Real Analysis and Infinity

Real Analysis and Infinity

H. SEDAGHAT

OXFORD
UNIVERSITY PRESS

Great Clarendon Street, Oxford, OX2 6DP,
United Kingdom

Oxford University Press is a department of the University of Oxford.
It furthers the University's objective of excellence in research, scholarship,
and education by publishing worldwide. Oxford is a registered trade mark of
Oxford University Press in the UK and in certain other countries

Published in the United States of America by Oxford University Press
198 Madison Avenue, New York, NY 10016, United States of America

British Library Cataloguing in Publication Data

Data available

Library of Congress Control Number: 2021946691

ISBN 978–0–19–289562–2

DOI: 10.1093/oso/9780192895622.001.0001

Printed and bound by
CPI Group (UK) Ltd, Croydon, CR0 4YY

Contents

Preface

This is an introductory book in real analysis. It covers the fundamental concepts and methods with the dual purpose of exposing the reader to both results (theorems and the like) and their proofs. As a textbook, it is designed for a one semester, first course in real analysis or advanced calculus.

Like other books in this category, this one also contains a substantial number of facts about real numbers and the functions that are defined on them. The standard facts include a detailed and precise description of real numbers and their properties and the concept of limit expressed primarily in terms of the convergence of sequences of real numbers; subsequently, using these two concepts, we explain the foundations of differential and integral calculus as well as the basics of the theory of infinite series of functions.

Also like other books in this category, proofs of most results are presented in detail. Some results are left to the reader to prove using arguments that are similar to what is presented in the text. This split is meant to help the reader gain a firmer grasp of the rules of mathematical reasoning through active participation.

What is different about this book. Every book has its own flavor, i.e., what distinguishes it from other books in the category. This book is different in some ways from other texts on the subject, too.

The major difference is its parallel goal of highlighting the crucial role that the concept of infinity plays in analysis. In no other branch of mathematics do we find infinity being an integral part of the foundations of the subject. Analysis contains elements of algebra and geometry at its core, but infinity is its defining characteristic.

Readers who have not been exposed to set theory or advanced mathematics can still recognize the footprint of infinity in the decimal expansions of irrational numbers. And after exposure to trigonometry and calculus, they also see it in the so-called transcendental functions like $\sin x$ or $\log x$, which at first seem rather mysterious things. A stiff course in calculus may make students comfortable with these ideas through repeated use but not necessarily through comprehension.

There are also the often popular infinity "paradoxes," like infinite regions with finite area or objects with infinite surface area but finite volume (so when filled up with paint, the infinite interior surface is covered with a finite amount of paint). These are intriguing and amusing concepts, though when we are first exposed to them, we often feel uncomfortable about their conflict with common experience.

One benefit of studying this book is that all these mysteries are fully explained. All the unpleasant confusion is taken out and replaced by actionable knowledge.

While there is no "definition of infinity" in this book, we clarify its meaning by discussing the manifestations of infinity that are relevant to the study of real analysis. The introductory chapter of this book explains what this means.

And it is infinity that spawns the old paradoxes that have intrigued numerous generations since Zeno, if not from even earlier times. After finishing with this book, the reader will have been exposed to enough knowledge to be able to tell which of these paradoxes are unavoidable (logical) consequences of using infinity and the extent to which their paradoxical nature derives from semantics.

The key concept that underlies and connects all other concepts in analysis is *convergence* (also known as the *limit*). Through convergence we explain the defining characteristic of real numbers, namely, *completeness*, the technical name for the "continuum." Through convergence, we also define continuity for functions, their derivatives and integrals, and the enigmatic (but not mysterious) infinite series.

The concept that underlies convergence itself is infinity.

In broader mathematics, infinity is introduced as a set theory axiom by the same name. Remarkably, there really is no other way of talking consistently about infinity without running into confusion or contradictions.

We don't start at such a basic level but jump a bit forward and begin from where most practitioners of analysis do: an acceptance of or agreement about what the set of all natural numbers (positive integers) are, including their ordering, their algebraic properties, and the idea that there are infinitely many of them that we get by adding 1 to itself repeatedly.

From this initial point, we construct all other types of numbers up to the complete system of real numbers, then proceed to define all other concepts of calculus that most readers of this book have already seen to a varying extent. We also discuss many topics that may be unfamiliar and in some cases technical, but are often useful, intriguing, and compelling.

A review of the content. To accomplish the goals set above, I start in Chapter 1 with an informal discussion of several topics that involve infinity in order to help motivate some of the concepts and methods that we study later in this book. If using this book as a textbook for a course in real analysis, then this chapter is mostly a self-reading assignment before perhaps the second class meeting.

In the second chapter, we establish some of the background concepts, including the existence of infinitely many different infinities, the smallest two of which are the most important in analysis. Much of the material in this chapter may be skipped or used for later reference; but the sections that are likely to be new, such as those on countability, may need to be considered for class discussion or be formally assigned as reading material.

I discuss sequences in Chapter 3, and this constitutes the usual start of an analysis course. Convergent sequences are the most important tools for our work in this book. I liken these sequences to *infinite-resolution lenses* that help us resolve the mysterious infinitesimals and explore their murky domain.

One way in which the coverage in this book differs from that in many textbooks is the use of sequences to develop all the basic concepts (continuity, the derivative, the integral, and infinite series).

This approach misses no essential facts of analysis at this level, and the equivalence to the traditional neighborhoods approach (ε, δ) is proved easily enough as we progress. We don't shy away from using the traditional approach, often alongside sequences and especially where it leads to easier or better comprehension.

In Chapter 4, we build up the real numbers from the rational numbers. The latter are fractions, i.e., ratios of two integers whose algebraic properties are familiar to anyone who has taken in course in elementary algebra.

A sequence of rational numbers may converge to (or approach) another number that is typically not rational. We discover that the totality of the so-called Cauchy sequences of rational numbers forms a complete set (no holes or gaps). This set turns out (essentially) to be the set of all real numbers.

Presenting this chapter before discussing the derivative or the integral is natural because of our sequence-based approach in this book; it is possibly also desirable given the interesting topics involving infinity that often interest many readers.

In most real analysis textbooks, this construction is entirely omitted in favor of a more economical approach where completeness is introduced as an axiom that often takes the form of the least upper bound property. Under time pressure, one simply adds this axiom to the field and order axioms, then moves on to proving theorems.

But as in most other things, expediency in this case comes with a price tag: beyond having to accept an extra axiom that is not mathematically required, most students complete their introductory analysis course without appreciating the role of infinity in the conceptualization of real numbers (especially the irrational numbers) and by extension, in the entire body of analysis and applied mathematics.

While students pursuing this course of action may develop good skill and significant knowledge, they will at best have an incomplete grasp of the foundations of analysis that could have been avoided by just a little extra investment of time and energy. In the course of my teaching analysis over so many years, I have heard students express confusion about completeness and dissatisfaction with their understanding of it.

As a bonus, students who move on to higher-level mathematics courses such as measure theory or functional analysis will discover that the construction of real numbers that we discuss here is a blueprint for the concept of "completion of a metric space," a fundamental concept in analysis.

In Chapter 5 on infinite series of constants (real numbers), we study a topic familiar from calculus, which sets the stage for the broader and more important study of series of functions in Chapter 8. In addition to the usual topics like definition of series as sequences of partial sums, Cauchy's criterion, and convergence tests, we also discuss interesting topics like expressing the real numbers as infinite series, and Riemann's rearrangement theorem for conditionally convergent series.

Chapters 6 and 7 discuss familiar topics from calculus, namely, continuity, differentiation, and integration.

The sequence-based approach in Chapter 6 is different from the traditional approach of taking limits based on neighborhoods (the ε, δ method). I discuss the connection between the two approaches in Chapter 6 and show that they are equivalent.

In Chapter 6, I also discuss continuous functions and their relation to the differentiable ones, rules for taking derivatives, l'Hospital's rule for calculating limits of indeterminate forms, and the Newton–Raphson method for estimating solutions of nonlinear equations. In various topics, I highlight occurrences of infinity and their consequences that are not discussed in most textbooks.

Chapter 7 is concerned with the development of the Riemann integral and its relationship to derivatives and continuity. We define integrable functions, derive their properties, and discuss their some of their applications.

We use the integral to define the natural logarithm and its properties, followed by the definition of their inverse—the exponential functions and their properties. Then we discuss the improper Riemann integral where infinity occurs in an explicit way.

In the closing Chapter 8, many of the earlier topics find powerful applications. We see that most functions can be expanded as infinite series of simpler, familiar functions, like power functions or trigonometric functions.

This knowledge helps us develop powerful tools with which to study unfamiliar functions that often appear as solutions of differential equations, including those in physics and engineering. In fact, the familiar "transcendental" functions that appear so often in introductory calculus (logarithm, exponential, trigonometric, hyperbolic, etc.) are tiny droplets in a sea of functions that may appear as solutions of important differential equations. Large classes of functions without closed forms are readily accessible using the ideas discussed in Chapter 8.

Needless to say, infinity lies at the core of most concepts in Chapter 8. In particular, the two most basic types of convergence for sequences of functions that I discuss in Chapter 8, namely, pointwise and uniform, can be distinguished by a less transparent occurrence of infinity in them. We discuss this issue as a way of differentiating between these types of convergence, and to explain how this occurrence of infinity is responsible for the shortcomings of pointwise convergence.

The style and layout. To help the reader identify the definitions of major concepts, I have enclosed them in easily identifiable boxes. The results with proofs are clearly identified as theorems, corollaries, etc. These important facts are connected by less formal prose intended to guide the reader from definitions to theorems and often to the next definition.

I have given priority to pedagogical descriptions over abstraction or elegance whenever a decision needed be made. I don't avoid repetition because efficient writing, so desirable from a professional mathematical perspective, is not the proper way of introducing new material to students. Most new readers tend to feel intimidated and lose interest prematurely if presented with a logically tight presentation that leaves the reader with too much to decipher.

Of course, even a pedagogically oriented textbook at this level contains more abstraction and leaves more to readers to work through than what they may be accustomed to. But there seems to be little point in expecting them to do more heavy lifting than is absolutely required.

I go to some length to provide context and ease the reader into each important topic. Analysis concepts often need to be read carefully and in many cases, multiple times; students must realize that not getting it at first reading is normal, not a sign of inability to comprehend.

Each major topic is motivated by informal discussion, examples, and diagrams before plunging into definitions, theorems, and proofs. Experienced readers may jump from definitions to theorems and skim through the rest. But I urge readers who are seeing this material for the first time to read through the informal discussions that connect major results to be better informed about what those results are and why they are important.

I frame important definitions in order to highlight them, and *include examples as part of the general prose* so as to help the reader get used to thinking in terms of specific cases—and hopefully, begin to ask questions that the theorems and proofs will subsequently answer.

After absorbing the theorems, more technical examples often follow to help the reader put general results to use in solving problems or answering questions. Finally, the exercises help readers gauge their understanding of the material and in many cases, add to what they read in the text of a chapter. They are not meant as tests of the reader's mastery of topics or methods but as opportunities to fine-tune the reader's understanding of the material in a chapter.

The more exercises the reader works through, the greater the level of understanding of the subject. If a problem presents difficulties, then it is usually good practice to move on to others. Later, sometimes even after the study of a new chapter is underway, a review of parts of the old chapter that relate to a problematic exercise may help resolve any issues that might have existed earlier.

Finally, of potential interest is an earlier, self-study version of this book named *Achieving Infinite Resolution*, which is a more relaxed, slow-paced reading in analysis.

Its content includes amusing stories, some history, anecdotes and paradoxes, as well as the major results that are discussed in this book (but not all of the proofs). Reading the self-study book, which is more focused on infinity than this book is, may add to the depth (and pleasure) of getting acquainted with analysis. You can check sample content from *Achieving Infinite Resolution* on Amazon.com.

Course suggestions. A variety of courses can be designed using this book, depending on the level of depth desired, the available time, etc. The instructor can omit some sections or cover a few sections out of order if desired. Here are some thoughts that may help in making such decisions.

Ideally, all eight chapters are covered, keeping in mind that the short introductory chapter may be left to students as a reading assignment; and Sections 2.1–2.3 of Chapter 2, along with the two appendices, are primarily for reference. If time is short or it is needed for emphasizing the essentials, one or more of the following sections may be omitted without loss of continuity:

5.6, 6.6, 6.7, 7.6, 7.7, 8.3.3, 8.5.4, 8.6, 8.7

Though I don't recommend it, a traditional first course in real analysis where the construction of real numbers is suppressed can still be extracted from this book by omitting Section 4.3 and subsection 4.2.1 and replacing this material with a "completeness axiom" to the list in Section 4.1. The first part of Section 4.2 (Cauchy sequences) may be added to Chapter 3.

Further, Chapter 5 may be covered after Chapters 6 and 7, for a more traditional. The list of sections that may be omitted when time is short is the same as the one above.

A more restricted course that emphasizes the foundations might cover the following chapters and sections:

Chapters 1–4, Sections 6.1–6.4, and 7.1–7.4

Acknowledgment and dedication. I would like to acknowledge with appreciation the role that my mentors played in bringing me up to speed. These were the few special teachers from the early years through college who put in the extra effort of getting through to me at a young age when it was not always easy or pleasant for them to do so.

I also acknowledge with appreciation the role of my many students, from freshmen to postgraduate, who over so many years of college teaching taught me a lot about communicating mathematics. I would like to thank those among them, especially Agnes Grocholski, who read and commented on some of the material that appears in this book.

I dedicate this book to these students and to all readers who wish to learn about analysis and infinity.

Finally, sincere thanks go to my editor Dan Taber without whose support and guidance this book would not exist.

H. Sedaghat

1

Manifestations of Infinity: An Overview

What we see on a television, computer, or phone screen looks smooth and continuous. But it is common knowledge that digital images on these screens are
composed of *pixels*: tiny squares that are arranged in a grid to collectively form
an image. The number of pixels per inch (or centimeter) is the *screen resolution*.
More pixels per inch means greater resolution and a sharper and more vivid image. But the screen resolution is always a *finite number*: no matter how sharp and
richly textured, every digital image turns into a jumble of colored or grey and white
squares when we zoom in on or magnify the image by a factor that is comparable
to the resolution number.

What about non-digital images? How does a drawing on a piece of paper stand
up to magnification, or for that matter, smooth-looking surfaces like that of a
bowling ball or polished marble?

A surface, like that of paper, marble, or a bowling ball, looks continuous to the
eyes; but, it is not actually continuous. These material objects are ultimately made
up of atoms, nature's three-dimensional pixels. All forms of matter have finite
resolution in this broader sense.

If matter, and even energy, are grainy or discrete at their core, *space* and *time*,
as we understand them, are not. To this day, even in quantum theory and its field
extensions where matter and energy are discrete, space and time are considered
continuous. Space is a "container" of fields, and time is a parameter by which
motion and transformation are tracked.

Space—in the sense that is understood in science since Newton—is a mathematical concept with continuous structure, i.e., a "continuum." Time is a similar
continuum, though of a scalar or numerical nature. Both are infinitely divisible;
any chunk of space or interval of time can be divided in two, with each piece having the same exact nature. So the dividing can go on indefinitely. Space and time
both seem to have *infinite resolution: there are no pixels of space or grains of time*
(sand-filled hourglasses notwithstanding to the contrary).

To deal with space and time, it is necessary to use appropriate tools that deal
with *infinity*. The significance of infinity in calculus and many other areas of mathematics, science, and engineering cannot be explained by conducting experiments
or making scientific measurements. Nothing that is infinite has been identified in
nature; even the cosmos as we know it is finite, based on the farthest that we see
by the light that is emitted by its matter constituents.

Real Analysis and Infinity. Hassan Sedaghat, Oxford University Press.
© H. Sedaghat, (2022). DOI: 10.1093/oso/9780192895622.003.0001

Equally subtle is the *infinitesimal*, a manifestation of infinity that sits at the core of calculus and every scientific discipline that uses it for modeling. Again, we have no physical way of detecting anything infinitesimal in nature using tools made of physical things (matter and radiation), all of which have *finite resolution*.

Calculus turned out to be a conceptual *infinite-resolution device* that could manage the infinite and the infinitesimal in a logically consistent way (though it didn't start out like that).

We focus on real analysis in this book. Using calculus (and analysis broadly), we resolve points (and instants) by essentially zooming in *an infinite number of times*. The tool for accomplishing this feat is the concept of *limit*. The connection between limits and infinity is most readily brought out using *sequences*, i.e., infinite strings of numbers.

As we see later in Chapter 3, a *converging sequence* is like an abstract zoom feature on an idealized camera or a microscope. By going down the sequence, we may zoom infinitely often to get to individual points in space. Learning the precise way in which this abstract mechanism works is key in gaining a thorough understanding of analysis.

In this introductory chapter, we consider some of the many ways in which infinity appears in our conceptual framework. This discussion is light on details and calculations, but it helps set the stage for a more in-depth study in later chapters.

There is no "definition" of infinity in this book. We think of infinity as a *property of sets* and as such, it is defined by an axiom in set theory. Rather than engaging in a discussion of set theory, I leave it to you, the reader, to consult a textbook on set theory where some build up is necessary to get to the root of the matter.

The set theoretic axiom is necessary for any consistent treatments involving infinity but not required for our pursuits in analysis. In fact, it tends to distract from our work; and knowing that a consistent treatment exists is enough for assuring us that we don't run into contradictions or unwanted mysteries.

1.1 Infinity within each number

A whole or natural number can be made as large as we wish by adding more digits to it, by concatenation, or by addition or multiplication of numbers. But such a number always has a finite number of digits. On the other hand, a ratio of two natural numbers, also called a *rational number* or fraction, *typically* has an infinite number of digits when written in decimal form, like

$$\frac{11}{27} = 0.407407407\cdots = 0.\overline{407}$$

How do we obtain this result? The short answer is by *long division*,[1] which we learn about before college. While we rarely calculate with long division nowadays, it is available to us as an option when needed. Unfortunately, when it comes to *irrational* numbers, this option is no longer available.

Notice that a knowledge of just the first 3 digits and the fact that these digits repeat in exactly the same order (i.e., periodically, with period 3) tells us everything about the decimal representation of 11/27. Because the first decimal place is a 4, we know that the decimal place $3k + 1$ is also 4 for every natural number k. For instance, if $k = 543$ then digit

$$3(543) + 1 = 1630$$

is a 4, while digit 1631 is a 0 and digit 1632 is a 7, and things go on repeating like that. So while 11/27 has an infinite number of digits, there are really only 3 different digits, and these appear in a fixed pattern.

We may contrast this situation with irrational numbers like $\sqrt{2}$ or π whose decimal forms consist of infinitely many digits that appear in no known pattern. It is quite difficult to tell what number the 1630th digit of $\sqrt{2}$ or of π is; and while decimal forms of these numbers are known to billions of digits, not all of their digits are (nor can be) known.

Such irrational numbers exhibit another occurrence or manifestation of infinity, one that is less transparent than the one for rational numbers like 11/27. The easiest way to characterize this deeper manifestation of infinity is by noticing that while a digital computer can store and recall all digits of 11/27 by using a few memory slots, it cannot do the same for all digits of $\sqrt{2}$ or π because the physical memory of a digital computer is always finite.

In calculus, however, this deeper manifestation of infinity in numbers is handled by *using the infinite itself* in different ways that we discuss in Chapter 5 and later. A common method is by *adding infinitely many numbers*; for example, π is such a sum of fractions:

$$\pi = \frac{4}{1} - \frac{4}{3} + \frac{4}{5} - \frac{4}{7} + \frac{4}{9} - \frac{4}{11} + \cdots$$

This *infinite series* was known to Leibniz (1646–1716) in the 17th century and to Indian mathematicians even earlier, in the 14th century. The modern derivation of this and similar series using the methods of calculus is rather simple, as we discover in Chapter 8.

---

[1] An alternative after-the-fact proof is as follows: let x be the decimal expansion $0.407407407\cdots$ and observe that $1000x - x = 407$. Solving this equation for x gives the rational number $x = 407/999$, which reduces to 11/27 after canceling the common factor 37 from the numerator and the denominator. Notice that this argument extends to all repeating decimal expansions and shows them to be rational numbers. Some rational numbers have a *finite* number of digits, like $2/5 = 0.4$. It is also valid to write $2/5 = 0.4000\cdots = 0.4\overline{0}$, which can be said to have period 1. An alternative form that is often more convenient in analysis discussions is $2/5 = 0.3999\cdots = 0.3\overline{9}$, also with period 1.

Notice that in the above series for π we can keep adding as many fractions as we want, thereby improving the accuracy of our knowledge of π. In fact, since the series is *alternating* in signs, we discover in Chapter 5 that we can even tell rather easily how many fractions will guarantee a desired level of accuracy. If we want to know what the 1630th digit is, then we add the required number of fractions in the infinite series until digit 1630 or any other desired digit is determined.[2]

Finally, worth mentioning is that when we approximate an irrational number using a sequence of rational numbers or fractions of integers, the fractions in the sequence have larger and larger denominators (and numerators) without any bounds. For example, the series (s) for π above gives a sequence of fractions that approach π as follows:

$$s_1 = \frac{4}{1}$$

$$s_2 = \frac{4}{1} - \frac{4}{3} = \frac{8}{3}$$

$$s_3 = \frac{4}{1} - \frac{4}{3} + \frac{4}{5} = \frac{52}{15}$$

and so on. For most values of n, the nth fraction in this list has a denominator

$$(1)(3)(5)(7)(9) \cdots (2n - 1)$$

and this becomes infinitely large as n does. Here, we have another way of seeing the infinite in an irrational number, namely, the rational numbers that get very close to it have denominators (and numerators) that are very large.

1.2 More than one infinity?

It may seem that the infinite must be unique, as it has no bounds and never ends. But there are actually *many* infinities in mathematics. Two of them play significant roles in analysis, the more familiar of which is the infinity of all natural numbers (positive integers) 1, 2, 3, etc. These make up a set of numbers that is commonly labeled $\mathbb{N}$ and needs no introduction.[3] The symbol $\aleph_0$ ("aleph-null") is used to denote the infinity of all natural numbers (the *cardinal number* of $\mathbb{N}$). $\aleph_0$ happens to be the *smallest* infinite (or the first non-finite) cardinal number in a very precise sense that is discussed in textbooks on set theory.

---

[2] Worth mentioning is that although simple and elegant, this alternating series expression for π converges too slowly to be of practical use. We discuss faster converging series later in this book.

[3] The precise definition of natural numbers is not trivial. They may be defined axiomatically (Peano's axioms) or in terms of sets in formal set theory.

There are greater infinities than $\aleph_0$. Consider the set of *all* numbers, rational and irrational, between 0 and 1, which we denote as the closed interval $[0,1]$. To appreciate how large the cardinal number of $[0,1]$ is, suppose that we pull numbers out of this set one at a time like drawing tiny glass beads out of a box.

Which numbers do we begin with? An easy choice is to take out the reciprocals of the natural numbers:

$$\frac{1}{1}, \frac{1}{2}, \frac{1}{3}, \frac{1}{4}, \frac{1}{5}, \frac{1}{6}, \ldots$$

Certainly all of these numbers are in $[0,1]$, they are rational, and there are infinitely many of them; looking at the denominators, you see that there are as many numbers in this list as in $\mathbb{N}$, the set of all natural numbers whose cardinality is $\aleph_0$. But still *infinitely many more* rational numbers remain in our $[0,1]$ box, like $0, 2/3$, and so on.

Let's be more clever in picking numbers; suppose that we take them out as follows:

$$\frac{0}{1}, \frac{1}{1}, \frac{1}{2}, \frac{1}{3}, \frac{2}{3}, \frac{1}{4}, \frac{3}{4}, \frac{1}{5}, \frac{2}{5}, \frac{3}{5}, \frac{4}{5}, \ldots \tag{1.1}$$

Here we take out all the fractions with a fixed denominator before moving on to the next set of fractions (discarding all repetitions along the way, like $2/4$, which equals $1/2$). This method takes out *all* rational numbers between 0 and 1 (including 0 and 1). For example, we reach the number $807/1102$ when removing all the fractions with denominator 1102.

The infinite list in (1.1) shows that there are a lot more numbers in $[0,1]$ than there are numbers in $\mathbb{N}$; but both sets are infinite, so can we really say that the set of all rational numbers in $[0,1]$ has a larger cardinal number than $\aleph_0$? Surprisingly, we discover in Chapter 2 that the cardinal number of the sum of all natural numbers is also $\aleph_0$! So while one set has infinitely more numbers in it than the other, they are both equally infinite in a sense that we define below.

So we haven't discovered a new infinity yet. But note that the infinite list in (1.1) leaves out all of the *irrational* numbers that cannot be written as fractions. We can prove easily that there are infinitely many numbers of this type too: choose any known irrational number between 0 and 1, say, $1/\sqrt{2}$, which is approximately equal to 0.707. Now, the list

$$\frac{1}{\sqrt{2}}, \frac{1}{2\sqrt{2}}, \frac{1}{3\sqrt{2}}, \frac{2}{3\sqrt{2}}, \frac{1}{4\sqrt{2}}, \frac{3}{4\sqrt{2}}, \frac{1}{5\sqrt{2}}, \frac{2}{5\sqrt{2}}, \frac{3}{5\sqrt{2}}, \frac{4}{5\sqrt{2}}, \ldots \tag{1.2}$$

that we get by multiplying every number in (1.1) by $1/\sqrt{2}$ is an infinite list of distinct irrational numbers between 0 and 1.

So there are no fewer irrational numbers between 0 and 1 than there are rational numbers. In fact, we can get even more irrational numbers if we replace $\sqrt{2}$ in (1.2) by other irrational numbers that are larger than 1, say $\sqrt{3}$, $\sqrt[3]{2}$, π, etc. For each of these we obtain equally large lists of *distinct* irrational numbers, none of which overlap.

Phrases like "as many numbers as" or "no fewer numbers than" used above are the kinds of words where we want to count something but don't have numerals for it. When we apply these to our infinite lists of rational and irrational numbers above, notice that we count *by pairing the entries in each list together*. This idea is important when dealing with infinite sets because we can't use statements like "each list has a trillion numbers in it."

Counting by pairing items in one list with items in another list predated the creation of numerals. In early human history, people would count things using references like the fingers of their hands. If they could identify as many heads of cattle as the fingers of one hand, then that meant "5 heads of cattle" whether or not they had a name for the number 5. When the numbers used were small, there was no need to invent symbols or names for them.

But as human civilization grew, so did the sizes of numbers that they used. The need for numbers with names and symbols, or *numerals,* became more urgent when merchants and governments needed to count large numbers of items, be they coins, bags of wheat, or heads of cattle. It would be rather hard for merchants to deal with hundreds or thousands of such items using fingers or other primitive counting methods.

As long as we are dealing with a finite (and not very large) number of items, the existing numerals are fine. In fact, the Romans found their cumbersome number system adequate for all their dealings. But Europe's needs eventually grew, along with its population, to the point that the Roman numerals were no longer adequate. So when the more practical Hindu numerals arrived by way of the Islamic civilization (with some modifications), they were quickly adopted for use in commerce and mathematics.

But of course, numerals of any kind are useless when dealing with infinity since it is pointless to develop infinitely many names. So, in a remarkable turnaround, we must go back to the primitive ancestors' way of counting by pairing things up. But instead of primitive objects like fingers, this time we use *sets* of numbers that are infinite.

The German mathematician Georg Cantor (1845–1918) used the pairing method creatively in the 1870s to show that the infinity of all numbers between 0 and 1 is a *larger* order of infinity than $\aleph_0$. In other words, we can never pair up *all* the real numbers with the natural numbers: no matter how cleverly we pair things up, we always run out of natural numbers before all the real numbers are accounted for. This is certainly not a transparent point; we explain it later when discussing Cantor's original arguments in Chapter 2. In that chapter, we also see how Cantor proves that there is an *infinite* hierarchy of *different infinities* using a short, clever argument involving *power sets.*

So, there are two different orders of infinity in just the set of all numbers between 0 and 1. More broadly, the collection of all rational and irrational numbers is called the *set of all real numbers* and denoted by $\mathbb{R}$. We often think of this set geometrically as points on a number line, like the familiar x-axis.

The irrational numbers in $\mathbb{R}$ turn out to be far more enigmatic than the rational ones; while they don't have interesting algebraic properties, they play an important role in making the x-axis the "continuous line" that it is supposed to be. In analysis, this continuity property of the number line is called *completeness*, and we discuss it fully in Chapter 4. The irrational numbers complete the rational numbers by filling all the "holes" that exist in the set of all rational points on the number line. For instance, $\sqrt{2}$ is irrational so without irrational numbers, there would be a hole or gap on the number line where $\sqrt{2}$ would be.

1.3 Infinite processes

Analysis isn't limited to the real numbers; it relies on *infinite processes*. We use infinite processes known as Cauchy sequences to construct the set of real numbers. Also derivatives and integrals are meaningless without reference to some infinite process of partitioning and convergence. The Greeks believed that the *continuum*, namely, a continuous object like a chunk of space, was infinitely divisible: *in principle*, we can divide an idealized continuous object infinitely many times; the same is true of the *distance* between two points or an *interval* of time.

After dividing something up, the pieces don't vanish, and the outcome of an infinite process of dividing the continuum is a set of objects that are no longer divisible; the Greeks called them *atoms* ("atomos" or indivisibles).

The nature of these indivisibles or how we get to them via an infinite process is far from obvious. To take a specific case, suppose we divide a piece of the x-axis like the unit interval [0,1] infinitely many times as follows: split [0,1] in half and discard the right half. What is left is the interval $[0, 1/2]$. Split this interval in half the same way and repeat to get smaller intervals; in the next step, which is 2nd in the process, the interval $[0, 1/4]$ is left; in the 3rd step, we are left with $[0, 1/8]$ and so on. Notice that

$$\frac{1}{4} = \frac{1}{2^2} \quad \frac{1}{8} = \frac{1}{2^3} \quad \text{etc.}$$

After n steps, we are left with the interval $[0, 1/2^n]$, which has length $1/2^n$. As we pick n progressively larger, the fraction $1/2^n$ rapidly goes to 0. If the division process does not terminate, then we expect it to result in the interval [0,0], which has length 0. In fact, it contains just one number, 0. We conclude that the set consisting of a single number 0 is the result of our infinite process of shrinking [0,1] by splitting intervals in half and discarding the right half. The single-element set $\{0\}$ cannot be further split in half, so we can say that it is indivisible.

To see that this conclusion is not without difficulties, let's consider the intervals in each step of the shrinking process:

$$[0,1] \rightarrow \left[0, \frac{1}{2}\right] \rightarrow \left[0, \frac{1}{2^2}\right] \rightarrow \left[0, \frac{1}{2^3}\right] \rightarrow \cdots \rightarrow \left[0, \frac{1}{2^n}\right] \rightarrow \cdots \{0\}$$

Notice that every one of these intervals has infinitely many numbers in it. To start with, every number x in $[0,1]$ can be uniquely paired with its half $x/2$; so $[0, 1/2]$ has as many numbers as does $[0,1]$, which is an infinite set.[4] Applying the same argument to the pair $[0, 1/2]$ and $[0, 1/2^2]$, we find that these intervals have the same number of elements, so $[0, 1/2^2]$ is as infinite as $[0,1]$. This is true about $[0, 1/2^n]$ for every $n = 1, 2, 3, \ldots$, so all the sets in the process have infinitely many elements in them.

But this process is supposed to end in the single-element set $\{0\}$, so how do the intervals lose the numbers in them in the end?

We may be tempted to conclude that the above argument is incorrect and wonder: what if dividing $[0,1]$ infinitely often did not result in $\{0\}$ but produced an indivisible object of "infinitesimal length" instead? This would remove a problem but replace it with another: it is necessary to define the *infinitesimal*. This alternative has in fact been considered in an area of mathematical logic called *smooth infinitesimal analysis* that was developed in the 1970s to give logical meaning to infinitesimals and use them to derive the basic results of calculus. But this line of reasoning has its cost: the abolishment of the "law of excluded middle" in this approach makes proving theorems difficult and in some cases impossible, since proof by contradiction is not allowed in this context.

A separate *non-standard* version of analysis that does not carry such a high price was introduced in the 1960s where infinity and infinitesimal both appear explicitly. This approach is more flexible than that of smooth infinitesimal analysis, but the level of abstraction needed to consistently handle infinities as numbers is still rather high; I discuss these matters in more detail in the self-study book *Achieving Infinite Resolution*.

In the standard version of analysis that we discuss in this book, infinity is not a number but a property or attribute of *sets*. Infinite processes are usually defined as *convergence* of sequences of objects to *limits*. These limits can be, and *often are*, markedly different from objects that are being transformed by the infinite process. The degree to which the limit differs from the sequentially transformed objects has a lot to say about the nature of the infinite process.

---

[4] If you are wondering how the set $[0, 1/2]$, which is half the size of $[0,1]$, can have as many numbers in it as $[0,1]$, then be advised that we discuss just this type of issue in Chapter 2. It happens that this oddity is a defining feature of infinite sets since it obviously does not occur for finite sets.

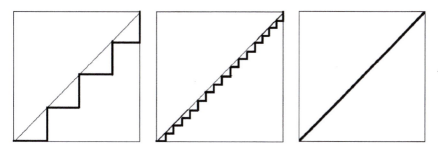

Fig. 1.1 Staircase curves converge uniformly to the diagonal

I discuss sequences and their convergence in Chapters 3 and beyond, but let's end this section with another, maybe more compelling, example that seems paradoxical.

Consider the squares in Figure 1.1, each of which has sides 1 unit long.

In the left and center squares, we see staircase curves (thick lines) right under a diagonal. The lengths of both of these curves is the same: *2 units*, because the treads add up to 1 going horizontally and the risers add up to 1 vertically. Adding more steps to the existing ones doesn't change their total length, although the size of each step does shrink.

As the number of steps grows infinitely large, the stairs smoothen out into a straight line, namely, the diagonal of the square on the right in Figure 1.1. Since all the while the length of the staircase remains exactly the same, it seems that the diagonal should also have length 2.

But the Pythagorean theorem states that the length of the diagonal is $\sqrt{1^2 + 1^2}$ or $\sqrt{2}$, which is definitely less than 2. At the very end, there is a sudden jump in the lengths of staircase curves from 2 down to $\sqrt{2}$. Even as the *shapes* of the staircase curves become good approximations to the diagonal of the square, their *lengths* do not approximate the length of the diagonal, no matter how large the number of steps. Since lengths are associated uniquely with curves, if the curves fade into a line, then shouldn't their lengths incrementally converge to the length of the line too?

I discuss this problem in Chapter 8 where we see that the lengths of the staircase curves is fixed at 2 by the increasing *variation* caused by the addition of more steps. However, the variations are absent in the limit, namely, the flat diagonal line. Put another way, infinite processes need not preserve the lengths of curves. This is reminiscent of objects called "fractals" that are typically the results of infinite iteration processes. However, the staircase curves above don't end in a fractal; they flatten out to a simple straight line.

Similarly, *areas* of surfaces that bound finite *volumes* can actually be infinite (and not fractal). As a simple example, we discuss a "*Stack of Boxes*" in Chapter 5 that

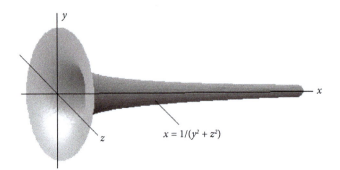

$$x = 1/(y^2 + z^2)$$

Fig. 1.2 Gabriel's Horn (a surface of revolution

has a finite volume but infinite surface area. Such a Stack of Boxes is in the same category as the famous *Gabriel's Horn* (or *Torricelli's Trumpet*), the horn-shaped surface of revolution in Figure 1.2 with infinite surface area but *finite volume*; see the discussion of improper Riemann integration in Chapter 7. The Stack of Boxes is a discrete analog that does not require integration.

1.4 Trigonometric functions and logarithms: Infinity built in

When we are presented with a power function like x^3 with an integer power or a sum like $x^3 + 2x$, we have no difficulty making sense of these: the first function multiplies a number x by itself three times; the second also adds twice x to the first.

These definitions and calculations involve a *finite number of algebraic operations*, like addition and multiplication. But when we come across trigonometric functions like $\sin x$ and $\cos x$ or the logarithm $\log x$, they seem enigmatic by comparison. How do we calculate the values of these quantities? How does a calculator with designated buttons come up with an answer that is usually accepted without doubt?

Strange as it may seem to modern students and most other people, trigonometric functions, logarithms, and the like were created to help us understand and solve important scientific, engineering, and even bureaucratic problems in government and business. Even through the 1970s extensive tables of values of these were supplied with most introductory mathematics textbooks as well as in reference books in science and engineering.

Trigonometric functions are defined with the aid of a right triangle. If x is one of the two acute angles (not 90 degrees), and we divide the length of the side in front of the angle x by the length of the hypotenuse (the side in front of the 90 degree angle), then we get the value of "sine of x" or $\sin(x)$ or more succinctly, $\sin x$.

Notice that this definition is *geometric* in nature, not algebraic. But when we need to calculate, say, $\sin 20^\circ$, we don't grab some prefabricated right triangle and measure its sides and angles; instead, we typically pick up a calculating device and punch in a few designated keys. A few digits pop up on the screen that we accept as sin $20°$. We rarely think about where these digits come from, let alone how they relate to the geometric definition of $\sin x$.

The situation with logarithms is similar. We define $\log x$ as the number y with the property that $10^y = x$. An equality like this routinely appears in scientific, engineering, and statistical calculations. For example, consider when a rumor "goes viral." Suppose that, of the people who hear about or view the rumor on their devices in the first hour, at least 10 broadcast it. If every hour, each broadcast is further broadcasted by at least 10 more people, then in 2 hours, no fewer than 100 or 10^2 people know about it; in 3 hours, 10^3 or $1,000$ people, and so on. How long does it take for at least three million people to hear this rumor?

To answer this question, we find y such that $10^y = 3,000,000$. The value of y that solves this equation is[5]

$$y = \log 3,000,000 = 6 + \log 3 \quad \text{hours}$$

(less than 6.5 hours). As with the trigonometric functions, to obtain the value of the quantity $\log 3$ (approximately 0.477), nowadays we just use a calculating device.

If all that we know of $\sin x$ or $\log x$ is their original definitions and the digital approximations of their values that are generated by machines, then it should come as no surprise that these functions seem so enigmatic.

Functions like $\sin x$, $\log x$, etc. are actually related to the nice and simple power functions through infinite processes. Think of them as irrational functions; a more common name for them is *transcendental functions*. We find in Chapter 8 that *each of these functions can be expressed as an infinite series of power functions*, or a *power series*. For instance,

$$\sin x = x - \frac{x^3}{6} + \frac{x^5}{120} - \frac{x^7}{5040} + \cdots$$

From this infinite series, we may calculate the value of $\sin x$ for any value of x. For example, to calculate $\sin 20°$, we first change 20 degrees to the radian measure (because in the infinite series that is given above x is in radians):

$$20° = 20 \times \frac{\pi}{180} = \frac{\pi}{9}$$

and then insert $x = \pi/9$ radian in the infinite series for $\sin x$ to get the numerical value for $\sin 20°$.

As you might imagine, the result has an infinite number of digits since it is an irrational number. But a nice feature of the infinite series is that a *finite number of*

---

[5] I discuss logrithms and their inverse, the exponential functions in Chapter 7.

fractions of the infinite series can give a close approximation. For instance, if we use just the above 4 explicitly listed fractions, then we get the approximation

$$\frac{\pi}{9} - \frac{(\pi/9)^3}{6} + \frac{(\pi/9)^5}{120} - \frac{(\pi/9)^7}{5040}$$

With $\pi/9$ being approximately 0.349, the above sum is approximately equal to 0.342 to three decimal places. This agrees rather well with what you might get for $\sin 20^\circ$ on a calculator with the "sine" button (in the degree mode, or $\sin(\pi/9)$ in the radian mode). We deal similarly with other transcendental functions such as $\log x$.

The important point about these calculations is that *they show us where the answers come from*. Other than the hidden presence of infinity, there is really nothing mysterious about transcendental quantities. If you were bothered in the past by the mysterious nature of transcendental functions, then hopefully after reading this book the mystery disappears, and you will never be bothered by it again!

1.5 Exercises

Exercise 1 *We found in this chapter that the decimal form of the rational number 11/27 had period 3.*

(a) Use ordinary long division to find the period of the number 87/112. Identify the first and last digits of each cycle. Feel free to use a calculator to check your result and note that the first cycle doesn't start with the first digit after the decimal point.

(b) Repeat (a) for 807/1102.

Exercise 2 *Irrational numbers are **not** algebraically interesting. Show that the sum, difference, product, or ratio of two irrational numbers may fail to be irrational. You can pick your favorite irrational numbers to play with or just calculate the following:*

$$\sqrt{2} - \sqrt{2} = \sqrt{2} + (-\sqrt{2}) = ?$$

$$\sqrt{12}\left(\frac{1}{\sqrt{3}}\right) = \frac{\sqrt{12}}{\sqrt{3}} = ?$$

$$\left(\sqrt{2}\right)\left(\sqrt{8}\right) = ?$$

Exercise 3 *We can approximate $\sqrt{2}$ using the "divide and average rule." Start with a convenient rational number r_1 as a first guess at $\sqrt{2}$, divide 2 by this number, and average the result to get a new approximation*

$$r_2 = \frac{r_1 + (2/r_1)}{2}$$

(a) Let $r_1 = 1$, and use the above rule to find $r_2 = (1 + 2)/2 = 3/2 = 1.5$, then repeat using the recursion formula:

$$r_{n+1} = \frac{r_n + (2/r_n)}{2} \tag{1.3}$$

Calculate r_3, r_4, and r_5 to get answers in the form of fractions.

(b) Round your fractional answers in (a) to at least 6 decimal places to see that they are getting closer to $\sqrt{2}$. Compare the value of r_5 with what a calculator gives for $\sqrt{2}$, and find the number of decimal places that they agree.

NOTE. The divide-and-average rule was known to ancient Babylonians some 4500 years ago. I show how it works in Chapter 6 as a special case of the Newton–Raphson method, a widely used fast algorithm for approximating the solutions of nonlinear equations.

2

Sets, Functions, Logic, and Countability

This chapter and Chapter 3 supply some of the background material to help make understanding the rest of the book easier. Some sections or topics are likely familiar and can be skipped as desired. The sections on Cantor's Theorem and *countable* and *uncountable* sets in this chapter are likely to be new, and as such, they deserve greater attention.

2.1 Sets and relations

The most basic definition of a *set* is just what comes to mind: a collection of objects. The set of all people working in an office, the set of all office furniture, the set of all water bottles in the office are all valid examples of sets. The objects that constitute a set are called its *elements* (or *members*). Sets that are too large to list all their elements explicitly are described by a shared property; for example, the set of all New Yorkers or the set of all stars in the Milky Way galaxy. The *set notation* often uses braces; for example, in the aforementioned office, we may have the following sets of workers and water bottles:

$$W = \{\text{Bob, Debbie, Mary, Josh}\},$$

$$B = \{\text{Debbie's bottle, Josh's bottle, spare bottle}\}$$

assuming that Bob drinks soda and Mary drinks tea. More generally, for sets with large numbers of elements, or sets whose elements are not fixed or fully known, the notation is based on the shared property of elements; for example,

$$N = \{P : P \text{ is a New Yorker}\}, \qquad M = \{S : S \text{ is a star in the Milky Way}\}$$

In the above set notation, read N as: *the set of all P (people or persons) such that P is a New Yorker*; M is read similarly.

Real Analysis and Infinity. Hassan Sedaghat, Oxford University Press.
© H. Sedaghat (2022). DOI: 10.1093/oso/9780192895622.003.0002

Membership and equality

Being an element of a set is usually indicated by the symbol $\in$. So if A is a set, and a is an element of A, then we write $a \in A$ and say that *a is a member of A* or that *a belongs to A*. For instance,

$$\text{Bob} \in W, \qquad \text{Debbie's bottle} \in B, \qquad \text{sun} \in M$$

which we read as Bob is an element of W and so on. We may list the elements of a set in any order; rearranging the elements of a set does not change it. Thus

$$W = \{\text{Debbie, Mary, Bob, Josh}\}$$

is exactly the same set of four people that I listed previously. Generally, we do not repeat the elements in set notation; but if Josh left and was replaced by another Bob, then W would be

$$W = \{\text{Debbie, Mary, Bob, Bob}\}$$

which is an awkward way of writing W. In this case, we must distinguish between the two Bobs in W. For example, we could use their middle names; or write OBob for the original Bob and NBob for the new Bob. Then we can write W more descriptively as

$$W = \{\text{Debbie, Mary, OBob, NBob}\}$$

We now define equality of sets.

Equal sets. Two sets A and B are equal if every member of A is a member of B and vice versa. More precisely,

$$A = B \quad \text{means:} \quad \text{for every } x, \text{ if } x \in A \text{ then } x \in B \text{ and conversely.}$$

Rearranging the elements of a given set evidently results in an equal set; for instance, in the original W above, listing the names alphabetically does not give a different set of office workers. A little less trivially, the following infinite sets are equal:

$$\{\pi n : n \text{ is an integer}\} = \{x : \sin x = 0\}$$

This follows from the facts that (a) sine of every integer multiple of π is 0, so every number in the set on the left-hand side is also in the set on the right; and (b)

$\sin x \neq 0$ if $x \neq \pi n$ for all integers n, so every number in the set on the right-hand side is also in the set on the left.

The number of elements in a (finite) set is called the *cardinality* of the set; we denote it by the symbol #.[1] Thus, $\#W = 4$ and $\#B = 3$. Notice that there are no precise values for the numbers of elements of N or M at any given time. If time *is* considered in the definitions of N or M, say,

$$N(t) = \{P : P \text{ is a New York resident on January 1 of year } t\}$$

then from year to year, $\#N(t)$ is a specific number at the start of each year; but it changes from year to year.

In working with sets, it is often necessary to consider a set that contains no elements: { }. This set, being unique and special, has been given a name and a symbol:

Empty set. The unique set containing no elements is the *empty set*. It is denoted by $\emptyset$. Note that $\#\emptyset = 0$.

In mathematics, the empty set is also used to indicate *non-existence*. For example, the equation $x^2 + 1 = 0$ doesn't have any solutions that are real numbers, so we say that the set of all *real* solutions of this equation is empty. On the other hand, the same equation has two imaginary (or complex) solutions $\pm i$ where $i = \sqrt{-1}$; so we say that the set of all solutions of this equation is $\{-i, i\}$.

Subsets

Any collection of elements of a given set S is a *subset* of S. In particular, S is considered a subset of itself, and the empty set is trivially contained in every set. These two exceptional subsets are called *improper*. Every other subset is *proper*.

We write $A \subset S$ if A is a subset of S whether proper or improper. For example, for every positive integer n

$$\{1, 2, ..., n\} \subset \mathbb{N}$$

and this is obviously a proper subset of $\mathbb{N}$. The subset relation $\subset$ is also called *set inclusion* or alternatively, the *containment relation*.

---

[1] We discuss this concept for *infinite* sets a little later. Cardinality or cardinal number of a set is an abstract concept whose general and precise definition can be found in texts on axiomatic set theory. Other symbols that are also used for cardinality include the bar notation, as in $|W| = 4$ and the Hebrew aleph $\aleph$ as in $\aleph(W) = 4$.

Set inclusion lets us express the equality of sets as *double containment*:

$$A = B \text{ means } A \subset B \text{ and } B \subset A$$

We discover later on that double containment turns out to be a convenient way of proving statements about sets.

Note that if a set is finite, then the total number of its subsets is also finite. For example, the set W of office workers above has the following subsets (proper and improper):

∅, {Bob}, {Debbie}, {Mary}, {Josh}, {Bob, Debbie}, {Bob, Mary}, {Bob, Josh},

{Debbie, Mary}, {Debbie, Josh}, {Mary, Josh}, {Bob, Debbie, Mary},

{Bob, Debbie, Josh}, {Bob, Mary, Josh}, {Debbie, Mary, Josh}, W

a total of 16, or 2^4.

Infinite sets

Sets of interest in analysis are typically *not* finite; for example, the set of all whole, or *natural numbers*

$$\mathbb{N} = \{1, 2, 3, \ldots\}$$

is an infinite set (it is not uncommon to include 0 in $\mathbb{N}$, but we leave it out). The *triple-dots notation (ellipsis) with one end open* in a mathematical context such as the definition of a set, or in an equation, usually means "so on, ad infinitum."

Here is one way to think of an infinite set:

An *infinite set* is *a set that is not contained in any finite set.*

I should point out that there is no reason to believe that such a set exists; certainly, none is exhibited in the natural world. However, analysis and especially applied mathematics as it is practiced today in science and engineering (calculus, differential equations, etc.) would not exist without them. The existence of infinite sets is guaranteed only axiomatically; a more useful definition is given a little later in this chapter.

We have already encountered one set that is not contained in any finite set, namely, $\mathbb{N}$. Strictly speaking, we have not defined $\mathbb{N}$, and the definition of this concept is a nontrivial matter. The best-known definition of natural number (including 0) is based on Peano's five axioms.[2]

---

[2] Named after the Italian mathematician Giuseppe Peano (1858–1932). But as the British logician and philosopher Bertrand Russell (1872–1970) explained in his book "Introduction to Mathematical

REAL ANALYSIS AND INFINITY   19

We may use a set theory construction based on the *empty set* $\varnothing$ to identify $\mathbb{N}$ with the set

$$\varnothing, \{\varnothing\}, \{\varnothing, \{\varnothing\}\}\ldots$$

where $\varnothing$ represents 0, the set $\{\varnothing\}$ that contains only the empty set represents 1, and so on.

Such constructions relate $\mathbb{N}$ to other objects that are mathematically defined via set theory, but they don't give us the meaning we normally associate with the integers 0,1,2, etc. The reason is that these numbers are themselves abstractions of everyday objects; e.g., one flower, two pebbles, five fingers, etc. that we discussed earlier.

We content ourselves here with the knowledge that $\mathbb{N}$ is a mathematically well-defined concept one way or the other and move on to larger sets of numbers.

The next set of interest is that of *all integers* $\mathbb{Z}$ (*zahlen* is German for numbers), which is infinite and consists of all natural numbers together with their negatives and zero:

$$\mathbb{Z} = \{\ldots, -2, -1, 0, 1, 2, \ldots\}$$

The cardinality of $\mathbb{N}$ is usually denoted by the first letter of the Hebrew alphabet:

$$\#\mathbb{N} = \aleph_0$$

which is *aleph null*. Unlike the symbol ∞, the number $\aleph_0$ has a precise meaning; using a simple argument of Cantor's, we will see later that there is a vast hierarchy of infinities in mathematics in which $\aleph_0$ happens to be the *smallest*. We will also see later that $\#\mathbb{Z} = \aleph_0$, which may seem surprising since $\mathbb{N}$ is a proper subset of $\mathbb{Z}$. We discuss this issue later on.

2.1.1 Set operations

Four *set operations* are of fundamental importance as they apply to *all sets*. To avoid confusion, we apply these operations to subsets of a fixed *universal set* U.[3]

The four operations are (a) *intersection* (or *meet*) of sets, (b) *union* (or *join*) of sets, (c) the *complement* of a subset, and (d) *product* of sets. Let's see what these concepts mean.

Intersection and union

Philosophy" (pp. 5–9.) even these axioms don't strictly identify $\mathbb{N}$ but all sets that are in one-to-one correspondence with it (what we later call sequences).

[3] This is no loss of generality. If we have any collection of sets of interest, then the set of all elements in all the sets of the collection is the universal set U. So every set in the collection is a subset of U.

Intersection. For each pair of sets A, B in U (subsets of U), their *intersection* $A \cap B$ is the set of all elements (of U) that are in both A and B.

For example, if A is the set of all boys in New York City, and B is the set of all fourth graders in NYC, then $A \cap B$ is the set of all NYC boys who are in fourth grade. Here, depending on the context of interest, U may be the set of all elementary school students in NYC, or the set of all students in NYC, or the set of all elementary school students in North America, or even the set of all human beings.

The intersection operation is visually illustrated in the right hand panel of Figure 2.1 using a *Venn diagram*.

Union. For each pair of sets A, B in U (subsets of U), their *union $A \cup B$ is the set of all elements (of U) that are in A or in B.*

Here the "or" is *inclusive*, meaning "or both." For example, if A is the set of all American citizens, and B is the set of all nurses, then $A \cup B$ is the set of all human beings who are either American, or a nurse, or both. Here, depending on the context of interest, U may be the set of all human beings, or the set of all primates, or the set of all mammals, or even the set of all living things on Earth. See the left panel of Figure 2.1 for a visual illustration of the union operation using a Venn diagram.

Complement. The *complement of A in U* is the set A^c of all elements (of U) that are *not* in A. In symbols:

$$A^c = \{x : x \notin A\}$$

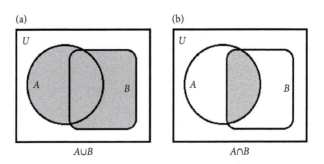

(a)   (b)

$A \cup B$ $A \cap B$

Fig. 2.1 Venn diagrams, union (left) and intersection (right)

For instance, the complement of the set of all even numbers in $\mathbb{N}$ is the set of all odd numbers

$$\{1, 3, 5, \ldots\} = \{2n - 1 : n \in \mathbb{N}\}$$

Note that $\left(A^c\right)^c = A$ since an element of U that is not in A^c has to be in A.

It is sometimes necessary to consider what is in a set A but not in another set B, that is, $A \cap B^c$; this is the *relative complement* of B in A, and it is often written as $A\backslash B$. In words, this set is what is left of A when we punch B out; it is the ordinary complement B^c if $A = U$ so that $B^c \subset A$.

The relative complement is useful when the underlying set is unclear. For instance, if E is the set of all even numbers, then the complement of E in the set of real numbers $\mathbb{R}$ is written $\mathbb{R}\backslash E$; note that this is not the set of all odd numbers.

The following result connects intersections, unions, and complements; it is named after the English mathematician Augustus De Morgan (1806–1871). The equalities are visually illustrated by Venn diagrams in Figure 2.2; the shaded regions in each panel show the equal sets in (2.1) following.

Theorem 4 (*De Morgan laws*) *The following equalities are true for all sets A and B*

$$(A \cup B)^c = A^c \cap B^c \tag{2.1}$$

Proof If $x \in (A \cup B)^c$, then x is not in $A \cup B$. Because the union contains both A and B, we conclude that x is not in A and x is not in B, which is the same as saying that x is in both A^c and B^c. It follows that $x \in A^c \cap B^c$. In this argument, x was not a specific member of $(A \cup B)^c$, so the argument applies to all members of $(A \cup B)^c$ and establishes that

$$(A \cup B)^c \subset A^c \cap B^c \tag{2.2}$$

Next, we prove the *reverse* containment. Let $x \in A^c \cap B^c$ and retrace the steps in the above argument but going in the opposite direction: x is in both A^c and

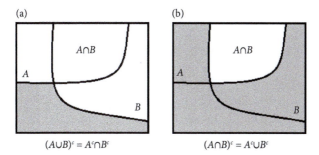

Fig. 2.2 Venn diagrams illustrating De Morgan's laws

B^c so that x is not in A and x is not in B. So x is not in the union $A \cup B$, which means that $x \in (A \cup B)^c$. Because x may be any member of $A^c \cap B^c$, we have shown that

$$A^c \cap B^c \subset (A \cup B)^c \tag{2.3}$$

Now (2.2) and (2.3) complete the proof by double containment.

You can prove the second equality in (2.1) similarly (see the Exercises for this chapter). ∎

The equalities in (2.1) represent *logical statements* in the following sense: replace the sets by propositions p and q, their union by "*or*," their intersection by "*and*," and complementation is the negation, or "*not*"; we discuss these more fully in the section on logic below. Then the first equality in (2.1) reads as:

"*not[p or q]*" is equivalent to "*not p and not q*"

This statement is rather self-evident in the usual context of daily life. For instance, if we know that a ball is not green *or* red then the ball is not green *and* it is not red. Similarly, the second equality reads as:

"*not[p and q]*" is equivalent to "*not p or not q*"

For instance, if someone is not a Girl Scout, then that person is not a scout or not a girl (though could be one or the other).

The above ideas of taking the union or intersection readily extend to any collection of sets in U. If $A_1, A_2, \ldots, A_n \subset U$, then

$$A_1 \cap A_2 \cap \ldots \cap A_n = \{x : x \in A_1 \text{ and } x \in A_2 \text{ and } x \in A_3 \text{ and } \cdots \text{ and } x \in A_n\}$$

A more concise (and useful) way of writing the above intersection is

$$\bigcap_{i=1}^{n} A_i = \{x : x \in A_i \text{ for } all \text{ values of } i = 1, 2, \ldots, n\} \tag{2.4}$$

For instance, if $A_i = \{1, \ldots, i\}$ for $i = 1, 2, \ldots, 10$, then $\bigcap_{i=1}^{10} A_i = \{1\}$ since 1 is the only number that is in all ten of the A sets.

The intersection of a number of nonempty sets may be empty. For example, if A_1 is the set of all even natural numbers, A_2 is the set of all odd natural numbers, and A_3 is the set of all multiples of 3, i.e., $A_3 = \{3k : k = 1, 2, 3, \ldots\}$; then $A_1 \cap A_2 \cap A_3 = \emptyset$, since $A_1 \cap A_2 = \emptyset$, even though both $A_1 \cap A_3$ and $A_2 \cap A_3$ are infinite sets.

If the intersection of a pair of sets is empty, then the sets are *disjoint*.

For example, the sets A_1 and A_2 mentioned above are disjoint.

Notice that since every set A is disjoint from its complement, we may write

$$A \cap A^c = \emptyset$$

Next, if $A_1, A_2, ..., A_n \subset U$, then

$$A_1 \cup A_2 \cup ... \cup A_n = \{x : x \in A_1 \text{ or } x \in A_2 \text{ or } x \in A_3 \text{ or } \cdots x \in A_n\}$$

A more concise (and useful) way of writing the above union is

$$\bigcup_{i=1}^{n} A_i = \{x : x \in A_i \text{ for } \textbf{some} \text{ value of } i = 1, 2, ..., n\} \qquad (2.5)$$

The above concepts of intersection and union of a collection of sets are valid for infinite collections, too. We discuss those later in this chapter.

Products of sets

There is a concept of product for sets that is important in all areas of mathematics, including analysis.

Product. The *product (direct, or Cartesian) of two sets A and B in U is the set* $A \times B$ *of all ordered pairs* (a, b) *with a from A and b from B.*

The aforementioned pair (a, b) is called *ordered* because it is not the same as (b, a), except trivially when $a = b$.

Note that if A (or B) is empty, then so is $A \times B$ because there are no ordered pairs with the first element in A (or second element in B).

When we talk about the Cartesian or coordinate xy-plane in algebra with the usual x-axis and y-axis, we are talking about a collection of ordered pairs (x, y) where both x and y are real numbers.

Like intersection and union, the product operation extends to any number of sets. If $A_1, A_2, ..., A_n$ are given sets, then

$$\prod_{i=1}^{n} A_i = A_1 \times A_2 \times \cdots \times A_n = \{(x_1, x_2, ..., x_n) : x_i \in A_i \text{ for } i = 1, 2, ..., n\}$$

The objects $(x_1, x_2, \ldots, x_n)$ are called *n-tuples* or *n-vectors* depending on the context. For example, in calculus when $n = 3$ and $A_i = \mathbb{R}$ for all three values of i, then

$$\prod_{i=1}^{3} A_i = \mathbb{R} \times \mathbb{R} \times \mathbb{R} = \{(x_1, x_2, x_3) : x_1, x_2, x_3 \in \mathbb{R}\}$$

is the familiar "three-dimensional space." The product of a set with itself n times is usually abbreviated as a power A^n (this is the case $A_i = A$ for $= 1, 2, \ldots, n$ in the definition of the product of n sets). Thus, $\mathbb{R} \times \mathbb{R} \times \mathbb{R} = \mathbb{R}^3$.

We discuss infinite products of sets in Chapter 3. These types of products are of special importance in analysis; after all, they involve infinity!

Intervals

Connected or single-piece subsets of $\mathbb{R}$ are especially important sets in real analysis; they are called *intervals*. For instance, the set of all real numbers, or points on the x-axis that are between -2 and 3.5, including -2 but excluding 3.5, is an interval that is written as $[-2, 3.5)$.

There is a variety of intervals depending on whether or which endpoint is included. If a and b are real numbers, and $a < b$, then we have the following types of intervals:

$$(a, b) = \{x \in \mathbb{R} : a < x < b\} \quad \text{(open interval)}$$
$$[a, b] = \{x \in \mathbb{R} : a \leq x \leq b\} \quad \text{(closed interval)}$$

In the case of an open interval, the two endpoints a and b are excluded; but in a closed interval, both endpoints are included. If only one of the endpoints is included, then we have the *half-open* (or *half-closed*) intervals:

$$[a, b) = \{x \in \mathbb{R} : a \leq x < b\}, \quad (a, b] = \{x \in \mathbb{R} : a < x \leq b\}$$

The above intervals are all *bounded intervals* because the numbers in them never drop below a nor exceed b.

Disconnected or multiple-piece sets of numbers can often be written as unions of intervals. For example, the set

$$\{x : 1 < x^2 < 4\}$$

consists of all numbers x whose squares are larger than 1 but smaller than 4. The number $3/2$ is such a number, as is $-3/2$ since $(-3/2)^2 = 9/4 = 2.25$. But 0 is not such a number because $0^2 = 0$ is not larger than 1; so the set above is not connected. You can readily check (by squaring) that the numbers that work are those in the following union of intervals:

$$(-2, -1) \cup (1, 2)$$

Notice that this disjoint union of intervals is not itself an interval because it misses points between -1 and 1, so it is not a single, connected piece.

Intervals that are not bounded, or *unbounded intervals*, are also routinely used in analysis, and they may be open or closed; for instance,

$$[a, \infty) = \{x \in \mathbb{R} : x \geq a\}, \quad (-\infty, a] = \{x \in \mathbb{R} : x \leq a\}$$

are closed unbounded intervals.

We use the symbol ∞ here to highlight the fact that $[a, \infty)$ contains all real numbers starting from a and including arbitrarily large numbers. We can't close the interval at ∞ because ∞ is not a number but a shorthand for *there is no endpoint on the right*. Similarly, $-\infty$ is shorthand for *there is no endpoint on the left*.

With this in mind, here are the open unbounded intervals:

$$(a, \infty) = \{x \in \mathbb{R} : x > a\}, \quad (-\infty, a) = \{x \in \mathbb{R} : x < a\}$$

The set of all real numbers can also be written as an interval: $\mathbb{R} = (-\infty, \infty)$.

It is worth noticing that every bounded interval is the intersection of two unbounded ones; for instance,

$$[a, b) = [a, \infty) \cap (-\infty, b)$$

For example,

$$(0, 2] = (0, \infty) \cap (-\infty, 2]$$

Infinite vs. unbounded

Notice that every finite set of numbers is bounded; all numbers are greater than the smallest number in the set and less than the largest. In sets like $\mathbb{N}$ or $\mathbb{Z}$, every infinite set is unbounded; for these sets, "infinite" and "unbounded" are the same things. This is *not* the case in sets like $\mathbb{R}$ or the set of rational numbers $\mathbb{Q}$. These sets contain subsets that are infinite and bounded at the same time; for example, the interval $[0,1]$, or even just the set of rational numbers in it, are infinite sets, and yet they are bounded, as they do not stretch to infinity the way an infinite subset of $\mathbb{Z}$ must. We may (loosely) think of $[0,1]$ as finite if it is clear from the context that we mean a finite *interval* and not a finite set.

Boundedness is a *metric* property, having to do with the concept of *length*. A bounded set in $\mathbb{R}$ is either finite or infinite but of *finite length*, like the interval

[0,1], whereas an unbounded set has infinite length and must stretch to infinity like $(-\infty, 1)$.

The interesting sets in analysis are typically bounded *and* infinite, like intervals. Finite sets are rarely of interest in themselves; in fact, with most concepts of length on the number line, *all* finite subsets of $\mathbb{R}$ have length 0. If all single-element sets $\{x\}$ have length zero in some metric (such as the usual Euclidean distance), then every finite set is a disjoint union of these points, so adding zero to itself a finite number of times returns 0 for the length of a finite set.

2.1.2 Relations

It is convenient to define relations as sets and deal with them using set theory. Specifically,

---

If S is a nonempty set, then *a relation R in S is any nonempty subset of* $S \times S = S^2$.

An important relation in $\mathbb{N}$ is "divisibility" $R_d \subset \mathbb{N} \times \mathbb{N}$, which is defined as "$(m, n) \in R_d$ if m divides n" (the remainder of the division is zero). Thus $(2, 2)$ and $(2, 6)$ are both in R_d since $2/2 = 1$ and $6/2 = 3$ with zero remainders in both cases. On the other hand, $(2, 5)$ is not in R_d because when we divide 5 by 2, there is a remainder of 1.

It is sometimes convenient to write aRb, which we read as "a is R-related to b" if $(a, b) \in R$.

For divisibility, the common symbol used is $m|n$ for "m divides n" or $(m, n) \in R_d$.

For a relation R among numbers, the ordered pairs that make up R can be plotted in the usual way in an xy-plane so we may also identify R with its *graph*. We discover later in this chapter that (numerical) functions are special types of relations, and their graphs are indeed ordered pairs of numbers.

Equivalence relations and classes

Among the most frequently encountered relations in mathematics is the *equivalence relation*.

Equivalence relation. Any relation $\sim$ having the following three properties is an equivalence relation: (i) $x \sim x$ for every $x \in S$ (*reflexive property*). (ii) If $x \sim y$, then $y \sim x$ for all $x, y \in S$ (*symmetric property*). (iii) If $x \sim y$ and $y \sim z$, then $x \sim z$ for all $x, y, z \in S$ (*transitive property*).

Equivalence relations often occur in everyday life. For example, when you go to the store and see two shirts that cost the same, then the shirts are "cost equivalent": $x \sim y$ if shirt x has the same price as shirt y (even if x and y have different colors, are not the same size, etc.). You can easily check that all three properties (a)–(c) above hold. Similarly, having "the same color" or having "the same size" also define equivalence relations among shirts, regardless of their prices.

Consider the cost equivalence relation above. For each shirt s, take all the shirts with the same price tag (say, p) as s. This group of shirts is a set E_s that we may call *equivalence class* (or cell) of the shirt s. If a shirt t has the same price p as s then $t \in E_s$.

If a shirt t is priced at q, and $q \neq p$, then *no shirt* in E_t (all of which are priced q) can be in E_s, and conversely. Therefore, the two sets E_s and E_t are disjoint: $E_s \cap E_t = \emptyset$ if s and t have different prices. Further, the *union* of all of these disjoint sets gives all the shirts in the store because every shirt has a price tag. This means that the sets E_s partition all the shirts into disjoint sets classified according to cost. Equivalence classes are sometimes called *cells*.

Every equivalence relation in a set S has its own collection of equivalence classes that partition S into disjoint subsets. If $\sim$ is an equivalence relation in S, then for each element of s, we define the equivalence class of s as the following subset of S:

$$[s] = \{x \in S : x \sim s\}$$

The brackets notation is common, and we stick to it from now on. As the above example of shirts illustrates, these equivalence classes have the following special properties:

If $\sim$ is an equivalence relation in a nonempty set S, then
(a) $[s] = [t]$ if $s \sim t$ and $[s] \cap [t] = \emptyset$ if $s \nsim t$ (s and t are not equivalent);
(b) the union of $[s]$ for all $s \in S$ is the whole set S.

Order relations

When we look at positive integers, we think of 1 as being less than 2, 2 less than 3, and so on. This arrangement of numbers is a way of *ordering* the set $\mathbb{N}$. The essential property of an ordering relation is that it be *transitive*; 4 is less than 6, and 6 is less than 13, so 4 is less than 13.

Order relations share the transitive property with equivalence relations discussed above. But unlike equivalence relations, an order relation can never be symmetric since that would remove the essence of ordering, namely, one element coming before, or after, another.

To properly define order relations, recall the set inclusion relation $\subset$ that we discussed earlier ($A \subset B$ if every element of the set A was also an element of the set B). If $A \subset B$ and $B \subset A$ then A and B are the same sets by double containment, that is, $A = B$. We use this as a template for our next definition.

A relation $\preceq$ is *antisymmetric* if "$x \preceq y$ and $y \preceq x$ imply $x = y$."

This leads to the following definition of order relation:

Order relation. A relation $\preceq$ is an *ordering* or an *order relation* on a set S if it is both transitive and antisymmetric. If $x \preceq y$ and $x \neq y$, then write $x \prec y$ and call $\prec$ a *strict ordering*.

We use the symbol $\succeq$ to mean the opposite or *reverse ordering* (but not the negation) of $\preceq$ in the sense that $x \succeq y$ whenever $y \preceq x$. Clearly, $\succeq$ is an order relation.

The usual ordering $\leq$ for integers is an antisymmetric order relation. An interesting observation about the strict relation $<$ is that it can be defined through algebra: $m < n$ if $n - m$ is positive, i.e., $n - m \in \mathbb{N}$. For instance, $2 < 7$ because $7 - 2 = 5$ is positive. This observation will be important in our construction of the set of real numbers.[4]

Notice that the above definition of ordering is *partial* in the sense that it does not preclude the possibility that for *some* pairs of elements x, y we have *neither $x \preceq y$ nor $y \preceq x$*. This may occur for set inclusion $\subset$, which orders the subsets of a set. For instance, consider the subsets $\{1, 2\}$ and $\{2, 3\}$ of $\mathbb{N}$. Note that neither $\{1, 2\} \subset \{2, 3\}$ holds nor $\{2, 3\} \subset \{1, 2\}$. Therefore, the above definition is often said to define a *partial ordering*.[5]

Set inclusion is then a partial ordering (on the set of *all* subsets of a given set). But the usual ordering of integers is *not* partial. For every pair m, n of integers, we can state that either $m \leq n$ or $n \leq m$ must be true. Because in this book we are typically interested in this type of ordering, we define:

---

[4] If 0 is included in $\mathbb{N}$, then the *non-strict* ordering can also be defined in this way: $m \leq n$ if $n - m$ is positive or zero.

[5] Partial orderings are also defined to be reflexive: $x \preceq x$ for every x. This does not follow from our definition above, even though whenever $x \preceq y$ and $y \preceq x$, we can infer via the transitive property that $x \preceq x$. It is possible that there are elements that are not related to themselves or any other elements by $\preceq$.

Total ordering. A relation $\leq$ is a *total ordering* on a set S if it is transitive, antisymmetric, and *total*; i.e., for all $x, y \in S$, either $x \leq y$ or $y \leq x$.

The divisibility relation $m|n$ is a partial ordering on $\mathbb{N}$, but it is not total. Notice that a total ordering $\leq$ is a reflexive relation via the transitive property. The totality property can also be expressed in terms of the *strict* relation $\prec$ slightly differently as follows:

The relation $\prec$ is a *strict total ordering* on a set S if it is transitive; and for every pair $x, y \in S$, *exactly one* of the following is true: $x \prec y$ or $y \prec x$ or $x = y$.

Because one of three possibilities must be the case, this observation means that $\prec$ has the *trichotomy property*.

It is worth mentioning here that for a *total* ordering, the symbol $\geq$ does in fact mean the negation of $\leq$; however, for the strict case, the negation of $<$ is not $>$ even when it is a total ordering: the negation of $x < y$ is $x > y$ or $x = y$ due to trichotomy, if $<$ is a total ordering.

Ordered sets

If a set has an order relation, then we call it an ordered set. In particular, we have the following:

A nonempty set S together with an order relation $\leq$ forms an *ordered set*. S is a *totally ordered set* if $\leq$ (or $<$) is a total ordering.

The set $\mathbb{N}$ of all natural numbers is an ordered set with its usual order relation $\leq$ (or $<$). We take this as given; but also notice that for every pair of numbers m, n in $\mathbb{N}$, we have

$$m < n \quad \text{whenever } n - m \in \mathbb{N}$$

We can use this fact to see that $<$ is a total ordering since exactly one of the following is true:

$$n - m \in \mathbb{N} \quad \text{or} \quad m - n \in \mathbb{N} \quad \text{or} \quad m = n$$

We will find later that the set of all real numbers, or the number line, is also totally ordered and that its ordering is an extension of the order relation on $\mathbb{N}$ (the two order relations agree on the subset $\mathbb{N}$ of $\mathbb{R}$). This total ordering property of the set of real numbers is one of their identifying characteristics and fundamentally important in analysis.

2.2 Functions and their basic properties

Here is a simple definition of function that will be enough for most of what we do.

A function $f : D \rightarrow S$ is an *assignment* of a *unique* element $f(x)$ in the set S to each element x of D. The set D on which f is defined is its *domain*, and the set S that contains the function values $f(x)$ is the *range* of f.

The word "unique" above works only one way: f can only pick one element in the range for each given element in its domain, but many (or even all) elements in the domain can be assigned a single element in the range.

A correspondence that assigns more than one element in the range to a single element in the domain is not a function, but it is a relation. For instance, in an auditorium that is not fully booked, people may use an extra chair for their coats or other belongings, so the assignment of chairs to people is not a function. In this case, if we reverse the arrow, so to speak, and consider the assignment of people to chairs, then we have a function because two people do not use the same chair.

Every function $y = f(x)$ can be thought of as a set of ordered pairs (x, y) where if the ordered pairs (u, v) and (u, w) are both related via f, then $v = w$ because $v = f(u)$ and $w = f(u)$). In function notation,

$$x = u \Rightarrow f(x) = f(u)$$

The aforementioned set of ordered pairs uniquely identifies the function f; in common usage, it is considered to be the *graph* of f

$$\{(x, y) : y = f(x),\ x \text{ in the domain of } f\} = \{(x, f(x)) : x \text{ in the domain of } f\}$$

since the ordered pairs $(x, f(x))$ or (x, y) can be plotted in the usual xy coordinate system.

The arrow symbol is widely used because functions are often thought of as *mappings*.

Odd-even symmetry

A function has *symmetry relative to the y-axis* if the half on the left of the y-axis is the mirror image (or reflection across the y-axis) of the half to the right. In symbols

$$f(-x) = f(x)$$

This type of function is called an *even function*. The square function x^2 is a simple example; another example is the cosine function $\cos x$.

There also *odd functions*; an odd function $f(x)$ satisfies the identity

$$f(-x) = -f(x)$$

The cube function x^3 is odd because $(-x)^3 = (-x)(-x)(-x) = -x^3$; an important odd function is $\sin x$.

The graph of an odd function is *symmetric relative to the origin*: every point on the graph to the left or below the origin is the reflection across the origin of a point on the graph to the right or above the origin.

The reason for using the terms odd and even in this context is that multiplying such functions works analogously to multiplying odd and even integers, in the sense that multiplying an even number by an odd one produces an odd number, while multiplying two odd or two even numbers yields an even number.

For functions, if $f(x)$ is odd and $g(x)$ is even, then

$$fg(-x) = f(-x)g(-x) = -f(x)g(x) = -fg(x)$$

so the product $fg(x)$ is an odd function. Similarly, if $f(x)$ and $g(x)$ are even functions, then

$$fg(-x) = f(-x)g(-x) = f(x)g(x) = fg(x)$$

so the product fg is an even function. A similar calculation shows that when f and g are both odd, then their product fg is an even function.

Functions as mappings

To better appreciate the mapping aspect of functions, we examine the square function's effect on *intervals* of numbers. Consider first the interval $[0,1]$ and the square function x^2. This function maps $[0,1]$ onto itself. But not every part of this interval is left intact. Compare the effect of the square function on each of the two half-intervals $[0,1/2]$ and $[1/2,1]$, each having length $1/2$. Since

$$f(0) = 0, \quad f\left(\frac{1}{2}\right) = \left(\frac{1}{2}\right)^2 = \frac{1}{4}, \quad f(1) = 1$$

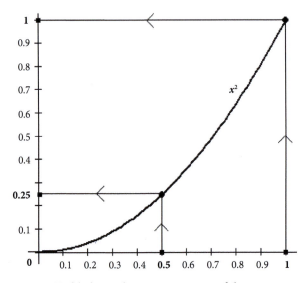

Fig. 2.3 Highlighting the mapping aspect of the square function

the left half $[0,1/2]$ is *compressed* by $f(x) = x^2$ to $[0,1/4]$ that has length $1/4$, while the right half $[1/2,1]$ is *stretched* to $[1/4,1]$ with length $3/4$; see Figure 2.3.

The square function stretches (or compresses) an interval of fixed length by different amounts.

2.2.1 Operations on functions

Functions can be combined in a variety of ways. Numerical functions can be added or multiplied because we can add and multiply numbers in the usual way: if $f(x) = x^2$ and $g(x) = 3x - 1$ then

$$f(x) + g(x) = x^2 + 3x - 1, \qquad f(x)g(x) = x^2(3x - 1) = 3x^3 - x^2$$

Here is another way of combining functions that doesn't require algebraic structure and is therefore more basic.

Composition of functions. Let $f : D \to S$ and $g : S \to T$ be two given functions (notice the shared set S). Then for each $x \in D$, the action of f gives $f(x) \in S$. Since g acts on the set S, the action of g following f gives $g(f(x)) \in T$. This combined action is called the *composition of f and g* and commonly written as $g \circ f$. It is a single function $g \circ f : D \to T$ from D into T.

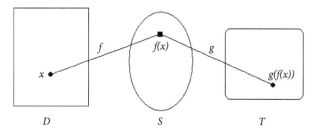

Fig. 2.4 Composition of functions

The above definition is illustrated schematically in Figure 2.4. For example, consider again the functions $f(x)$ and $g(x)$ above. Then

$$g \circ f(x) = g(f(x)) = 3f(x) - 1 = 3x^2 - 1$$

Function composition is *not* commutative; that is, if done in *reverse order*, then we generally get different results.

To begin with, $f \circ g$ need not even be defined just because $g \circ f$ is; see Exercise 35 at the end of the chapter. But even when $g \circ f$ and $f \circ g$ are both defined, they are usually not equal. Going back to the functions $f(x)$ and $g(x)$ above, notice that since

$$D = S = T = \mathbb{R}$$

both $g \circ f$ and $f \circ g$ are defined with the same domain, $\mathbb{R}$. But the latter is not the same as the former:

$$f \circ g(x) = f(g(x)) = g(x)^2 = (3x - 1)^2 = 9x^2 - 6x + 1$$

Function iteration. A function $f : D \to S$ may be composed with itself any number of times to give $f \circ f$, $f \circ f \circ f$, and so on provided that $S \subset D$; see Exercise 35. It is common to use the notation f^n for $f \circ f \circ \cdots \circ f$ (n times) to denote this *iteration* of the function. This iteration describes the repetition of a process.

For instance, suppose that a form of bacteria doubles its numbers (by splitting in half) every so often. Thus, $f(P) = 2P$ for any given population size P. If the original population size is P_0, then the population at first splitting of each bacterium is $f(P_0) = 2P_0$. Later, when splitting occurs again,

$$f^2(P_0) = f(f(P_0)) = f(2P_0) = 2^2 P_0$$

and so on:

$$f^3(P_0) = f(f(f((P_0)))) = f(2^2 P_0) = 2^3 P_0$$

$$\vdots$$

$$f^n(P_0) = 2^n P_0$$

2.2.2 Image and inverse image sets

Consider a function $f : D \longrightarrow S$. This means that f maps each point x of the domain D into an element $y = f(x)$ of the range S. We call y the *image* of x (under f). For instance, if f is the square function $f(x) = x^2$, then the image of $x = 3$ is $y = f(3) = 9$.

A point to keep in mind is that the entire range S may not be covered in this way even if all elements of D are exhausted. For instance, under the square function $f(x) = x^2$, only the non-negative numbers are covered by mapping all of the real numbers. These non-negative numbers are the images of all real numbers (the domain).

For each nonempty subset A of the domain D, the *image set* of A is the subset of S (range) consisting of those elements that are images of some element in A. In symbols,

$$f(A) = \{y \in S : y = f(x) \text{ for some } x \in A\}$$

For instance, the image of the set $A = \{-3, 0, 2, 3\}$ under the square function $f(x) = x^2$ is

$$f(A) = \{f(-3), f(0), f(2), f(3)\} = \{0, 4, 9\}$$

The image set $f(D)$ is a subset of the range S, but it need not be all of S. In the case of the square function, the image set of its entire domain $D = \mathbb{R}$ is

$$f(\mathbb{R}) = \{y \in \mathbb{R} : y = x^2 \text{ for some } x \in \mathbb{R}\} = \{y \in \mathbb{R} : y \geq 0\}$$

where the last inequality holds because every non-negative real number is the square of some real number, possibly negative, like $3 = (-\sqrt{3})^2 = (\sqrt{3})^2$. The interval notation gives a more succinct expression:

$$f((-\infty, \infty)) = [0, \infty)$$

If $f : D \longrightarrow S$ is a given function, then for some y in S there may not be any x in D such that $f(x) = y$; and if such an x exists, then it need not be unique. This necessitates defining a backward mapping.[6]

For a function $f : D \longrightarrow S$ and each element y in S, the *inverse image set* of y is the set $f^{-1}(y)$ of *all $x \in D$ for which $f(x) = y$*. In symbols:

$$f^{-1}(y) = \{x \in D : f(x) = y\}$$

If we think of an element y as a person, then $f^{-1}(y)$ may indicate y's parents or ancestors, depending on how f is defined.

Here's a numerical example: if $f(x) = x^2$ is the square function, then the inverse image of 4 is $f^{-1}(4) = \{-2, 2\}$ because $f(2) = f(-2) = 4$.

Note that $f^{-1}(y)$ is a *subset* of the domain D for every y in S. This idea is illustrated in Figure 2.5.

If f is the square function $f(x) = x^2$, and $D = S = (-\infty, \infty)$, then for $y < 0$ there is no x such that $f(x) = y$ because this would mean that $x^2 < 0$, which is impossible for a real number x. We express this by writing $f^{-1}(y) = \emptyset$. On the other hand, if $y > 0$, then $f(x) = y$ means $x^2 = y$. Taking square roots we find two possible values $x = \pm\sqrt{y}$ that work. Finally, if $y = 0$, then the only x that gives $f(x) = 0$ is $x = 0$. These observations are summarized below:

$$f^{-1}(y) = \{-\sqrt{y}, \sqrt{y}\} \text{ if } y > 0$$
$$f^{-1}(0) = \{0\} \text{ and } f^{-1}(y) = \emptyset \text{ if } y < 0$$

Like the image sets, inverse image sets can be defined of sets, too.

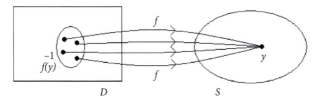

Fig. 2.5 Illustrating the inverse image of a point

---

[6] The notation $f^{-1}(y)$ should not be read as the value of the inverse *function* at the point y. In fact, the inverse image $f^{-1}(y)$ is generally a set containing multiple elements or numbers. We discuss inverse functions later.

For a function $f : D \longrightarrow S$ and each subset B of S, the *inverse image set* of B is the set $f^{-1}(B)$ of *all* $x \in D$ for which $f(x)$ is in B. In symbols

$$f^{-1}(B) = \{x \in D : f(x) \in B\}$$

If $f(x) = x^2$, then the inverse image of the interval $[0,1]$ in the range is the interval $[-1, 1]$. The logical argument is as follows: Let $B = [0, 1]$ and $y \in B$ and look for all real numbers x such that $y = f(x) \in B$. We require all x with the property that $x^2 = y \in [0, 1]$, or equivalently, $0 \le x^2 \le 1$. Since if $-1 \le x \le 1$, then $x^2 \le 1$ and further, $x^2 \ge 0$ (the square of a real number cannot be negative), it follows that $0 \le x^2 \le 1$ whenever $-1 \le x \le 1$; or equivalently, $f(x) \in [0, 1]$ whenever $x \in [-1, 1]$. The definition of inverse image now states that $f^{-1}([0, 1]) = [-1, 1]$.

It is worth mentioning that $f^{-1}([-1, 1]) = [-1, 1]$ is also true. The reason is that the negative part $[-1, 0)$ does not contain the image of any points after squaring, so we conclude that $f^{-1}([-1, 1]) = f^{-1}([0, 1])$.

Image sets and inverse image sets follow specific rules with regard to unions and intersection that are listed next.

Theorem 5 *Let $f : D \to S$ be a given function, and consider the subsets $A_1, A_2 \subset D$ and $B_1, B_2 \subset S$. Then the image sets satisfy*

$$(a) \quad f(A_1 \cup A_2) = f(A_1) \cup f(A_2)$$
$$(b) \quad f(A_1 \cap A_2) \subset f(A_1) \cap f(A_2)$$

and the inverse image sets satisfy

$$(c) \quad f^{-1}(B_1 \cup B_2) = f^{-1}(B_1) \cup f^{-1}(B_2)$$
$$(d) \quad f^{-1}(B_1 \cap B_2) = f^{-1}(B_1) \cap f^{-1}(B_2)$$

Proof I prove (a) and (d) and leave (b) and (c) as exercises. The argument for (a) is basic double-containment: choose any $y \in f(A_1 \cup A_2)$ so that $y = f(x)$ for some $x \in A_1 \cup A_2$. Now, $x \in A_1$ or $x \in A_2$, so $y \in f(A_1)$ or $y \in f(A_2)$, i.e., $y \in f(A_1) \cup f(A_2)$. This conclusion applies to every $y \in f(A_1 \cup A_2)$, so we have shown that

$$f(A_1 \cup A_2) \subset f(A_1) \cup f(A_2)$$

For the reverse inclusion, we start with an arbitrary $y \in f(A_1) \cup f(A_2)$. If $y \in f(A_1)$, then $y = f(x_1)$ for some $x_1 \in A_1$, which also implies that $x_1 \in A_1 \cup A_2$.

Therefore, $y \in f(A_1 \cup A_2)$. If $y \in f(A_2)$, then $y = f(x_2)$ for some $x_2 \in A_2$, which also implies that $x_2 \in A_1 \cup A_2$. So again $y \in f(A_1 \cup A_2)$, and it follows that

$$f(A_1 \cup A_2) \supset f(A_1) \cup f(A_2)$$

and (a) is proved. It is worth mentioning that we could prove the reverse containment without considering the elements. We might notice that since $A_1 \subset A_1 \cup A_2$, it is true that $f(A_1) \subset f(A_1 \cup A_2)$. Similarly, $f(A_2) \subset f(A_1 \cup A_2)$. Since the set $f(A_1 \cup A_2)$ contains both of the smaller sets $f(A_1)$ and $f(A_2)$, it contains their union $f(A_1) \cup f(A_2)$.

To prove (d), we use double containment again. For one direction we don't need to consider elements. Note that $B_1 \cap B_2$ is a subset of both B_1 and B_2, so $f^{-1}(B_1 \cap B_2)$ is contained in both $f^{-1}(B_1)$ and $f^{-1}(B_2)$ and it follows that

$$f^{-1}(B_1 \cap B_2) \subset f^{-1}(B_1) \cap f^{-1}(B_2)$$

For the reverse inclusion, let $x \in f^{-1}(B_1) \cap f^{-1}(B_2)$. Then $x \in f^{-1}(B_1)$, so $y = f(x) \in B_1$; and $x \in f^{-1}(B_2)$, so $y = f(x) \in B_2$; and we conclude that $f(x) = y \in B_1 \cap B_2$. By the definition of inverse image, this means that $x \in f^{-1}(B_1 \cap B_2)$. Since this works for every $x \in f^{-1}(B_1) \cap f^{-1}(B_2)$, we conclude that

$$f^{-1}(B_1 \cap B_2) \supset f^{-1}(B_1) \cap f^{-1}(B_2)$$

and the proof is complete.   ∎

The proper inclusion in Theorem 5(b) is *not a typo*. In this case, equality may not hold for all subsets. Here is a simple example that refutes the equality. Consider the square function $f(x) = x^2$ and the intervals $A_1 = [-1, 0]$ and $A_2 = [0, 1]$. Then

$$f(A_1 \cap A_2) = f(\{0\}) = \{0^2\} = \{0\}$$

which is definitely not equal to

$$f(A_1) \cap f(A_2) = [0, 1] \cap [0, 1] = [0, 1]$$

2.2.3 One-to-one functions and bijections

As we saw in the example of the square function above, the inverse image set of a single element may contain more than one element (if not empty). In fact, for the square function x^2, the *only* number whose inverse image is just one number is 0. But some functions have the special property that $f^{-1}(y)$ is a single element (or possibly empty) for *every y* in the range of *f*. This means that such *f* can only map a *single* element in their domain to each element in their range. Such functions are reasonably called *one-to-one*.

One-to-one function. A function $f : D \longrightarrow S$ with domain D is *one-to-one* or *injective* on D if for each pair of elements u, v in D the following is true:

$$f(u) = f(v) \Rightarrow u = v \tag{2.6}$$

For a simple illustration of how to use (2.6), consider $f(x) = 2x - 1$. Here D and S are both the entire number line $(-\infty, \infty)$. If u, v are any pair of numbers that satisfy $f(u) = f(v)$, then we quickly derive

$$2u - 1 = 2v - 1$$
$$2u = 2v$$
$$u = v$$

This shows that (2.6) is true for arbitrary pairs of numbers in D, so $2x - 1$ is a one-to-one function. If we try this calculation with the square function $f(x^2)$, then we obtain

$$u^2 = v^2$$
$$u = \pm v$$

Without further restriction (e.g., $x \geq 0$), the square function is not one-to-one. An equivalent way of stating the above definition is sometimes useful:

$$u \neq v \Rightarrow f(u) \neq f(v) \tag{2.7}$$

This is the contrapositive of the original definition above and logically equivalent to it (see section 2.3). On the other hand, the converse statement

$$u = v \Rightarrow f(u) = f(v) \tag{2.8}$$

that we get by simply reversing the implication in (2.6) is not equivalent to the original. In fact, (2.8) is true for *every* function f (one-to-one or not) since a function can only assign one value to each element of its domain.

Monotone functions

Many numerical functions have the useful feature that their graphs are always rising or always falling. The horizontal line test shows, intuitively, that all such functions are one-to-one. Here's the formal definition.

Increasing, decreasing, and monotone functions

Let $f : D \rightarrow S$ be a function where D is totally ordered by a relation $<$ and S is totally ordered by a relation $\prec$. Then f is an *increasing* function on the set D if for every pair of elements x, y in D

$$x < y \Rightarrow f(x) \prec f(y) \tag{2.9}$$

Similarly, f is a *decreasing* function on D if

$$x < y \Rightarrow f(y) \prec f(x) \tag{2.10}$$

If the relation $<$ is replaced by its *non-strict* form $\leq$ in the above implications, then we say that f is *nondecreasing* (or *non-increasing*). We call a function *monotone* on D if it is either nondecreasing or non-increasing on D.

The function $f(x) = x^3$ is increasing on the entire number line $(-\infty, \infty)$, and its negative $-x^3$ is decreasing on $(-\infty, \infty)$. Both are monotone functions.

The square function is neither increasing nor decreasing on $(-\infty, \infty)$; therefore, x^2 is not monotone on $(-\infty, \infty)$. However, x^2 is increasing on the interval $[0, \infty)$ and decreasing on $(-\infty, 0]$, so it is monotone on each of these latter two intervals.

A monotone function is basically one that does not oscillate; i.e., its direction does not change. A function may be monotone but not strictly; for example,

$$f(x) = \begin{cases} x^2, & \text{if } x < 0 \\ 0, & \text{if } x \geq 0 \end{cases}$$

is decreasing on $(-\infty, \infty)$, but not strictly since $f(x) = 0$ is constant for all positive numbers x. An extreme example of a monotone function is a *constant function*

$$f(x) = c$$

where c is any fixed number. This function is trivially non-increasing *and* nondecreasing everywhere.

The idea of increasing or decreasing functions is not limited to numerical functions and applies to functions on any ordered set. For example, consider the following collection of subsets of $\mathbb{N}$:

$$\{1\}, \{1, 2^2\}, \{1, 2^2, 3^2\}, \{1, 2^2, 3^2, 4^2\}, \ldots$$

Note that this collection is totally ordered by set inclusion $\subset$. The function $\#$ that assigns the number of elements in a (finite) set to the set itself is a function whose domain includes this collection of sets, and its range is contained in $\mathbb{N}$. For instance,

$$\#\{1, 2^2, 3^2\} = 3$$

This function (whose domain consists of sets instead of numbers) is increasing in the sense that if A and B are any two distinct sets in the collection, and $A \subset B$, then $\#A < \#B$.

The following result is general and quite useful.

Theorem 6 *A function $f : D \rightarrow S$ that is either increasing or decreasing on D is one-to-one.*

Proof Suppose that f is increasing on D, and let u, v be an arbitrary pair of elements in D. If $u \neq v$, then either $u < v$ or $v < u$ since $<$ is a total ordering on D. If $u < v$, then by (2.9) $f(u) < f(v)$; so in particular, $f(u) \neq f(v)$. This shows that (2.7) is true; the same argument works if $v < u$ since we can just switch u and v. So f is one-to-one. A similar argument works if f is decreasing. ∎

The *converse* of the above theorem is not true. It is possible for a one-to-one function to *not* be monotone, and a simple example of this type of function is $f(x) = 1/x$. This is one-to-one on its domain, the set of all nonzero real numbers, because if u, v are two nonzero numbers, then

$$f(u) = f(v) \Rightarrow \frac{1}{u} = \frac{1}{v} \Rightarrow u = v$$

The familiar graph of this function is a hyperbola that is decreasing on both the set of negative numbers and the set of positive numbers. However, $1/x$ is not monotone because as the sign of x changes from negative to positive, the value of $f(x)$ jumps *up* from negative to positive, causing a change in its otherwise downward form.

It may seem that for every function on the number line $(-\infty, \infty)$ it must be possible to split the line into small enough intervals such that on each interval the function is monotone. This is possible to do for the functions that typically appear in a standard calculus textbook, including trigonometric functions of very short periods (or high frequency). But we see later that there are functions that are not monotone on *any* interval, no matter how short. We cannot visualize such functions, but they appear quite naturally in analysis.

Bijections

Suppose that a function covers its entire range, which means that *every* element of the range is assigned to some element of the domain. Such a function is *onto* its range, or more technically, the function is *surjective*. This feature can be combined with one-to-one to give a special type of function.

A function $f : D \longrightarrow S$ is onto S, or surjective, if the image set of D is all of S, that is, $f(D) = S$. In words, for every y in S, there is (at least) one x in D such that $y = f(x)$. A function that is *both* one-to-one and onto is called a *bijection*.

As a simple example of a bijection, consider $f(x) = 2x - 1$. The domain and range are the same set, namely, the entire number line, so $D = S = (-\infty, \infty)$. We previously saw that this function is one-to-one; to show that it is also onto $(-\infty, \infty)$, pick an arbitrary y in this set; we need to find a value of x such that $y = 2x - 1$. The obvious thing to do is to solve the last equation for x to get $x = (y + 1)/2$. This value satisfies the definition of onto functions for our chosen y. So we may justifiably claim that for every y in $S = (-\infty, \infty)$, there is $x = (y + 1)/2$ in $D = (-\infty, \infty)$ such that $y = f(x)$. Therefore, our function is onto as well as one-to-one, so it is a bijection.

This argument can be extended to all affine functions

$$f(x) = ax + b \tag{2.11}$$

to prove that these functions are bijections of the whole number line $(-\infty, \infty)$ for each given pair of real numbers a, b provided that $a \neq 0$.

Bijections are important in mathematics, as they represent a common feature of maps that express the equivalence of mathematical structures, like group isomorphisms in algebra or isometries of metric spaces—namely, transformations of metric spaces that preserve distances between points, such as rotations or translations of the xy-plane.

An essential feature of bijections is their role as counting tools.

Counting with bijections. *One of the most important uses of bijections is to count elements of sets by setting up a unique matching between elements of a set whose cardinality is known and another less familiar set.*

If A and B are finite sets with $\#A = m$ and $\#B = n$ then a bijection from A onto B exists only when $m = n$.

To see why, note that if $f : A \longrightarrow B$ is a bijection, then every element a in A is assigned to exactly one b in B, namely, $b = f(a)$. So B has at least as many elements as A, which means that $m \leq n$. On the other hand, f is also onto B so that for each *element b in B, there is some a in A that is assigned to b. This means that A has at least as many elements as B, so m ≥ n. Therefore, it must be true that m = n.*

This type of reasoning does not extend immediately to infinite sets since we do not have the sizes m and n that can be readily compared to each other. In particular, if a finite set is a proper subset of another, then these sets cannot have the same number of elements. However, this is not the case for infinite sets: it is often easy to find bijections between an infinite set and proper infinite subsets of it. I will discuss this and related issues later in this chapter.

Inverse functions

If f is a bijection, then as we saw above the inverse image $f^{-1}(y)$ of every element in S is a single element set $\{x\}$ in D. This means that f^{-1} can be viewed as a function $f^{-1} : S \longrightarrow D$ called the *inverse function* of f. Rather than writing $f^{-1}(y) = \{x\}$, we drop the braces and write $f^{-1}(y) = x$.

Calculating inverse functions of bijections is not an easy task generally and often requires advanced methods. However, simple examples do exist. We saw earlier that $f(x) = 2x - 1$ is a bijection. For each y, there is a number $x = (y + 1)/2$ that is mapped to y by the function $f(x)$. So we conclude that $f^{-1}(y) = (y + 1)/2$.

The following is a useful result about inverse functions that we use later on. The proof is left as an exercise.

Theorem 7 *If $f : D \to S$ is a bijection, then the following are true:*
  (i) $f^{-1} : S \to D$ is also a bijection
  (ii) $f^{-1}(f(x)) = x$ and $f(f^{-1}(y)) = y$ for every $x \in D$ and $y \in S$

Important examples of inverse functions in analysis are the logarithmic and exponential functions that we discuss in Chapter 7.

2.3 Basics of logic concepts and operations

In this section, we discuss a few basic logical concepts that are essential to a meaningful discussion of some mathematical topics in later chapters. It is true that many ideas in calculus, including some that involve infinity, can be illustrated by diagrams and concrete examples, and we put this thought to good use. But some ideas, especially those involving deeper manifestations of infinity, can be properly understood only through the use of logical deduction.

Analysis, like other branches of mathematics, uses *binary logic* (every statement is either *true* or *false*) to deduce its results and communicate them. The advantage of this approach is twofold: not only do we obtain *precise* results in the sense that once a statement is proven true it is never false, but it also makes *proof by contradiction* possible. A substantial number of results in analysis that involve infinity would not be provable (i.e., it is not possible to prove them true or false) without using proof by contradiction.

The discussion in this section is largely informal, relying on ordinary language rather than the usual precise and economical symbolism of mathematical logic. If you are curious about the details and value purity and precision, then you should definitely check a textbook on mathematical logic and preferably *also* one on axiomatic set theory. The presentation here is focused on logic's role as a language and communication tool for analysis.

2.3.1 Fundamental connectives and truth tables

Mathematical statements can be combined, as logical *propositions*, to create more complex statements using the *conjunction* "*and*," the *disjunction* "*or*" (inclusive), and the *implication* "$\Rightarrow$."

If p and q are two statements or propositions, then
(a) the conjunction [p *and* q] is true only when both p is true and q is true; otherwise, it is false;
(b) the disjunction [p *or* q] is false only when both p is false and q is false; otherwise, it is true;
(c) the implication [$p \Rightarrow q$] is false only when p is true and q is false; otherwise, it is true;
(d) p is (logically) *equivalent* to q, denoted [$p \Leftrightarrow q$] if [$p \Rightarrow q$ *and* $q \Rightarrow p$]. Where confusion is unlikely, we drop the brackets from the implication and just write $p \Rightarrow q$.

You may have noticed from the above description that *a false statement implies anything*. Here is an amusing anecdote:

Bertrand Russell wins an argument: *While attending a party, Russell tried to explain the point about false statements to someone who kept rejecting it initially but who eventually agreed to accept it if Russell could prove that "1 = 0" implies "Russell was the Pope." After a moment's reflection, Russell said if 1 = 0 then 2 = 1." Pope and I are two, so Pope and I are one.*

You can extend Russell's idea to show that "1 = 0" implies that all human beings are 5 feet tall, or that all Americans are vegetarians. Just show that 1 = 0 implies $N = 1$ for any positive integer N and the rest should be easy!

Proving necessity and sufficiency

In mathematical statements of theorems and their proofs, $p \Rightarrow q$ as "p implies q," by which we mean "if p is true, then q is true." Also, $p \Leftrightarrow q$ is written as "p (is true or false) if and only if q (is true or false)". The combination *if and only if* is often abbreviated as *iff*. To prove a statement containing "if and only if," we must prove the truth of *both* the "if" part $p \Rightarrow q$ and the "only if" part $q \Rightarrow p$. The order in which this is done is not important and usually dictated by pedagogy. In mathematics, an iff statement is also called a *necessary and sufficient condition*.

Truth tables

A simple way of checking the validity of abstract logical statements is provided by means of their *truth tables*. Each statement p is assigned one of two possible "truth values": T for true and F for false. The definitions of fundamental connectives are more comprehensively illustrated using their truth tables.

For conjunction (and), we have the following table:

p	q	p and q
T	T	T
T	F	F
F	T	F
F	F	F

This shows what (a) at the beginning of section 2.3.1 stated as well as other possible combinations that lead to the false outcome. For disjunction (inclusive or), we have the table

p	q	p or q
T	T	T
T	F	T
F	T	T
F	F	F

The only combination that yields a false outcome is when both p and q are false. For implication, the truth table is as follows:

p	q	$p \Rightarrow q$
T	T	T
T	F	F
F	T	T
F	F	T

REAL ANALYSIS AND INFINITY 45

Using combinations of just these three truth tables, we can create many more new truth tables. For example, consider the compound statement

$$[p \Rightarrow q \text{ and } p] \Rightarrow q \qquad (2.12)$$

In addition to the two columns for p and q, we need to have columns for the conjunction and the implications:

p	q	$p \Rightarrow q$	$p \Rightarrow q \text{ and } p$	$[p \Rightarrow q \text{ and } p] \Rightarrow q$
T	T	T	T	T
T	F	F	F	T
F	T	T	F	T
F	F	T	F	T

To see why the entry in first row of truth values and fourth column is a T, we check the entries in the first row for $p \Rightarrow q$ and p: they both have the value T, so their conjunction also comes up T. In the second, third, and fourth rows of the same column, one of $p \Rightarrow q$ or p has the value F, so their conjunction also has value F. In the final column for the outermost implication, we consider the truth values for $[p \Rightarrow q \text{ and } p]$ and those for q, in that order, and find that the combination. T $\Rightarrow$ F never occurs. So according to the truth table for implication, the final column can only have T as valid entries.

A truth table such as the last one that we considered with all entries of its final column being T shows that the statement in the final column is always true. In this case, we have a *tautology*.

Quantifiers

Also essential are the two *quantifiers* for statements that contain *variables* like x (x may be a number, vector, or any other valid concept):

Quantifiers

The *universal quantifier*: [*for all* x]$p(x)$ means that the statement or proposition $p(x)$ is *true for all values* of the variable x. The *existential quantifier*: [*there exists* x]$p(x)$, or [*for some* x]$p(x)$, means that the statement or proposition $p(x)$ is *true for some value* of the variable x. The notation [*there exists* x] is usually read as "there exists (or there is) an x such that." The symbolic abbreviations $\forall$ for "for all" and $\exists$ for "there exists" are often used in more advanced or more formal contexts, but we will not use them.

Propositions or statements containing more that one variable often occur in math. If $p(x, y)$ is a statement with two variables, then we may use more than one quantifier to "bound" the variables, as in [*for all x*][*there exists y*]$p(x, y)$. A concrete example is

"for all x, there exists a y such that $x + y > 0$"

Note that any $y > -x$ will do in this case; so in general, no assumption of uniqueness is involved for the "there exists" quantifier. Of course, if I replaced the inequality $>$ with equality $=$, then only a unique $y = -x$ would work.

Negation

The definition is straightforward:

The *negation* of a statement p is [*not p*].

It is evident that the negation of p is true when p is false and vice versa.

The negations of conjunctions, disjunctions, implications, and the quantifiers are obtained as follows:

---

not[*p and q*]$\Leftrightarrow$[*not p or not q*];
not[*p or q*]$\Leftrightarrow$[*not p and not q*];
not[*p* $\Rightarrow$ *q*]$\Leftrightarrow$[*not p or q*];
not[*for all x*]$p(x)$ $\Leftrightarrow$[*there exists x*]*not* $p(x)$; or perhaps easier to read, [*for some x*]*not* $p(x)$
not[*there exists x*]$p(x)$ $\Leftrightarrow$[*for all x*]*not* $p(x)$; or perhaps easier to read, *not*[*for some x*]$p(x)$$\Leftrightarrow$[*for all x*]*not* $p(x)$

The negation of implication can be stated more informally as: $p \nRightarrow q$ or "p does not imply q" if "p is true and q is false."

We sometimes need to consider the negation of composite statements where the above list comes in very handy. For instance, consider the following mathematical statement:

For all x, there is y such that $p(x, y)$ is true.

The negation of this may be worked out as follows:

not[*for all x*[[*there exists y*]$p(x, y)$]]$\Leftrightarrow$*there exists x*[*not*[*for some y*]$p(x, y)$]]
$\Leftrightarrow$*there exists x*[*for all y*]$p(x, y)$

The last statement can be written less formally as

there is x such that $p(x, y)$ is true for all y.

Notice the absence of the word "not" in the above negation. Switching between quantifiers is commonplace in analysis arguments.

2.3.2 Converse, contrapositive, and contradiction

Related to the implication are the following two concepts that arise often in mathematical writing:

The *converse* of $p \Rightarrow q$ is $q \Rightarrow p$; the *contrapositive* of $p \Rightarrow q$ is $[not\ q \Rightarrow not\ p]$

The truth of $p \Rightarrow q$ has no bearing on its converse $q \Rightarrow p$, which may be true or false; for instance, assuming that "if we eat tainted food, then we get sick" is a true statement, its converse "if we are sick, then we ate tainted food" is certainly open to question. However, the contrapositive of $p \Rightarrow q$ is logically equivalent to it:

$$[not\ q \Rightarrow not\ p] \Leftrightarrow [p \Rightarrow q]$$

This is easy to show using the truth table method:

p	q	*not q*	*not p*	*not q* $\Rightarrow$ *not p*
T	T	F	F	T
T	F	T	F	F
F	T	F	T	T
F	F	T	T	T

Notice that the rightmost column has the same truth values as the the truth table for $p \Rightarrow q$ regardless of the truth values of p or q. Therefore, we have equivalence.

Contradiction. A *contradiction* is a statement that is always false. The most basic contradiction is the statement

$$p\ and\ [not\ p] \tag{2.13}$$

Let's check the truth table for this statement:

p	*not p*	*p and [not p]*
T	F	F
F	T	F

Notice that the only truth value in the rightmost column is F. This is generally true of contradictory statements that turn out to be the opposites of tautologies.

Proof by contradiction, or indirect proof. *This usually involves showing that negating the conclusion to be proved implies the negation of some accepted fact or hypothesis p so we obtain p and [not p]. Since this false statement results from negating the conclusion, the conclusion has to be valid. A proof done by contradiction is also called an indirect proof.*

Proofs by contradiction appear *routinely* in analysis and are often unavoidable when working with infinity. To give a flavor of such a proof, we prove Euclid's theorem on primes.

Consider the set of all prime numbers: $P = \{2, 3, 5, 7, 11, \ldots\}$. Recall that a positive integer is a *prime* if it is not 1, and the only integers that divide are 1 and itself. Are there infinitely many primes?

Euclid proved that the answer is yes:

Euclid's Theorem *There are infinitely many prime numbers.*

Proof   Here the statement p is "There are infinitely many prime numbers." By way of contradiction, assume [*not p*], i.e., "there are finitely many prime numbers." Then we may list all of them in the order of increasing magnitude as follows:

$$2 = p_1 < 3 = p_2 < 5 = p_3 < \cdots < p_k \text{ (the largest prime)}$$

Note that the product $p_1 p_2 \cdots p_k$ is divisible by every one of these prime numbers. Further, the number $n = p_1 p_2 \cdots p_k + 1$ is not prime by our assumption since it is larger than p_k. So there is a prime (at least one) that divides n, and this prime must be one of $p_1, \ldots, p_k$ (assumed to be all of the primes). Suppose that it is p_j, and note that p_j also divides the product $p_1 p_2 \cdots p_k$. Since p_j also divides n, it must divide the difference $n - p_1 p_2 \cdots p_k = 1$. But this negates the fact that no integer divides 1 (other than 1) and implies a contradiction. This contradiction occurred because we assumed that there are finitely many primes, so this assumption is false. We conclude that there are infinitely many primes.   ∎

We encounter may more instances of proofs by contradiction in later chapters.

2.3.3 Rules of inference

Proofs in mathematics typically involve "deduction" and proceed via *rules of inference*. The most fundamental is the following:

> **Modus ponens** $[p \Rightarrow q \text{ and } p] \Rightarrow q$,
> in words: "if p implies q, and p is true, then q is true."

For instance, suppose that "if it is raining, then we need an umbrella." This implication can be written as $p \Rightarrow q$ if p is the statement "it is raining" and q is "we need an umbrella." Now the modus ponens says something rather obvious, that is, if we grant the implication and also grant that it is raining, then it follows that we need an umbrella. Indeed, modus ponens is always true (or a *tautology*) as we saw in the truth table for (2.12) above.

Many mathematical theorems are proved by applying the rules of inference multiple times in the form of a chain if necessary, along with the other facts that are mentioned above.

A specific example from calculus is the following: we show later that "if $f(x)$ is a continuous function, then it is integrable." We also find that "the function $f(x) = \sin(x^2)$ is continuous, since it is a composition of two continuous functions," namely, $\sin x$ and x^2. Therefore, by the modus ponens, $\sin(x^2)$ is integrable. This is mathematically useful information since there is no known elementary formula for the integral of $\sin(x^2)$ nor a simple proof without using continuity that $\sin(x^2)$ is integrable.

A similar statement that is always true (can be verified using a truth table) is the *transitivity of implication*. This is another rule of inference, which is often called the *(hypothetical) syllogism:*[7]

> **Syllogism (hypothetical)** $[p \Rightarrow q \text{ and } q \Rightarrow r] \Rightarrow [p \Rightarrow r]$

Here is an example from calculus: we prove later that "if a function $f(x)$ has a derivative then it is continuous"; and "if $f(x)$ is continuous, then it is integrable". From these statements we infer that "if $f(x)$ has a derivative, then it is integrable." Again it is easier to find the derivative of a function than its integral, so the information here is quite useful. For instance, using basic derivative rules, not only can we show $\sin(x^2)$ has a derivative but in fact calculate its derivative as $2x \cos(x^2)$. We can tell that $\sin(x^2)$ is integrable without knowing what its integral is. This

---

[7] The term "hypothetical" is used because the syllogism involves a hypothetical statement, if p then q. This is a little different from the more familiar "categorical syllogism" consisting of a trio of statements like "All mammals are apes," and "Uncle Bob is a mammal," from which it follows that "Uncle Bob is an ape."

knowledge comes to us indirectly through logical relations from a relatively simple derivative formula.

2.4 Mathematical induction

Suppose that we want to prove a statement like

$$2^n > 2n + 1 \quad \text{for every positive integer } n \geq 3$$

We can certainly insert specific values for n like $n = 3$ and calculate $2^3 = 8$ while $2(3) + 1 = 7$ to verify the validity of the statement for $n = 3$. But doing so is not enough to prove the statement since we cannot substitute all the infinitely many integers that the statement refers to.

Proving a statement or property $P(n)$ for infinitely many integer values of n is done using *mathematical induction on n*. The basic idea is the fact that $n + 1$ is the next integer (or the least integer larger than n) for every given integer n.

Theorem 8 *(Mathematical induction, or the induction principle) Let m be a fixed integer (often m = 1) and let P(n) be a statement or property for every n ≥ m. If the following two conditions are satisfied, then P(n) is true for n ≥ m:*
  (a) P(m) is true,
  (b) If we assume that P(k) is true for some k ≥ m, then P(k + 1) is also true.

We prove this theorem shortly but first some background. The induction principle essentially states that if we *start* with a valid statement and show that for every valid statement, the next statement is valid too, then all the subsequent statements will be valid. It is easy to see the importance of (b), often called the *induction step*; to see that the *initial step* or starting point (a) is also important, consider the false statement:

$$FP(n) : n + 1 = n \quad \text{for every positive integer } n$$

Note that $FP(1)$ is false since $1 + 1 \neq 1$, so (a) is not true. However, if as stated in (b), we *assume* (induction hypothesis) that $FP(k)$ is true for some positive integer $k \geq 1$, i.e., $k + 1 = k$, then $FP(k + 1)$ is also true since

$$(k + 1) + 1 = k + 1$$

and it follows that (b) is valid under the induction hypothesis alone.

While the induction principle seems self-evident, it is not an axiom; and since it involves infinity, it is not something that we can intuitively accept. The principle of mathematical induction is rooted in the *Well-ordering Property* (often taken as an axiom) for the set of positive integers. This states that the set $\mathbb{N}$ is not only totally ordered by the usual order relation $\leq$, but it actually has the following additional property:

The well-ordering principle: *Every nonempty subset of $\mathbb{N}$ has a least element.*

We say that $\mathbb{N}$ is well-ordered. Note that the set $\mathbb{Z}$ of all integers is totally ordered but not well-ordered. But if m is any *negative* integer, then the set

$$\mathbb{N}_m = \{m, m + 1, \ldots, -1, 0\} \cup \mathbb{N}$$

is also well-ordered since if a subset S of $\mathbb{N}_m$ contains numbers that are not in $\mathbb{N}$, then the least element of S is simply one of the numbers $m, m + 1, \ldots, -1, 0$.

The interval $[0, 1]$ of real numbers is also totally ordered under its usual order relation, but it is not well-ordered because, e.g., the subset $(1/2, 1]$ has no least element ($1/2$ is not in this interval).

We can now prove the induction principle using the well-ordering property.

Proof (of the induction principle) We prove by contradiction: assume that (a) and (b) are true for some $P(n)$ but that $P(n)$ is false for some integer $n_0 \geq m$. Then the set

$$F = \{j \in \mathbb{Z} : j \geq m, \ P(n_0) \text{ is false}\}$$

contains the number n_0 and therefore is nonempty. Because F is a subset of the well-ordered set $\mathbb{N}_m$, it must have a least element, say j_0.

Note that $P(j_0)$ is false and further, $j_0 > m$ because $P(m)$ is true. But then $j_0 - 1 \geq m$, and $j_0 - 1$ is not in F, so $P(j_0 - 1)$ is true. Thus, (b) implies that

$$P(j_0 - 1 + 1) = P(j_0)$$

is true, which is a contradiction. So the set F must be empty, and for that to happen, the integer n_0 above cannot exist. ∎

To illustrate mathematical induction, we prove the statement:

$$P(n) : 2^n > 2n + 1 \quad \text{for } n \geq 3$$

Take $m = 3$ and observe that

$$P(3) : 2^3 = 8 > 7 = 2(3) + 1$$

So $P(m)$ is true for $m = 3$, and (a) is verified. Now we must show that (b) also holds. Assume that $k \geq 3$ and that $P(k)$ is true, i.e.,

$$2^k > 2k + 1 \tag{2.14}$$

With this as given, we proceed to show $P(k + 1)$ must be true, i.e.,

$$2^{k+1} \overset{?}{>} 2(k + 1) + 1 = 2k + 3 \tag{2.15}$$

Note that

$$2^{k+1} = (2)2^k = 2^k + 2^k > (2k + 1) + 2^k \tag{2.16}$$

The inequality is valid by the induction hypothesis (2.14). Further, since $k > 1$, we have

$$2^k > 2^1 = 2$$

and using this in (2.16) gives

$$2^{k+1} > 2k + 1 + 2 = 2k + 3$$

which verifies (2.15). Since we have now shown that (b) is also true, mathematical induction completes the proof that the statement $P(n)$ above is valid.

We close this section with another illustration of the induction principle by proving the following useful equality:

$$1 + 2 + 3 + \cdots + n = \frac{n(n + 1)}{2} \quad \text{for all } n \in \mathbb{N}$$

Let $P(n)$ be the label for the above statement with $m = 1$. Then $P(1)$ states that

$$1 = \frac{1(1 + 1)}{2}$$

which is true. So condition (a) above is verified. Next, assume that $P(k)$ is true for some $k \geq 1$, i.e.,

$$1 + 2 + 3 + \cdots + k = \frac{k(k + 1)}{2}$$

To verify condition (b), we need to show that $P(k + 1)$ is true, i.e.,

$$1 + 2 + 3 + \cdots + k + (k + 1) \overset{?}{=} \frac{(k + 1)(k + 2)}{2} \tag{2.17}$$

By the induction hypothesis,

$$(1 + 2 + 3 + \cdots + k) + (k + 1) = \frac{k(k + 1)}{2} + (k + 1)$$

$$= \frac{k(k + 1) + 2(k + 1)}{2}$$

$$= \frac{(k + 1)(k + 2)}{2}$$

The last equality verifies (2.17) and proves (b). The induction principle now completes the proof.

2.5 Bijections and cardinality

Infinity is not an intuitive concept, but intuition *does* help us distinguish between infinite things that *look different* to the mind, like $\mathbb{N}$ and $\mathbb{R}$, both of which are actually *idealizations* of what our intuition tells us. We visualize $\mathbb{R}$ as a straight line (the x-axis), a sort of idealized long, thin rod. But we think of $\mathbb{N}$ as a discrete collection of points, or numbers 1, 2, 3, etc. usually placed on the x-axis. Both $\mathbb{R}$ and $\mathbb{N}$ are infinite, yet they look different. Are they *actually* different? In what precise sense?

The above question strikes at the heart of the matter as far as analysis is concerned. The description that I gave used imprecise language (idealized rod, discrete points) that does not stand up to scrutiny. For example, the numbers in $\mathbb{N}$ are also points in $\mathbb{R}$, so we may conclude that the latter is a larger collection of objects than the former. But if they are both infinite, then what does it mean to call one larger than the other? In fact, the same question arises regarding other special numbers, like are there more positive integers than even positive integers?

To understand the difference between $\mathbb{N}$ and $\mathbb{R}$, we need to go beyond intuition and develop sufficiently precise concepts and language that stand up to scrutiny.

Set versus magnitude

We begin by considering the distinction between set and *magnitude*. It is not necessary to define "infinite magnitude" (which is also intuitively incomprehensible) in order to define the concept of an infinite set. This is as a *single* object that contains an infinite number of objects.

For example, imagine a bucket of sand, which is thought of as a single object (one bucket) that contains a very large number of sand grains; we can think of the bucket without having to visualize every single grain of sand. In a similar vein, we can think of sets as being "actually infinite" without having to imagine, or work with, infinite numbers or magnitudes. Physical reality is inconceivable without physical measurements, something that we cannot apply to infinite (or infinitesimal) magnitudes.

The above idea suggests that we may work with infinity in a consistent way using sets without having to define infinite magnitudes. The level of precision in this section falls far short of what we may find in a book on axiomatic set theory, but it is precise enough to clarify the issues that I mentioned above. Let's begin with a popular example to illustrate counting infinite sets.

There is always a vacancy at Mr. Hilbert's *Hotel Infinity*

You may have had the unpleasant experience of wanting to book a room at a hotel or a seat on an airliner during the peak travel season but finding that no vacant rooms or empty seats are available. These mundane problems belong to our finite world; in a realm where the infinity is a thing, these issues aren't!

Consider a hotel with an infinite number of rooms and call it *Hotel Infinity*.[8] Even with a firm policy of allowing only one guest per room, the proprietor Mr. Hilbert discovered that he could always find vacancies in his unusual place without throwing anyone out; here is how: If all rooms are full when a new guest arrives then he moves the guest in Room 1 to Room 2, the guest in Room 2 to Room 3 and so on. In this way Room 1 is made available and all existing guests still have rooms of their own. There is no *last room* whose occupant might either have to be kicked out or be forced to share it with another guest (or contrary to the hotel's own policy).

It is also remarkable that this shifting of guests can be done over and over again, so even more guests can be accommodated, should more of them arrive. Throughout this process, the hotel remains the same (no new construction necessary), and the policy of one guest per room is strictly enforced. Hilbert's hotel can be more paradoxical yet: it can accommodate infinitely many new guests not just any (finite) number of them (see the Exercises).

This story shows that subsets of infinite sets can have as many elements as their parent sets. Let's see what this means next.

Counting things by pairing them up

Important in studying infinite sets is to find a consistent way to compare the cardinalities (or sizes) of infinite sets. This is done using bijections, which we discussed previously.

---

[8] This hotel was introduced as a thought experiment by the German mathematician David Hilbert (1862–1943), who called it the "Grand Hotel" in a 1924 lecture on infinity. Hilbert's lecture was not published, but it was popularized by the famous physicist George Gamow in his classic 1947 book *One Two Three ... Infinity* and since then by many other authors.

> A pair of sets A and B have the same cardinal number, or #A = #B, if there is a bijection from A onto B.

For finite sets, this amounts to saying that A and B have the same number of elements. But for infinite sets, the situation is more complex. In particular, an infinite set can have the same cardinal number as some of its proper subsets. In fact, following Cantor, we may define infinite sets as just such sets![9]

> **Infinite set** (Cantor) *A set is infinite if it can be put in a one-to-one correspondence (bijection) with a proper subset of itself.*

Consider Hilbert's hotel again. The set of guests in the original fully booked hotel is a *proper* subset of the set of guests in the hotel after new guests arrive. If we label the first guest who arrives after the hotel is "fully booked" as g_0 and let $g_1, g_2, g_3, \ldots$ be the guests already in rooms 1,2,3, etc., then we can set up a matching $f : \{0, 1, 2, 3, \ldots\} \to \mathbb{N}$ between the new number of guests and the set of rooms:

$$f(g_n) \to n + 1, \quad n = 0, 1, 2, 3, \ldots$$

This f is a bijection with $f(g_0) = 1$ with $n = 0$ (the new guest is placed in room 1), $f(g_1) = 2$ with $n = 1$ (the guest already in room 1 is moved to room 2), and so on. We see that the set $\{0, 1, 2, 3, \ldots\}$ has the same cardinality as its proper subset $\mathbb{N}$; that is, the now greater number of guests can be matched evenly with the same number of rooms.

Next, let's show that the sets $\mathbb{N}$ and $\mathbb{Z}$ (all integers) have the same cardinality $\aleph_0$. Note that $\mathbb{N}$ is a disjoint union of the two sets of even numbers and odd numbers. These two subsets of $\mathbb{N}$ can be evenly matched with the positive and non-positive integers as shown below:

$$
\begin{array}{ccccccccc}
\cdots & & -2 & -1 & 0 & 1 & 2 & & \cdots \\
\text{odd numbers} & & & & & & & & \text{even numbers} \\
\longleftarrow & & 5 & 3 & 1 & 2 & 4 & & \longrightarrow
\end{array}
$$

---

[9] Defining infinite sets is no simple task in set theory, and Cantor's definition is not fault-proof. A more precise definition and further discussions can be found in a text on axiomatic set theory.

This assignment can be coded as a function $f : \mathbb{Z} \to \mathbb{N}$ in the following way:

$$f(n) = \begin{cases} 2n, & \text{if } n > 0 \\ -2n + 1, & \text{if } n \le 0 \end{cases}$$

For instance, since $0 \le 0$, we have $f(0) = -2(0) + 1 = 1$ as shown in the table; similarly, $f(1) = 2(1) = 2$ since $1 > 0$, and $f(-1) = -2(-1) + 1 = 3$ since $-1 \le 0$. The existence of a bijection between $\mathbb{N}$ and $\mathbb{Z}$ proves that

$$\#\mathbb{Z} = \#\mathbb{N} = \aleph_0$$

I emphasize that the above bijection is not unique; see Exercise 46. However, the existence of *any one bijection* is enough to show that two sets have the same cardinality.

Equipollent sets

The existence of a bijection between two sets A and B gives an equivalence relation $A \sim B$ as you can quickly verify by checking the three conditions of reflexivity, symmetry, and transitivity. This relation is called the *equipollence relation*, so we say that A and B are *equipollent sets*.

In particular, the symmetry condition $A \sim B$ implies $B \sim A$ because the inverse function $f^{-1} : B \to A$ is also a bijection. As an example, the function $f : \mathbb{N} \to \mathcal{E}$ defined as $f(n) = 2n$ is a bijection onto the set $\mathcal{E}$ of all even numbers. Its inverse is obtained by solving the equation $2n = k$ for n to get $n = k/2$; this gives the inverse $f^{-1}(k) = k/2$, which is a bijection from $\mathcal{E}$ onto $\mathbb{N}$.

I end this section on the note that sometimes it is not easy to find an explicit bijection between sets that we know by *other* means to be equipollent. As an example, consider the set of prime numbers: $P = \{2, 3, 5, 7, 11, \ldots\}$. As we saw earlier, Euclid proved that this is an infinite set. However, an explicit bijection $f : \mathbb{N} \to P$ is not known in this case since we know of no formula that gives every element of P. If such a bijection $f(n)$ were known, then it would serve as a formula that determines every prime number—a number theorist's dream come true!

2.6 An infinity of infinities: Cantor's theorem

It is time to answer the question, are there infinite cardinal numbers other than $\aleph_0$? Cantor showed that there were, using the set of all subsets, or the *power set* of $\mathbb{N}$. Let us use the notation $\mathcal{P}(S)$ for the power set of S.

In Exercise 18, we see that if S has n elements; then it has 2^n subsets, including S itself and the empty set. Therefore, $\#\mathcal{P}(S) = 2^n$. In particular, $\#\mathcal{P}(S) > \#S$ for every finite set S. Therefore, the number of subsets becomes quite large even when the number of elements is quite modest. But what about $\#\mathcal{P}(\mathbb{N})$?

Based on the above discussion, it's reasonable to expect that $\#\mathcal{P}(\mathbb{N})$ is not only greater than $\#\mathbb{N}$ but *vastly greater*. But in what sense? Here we are dealing with infinite sets, so we cannot even assume that $\#\mathcal{P}(\mathbb{N}) > \#\mathbb{N}$ just because this inequality is true for finite sets; after all, if S is a finite set, then $\#\mathcal{P}(S)$ may be unimaginably large in numerical terms, but it is still a finite integer. Furthermore, while we have a well-defined order relation $>$ for *finite* cardinal numbers, it is not obvious that a similar relation exists for infinite cardinals.

These issues are complicated matters, requiring axiomatic set theory for their proper resolution, including the definition of an order relation for infinite cardinals. However, our goal here is more modest, so we proceed by reasoning as follows: for every set S there is an obvious one-to-one function $S \to \mathcal{P}(S)$, namely, the function that assigns to each element $a \in S$ the single-element set $\{a\}$ in $\mathcal{P}(S)$. Of course, this function is not onto since it misses many other elements of $\mathcal{P}(S)$, including $\varnothing$ and S. So $\#S$ must be less than or equal to $\#\mathcal{P}(S)$; the next result shows that it is in fact *less than*, not equal to.

Theorem 9 *(Cantor): There are no functions $f : S \to \mathcal{P}(S)$ that are onto $\mathcal{P}(S)$ (or surjective) for any set S. In particular, there are no bijections between S and its powerset.*

Proof The proof is by contradiction. Suppose that there is a function $f : S \to \mathcal{P}(S)$ that is onto. Notice that the image of each element $a \in S$ is a set $f(a) \in \mathcal{P}(S)$ since $f(a)$ is a *subset* of S. Since we are assuming that f is onto, it must be that *every* subset of S must be assigned some element of S. Consider the subset of S containing *all elements that are not in their images*; i.e., the set

$$N = \{a \in S : a \notin f(a)\}$$

Recall that by assumption f is onto $\mathcal{P}(S)$, so there is an element $t \in S$ whose image $f(t) = N$ is in $\mathcal{P}(S)$. Is $t \in N$ or $t \notin N$?

If $t \in N$, then the definition of N implies that $t \notin f(t) = N$, which is a contradiction. If $t \notin N$, then the definition of N implies that $t \in f(t) = N$, and we reach a contradiction again. These contradictions are unavoidable if we assume that a function exists from S onto $\mathcal{P}(S)$. It follows that an onto function cannot exist.   ∎

Cantor's Power-set Theorem implies that $\#\mathcal{P}(\mathbb{N})$ *is an infinity that is greater than* $\aleph_0$. Further, our discussion of finite sets suggests that $\#\mathcal{P}(\mathbb{N})$ is, in some sense, vastly greater than $\aleph_0$. We can't quantify this here as easily as we did the finite case, but you can explore the difference in size a little more in the exercises. Borrowing the notation for finite sets, it is common to write

$$\#\mathcal{P}(\mathbb{N}) = 2^{\aleph_0}$$

I emphasize that here $2^{\aleph_0}$ is just notation and *not* "2 raised to the power $\aleph_0$."

We have proved now that the infinite cardinal $\aleph_0$ is less than the infinite cardinal $2^{\aleph_0}$ in a precise sense. But are there any cardinal numbers *between* these two infinite cardinals?

For a large finite integer n, there are many integers between n and 2^n, so the question is not without meaning. But it turns out that it is not easy to come up with an infinite set whose cardinal number is larger than $\aleph_0$ and smaller than $2^{\aleph_0}$. Indeed, a well-known hypothesis of set theory, namely, the *continuum hypothesis*, actually posits that there are no *distinct* cardinals between $\aleph_0$ and $2^{\aleph_0}$.

The continuum hypothesis is an *undecidable* proposition in set theory, which means that it can be neither proved nor disproved using the common axioms of set theory (the Zermelo–Fraenkel axioms). It may be added to the list of axioms for which additional mathematical statements can be proved or disproved with its aid.

As far as our work in this book is concerned, whether we accept the continuum hypothesis or ignore it makes no difference because the sets that we are interested in either have cardinal number $\aleph_0$ or $2^{\aleph_0}$. We will see later that $2^{\aleph_0}$ is in fact the cardinality of the set of all *real* numbers.

Note that the power-set theorem applies to arbitrary sets, not just $\mathbb{N}$. So if we pick $S = \mathcal{P}(\mathbb{N})$, then we may conclude that $\#\mathcal{P}(\mathbb{N}) < \#\mathcal{P}(\mathcal{P}(\mathbb{N}))$. This may be repeated indefinitely to yield an *infinite chain of ever larger infinities*:

$$\#\mathbb{N} < \#\mathcal{P}(\mathbb{N}) < \#\mathcal{P}(\mathcal{P}(\mathbb{N})) < \#\mathcal{P}(\mathcal{P}(\mathcal{P}(\mathbb{N}))) < \cdots$$

or equivalently,

$$\aleph_0 < 2^{\aleph_0} < 2^{2^{\aleph_0}} < 2^{2^{2^{\aleph_0}}} < \dots \tag{2.18}$$

Sets having cardinal number $2^{2^{\aleph_0}}$ do arise often enough in analysis as structured sets of functions. Our work here rarely involves these larger sets; they are encountered more often in more advanced treatments of analysis.

2.7 Countable or uncountable?

The very first two cardinals in (2.18), namely, $\aleph_0$, and $2^{\aleph_0}$ are the most common infinite cardinals because many familiar sets including $\mathbb{N}$ and $\mathbb{R}$ have these cardinalities. We now discuss the main difference between $\aleph_0$ and $2^{\aleph_0}$.

Every finite set is *countable* in the sense that we can start with any element and keep counting till we run out of elements. Of course, we may run out of names long before we run out of elements, but we can use mathematical notation such as 10^n and the like.

We think of the set $\mathbb{N}$ as countable too; not literally of course, but in the sense that its elements can be *listed one after another*. The same listing of consecutive elements is possible for any set that is equipollent to $\mathbb{N}$; for if $f: \mathbb{N} \to S$ is a bijection, then $f(1), f(2), f(3), \ldots$ is a listing of elements of S. So, since $\mathbb{N}$ has cardinality $\aleph_0$, we define the following:

Countable and uncountable. A set S is *countably infinite* or *denumerable* if S has cardinality $\aleph_0$, i.e., it is equipollent to $\mathbb{N}$. If S is either finite or countably infinite, then S is *countable*. If S is infinite but *not* equipollent to $\mathbb{N}$, then such a set is not countable, or *uncountable* for short.

Saying that $2^{\aleph_0}$ is greater than $\aleph_0$ is not exactly the same as saying 2^n is larger than n. We can get from n to 2^n in a finite number of steps through a string of integers. But how do we get from $\aleph_0$ to $2^{\aleph_0}$?

There are ways, and we discuss one after first examining a few related concepts. We discuss some of these concepts in this chapter and finish our quest in Chapter 3.

Intersections and unions of infinitely many sets.

The concepts of union and intersection of sets that we defined earlier for finite sets in (2.5) and (2.4), respectively, extend to infinite collections of sets in a straightforward way.

Consider a nonempty set $\mathcal{I}$ (which may be countable like $\mathbb{N}$ or uncountable like $\mathbb{R}$) and a collection of sets A_i *indexed* by $\mathcal{I}$; that is, there is a set A_i for each element $i \in \mathcal{I}$. We define the union of this indexed collection of sets as

$$\bigcup_{i \in \mathcal{I}} A_i = \{x : x \in A_{i_0} \text{ for some } i_0 \in \mathcal{I}\} \tag{2.19}$$

The "for some" here is the logical quantifier we discussed earlier; it means for *at least one* A_i but possibly more, and maybe even all sets A_i. The intersection is defined similarly:

$$\bigcap_{i \in \mathcal{I}} A_i = \{x : x \in A_i \text{ for all } i \in \mathcal{I}\} \tag{2.20}$$

These concepts may seem abstract and impractical, but they appear routinely in analysis. We use it soon in Chapter 4 when discussing the *nested intervals property*.

De Morgan laws hold for arbitrary unions and intersections too and as such, serve as basic results in analysis.

Theorem 10 *De Morgan laws (generalized): Let $\{A_i : i \in \mathcal{I}\}$ be a collection of sets indexed by $\mathcal{I}$. The following equalities are true:*

$$\left(\bigcup_{i \in \mathcal{I}} A_i\right)^c = \bigcap_{i \in \mathcal{I}} A_i^c \quad and \quad \left(\bigcap_{i \in \mathcal{I}} A_i\right)^c = \bigcup_{i \in \mathcal{I}} A_i^c \qquad (2.21)$$

Proof I prove the first of these; the second uses a similar argument that you can write down as an exercise. The argument is a compressed form of double containment where we use *iff* (if and only if) instead of $\Rightarrow$ in two directions—first to prove $\subset$ and then to prove the reverse containment $\supset$.

Note that an element $x \in (\bigcup_{i \in \mathcal{I}} A_i)^c$ iff $x \notin \bigcup_{i \in \mathcal{I}} A_i$ iff $x \notin A_i$ for all $i \in \mathcal{I}$ (recall that the negation of "for some" is "for all"). Therefore, $x \in A_i^c$ for all $i \in \mathcal{I}$, and this is true iff $x \in \bigcap_{i \in \mathcal{I}} A_i^c$. ∎

An easy way to remember the De Morgan equalities is to notice that distributing the complementation symbol over a union switches it to intersection and vice versa.

The following result, often useful in analysis and topology, generalizes Theorem 5 to arbitrary indexed sets. The proof, left as an exercise, uses arguments that are similar to those used in the proofs of Theorems 5 and 10.

Theorem 11 *Let $\{A_i : i \in \mathcal{I}\}$ be a collection of subsets of a set A, and $\{B_j : j \in \mathcal{J}\}$ be a collection of subsets of a set B. The following are true for every function $f : A \to B$:*

$$(a) \quad f\left(\bigcup_{i \in \mathcal{I}} A_i\right) = \bigcup_{i \in \mathcal{I}} f(A_i)$$

$$(b) \quad f\left(\bigcap_{i \in \mathcal{I}} A_i\right) \subset \bigcap_{i \in \mathcal{I}} f(A_i)$$

And the following are true for the inverse images:

$$(c) \quad f^{-1}\left(\bigcup_{j \in \mathcal{J}} B_j\right) = \bigcup_{j \in \mathcal{J}} f^{-1}(B_j)$$

$$(d) \quad f^{-1}\left(\bigcap_{j \in \mathcal{J}} B_j\right) = \bigcap_{j \in \mathcal{J}} f^{-1}(B_j)$$

In calculus-related discussion, the index set $\mathcal{J}$ is usually $\mathbb{N}$ or some set equipollent to $\mathbb{N}$. In this case, since $\mathbb{N}$ is countable and its elements can be listed sequentially, the notation is less abstract:

$$\bigcup_{i \in \mathbb{N}} A_i = \bigcup_{i=1}^{\infty} A_i = A_1 \cup A_2 \cup A_3 \cup \cdots \qquad (2.22)$$

$$\bigcap_{i \in \mathbb{N}} A_i = \bigcap_{i=1}^{\infty} A_n = A_1 \cap A_2 \cap A_3 \cap \cdots \qquad (2.23)$$

In practice, we often run into situations when $\mathcal{J}$ is some set equipollent to $\mathbb{N}$. For example, the starting index may be 0 or some other integer less than 1; and indices may skip integers, as in $\mathcal{J} = \{-1, 1, 3, 5, 7, \ldots\}$; or the starting index may be greater than 1, as in $\mathcal{J} = \{3, 4, 5, 6, \ldots\}$. As we saw earlier, these index sets are equipollent to $\mathbb{N}$, so there are no essential differences between them and $\mathbb{N}$ as far as indexing goes.

Infinite unions and intersections often give surprising answers. For example, consider the subsets $A_i = \{i, i+1, i+2, \ldots\}$ of $\mathbb{N}$, each an infinite set. What is the intersection of all of these sets?

$$\bigcap_{i \in \mathbb{N}} A_i = \{1, 2, 3, \ldots\} \cap \{2, 3, 4, \ldots\} \cap \{3, 4, 5, \ldots\} \cap \cdots$$

To answer, we use the previous definition of infinite intersection. An element of the intersection must be in all of constituent sets; so if this infinite intersection has a number n in it, then $n \geq i$ for every i in $\mathbb{N}$. But since there is no such n in $\mathbb{N}$, it follows that

$$\bigcap_{i \in \mathbb{N}} A_i = \varnothing$$

even though each of the sets A_i is an infinite set.

Let $A_1, A_2, A_3, \ldots$ be given sets. If $\#A_i$ is a finite cardinal number for every index $i \in \mathbb{N}$, then we expect that the cardinality of $\bigcup_{i \in \mathbb{N}} A_i$ is $\aleph_0$ since we can just list all elements of all of the sets one after another and get a countable list. Now, what if each A_i is equipollent to $\mathbb{N}$ (or copies of $\mathbb{N}$)?

It turns out that the cardinality of $\bigcup_{i \in \mathbb{N}} A_i$ is still $\aleph_0$; in fact, the following stronger statement is true since it implies that possible overlaps between the sets A_i are not to blame:

Theorem 12 *A countably infinite union of mutually disjoint, countably infinite sets is countably infinite.*

Proving Theorem 12 will be simpler after we first prove a lemma that is interesting by itself.

Lemma 13 The direct product $\mathbb{N} \times \mathbb{N}$ is equipollent to $\mathbb{N}$ and has cardinality $\aleph_0$.

Proof Notice that we are showing in effect that $\aleph_0^2 = \aleph_0$. To see that $\mathbb{N} \times \mathbb{N}$ is equipollent to $\mathbb{N}$, let's use an idea of Cantor's. We list the elements of $\mathbb{N} \times \mathbb{N}$, i.e., ordered pairs of integers, in a layered pattern like a matrix; then we count them in a zigzag fashion:

$$
\begin{array}{cccc}
(1,1) \longrightarrow & (1,2) & (1,3) \longrightarrow & (1,4) \cdots \\
 & \swarrow \quad \nearrow & \swarrow & \nearrow \\
(2,1) & (2,2) & (2,3) & (2,4) \cdots \\
\downarrow \quad \nearrow & & \swarrow \quad \nearrow & \swarrow \\
(3,1) & (3,2) & (3,3) & (3,4) \cdots \\
 & \swarrow \quad \nearrow & \swarrow & \nearrow \\
(4,1) & (4,2) & (4,3) & (4,4) \cdots \\
\downarrow \quad \nearrow & \vdots \quad \swarrow & \vdots \quad \nearrow & \vdots
\end{array}
$$

Each row is just a copy of $\mathbb{N}$; the arrows define a path starting from (1,1) and proceed along the diagonals in such a way that the sum of the two coordinates along each diagonal equals a fixed integer. The first diagonal is just (1,1), and its coordinates add to 2; then the second diagonal has two pairs (1,2) and (2,1), and the coordinates of each pair add to 3; the third diagonal has three pairs with coordinates of each adding to 4, and so on. If we list the consecutive pairs along the zigzag path, we get

$$(1, 1), (1, 2), (2, 1), (3, 1), (2, 2), (1, 3), (1, 4), \ldots$$

This list of ordered pairs contains all of $\mathbb{N} \times \mathbb{N}$ and, as listed, is in a one-to-one correspondence with $\mathbb{N}$. Thus, $\mathbb{N} \times \mathbb{N}$ is equipollent to $\mathbb{N}$ and has cardinality $\aleph_0$. ∎

In the exercises you can prove the equipollence of $\mathbb{N} \times \mathbb{N}$ and $\mathbb{N}$ more efficiently using a *direct* bijection.

Now that we know $\mathbb{N} \times \mathbb{N}$ is equipollent to $\mathbb{N}$, the proof the theorem is straightforward.

Proof (of Theorem 12) Since each A_i is countably infinite, its elements can be put into one-to-one correspondence with $\mathbb{N}$, and we can write out its elements as a sequence:

$$A_i = \{a_{i,1}, a_{i,2}, a_{i,3}, a_{i,4}, \ldots\}$$

This way of writing a countable set A is called an *enumeration* of A (using $\mathbb{N}$ for the index set). We do this for every i and stack up the sets as follows:

$$A_1 = \{a_{1,1}, a_{1,2}, a_{1,3}, a_{1,4}, \ldots\}$$
$$A_2 = \{a_{2,1}, a_{2,2}, a_{2,3}, a_{2,4}, \ldots\}$$
$$A_3 = \{a_{3,1}, a_{3,2}, a_{3,3}, a_{3,4}, \ldots\}$$
$$A_4 = \{a_{4,1}, a_{4,2}, a_{4,3}, a_{4,4}, \ldots\}$$
$$\vdots$$

Now compare the indices of the elements $a_{m,n}$ with the entries of $\mathbb{N} \times \mathbb{N}$ listed above to notice that the assignment $(m, n) \rightarrow a_{m,n}$ defines a bijection (recall also that the sets A_i are mutually disjoint, so no two $a_{m,n}$ are identical). This completes the proof. ∎

Theorem 12 can be readily extended to the following more general statement that you can prove in Exercise 52.

Corollary 14 *A countable union of countable sets is countable.*

We now see that a countable unions of sets with cardinality $\aleph_0$ again has cardinality $\aleph_0$. A consequence of this fact is that by adding countable sets to existing countable sets, we only get countable sets again. But is there any *countable* operation that gets us from $\aleph_0$ to $2^{\aleph_0}$?

It turns out that there is: by taking countable *products* of countable sets. We discuss the details in Chapter 3 after getting acquainted with *sequences*.

Finally, consider the question: If a set S is countable and $A \subset S$ then must A be also countable? This makes sense and it is true.

Theorem 15 *Every subset of a countable set is countable.*

Proof Let S be a countable set and $A \subset S$. We can assume that A is nonempty since the empty set is (trivially) countable as a finite set. Pick any fixed element m in A and define a function $f : \mathbb{N} \rightarrow A$ as follows[10]:

$$f(n) = \begin{cases} n & \text{if } n \in A \\ m & \text{if } n \notin A \end{cases}$$

[10] Note that f here is just the identity function on A that has been extended to all of $\mathbb{N}$ by defining it as constant outside A.

We may now write

$$A = \bigcup_{n \in \mathbb{N}} \{f(n)\} \qquad (2.24)$$

This equality expresses A as a countable union of single element sets, so A is countable by Corollary 14.   ∎

As an obvious corollary, we also have the following equivalent contrapositive version of Theorem 15:

Corollary 16 *If a set S has an uncountable subset, then S is uncountable.*

Using Theorem 15, the following corollary can be proved (left as an exercise).

Corollary 17 *Let A and B be given nonempty sets, and let $f : A \to B$ be a function.*
(a) If A is countable and f is surjective (onto), then B is countable.
(b) If f is injective (one-to-one) and B is countable, then A is countable.

2.8 Exercises

Exercise 18 *(a) How many subsets does the set $A = \{1, 2, 3, 4, 5\}$ have? Consider the following idea: think of five blank places where you can either put a number in A or leave blank. For example, _ , 2, 3, _ , 5 is the subset $\{2, 3, 5\}$, and _ , _ , _ , _ , _ is the empty set Ø. Count the number of ways can you fill in the black spaces starting from Ø. Note that we cannot put, say, 4 in the space for 5.*
(b)Suppose that a set S has n elements where n is any unspecified positive integer. Use the idea in (a) to prove that the number of subsets (proper and improper) of S is 2^n.

Exercise 19 *(a) Draw a Venn diagram for $(A \backslash B) \cup (B \backslash A)$. This union is also called the "symmetric difference" of A and B; it is sometimes denoted $A \triangle B$ and represents the "exclusive or" in the sense that it is the set of all elements that are either in A or in B but not in both.*
(b) What is the relative complement $A \backslash B$ if A and B do not intersect? What is $A \triangle B$?
(c) What is the relative complement $A \backslash B$ if $A \subset B$? What is $A \triangle B$?

Exercise 20 *Prove the **second** De Morgan law using the same double-containment approach we used for the first law.*

Exercise 21 *Prove the distributive properties for intersections and unions (use double-containment arguments):*

$$A \cap (B \cup C) = (A \cap B) \cup (A \cap C)$$
$$A \cup (B \cap C) = (A \cup B) \cap (A \cup C)$$

Exercise 22 *Consider the following subsets of* $\mathbb{Z}$:

$$A = \{2k : k \in \mathbb{Z}\} = \{..., -6, -4, -2, 0, 2, 4, 6, ...\}$$
$$B = \{k \in \mathbb{Z} : -9 \le k < 9\} = \{-9, -8, ..., 7, 8\}$$

Find each of the following sets:

$$A \cap B, \quad A^c = \mathbb{Z} \backslash A, \quad \mathbb{N} \backslash A, \quad B \cap \mathbb{N},$$
$$B \cup \mathbb{N}, \quad (B \cup \mathbb{N})^c, \quad B^c \cap \mathbb{N}^c$$

Exercise 23 *Write the set B in Exercise 22 as the intersection of two infinite subsets of* $\mathbb{Z}$.

Exercise 24 *(a) Let* $A = \{0, 1, 2\}$ *and* $B = \{x, y\}$. *Find* $A \times B$ *and* $B \times A$ *and* $A^2 = A \times A$. *How many elements does each product have?*

(b) Suppose that A is a set with m elements, and B is a set with n elements. Prove that $\#A \times B = mn$ *for all integers m, n.*

Exercise 25 *Write each of the bounded intervals below as the intersection of two unbounded intervals:*

$$[3, 8], \quad \{a\} = [a, a], \quad (-\pi, \pi), \quad \left(-\frac{1}{2}, 0\right]$$

Exercise 26 *(a) Write the set* $\{x : \sin x < 1/2\}$ *as a union of (infinitely many) disjoint intervals. How does this set compare to* $\{x : \sin x \le 1/2\}$?

(b) Write the set $\{x : \sin x < 1\}$ *as a union of (infinitely many) disjoint intervals. How does this set compare to* $\{x : \sin x \le 1\}$?

Exercise 27 *(a) Is the divisibility relation* $m|n$ *an equivalence relation in* $\mathbb{N}$? *If not, specify which properties hold and which do not.*

(b) Is divisibility a partial ordering in $\mathbb{N}$?

(c) If the answer in (b) is yes, then is divisibility a total ordering in $\mathbb{N}$?

Exercise 28 *Use mathematical induction to prove each of the following statements:*

(a) The sum of the first n odd numbers is n^2; i.e.,

$$1 + 3 + 5 + 7 + \cdots + 2n - 1 = n^2$$

(b) For all integers greater than 4

$$2^n > n^2$$

Exercise 29 *Explain why the following sets are not well-ordered:*

$$\left\{1, \frac{1}{2}, \frac{1}{3}, \ldots, \frac{1}{n}, \ldots\right\}$$

All rational numbers in the interval $[0, 1]$

Exercise 30 *Specify whether the following functions are even, odd, or neither:*

(a) x^9 *(b)* $x^9 + 3$ *(c)* $x^6 + 3$ *(d)* $x^6 + x^9$

Exercise 31 *Show that if f and g are both odd functions, then their product fg is an even function.*

Exercise 32 *Consider the affine function $f(x) = 2x - 1$.*
   (a) Show that the interval $[0, 1]$ is mapped to the interval $[-1, 1]$, which is twice as long as $[0, 1]$. Extend this observation to an arbitrary interval $[a, b]$ where $a < b$ is mapped to an interval twice as large. Note that the length of $[a, b]$ is $b - a$; what is $f(b) - f(a)$?
   (b) Repeat the above calculations for $f(x) = -x/3 + 1$.
   (c) What can you conclude about the general $f(x) = mx + c$ where m and c are arbitrary numbers?

Exercise 33 *Consider the exponential function $f(x) = 2^x$.*
   (a) Calculate the values of $f(x)$ for $x = -1, 0, 1, 2, 3$. Compare the effect of f on the intervals $[-1, 0]$, $[0, 1]$, $[1, 2]$ and $[2, 3]$; which ones are compressed (to shorter length), and which ones are stretched (to longer length)? Are any of these intervals left intact?
   (b) Explain why in (a) f stretches each interval by double the amount that it stretches the previous interval. Note that $2^{x+1} = 2^x 2^1 = 2(2^x)$ for every value of x.

Exercise 34 *For the square function $f(x) = x^2$, find $f \circ f$ and $f \circ f \circ f$. Find a formula for f^n for arbitrary $n \in \mathbb{N}$.*

Exercise 35 *Composition of functions $f \circ g$ requires that the range of g be contained in the domain of f. Otherwise, $f \circ g$ is not defined for every point in the domain of g.*

For example, let $f(\theta) = \cot\theta = (\cos\theta)/(\sin\theta)$ be the cotangent function, which is defined on the set $D = \{\theta : 0 < \theta < \pi\}$. Notice that $f(\pi/2) = \cot(\pi/2) = 0$, which is not in D (indeed, $f \circ f(\pi/2)$ is not defined). Note that with D as its domain, the range of $\cot\theta$ is all of the real numbers.

Exercise 36 Let $f : D \to S$ be a function. Define a relation in D as follows: $x \sim y$ if $f(x) = f(y)$. For example, if $f(x) = x^2$, then $2 \sim -2$ because $f(-2) = 4 = f(2)$.
   (a) Verify that $\sim$ is an equivalence relation.
   (b) For each $x \in D$, explain why the equivalence class of x is $f^{-1}(f(x))$.
   (c) Consider the function $f(x) = x^3 - x$. What is the equivalence class of 0? What is the equivalence class of 6? Sketching a graph of $f(x)$ is helpful but not required.

Exercise 37 Explain why for every y in the range of a one-to-one function f, the inverse image set $f^{-1}(y)$ contains either a single element or is empty.

Exercise 38 Prove Parts (b) and (c) in Theorem 5.

Exercise 39 Suppose that a function f is decreasing everywhere on its domain D, and prove that the function must be one-to-one.

Exercise 40 Prove that the function $f(x) = ax + b$ is a bijection of $\mathbb{R}$ for all fixed (constant) real numbers a, b as long as $a \neq 0$.

Exercise 41 Consider the finite sets $A = \{1, 2\}$ and $B = \{a, b, c\}$.
   (a) Specify a one-to-one function that maps A into B; can you specify any function, one-to-one or not, from A onto B? Why not?
   (b) Specify a function that maps B onto A; can you specify any one-to-one function from B into A? Why not?
   (c) Can there be a bijection from A to B? Why not?

Exercise 42 **Equipollence.** Consider a relation $\sim$ (among subsets of a given set U) defined as $A \sim B$ if there is a bijection $f : A \longrightarrow B$ where A and B are subsets of U. Verify that $\sim$ is an equivalence relation. Equipollent sets have the same cardinal number.

Exercise 43 Use a truth table to show that the following statement is a tautology:

$$p \text{ or } [not\ p] \tag{2.25}$$

This statement is often called the "law of excluded middle" because it says that either a statement or its negation must be always true; there is no other alternative or "middle ground."

Exercise 44 *Prove each of the following equivalences:*

$$[p \Rightarrow q] \Leftrightarrow not[p \, and \, [not \, q]]; \qquad [[p \, or \, q] \Leftrightarrow not[not \, p \, and \, not \, q] \qquad (2.26)$$

These statements show that implication and disjunction may be defined in terms of negation and conjunction.

Exercise 45 *(a) Suppose that in response to endless complaints by inconvenienced guests, Hotel Infinity proprietor D. Hilbert decides on a less invasive plan to accommodate new guests. He moves the guest in room 1 to room 10, the guest in room 10 to room 100, the guest in room 100 to room 1000 and so on and offers a discount on rooms that are powers of 10. What function f describes this reassignment of guests? Notice that positive integer powers of 10 can be matched in a one-to-one way with the positive integers!*

(b) Later on, a huge bus with infinitely many passengers arrives at the hotel. Unfazed, Hilbert moves the guest in room 1 to room 2, moves the guest in room 2 to room 4, the guest in room 3 to room 6, and so on. In this way, he relocates all the guests to even-numbered rooms and makes all the odd-numbered rooms vacant and ready to receive the infinitely many new guests! What function f describes this reassignment of guests?

Exercise 46 *Show that the following function is a bijection from $\mathbb{Z}$ to $\mathbb{N}$:*

$$f(n) = \begin{cases} 2n+1, & if \, n \geq 0 \\ -2n, & if \, n < 0 \end{cases}$$

Which integers does this function map onto the odd numbers?

Exercise 47 *Prove that the set of all odd positive integers $\mathcal{O}$ is equipollent to $\mathbb{N}$ by finding a bijection $f : \mathbb{N} \to \mathcal{O}$. What is f^{-1}?*

Exercise 48 *Convince yourself that the power set $\mathcal{P}(\mathbb{N})$ contains all of the following types of sets:*
(i) all finite subsets of $\mathbb{N}$ (of which there are infinitely many)
(ii) all "co-finite" subsets of $\mathbb{N}$, namely, the complements of finite sets, like

$$\{100, \ 101, \ 102, ...\} = \{1, 2, ..., 99, 100\}^c$$

(iii) all infinite sets that are not co-finite, like the sets of all even numbers or all prime numbers

(iv) all sets of multiples of fixed integers, like

$$\{2, 4, 6, 8, ..., 2n, ...\}, \quad \{3, 6, 9, 12, ..., 3n, ...\}, \quad etc.$$

(v) all sets of powers of integers, like

$$\{1, 4, 9, 16, ..., n^2, ...\}, \quad \{1, 8, 27, 64, ..., n^3, ...\}, \quad etc.$$

Can you think of some other infinite sets in $\mathcal{P}(\mathbb{N})$? How about if you add 1 (or 2 or any other fixed positive integer) to every number in each of the sets in multiples or powers? And how about the complements of these sets? You can indeed go far creating distinct sets this way (distinct doesn't mean disjoint, so overlapping of sets is not a problem). Doing so will enhance your grasp of just how vastly larger the set $\mathcal{P}(\mathbb{N})$ and its cardinal number $2^{\aleph_0}$ are compared to $\mathbb{N}$ and $\aleph_0$.

Exercise 49 *Prove each of the following equalities about infinite intersections and unions of intervals:*

$$(a) \bigcap_{n=1}^{\infty} \left(0, \frac{1}{n}\right) = \varnothing \quad (b) \bigcap_{n=1}^{\infty} \left(-\frac{1}{n}, 1 + \frac{1}{n}\right) = [0, 1] \quad (c) \bigcup_{n=1}^{\infty} \left[\frac{1}{n}, 1 - \frac{1}{n}\right] = (0, 1)$$

Exercise 50 *Prove Theorem 11.*

Exercise 51 *(a) An efficient way to prove that $\mathbb{N} \times \mathbb{N}$ is equipollent to $\mathbb{N}$ is by defining a bijection explicitly. Verify that the function $f : \mathbb{N} \times \mathbb{N} \to \mathbb{N}$ defined below is a bijection:*

$$f((m, n)) = 2^{m-1}(2n - 1) \quad m, n \in \mathbb{N}$$

(b) Similarly, show that $\mathbb{Z} \times \mathbb{N}$ is equipollent to $\mathbb{Z}$ by verifying that the following function is a bijection:

$$f((m, n)) = 2^{m-1}(2n - 1), \quad m \in \mathbb{N}, n \in \mathbb{Z}$$

Exercise 52 *Prove Corollary 14. In this corollary, some of the countable sets may be finite and some of them may overlap; recall that a countable union may also mean a finite union. The difficult case has already been proved in Theorem 12.*

Exercise 53 *Prove Corollary 17.*

3
Sequences and Limits

In this chapter, we discuss the basics of sequences of numbers. In particular, we focus on convergent sequences, our primary tools for exploring the concept of limit on which the entire calculus rests. In addition, in Chapter 4, we use convergent sequences of rational numbers to construct the real numbers, without which analysis would be impossible. Then in Chapter 5, we use convergent sequences to define infinite series of numbers; and later in Chapter 8, we move up to sequences and series of functions, concepts that form the foundations for much of applied mathematics.

3.1 Infinite lists of numbers

In its most basic form, a sequence is an infinite list of objects. There are actually no proper (infinite) sequences of objects in the universe as we know it since the number of all known particles of matter or quantity of radiation is finite. If we list all subatomic particles in the observable universe in a sequence, the list stops when we run out of particles. It will certainly be a long list but still far from infinite.

Here's a formal definition of a sequence.

Sequence

A sequence is a function whose domain is either the set $\mathbb{N}$ or a well-ordered set N equipollent to $\mathbb{N}$ (e.g., not like $\mathbb{Z}$). If S is a nonempty set, then a function

$$s : N \longrightarrow S$$

defines a *sequence in S*, and S is the *range* of s. The values $s(1), s(2), s(3), \ldots$ are called the *terms* of the sequence. Commonly a sequence s is denoted by its generic nth term $s(n)$ with unspecified n. We use the notation s_n rather than the function notation $s(n)$ in order to emphasize the discrete, list-like nature of the sequence.

Real Analysis and Infinity. Hassan Sedaghat, Oxford University Press.
© H. Sedaghat (2022). DOI: 10.1093/oso/9780192895622.003.0003

For example, let S be the set $\mathbb{Z}$ of all integers so that sequence $s : \mathbb{N} \to \mathbb{Z}$ is a sequence of integers; for instance, $s(n) = n^2 - 5n$ gives an integer for each value of n in $\mathbb{N}$. The first few terms of this sequence are listed explicitly below:

n	1	2	3	4	5	6	7	$\cdots$
$s(n)$	-4	-6	-6	-4	0	6	14	$\cdots$

Notice that $s(n)$ can never take values that are smaller than -6; i.e., $s(n) \geq -6$ for every $n \in \mathbb{N}$. So $\mathbb{Z}$ contains numbers that are not the image of any n under the function s. In contrast to functions of real numbers, the *image* of $\mathbb{N}$ under s, i.e., $s(\mathbb{N})$, is a subset of $\mathbb{Z}$ that is not easy to determine in this case and, indeed, typically.

In this book, the range S of a sequence is often the set $\mathbb{R}$ of all real numbers, but the concept of sequence is more general. For example, we can talk about a sequence of sets such as

$$s(n) = \left[0, \frac{1}{n}\right] \quad \text{(a sequence of intervals)}$$

that we encounter in Chapter 4 when discussing the nested intervals property and the completeness of the set of real numbers (here S is the power set $\mathcal{P}(\mathbb{R})$, or a suitable subset of it). Sequences of intervals or other types of sets appear routinely in measure and integration theory; we get a brief exposure in Chapter 7.

Also important are sequences of functions, like

$$s(n) = x^n \quad \text{or} \quad s(n) = \sin(nx) \quad \text{(for all } x \text{ in a given interval)}$$

that arise in the discussions of power series or Fourier series; see Chapter 8.

Even though sequences are special types of functions, it is generally more useful to think of them as *strings of numbers or points* rather than as mappings. We often denote a sequence s_n by listing its first few values, including the nth term if needed for clarity or completeness:

$$s_1, s_2, s_3, \ldots, s_n, \ldots \tag{3.1}$$

Here, s_1 is the first term of the sequence, s_2 is the second term, and so on. To see that the representation (3.1) can be useful, consider the complicated looking sequence defined by

$$s_n = \frac{(-1)^n - 1}{2\cos(\pi n)}, \quad n \in \mathbb{N} \tag{3.2}$$

The first four terms are calculated easily using the facts that $(-1)^n = -1$ for odd n (multiplying -1 by itself an odd number of times), and $\cos(\pi n) = (-1)^n$:

$$s_1 = \frac{(-1)^1 - 1}{2\cos\pi} = \frac{-1 - 1}{-2} = 1, \quad s_2 = \frac{(-1)^2 - 1}{2\cos(2\pi)} = \frac{1 - 1}{2} = 0$$

$$s_3 = \frac{(-1)^3 - 1}{2\cos(3\pi)} = \frac{-1 - 1}{-2} = 1, \quad s_4 = \frac{(-1)^4 - 1}{2\cos(4\pi)} = \frac{1 - 1}{2} = 0$$

We can now see that the sequence in (3.2) is actually simple: the image of $\mathbb{N}$ is just the two-element set $\{0, 1\}$ while, of course, the sequence itself is a function on $\mathbb{N}$, or an infinite string of alternating 0's and 1's

$$1, 0, 1, 0, 1, 0, \ldots$$

A sequence that consists of just one element, i.e., $s_n = c$, it is called a *constant sequence* and may also be written as

$$c, c, c, \ldots$$

This is sometimes called a *trivial* sequence. A sequence is *eventually* constant if it doesn't change its value beyond a certain index. An example is

$$s_n = \left\lceil \frac{5}{n} \right\rceil, \quad n = 1, 2, 3, \ldots \tag{3.3}$$

where the *ceiling function* $\lceil x \rceil$ gives the least integer that is greater than or equal to the number x. Here s_n is just the sequence $5, 3, 2, 2, 1, 1, \ldots$, which is eventually constant at 1.

Note that for each given c, there is precisely one constant sequence, namely, the one shown above. How about sequences like the one in (3.2) with just two distinct terms, like 0 and 1? We expect to have many more of those since we can rearrange the terms in different orders to come up with different sequences. To find out just how much more, we need to first introduce the concept of infinite product of sets. We do that later in this chapter and prove there are not only infinitely many sequences of 0's and 1's, but in fact there are $2^{\aleph_0}$ of them!

Sets versus sequences

When we write the set $\mathbb{N}$ as $\{1, 2, 3, \ldots\}$ using braces, it *looks like* the sequence $s_n = n$, which we write as $1, 2, 3, \ldots$ without the braces. But there is an important distinction: the elements of a set can be written in any order, so that, say, $\{2, 1, 3, \ldots\}$ is identical to $\mathbb{N}$. However, the sequence $2, 1, 3, \ldots$ (call it s_n') is *not* the same as the sequence s_n because $s_1' = 2$, while $s_1 = 1$ and $s_2' = 1$, while $s_2 = 2$. So $s_n' \neq s_n$.

Also important to remember is the fact that *a sequence (always infinite) may well have a finite range;* for instance, $1, 0, 1, 0, \ldots$ is an infinite repeating sequence while its range is a set consisting of just two numbers: 0 and 1. Different sequences like $0, 1, 0, 1, \ldots$ that start with 0 (all terms are different from $1, 0, 1, 0, \ldots$) or one like $1, 0, 1, 1, 0, 1, \ldots$ with the pattern $1,0,1$ repeating (different in infinitely many terms) both have the exact same range $\{0, 1\}$.

3.2 Sequence types and plots

Sequences of real numbers may be plotted on a real number line like the x-axis. This is sometimes helpful but visually difficult to assess because we lose track of which point comes before another; for instance, if we plot the sequence in (3.2) on the x-axis, we only see two points: one at 0 and the other at 1.

A more informative plot is a diagram that plots the sequence as a function. Time indices, which are positive integers in $\mathbb{N}$, are listed horizontally; and above each integer n, the value of s_n is marked vertically. To make the plot easier to understand, we also connect consecutive points of the sequence plot using line segments (these connecting lines are visual aids, not parts of the sequence). For example, Figure 3.1 shows the plot of the following sequence up to $n = 12$:

$$s_n = \cos\frac{\pi n}{3}: \quad \frac{1}{2}, -\frac{1}{2}, -1, -\frac{1}{2}, \frac{1}{2}, 1, \frac{1}{2}, -\frac{1}{2}, -1, -\frac{1}{2}, \frac{1}{2}, 1, \ldots \qquad (3.4)$$

The figure clearly illustrates the behavior of the sequence and provides a good understanding of its nature.

Periodic sequences

Notice from Figure 3.1 as well as the list of the first 12 elements of (3.4) that there is a pattern: the first six terms repeat. We say that this *sequence is periodic with period* 6.[1] The sequence in (3.2) is periodic with period 2. You can easily plot this sequence by hand to see a sawtooth pattern.

The general definition of periodic sequences is the following.

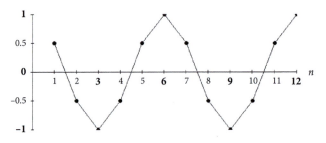

Fig. 3.1 Plot of a periodic sequence with period 6

---

[1] It is worth mentioning that if we make a simple plot of the values s_n on the x-axis without relating them to their indices, the plot would consist of just four points: $\pm 1, \pm 1/2$. Using only this plot might lead to the hasty conclusion that the sequence has period 4, which is wrong.

> **Periodic sequence** A sequence s_n is *periodic with period k* if $s_{n+k} = s_n$ for every
> n. If $k = 1$, then we have a *constant sequence*. A sequence that does not start off
> periodic but becomes periodic beyond a certain index is *eventually periodic*.

There is no actual or physical time involved in the definition of a sequence, and
"eventually" means "beyond a certain index." For example, the sequence in (3.3) is
eventually constant (period 1) starting with $n = 5$ (the fifth term).

In some applications, the index does track time (like seconds, hours, days, etc.),
and in those contexts, the term "eventually" has its usual temporal meaning.

We may call the initial segment that is not periodic the *transient part*; for instance, the transient part of the sequence in (3.3) consists of its first four terms.
Again, physical time is not implied in this definition of "transient" outside of
applications.

Is sin n periodic?

We close this section by taking a closer look at the sequence $s_n = \sin n$. Is this
a periodic sequence? This question is sensible given that for a real variable x, the
function $\sin x$ is periodic and has period 2π. But it turns out that the same is not
true if the variable is an integer.

To see why the *sequence* $\sin n$ is not periodic, we use a simple proof by contradiction. Assume that $\sin n$ is periodic. Then, in particular, the value $\sin 0 = 0$ must
repeat regularly. Recall that $\sin x = 0$ only if $x = k\pi$ for some integer k. But if n
is an integer, then $n = k\pi$ only when $k = 0$ because π is irrational. So $\sin n = 0$
only when $n = 0$, contradicting our earlier conclusion that the value 0 repeats. It
follows that $\sin n$ is not a periodic sequence.

Notice that this argument also shows that $\sin q$ is not periodic when the variable
q is restricted to the set of rational numbers.

The fact that $\sin n$ oscillates in the bounded interval $[-1, 1]$ suggests that $\sin n$
does have (nontrivial) *subsequences* that converge in $[-1, 1]$.

This is not an obvious statement; we prove it in Chapter 4 using the Bolzano–
Weierstrass theorem (a fundamental result in analysis) after discussing subsequences later in this chapter.

3.3 Monotone sequences and oscillating sequences

If the range S of a sequence is an ordered set with an ordering $\leq$ then some
sequences never reverse their direction. These sequences are simple but important.

I start with the basic definitions. Sequences are functions, and recalling the definition of monotone functions earlier, we already know what a monotone sequence is. But it is useful to see it in the sequence-specific notation too.

Let S be a set ordered by a relation $\leq$ (not necessarily total ordering). A sequence s_n in S is *nondecreasing* if $s_{n+1} \geq s_n$ for every n. It is *non-increasing* if $s_{n+1} \leq s_n$ for every n. With strict inequalities, s_n is *increasing* if $s_{n+1} > s_n$ for every n or *decreasing* if $s_{n+1} < s_n$ for every n. If a sequence in S is either nondecreasing or non-increasing, then it is a *monotone sequence*.

The sequence $s_n = 2n - 1$ is increasing because

$$s_{n+1} = 2(n+1) - 1 = 2n + 1 > 2n - 1 = s_n$$

for every n.

The range S of a sequence need not be a totally ordered set. For example, if S is the power set $\mathcal{P}(\mathbb{R})$, then S is partially ordered by set inclusion $\subset$. The sequence of intervals

$$I_n = \left[-\frac{1}{n}, \frac{1}{n} \right]$$

is a decreasing sequence in $\mathcal{P}(\mathbb{R})$ relative to $\subset$ since

$$I_{n+1} = \left[-\frac{1}{n+1}, \frac{1}{n+1} \right] \subset \left[-\frac{1}{n}, \frac{1}{n} \right] = I_n$$

The first few terms of the sequence are

$$I_1 = [-1, 1], \quad I_2 = \left[-\frac{1}{2}, \frac{1}{2} \right], \quad I_3 = \left[-\frac{1}{3}, \frac{1}{3} \right], \dots$$

clearly showing the decreasing nature of the sequence as the index n grows. This sequence is also called *a sequence of nested intervals* for the obvious reason.

Note that every constant sequence is simultaneously nondecreasing and non-increasing, hence (trivially) monotone. A sequence that is not monotone from the start may become monotone after a finite number of terms.

If there is a positive integer N such that a sequence s_n is nondecreasing (nondecreasing, monotone) for $n > N$, then s_n is *eventually nondecreasing (respectively, eventually non-increasing, eventually monotone)*.

Sequences that are not monotone have been given a name, too.

A sequence that is *not* eventually monotone is an *oscillating sequence*.

We have already seen a type of oscillating sequence: periodic sequences with period 2 or greater are certainly oscillating.

But oscillating sequences are generally not periodic. For example, the sequence

$$s_n = (-1)^n n^2 \tag{3.5}$$

is oscillating due to repeated sign changes but not periodic because the magnitudes of its terms get larger as n does, so there is no repetition:

n	1	2	3	4	5	6	$\cdots$
s_n	-1	4	-9	16	-25	36	$\cdots$

Another example of a non-periodic, oscillating sequence is

$$s_n = \frac{1 + \cos(\pi n)}{n} = \frac{1 + (-1)^n}{n}$$

The odd terms of this sequence are all 0, while the even terms are all positive and with decreasing values.

A more subtle example of an oscillating sequence that is not periodic is $\sin n$, which we discussed earlier.

Eventually monotone sequences and eventually periodic sequences are special types of sequences. Most sequences of real numbers are oscillating but not eventually periodic.

The so-called *chaotic sequences* that are generated by nonlinear recursions are among well-known oscillating sequences that are not eventually periodic. Familiar examples are generated by the following quadratic polynomial recursion, known as the discrete "logistic equation":

$$x_{n+1} = ax_n(1 - x_n), \quad n = 1, 2, 3, \ldots \tag{3.6}$$

with the value of the fixed parameter a between 3.7 and 4. If the initial value x_1 is in the interval $[0, 1]$, then all generated numbers x_n are also in the same interval. For most values of x_1 in $[0,1]$, the generated sequence x_n oscillates unpredictably in the sense that slightest changes in the initial value quickly spread to the entire interval. This feature is known as *sensitivity to initial values* in the literature, and popularized as the "butterfly effect."

Figure 3.2 shows the space-time plots of the first 50 terms of two sequences generated by the iteration of the logistic equation above with parameter value

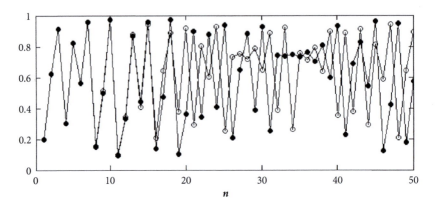

Fig. 3.2 Divergence of chaotic sequences starting near each other

$a = 3.9$. The starting values of these sequences are close to each other: $x_1 = 0.2$ (hollow circles), and $x_1 = 0.2001$ (filled circles).

We see in Figure 3.2 that for about a dozen terms, the two sequences are in close agreement, but then they begin to diverge from each other, and by the 20th term, the two sequences oscillate independently and apart from each other: they have completely forgotten how close they were in the beginning!

If we had chosen x_1 a lot closer to 0.2, say, $x_1 = 0.20000001$, then it would take only a few more steps before the two sequences diverge; but the separation happens eventually for most initial values of x_1 in $[0,1]$.

3.4 Convergent sequences and limits

We can get an idea as to what convergence and limit mean for sequences using an example. Consider the sequence

$$s_n = 1 + \frac{1}{n} \tag{3.7}$$

Notice that as the value of n gets larger the fraction $1/n$ approaches 0; therefore, s_n approaches 1. For example, if $n = 10$, then we find that $s_{10} = 1 + 1/10 = 1.1$; but for $n = 10,000$, we get $s_{10000} = 1.0001$. We say that s_n *converges* to 1 as n goes to infinity.

The word "converge" in this context is synonymous with "approaches" or "gets arbitrarily close to." It doesn't mean that s_n eventually equals 1; in fact, s_n *never* equals 1 exactly. But since we never run out of indices n, the point about convergence of sequences is a subtle one that needs a proper logical explanation, which we turn to shortly.

Before getting into specifics, I should point out that other issues arise when dealing with non-numerical sequences that are important in real analysis. For instance, what happens to the sequence of intervals

$$I_n = \left[0, \frac{1}{n}\right]$$

as n goes to infinity? The issues of approximation or the sense of getting close that I mentioned above for the numerical sequence s_n no longer seem relevant. Notice that each interval I_n is an infinite set no matter how large the index n, so it seems reasonable to expect that this sequence will converge to an infinite set.

This isn't the case. The right endpoint $1/n$ approaches 0 as n goes to infinity. While for every *finite* value of n the interval I_n is an infinite set, it is also true that there is no real number that is non-negative *and* less than $1/n$ for *every n*, except for 0. The only possible answer that doesn't end in a contradiction is that the sequence I_n converges to the set $\{0\}$. But this set has only one number in it, not infinitely many.

All is not lost, however; the intervals I_n do converge in a numerical sense: their *lengths* (which are numbers) approach 0, which is the length of the set $\{0\}$.[2] This aspect of convergence of sets turns out to be important in characterizing the completeness of the set of real numbers, one of the most important tenets of analysis.

In this section, we discuss the concepts of convergence and limit for numerical sequences. Much of the theory is concerned with answering (in a logically consistent way) what it means for a sequence to converge a number. A carefully developed theory of convergence for sequences of numbers can be extended to more general sequences when required. In Chapter 8, we actually discuss such extensions to sequences of functions.

Limits of sequences

We now turn to defining what it means exactly for a sequence of numbers to converge. Consider again the sequence s_n in (3.7), which we found to get as close to the number 1 as desired when the index n goes to infinity. There are two related components to this statement: one is "getting as close to 1 as desired," and the other is "as n goes to infinity."

To capture the first component, consider a positive number ε (the Greek letter *epsilon*) that can be arbitrarily small. To place the terms of the sequence within ε distance of 1, we need the inequalities

---

[2] The ordinary length of the interval $[a, b]$ is, of course, the number $b - a$. If $b = a$, then the length of $[a, a] = \{a\}$ is 0 for every number a.

$$1 - \varepsilon < s_n = 1 + \frac{1}{n} < 1 + \varepsilon \tag{3.8}$$

But what about the index n that is not specified? These inequalities are not true if n is not large enough to bring $1 + 1/n$ as close to 1 as is indicated in (3.8); for instance, if $\varepsilon = 0.1$ and $n = 5$, then $1 + 1/5 = 1.2$, whereas $1 + \varepsilon = 1.1$. Now is the time to consider the "as n goes to infinity" component of the aforementioned statement. This component means that we may pick the index n as large as is necessary to make the inequalities in (3.8) true.

So suppose that we choose a positive integer N large enough that

$$1 - \varepsilon < s_n < 1 + \varepsilon \quad \text{if} \quad n > N \tag{3.9}$$

The integer N is a "threshold" that once the index n crosses it, the inequalities (3.9) are guaranteed thereafter. They ensure that $1 + 1/n$ is at most ε units away from 1 from either side of 1.[3] Our task is to find the threshold N, and at this point there is no guarantee that such N even exists!

If it does exist, then N depends on our choice of ε because the smaller the value of ε, the larger N needs to be. Given $\varepsilon > 0$, we look for N by solving the inequalities in (3.8) for n. Adding a -1 to these inequalities to cancel out the 1's, we get

$$-\varepsilon < s_n - 1 < \varepsilon \tag{3.10}$$

or equivalently,

$$-\varepsilon < \frac{1}{n} < \varepsilon$$

This implies that

$$n > \frac{1}{\varepsilon}$$

Now any *integer* N that is larger than $1/\varepsilon$ works for any choice of ε (it is not necessary that N be uniquely defined). For example, if $\varepsilon = 0.03$, then $1/\varepsilon = 33.33$. Any integer greater than this number is a valid choice for N; we could pick $N = 34$ or larger to ensure that

$$1 - 0.03 < s_n < 1 + 0.03 \quad \text{if} \quad n > N$$

Of course, $N = 33$ will not do, but our calculation gives some value of N no matter what ε is chosen or given. When this is possible, we say that the limit exists, and the sequence s_n converges to the limit.

---

[3] This may seem overstated: since $1 + 1/n$ is always greater than 1, why bother with $1 - \varepsilon$? The generality is needed to take care of similar cases like $1 - 1/n$, which is always less than 1; or $1 + (-1)^n/n$, which can be on either side of 1 depending on whether n is odd or even.

You can use a similar argument in Exercise 78 to also prove that the sequence $1/n$ converges to 0.

Before stating the formal definition of limit for a converging sequence, I should point out that the two inequalities in (3.10) can be written succinctly as one inequality using the *absolute value*:[4]

$$|s_n - 1| < \varepsilon$$

Convergence and limit

A sequence s_n of real numbers *converges* to a real number s if for every $\varepsilon > 0$, we can find a positive integer $N = N(\varepsilon)$ such that for all indices $n > N$

$$|s_n - s| < \varepsilon \qquad (3.11)$$

We call s the *limit* of s_n and write

$$\lim_{n \to \infty} s_n = s$$

If for a given ε we find any N that works, then all larger numbers, like $N + 1$ or $2N$, also work because once s_n reaches within an ε of s, it stays within the ε distance of s for all larger indices n.

We can be more specific since the set $\mathbb{N}$ of all positive integers is well-ordered, i.e., there is *a least or smallest value* of N that works. For a fixed ε, such a least index is uniquely defined. For instance, for the sequence $s_n = 1 + 1/n$ above, $N = 34$ is the one and only *least index* that works for $\varepsilon = 0.03$. Or if $\varepsilon = 0.0002$, then the least index that works is

$$N = \frac{1}{0.0002} = \frac{10000}{2} = 5000$$

We use a standard notation for this least value: $\lceil x \rceil$ denotes *the least integer that is greater than or equal to x* for each positive number x. The notation $\lceil \; \rceil$ actually defines the ceiling function whose domain is the set of all real numbers, and its range is the set $\mathbb{Z}$ of all integers. For example, $\lceil 1/0.03 \rceil = 34$; in short, the least index that works for an arbitrary ε in this example is $\lceil 1/\varepsilon \rceil$.

We are now ready to define a concept that becomes more important when studying sequences of functions in Chapter 8.

---

[4] Recall that the absolute value of a number x is just the magnitude of x stripped of its sign; formally, $|x| = x$ if $x \geq 0$ and $|x| = -x$ if $x < 0$. For instance, $|-2| = -(-2) = 2$.

The ε-index of a convergent sequence

Assume that a sequence s_n converges to a real number. For each $\varepsilon > 0$, *the least index N_ε for which (3.11) holds is the ε-index of the sequence s_n.*

Note that $N_\varepsilon < \infty$ for each $\varepsilon > 0$ when the sequence converges, although it is possible (and generally true) that $N_\varepsilon \to \infty$ as $\varepsilon \to 0$. For example, $N_\varepsilon = [1/\varepsilon]$ for the sequence $s_n = 1 + 1/n$ and $1/\varepsilon$ goes to infinity as ε goes to 0.

The algebra of limits

Sequences are functions and as such, they can be added, multiplied, etc.; in short, the algebra of functions applies to sequences as well. If we have two sequences $s_1, s_2, s_3, \ldots$ and $s_1', s_2', s_3', \ldots$ then we can perform the following algebraic operations:

$$s_1 + s_1', s_2 + s_2', s_3 + s_3', \ldots \quad \textit{addition term by term, abbreviated as} \quad s_n + s_n'$$

$$s_1 s_1', s_2 s_2', s_3 s_3', \ldots \quad \textit{multiplication term by term, abbreviated as} \quad s_n s_n'$$

Subtraction and division are defined using addition and multiplication, respectively, as $s_n + (-s_n')$, and $s_n(1/s_n')$ if $s_n' \neq 0$ for all n.

Now, suppose that both s_n and s_n' converge to real numbers. It seems reasonable that their sums, products, etc. should converge the sum, product, etc. of their limits. The next result verifies this statement. The arguments that prove the next theorem are based on the definition of convergence above, even though the cases of products and quotients are a bit technical.

An essential tool in the proof is the "triangle inequality," a simple yet fundamental result about all metric spaces, including the real numbers. We can prove this result before constructing the real numbers since the idea is simply a property of absolute value as a metric, and the essential completeness property of real numbers is not invoked.

Lemma 54 *(Triangle inequality) If a and b are real numbers, then*

$$|a + b| \leq |a| + |b| \tag{3.12}$$

and

$$||a| - |b|| \leq |a - b| \tag{3.13}$$

Proof Proving (3.12) is easy on a case-by-case basis: either a and b have the same signs, both negative or both non-negative, or one is non-negative and the other negative. If they are both non-negative, then we can remove all absolute value signs and get the equality $a + b \leq a + b$, which is true.

If a, b are both negative, then all absolute value signs can be replaced with negative signs, which again gives a true statement

$$-(a + b) \leq -a - b$$

Now suppose that $a < 0 \leq b$. Then $|a + b|$ is a (positive) quantity that is less than b which is in turn less than $|a| + |b|$ because $|a| > 0$; symbolically,

$$|a + b| < b = |b| < |a| + |b|$$

which is consistent with (3.12). The last remaining case, $b < 0 \leq a$, is argued similarly with the roles of a and b interchanged.

To prove (3.13), let $c = a - b$, and use (3.12) to get

$$|a| = |b + c| \leq |b| + |c|$$

so that

$$|a| - |b| \leq |c| = |a - b| \tag{3.14}$$

Similarly, with $c = b - a$ and using (3.12), we get

$$|b| - |a| \leq |c| = |b - a| = |a - b| \tag{3.15}$$

Putting (3.14) and (3.15) together gives (3.13) since one of the two differences on the right-hand sides must be equal to $||a| - |b||$.   ∎

Theorem 55 (*Convergence and algebraic operations*) *Assume that $s_n \to s$ and $s'_n \to s'$ as $n \to \infty$. Then*

(a) The sum sequence $s_n + s'_n$ converges and

$$s_n + s'_n \to s + s' \tag{3.16}$$

(b) The product sequence $s_n s'_n$ converges and

$$s_n s'_n \to s s' \tag{3.17}$$

(c) The quotient sequence s_n / s'_n converges and

$$\frac{s_n}{s'_n} \to \frac{s}{s'} \tag{3.18}$$

provided that $s' \neq 0$ and $s'_n \neq 0$ for all n.

Proof   Because the sequences s_n and s'_n converge to s and s', the definition of convergence implies that for every number $\varepsilon_1 > 0$, there are positive integers N and N' such that

$$|s_n - s| < \varepsilon_1 \quad \text{for all } n \geq N \tag{3.19}$$

$$|s'_n - s'| < \varepsilon_1 \quad \text{for all } n \geq N' \tag{3.20}$$

If N'' is the larger of N and N', then both of the above inequalities hold whenever the index n exceeds N''.

(a) Notice that

$$|s_n + s'_n - (s + s')| = |s_n - s + s'_n - s'| \leq |s_n - s| + |s'_n - s'|$$

The last inequality is just the triangle inequality. According to (3.19) and (3.20), each of the two quantities on the far right is less than ε_1 if $n \geq N''$; so if ε is any positive number, and we set $\varepsilon_1 = \varepsilon/2$, then

$$|s_n + s'_n - (s + s')| < \frac{\varepsilon}{2} + \frac{\varepsilon}{2} = \varepsilon \quad \text{for all } n \geq N''$$

This proves (3.16), the limit theorem for sums.

(b) For the product, we must show that

$$s_n s'_n \to s s' \quad \text{as } n \to \infty$$

To this end, let ε be an arbitrary (unspecified) positive number and notice that

$$|s_n s'_n - s s'| = |s_n s'_n - s_n s' + s_n s' - s s'| = |s_n(s'_n - s') + (s_n - s)s'|$$

The simultaneous addition and subtraction of the term $s_n s'$ inside the absolute value causes no change in value on the left-hand side, but it allows us to use our hypotheses in (3.19) and (3.20). Using the triangle inequality gives

$$|s_n s'_n - s s'| \leq |s_n||s'_n - s'| + |s_n - s||s'|$$

The sequence s_n is convergent and therefore, it is bounded (see Appendix 9.1); that is, there is a positive number M such that $|s_n| \leq M$ for all n. Now, let

M' be the larger of M and $|s'|$ so that

$$|s_n s'_n - ss'| \leq M'(|s'_n - s'| + |s_n - s|)$$

The final step is to set $\varepsilon_1 = \varepsilon/(2M')$ so that $M' = \varepsilon/(2\varepsilon_1)$, and the last inequality above can be written as

$$|s_n s'_n - ss'| \leq \frac{\varepsilon}{2\varepsilon_1}(|s'_n - s'| + |s_n - s|) < \frac{\varepsilon}{2\varepsilon_1}(\varepsilon_1 + \varepsilon_1) = \varepsilon$$

This now proves (3.17), the limit theorem for products.
(c) Proceeding similarly to the product case, we obtain

$$\left|\frac{s_n}{s'_n} - \frac{s}{s'}\right| = \left|\frac{s_n s' - s'_n s}{s'_n s'}\right| = \left|\frac{s_n s' - ss' + ss' - s'_n s}{s'_n s'}\right| = \left|\frac{(s_n - s)s' + s(s' - s'_n)}{s'_n s'}\right|$$

The last fraction is split using the triangle inequality as follows:

$$\left|\frac{s_n}{s'_n} - \frac{s}{s'}\right| \leq \left|\frac{s_n - s}{s'_n}\right| + \left|\frac{s(s' - s'_n)}{s'_n s'}\right| = \frac{1}{|s'_n|}|s_n - s| + \left|\frac{s/s'}{s'_n}\right||s'_n - s'|$$

Next, we need to examine the coefficient $|1/s'_n|$. Intuitively, since $s'_n \to s'$ and $s' \neq 0$, we see that s'_n cannot get arbitrarily close to 0 and from this observation conclude that $|1/s'_n|$ is bounded, that is, there is a positive number M such that $|1/s'_n| \leq M$ for all large values of n. To be precise, we know that as s'_n gets close to s', there is a positive integer, say, N_1, such that $|s'_n - s'| < |s'|/2$ for every $n \geq N_1$. Now the triangle inequality implies that (see Appendix 9.1)

$$\frac{|s'|}{2} > |s'_n - s'| = |s' - s'_n| > |s'| - |s'_n|$$

so we can write

$$|s'_n| > |s'| - \frac{|s'|}{2} = \frac{|s'|}{2} \quad \text{for all } n \geq N_1$$

Thus we see that

$$\left|\frac{1}{s'_n}\right| < \frac{2}{|s'|} \quad \text{for all } n \geq N_1$$

From this argument, we see that the upper bound can be chosen as $M = 2/|s'|$ as long as $n \geq N_1$. The process now follows the same path as that of the products;

we choose N_2 to be the larger of N_1 and N'', and given an arbitrary $\varepsilon > 0$ and notice that for all $n \geq N_2$

$$\left| \frac{s_n}{s'_n} - \frac{s}{s'} \right| < \frac{2}{|s'|} |s_n - s| + \left| \frac{2s}{s'2} \right| |s'_n - s'| < \frac{2}{|s'|} \varepsilon_1 + \left| \frac{2s}{s'2} \right| \varepsilon_1 = A\varepsilon_1$$

where

$$A = \frac{2}{|s'|} \varepsilon_1 + \left| \frac{2s}{s'2} \right|$$

So if we define $\varepsilon_1 = \varepsilon / A$, then we see that

$$\left| \frac{s_n}{s'_n} - \frac{s}{s'} \right| < \varepsilon \quad \text{for all } n \geq N_2$$

This proves (3.18), the limit theorem for quotients. ■

A special case of (3.17) is when one of the sequences is a constant sequence, say, $s'_n = c$ for all n, where c is some fixed real number. For this case, (3.17) reads

$$cs_n \to cs \quad \text{as } n \to \infty$$

This is often called the *constant multiple rule* and is a useful result. As a quick application, we see that if s'_n converges to s', then with $c = -1$

$$s_n - s'_n = s_n + (-1)s_n' \to s + (-s') = s - s'$$

If we set $s = \lim_{n \to \infty} s_n$ and $s' = \lim_{n \to \infty} s'_n$, then the statement of the above theorem can be written more compactly as a list of *limit properties*.

Corollary 56 (*The algebraic properties of limits*) *If* $\lim_{n \to \infty} s_n$ *and* $\lim_{n \to \infty} s'_n$ *both exist and are real numbers, then*

(a) $\lim_{n \to \infty} (s_n + s'_n) = \lim_{n \to \infty} s_n + \lim_{n \to \infty} s'_n$

(b) $\lim_{n \to \infty} (s_n s'_n) = \left(\lim_{n \to \infty} s_n \right) \left(\lim_{n \to \infty} s'_n \right)$

(c) $\lim_{n \to \infty} \frac{s_n}{s'_n} = \frac{\lim_{n \to \infty} s_n}{\lim_{n \to \infty} s'_n}$ *if* $\lim_{n \to \infty} s'_n \neq 0$ *and* $s'_n \neq 0$ *for all* n

These properties help us calculate limits. For example,

$$\lim_{n\to\infty}\left(2-\frac{7}{n}\right)^4 = \lim_{n\to\infty}\left[\left(2-\frac{7}{n}\right)\left(2-\frac{7}{n}\right)\left(2-\frac{7}{n}\right)\left(2-\frac{7}{n}\right)\right]$$

$$= \lim_{n\to\infty}\left(2-\frac{7}{n}\right)\lim_{n\to\infty}\left(2-\frac{7}{n}\right)\lim_{n\to\infty}\left(2-\frac{7}{n}\right)\lim_{n\to\infty}\left(2-\frac{7}{n}\right)$$

$$= \left[\lim_{n\to\infty}\left(2-\frac{7}{n}\right)\right]^4$$

$$= \left[\lim_{n\to\infty}2 - \lim_{n\to\infty}\frac{7}{n}\right]^4$$

$$= (2-0)^4 = 16$$

Can you tell which of the properties in Corollary 56 are used to go from the first line to the second and from the third to the fourth?

We can also use the idea in this example to show that for any real number c and any positive integer m,

$$\lim_{n\to\infty}\frac{c}{n^m} = 0 \tag{3.21}$$

since we can write

$$\lim_{n\to\infty}\frac{c}{n^m} = \lim_{n\to\infty}\left[c\left(\frac{1}{n}\right)^m\right] = \lim_{n\to\infty}c\lim_{n\to\infty}\left(\frac{1}{n}\right)^m = c\left(\lim_{n\to\infty}\frac{1}{n}\right)^m = c(0)^m = 0$$

You may wonder, what if m is not a positive integer but, say, a positive rational or even irrational number? In such a case, the above reasoning does not work because we cannot write the power as repeated multiplication. However, *(3.21) is true if m is any positive real number*; and to show it, we may use the *definition* of convergence; see Exercise 78.

Two other important properties of limits are proved next.

Theorem 57 (a) *If a sequence converges, then its limit is unique.*

(b) *If x_n and y_n are convergent sequences of real numbers, and there is a positive integer K such that $x_n \leq y_n$ for all $n \geq K$, then*

$$\lim_{n\to\infty}x_n \leq \lim_{n\to\infty}y_n \tag{3.22}$$

That is, limits preserve the inequality $\leq$.

Proof (a) Suppose that s_n converges to a and to b, and $a \leq b$. If $a \neq b$, then we can set $\varepsilon = (b - a)/2 > 0$ and then find a positive integer N large enough that

$$|s_N - a| < \varepsilon \quad \text{and} \quad |s_N - b| < \varepsilon$$

From these inequalities, we conclude that

$$s_N < a + \varepsilon = a + \frac{b - a}{2} = \frac{a + b}{2}$$

$$s_N > b - \varepsilon = b - \frac{b - a}{2} = \frac{a + b}{2}$$

But this is a contradiction since the number s_N cannot be both less than and greater than the same number $(a + b)/2$. Therefore, $a = b$ and the limit of s_n is unique.

(b) First, we prove that if s_n converges, and $s_n \geq 0$ for all $n \geq K$, then $\lim_{n\to\infty} s_n \geq 0$. If this is true, then we may complete the proof of the theorem by setting $s_n = y_n - x_n$ and using Corollary 56(a).

So let's assume that $\lim_{n\to\infty} s_n = s$, and $s_n \geq 0$ for all $n \geq K$. If $s < 0$, then choose $\varepsilon = -s/2 > 0$ and find an integer N so large that

$$|s_n - s| < \varepsilon \quad \text{for } n \geq N$$

In particular (by choosing $N = K$ if necessary), we may conclude that

$$s_N < s + \varepsilon = s - \frac{s}{2} = \frac{s}{2} < 0$$

But this contradicts the hypothesis that $s_n \geq 0$ for all $n \geq K$, so s cannot be negative; i.e., $s \geq 0$, and the proof is complete. ∎

The nonzero conditions in Theorem 55(c) are necessary since we don't want to divide by 0 on either side of the equation, which would be algebraically problematic. But the case where $s'_n \neq 0$ for all n but

$$\lim_{n\to\infty} s_n = \lim_{n\to\infty} s'_n = 0 \tag{3.23}$$

is especially important for the sake of calculus. In a sense, it includes the concept of *derivative*, which, as we discuss later, is defined by the limit of a difference ratio

$$\frac{f(x_n) - f(a)}{x_n - a}$$

where x_n is a sequence that converges to the number a. If we take $s_n = f(x_n) - f(a)$ and $s'_n = x_n - a$ then

$$\text{derivative of } f(x) \text{ at } x = a \text{ is: } \lim_{n \to \infty} \frac{f(x_n) - f(a)}{x_n - a} = \lim_{n \to \infty} \frac{s_n}{s'_n}$$

Although we can't use Theorem 55(c) to calculate the last limit, it often has a well-defined value, namely, the derivative. For instance, if $f(x) = x^2$ is our reliable square function, and $x_n \to a$ as $n \to \infty$, then the derivative of x^2 at $x = a$ turns out to be

$$\lim_{n \to \infty} \frac{x_n^2 - a^2}{x_n - a} = \lim_{n \to \infty} \frac{(x_n - a)(x_n + a)}{x_n - a} = \lim_{n \to \infty} (x_n + a) = a + a = 2a$$

This type of limit calculation can be done in far more general situations as we discover in Chapter 6, so it is fair to say

The concept of derivative can be defined to develop calculus because (c) fails in Theorem 55 when (3.23) holds.

Failure of theorems or properties often suggests the existence of new structures or possibilities. Another famous example is *quaternions*, the numbers introduced by Hamilton[5] that have all the properties of the real numbers *except* for the commutative property.

Important mathematical operations in fact fail the commutative property; the most prominent among these operations is function composition that we encountered earlier in Chapter 2. This general operation is defined not only among ordinary functions but also among operators on spaces of functions. As such it plays a fundamental role in various areas of mathematical physics, including quantum theory where the non-commutative nature of operator composition (also known as operator multiplication) is tied to such important results as the uncertainty principle.

The squeeze theorem

I end this section with a very useful theorem with a funny name. Its name is quite apt; to illustrate, imagine three cars that are moving in a narrow alley. To avoid collisions, the car in the middle must keep ahead of the car that is following it but cannot move past the car that is in front. Now, suppose that the car ahead and the car behind gradually slow down so as to come to a stop in front of a building.

---

[5] William Rowan Hamilton (1805–1865) was an Irish mathematician and astronomer. He is known in physics for developing the Hamiltonian mechanics.

Then the car in the middle has to slow down too and eventually stop in front of the same building.

The cars ahead and behind essentially "squeeze" or "sandwich" the one in the middle, forcing it to follow them. If we take snapshots of the three cars in, say, 1-second intervals and line up the photos in a row, then we find three interlaced sequences (images of the three cars). One of the sequences is squeezed in between the other two. The inevitable outcome may be stated as a theorem.

Theorem 58 *(Squeeze theorem) Assume that a_n, b_n, and c_n are sequences of real numbers such that*

$$a_n \leq b_n \leq c_n$$

for every index n. If the flanking sequences a_n and c_n both converge to the same limit L, i.e.,

$$\lim_{n \to \infty} a_n = \lim_{n \to \infty} c_n = L$$

then the middle sequence b_n also converges to L:

$$\lim_{n \to \infty} b_n = L$$

Proof   We begin by collecting some facts: the sequences a_n and c_n both converge to the same limit L; so for every $\varepsilon > 0$, there is a positive integer N such that for all $n > N$,

$$L - \varepsilon < a_n < L + \varepsilon \quad \text{and} \quad L - \varepsilon < c_n < L + \varepsilon$$

Therefore,

$$b_n \leq c_n < L + \varepsilon \quad \text{and} \quad b_n \geq a_n > L - \varepsilon$$

from which it follows that for all $n > N$,

$$L - \varepsilon < b_n < L + \varepsilon$$

Since the above argument holds for every $\varepsilon > 0$, it follows that b_n converges to L, and the squeeze theorem is proved.   ∎

To illustrate, we find the limit of the following sequence as $n \to \infty$:

$$s_n = \frac{\sin n}{n}$$

The squeeze theorem provides a quick answer once we find suitable flanking or sandwiching sequences. To do this, recall that the sine function oscillates between

1 and −1, so

$$-1 \le \sin n \le 1$$

for every value of n; when we divide these inequalities by n, we obtain

$$-\frac{1}{n} \le \frac{\sin n}{n} \le \frac{1}{n}$$

The flanking sequences are $-1/n$ and $1/n$, and both of them converge to 0 as $n \to \infty$. Therefore, s_n converges to 0 too.

We close this section with the following useful topological fact about limits.

Theorem 59 *Consider a sequence s_n in a closed interval $[a, b]$ of real numbers, i.e., $s_n \in [a, b]$ for all n. If s_n converges to a limit s, then $s \in [a, b]$.*

Proof We prove this theorem by a contradiction argument. Assume that $s \notin [a, b]$. Then either $s < a$ or $s > b$; the argument is similar for both cases, so let $s < a$. Then $d = a-s > 0$ so there is a positive integer N such that $|s_N-s| < a-s$. In particular,

$$s_N - s < a - s$$

so that $s_N < a$. This contradicts the hypothesis that $s_n \in [a, b]$, so the assumption $s \notin [a, b]$ is false; i.e., $s \in [a, b]$, and the proof is complete. ∎

It is important that the interval in Theorem 59 be closed; for example, if s_n is in the open interval (a, b) or the half-open interval $(a, b]$ and converges to a, then the limit of s_n is *not* in the interval.

3.5 Bounded, unbounded, and divergent sequences

In this section, we briefly discuss sequences that don't converge. A sequence that can be contained in a bounded interval is a bounded sequence. More precisely

Bounded sequence A sequence s_n is *bounded* if there are real numbers a and b such that $a \le s_n \le b$ for all indices n. An *unbounded sequence* is then a sequence that is not bounded.

For instance, the sequence in (3.5) is unbounded because it can't be bounded by two real numbers, while the sequence in (3.4) is clearly bounded. *The latter is actually a bounded sequence that doesn't converge.* On the other hand, every convergent sequence of real numbers is bounded since it approaches a limit and in particular, isn't going far away. Here is how to prove this fact.

Theorem 60 *Every convergent sequence of real numbers is bounded.*

Proof Suppose s_n is a convergent sequence with limit s. Then in particular with $\varepsilon = 1$, there is a positive integer N such that

$$|s_n - s| < 1 \quad \text{for all } n > N$$

Therefore,

$$s - 1 < s_n < s + 1 \quad \text{for all } n > N$$

So all terms $s_{N+1}, s_{N+2}, \ldots$ with indices greater than N are bounded between the real number $s - 1$ and $s + 1$. That leaves a finite number of terms $s_1, s_2, \ldots, s_N$, which can't get infinitely large. In fact, define

$$M = \max\{|s_1|, |s_2|, \ldots, |s_N|\}$$

and note that

$$-M \le s_n \le M$$

for $n = 1, 2, \ldots, N$. It follows that all terms of the sequence are within a bounded interval; we can define this interval in many ways, e.g.,

$$-M + s - 1 \le s_n \le M + s + 1 \quad \text{for all } n \in \mathbb{N} \qquad \blacksquare$$

We now consider sequences that are not bounded or not convergent.

Divergent sequence A sequence s_n *diverges* or is *divergent* if it does not converge.

Unbounded sequence A sequence s_n is *unbounded* if it is not bounded. Specifically, if for every real number $M > 0$, there is an index N such that $|s_N| \ge M$.

By Theorem 60, unbounded sequences diverge. However, some unbounded sequences have a more constrained behavior as in the next definition.

> **Divergence to infinity** A sequence s_n of real numbers *diverges to* ∞ if for every positive real number M there is a positive integer N such that $s_n \geq M$ for all $n \geq N$. *Further,* s_n *diverges to* $-\infty$ *if* $-s_n$ *diverges to* ∞.

In other words, "s_n diverges to ∞" if its terms eventually exceed every *positive* real number; we sometime say that s_n "approaches" ∞ and though technically incorrect, this usage is intuitively appealing as long as it doesn't mislead.

Examples of sequences that diverge to ∞ or $-\infty$ are $2n-7, n^2-n, 3-\sqrt{n}$. These sequences are monotone (increasing or decreasing), but non-monotone sequences may diverge to ∞ or $-\infty$ also; see Exercise 83.

3.6 Subsequences

Consider the periodic sequence in (3.4)

$$s_n = \cos \frac{\pi n}{3}$$

Notice that for $n = 6k$ with $k = 1, 2, 3, \ldots$ or $n = 6, 12, 18, \ldots$ we get a sequence of constants: $s_{6k} = \cos(2\pi k) = 1$. The sequence s_{6k} is a part of the original s_n but unlike the latter, s_{6k} converges (trivially) to 1 as a constant sequence. You can check that there are five other parts of s_n that also converge as constant sequences; they are $s_{6k-1}, s_{6k-2}, s_{6k-3}, s_{6k-4},$ and s_{6k-5}.

Each of the six parts of the sequence s_n above is called a *subsequence* of s_n. Although a nontrivial periodic sequence of period m doesn't converge, it has m convergent subsequences. More generally, other types of non-convergent sequences can also have convergent subsequences, although the indices of the latter may not be evenly spaced. For instance, if

$$s_n = \cos(\pi\sqrt{n})$$

then $s_{4k^2} = \cos(2\pi k) = 1$ is a convergent (constant) subsequence for $k = 1, 2, 3, \ldots$ or $n = 4k^2 = 4, 16, 36, \ldots$ A less transparent example is $\sin n$ that has convergent subsequences; for this sequence, we need more powerful tools that are discussed in Chapter 4.

These results suggest that the concept of subsequence is important and useful in analysis, which is concerned with convergence and limits. We now give a formal definition.

Subsequence Let s_n be a given sequence. For every increasing sequence of indices

$$n_1 < n_2 < n_3 < \cdots$$

the sequence s_{n_k}, i.e.,

$$s_{n_1}, s_{n_2}, s_{n_3}, \ldots$$

is a *subsequence* of s_n.

Although every sequence has an infinity of subsequences, not all of them are convergent. In fact, *there are sequences that have no convergent subsequences*, like $s_n = (-1)^n n^2$ in (3.5). On the other hand, if a sequence itself converges, then so do all of its subsequences, and they all converge to the same limit.

Theorem 61 *If s_n converges to a limit s, then all subsequences of s_n converge to s.*

Proof Let s_{n_k} be an arbitrary subsequence of s_n. We begin with the observation that since the index sequence n_k is increasing in the set $\mathbb{N}$, it must be that $n_k \geq n_{k-1} + 1$ for all $k = 1, 2, 3, \ldots$ This leads to the chain

$$n_k \geq n_{k-1} + 1 \geq n_{k-2} + 1 + 1 = n_{k-2} + 2 \geq n_{k-3} + 3 \geq \cdots \geq n_1 + k - 1$$

Therefore, as $k \to \infty$ the sequence n_k diverges to ∞.

Now, by assumption, s_n converges to s; so given $\varepsilon > 0$, there is a positive integer N such that $|s_n - s| < \varepsilon$ for all $n \geq N$. We showed that n_k goes to ∞ in a strictly increasing fashion, so there is some integer K such that $n_k \geq N$ for all $k \geq K$. It follows that

$$\left| s_{n_k} - s \right| < \varepsilon \quad \text{for all } k \geq K$$

This means that s_{n_k} converges to s, i.e.,

$$\lim_{k \to \infty} s_{n_k} = s$$

Since the above argument applies to an arbitrary subsequence, it is true for all of them. ∎

A divergent sequence may have infinitely many subsequences that converge to the same limit. An example of such a sequence is easy to construct; here is one:

$$s_n = \begin{cases} 1/n & \text{if } n \text{ is even} \\ n & \text{if } n \text{ is odd} \end{cases}$$

Since $1/n$ converges to 0, all of *its* subsequences (infinitely many; see Exercise 85) converge to the same limit 0, but s_n itself diverges. Note that s_n in this case also has infinitely many subsequences that diverge to ∞.

The contrapositive version of Theorem 61 can be useful in proving divergence, especially of bounded sequences:

If two subsequences of a sequence s_n converge to different limits, then the sequence diverges.

In particular, this statement makes it obvious why nontrivial periodic sequences don't converge.

3.7 Limit supremum and limit infimum

In this section, we introduce concepts that help us gain a deeper understanding of the behavior of sequences. We begin with bounded sets.

Supremum and infimum of sets

Let A be a nonempty subset of $\mathbb{R}$. An *upper bound* of A is a number $u \in \mathbb{R}$ such that $x \leq u$ for all $x \in A$. If A has an upper bound, then A is *bounded from above*.

The *supremum of A* or the *least upper bound* is an upper bound that is less than or equal to all upper bounds of A. If this number exists, then it is denoted $\sup A$.

Similarly, a *lower bound* of A is a number $l \in \mathbb{R}$ such that $l \leq x$ for all $x \in A$. If A has a lower bound, then A is *bounded from below*. The *infimum of A* or the *greatest lower bound* is a lower bound that is greater than or equal to all lower bounds of A. If this number exists, then it is denoted $\inf A$.

If A is bounded from above and from below, then we say that A is *bounded*.

Upper bounds and lower bounds are not unique; for example, consider the half-open interval $A = [-1, 1)$ of real numbers. Then any number $u \geq 1$ is an upper bound of A since every number in $[-1, 1)$ is at most 1. In particular, 1 is the least upper bound since no upper bound of $[-1, 1)$ can be smaller than 1 without being less than some number in $[-1, 1)$. For instance, 0.999 is not an upper bound because it is less than, e.g., 0.9991, which is in $[-1, 1)$. We conclude that $\sup[-1, 1) = 1$.

Similarly, every number $l \le -1$ is a lower bound for A and $\inf[-1, 1) = -1$. In this example, we also see that the supremum or the infimum may or may not be in A (when they exist). Indeed, $[-1, 1)$ contains not a single one of its upper bounds!

Another example is that the countable set $S = \{1, 1/2, 1/3, 1/4, \ldots\}$ has sup $S = 1$, which is in S, and $\inf S = 0$, which is not in S.

Some sets don't have upper bounds or lower bounds. If A is $\mathbb{N}$ or $[1, \infty)$, then $\inf A = 1$; lower bounds for A include 0 and all negative numbers. But A doesn't have upper bounds because there is no real number u that is larger than all numbers in $\mathbb{N}$ or $[1, \infty)$. In particular, sup A doesn't exist. Going further, the set $\mathbb{Z}$ of integers has no upper bounds and no lower bounds.

In Chapter 4, we discover that every *bounded* subset of the real number set $\mathbb{R}$ has both a supremum and an infimum; this is a consequence of the completeness of $\mathbb{R}$. But bounded sets in $\mathbb{Q}$, which is not complete, may fail to have a supremum or an infimum due to the existence of gaps or holes; see Exercise 87.

However, whenever sup A or $\inf A$ exist, they are unique.

Lemma 62 *(a) If A has a supremum or least upper bound, then it is unique. Also, a greatest lower bound is unique, if it exists.*

(b) There is a sequence a_n in A that converges to sup A. Also there is a sequence in A that converges to $\inf A$.

Proof (a) Let $r = \sup A$. If r' is also a least upper bound of A, then in particular, r' is an upper bound, so $r \le r'$. Similarly, $r' \le r$ since r is also an upper bound of A, and r' is least by assumption. It follows that $r' = r$. The proof that the greatest lower bound is unique is essentially the same.

(b) We prove the assertion about sup A and leave the one about $\inf A$ as an exercise. First, if $s = \sup A$ is in A (e.g., if A is a finite set), then the constant sequence $a_n = s$ converges to s (trivially), and the proof is finished.

Next, suppose that s is not in A (hence, A is infinite). If $\sup A = \infty$, then for every positive integer n there is an element a_n of S such that $a_n \ge n$. It follows that a_n diverges to ∞ and thus to sup A. Finally, let $\sup A = s < \infty$. Then there is an element of $a_1 \in A$ such that $s - a_1 < 1$; if not, then $s \ge a + 1$ for all $a \in A$, and thus s is not the least upper bound. Therefore,

$$s - 1 < a_1 < s$$

Similarly, there is $a_2 \in A$ such that $a_2 > s - 1/2$ and so on; for every n, there is $a_n \in A$ such that

$$s - \frac{1}{n} < a_n < s$$

Since $\lim_{n \to \infty}(s - 1/n) = s$, the squeeze theorem implies that $\lim_{n \to \infty} a_n = s$. ∎

Let s_n be a sequence of real numbers and define the sequences

$$u_k = \sup\{s_k, s_{k+1}, s_{k+2}, \ldots\} = \sup_{n \geq k} s_n$$

$$l_k = \inf\{s_k, s_{k+1}, s_{k+2}, \ldots\} = \inf_{n \geq k} s_n$$

For a simple example, consider the sequence $s_n = 1/n$. Then for each index k

$$u_k = \sup\left\{\frac{1}{k}, \frac{1}{k+1}, \frac{1}{k+2}, \ldots\right\} = \frac{1}{k}$$

because $1/k$ has the smallest denominator of all the fractions inside the braces, thus it must be the largest fraction. On the other hand,

$$l_k = \inf\left\{\frac{1}{k}, \frac{1}{k+1}, \frac{1}{k+2}, \ldots\right\} = 0$$

because the fractions inside the braces get smaller and smaller, approaching 0.

For another example, consider the alternating sequence $s_n = (-1)^n$. In this case,

$$u_k = \sup\{(-1)^k, (-1)^{k+1}, (-1)^{k+2}, \ldots\} = 1$$

because the numbers inside the braces are always either -1 or 1. Similarly,

$$l_k = \inf\{(-1)^k, (-1)^{k+1}, (-1)^{k+2}, \ldots\} = -1$$

Here is one more example that in a sense combines the previous two: $s_n = (-1)^n/n$. For every index k

$$u_k = \sup\left\{\frac{(-1)^k}{k}, \frac{(-1)^{k+1}}{k+1}, \frac{(-1)^{k+2}}{k+2}, \ldots\right\} = \begin{cases} 1/k & \text{if } k \text{ is even} \\ 1/(k+1) & \text{if } k \text{ is odd} \end{cases}$$

while

$$l_k = \inf\left\{\frac{(-1)^k}{k}, \frac{(-1)^{k+1}}{k+1}, \frac{(-1)^{k+2}}{k+2}, \ldots\right\} = \begin{cases} -1/(k+1) & \text{if } k \text{ is even} \\ -1/k & \text{if } k \text{ is odd} \end{cases}$$

There are also sequences for which u_k or l_k may be undefined. Consider the simple sequence $s_n = 2n$ for which we get

$$u_k = \sup\{2k, 2k+2, 2k+4, \ldots\}$$

The supremum doesn't exist because the set in braces has no upper bounds; thus the sequence u_k is undefined in this case. On the other hand,

$$l_k = \inf\{2k, 2k + 2, 2k + 4, \ldots\} = 2k$$

is well-defined for all k.

The following lists the basic properties of the two sequences u_k and l_k.

Lemma 63 *Let s_n be a given sequence of real numbers.*

(a) The sequences u_k and l_k bound the sequence s_n in the following sense:

$$l_k \leq s_k \leq u_k \tag{3.24}$$

(b) u_k is a nonincreasing sequence, and l_k is a nondecreasing sequence.

Proof (a) is clear from the definition of supremum and infimum of sets.

To prove (b), we show that for every index k

$$u_k \geq u_{k+1} \quad \text{and} \quad l_k \leq l_{k+1} \tag{3.25}$$

For the first inequality, recall that for every k,

$$u_k = \sup\{s_k, s_{k+1}, s_{k+2}, \ldots\}$$

$$u_{k+1} = \sup\{s_{k+1}, s_{k+2}, s_{k+3}, \ldots\}$$

The only difference between the two quantities is that the second set doesn't contain s_k. If s_k is less than or equal to one of the other numbers $s_{k+1}, s_{k+2}, \ldots$ inside the braces, then the supremum isn't affected by dropping it, and we have $u_{k+1} = u_k$. But if s_k is greater than all the other numbers inside the braces, then dropping it will reduce the supremum: $u_{k+1} < u_k$. This proves the statement about u in (3.25). The statement about l is proved similarly and is left as an exercise. ∎

Lemma 63(b) shows that the bounding sequences u_k and l_k are monotone sequences. As such, each can either have a real number for a limit or diverge to ∞ or $-\infty$. Because u_k is nonincreasing, if its limit is a real number, then it must be the greatest lower bound or infimum of the sequence u_k and can thus be represented as

$$\lim_{k \to \infty} u_k = \inf_{k \geq 1} u_k = \inf_{k \geq 1} \sup\{s_k, s_{k+1}, s_{k+2}, \ldots\} = \inf_{k \geq 1} \sup_{n \geq k} s_n$$

Similarly, for l_k, which is nondecreasing, we can write

$$\lim_{k \to \infty} l_k = \sup_{k \geq 1} l_k = \sup_{k \geq 1} \inf\{s_k, s_{k+1}, s_{k+2}, \ldots\} = \sup_{k \geq 1} \inf_{n \geq k} s_n$$

If the limits are $\pm\infty$ instead of real numbers, then we use those symbols to indicate the limits. With this in mind, we have the following definition.

Limit supremum and limit infimum

Let s_n be a given sequence of real numbers. If the sequence u_k converges to a real number, then its limit is the *limit supremum* (or *limit superior*) of s_n and denoted by

$$\limsup_{n\to\infty} s_n = \inf_{n\geq 1} \sup\{s_n, s_{n+1}, s_{n+2}, \ldots\} = \lim_{k\to\infty} u_k$$

If u_k diverges to ∞ or $-\infty$, then we use these symbols to denote the limit supremum. Similarly, the *limit infimum* (or *limit inferior*) of s_n is

$$\liminf_{n\to\infty} s_n = \sup_{n\geq 1} \inf\{s_n, s_{n+1}, s_{n+2}, \ldots\} = \lim_{k\to\infty} l_k$$

or ∞ or $-\infty$ as appropriate.

For example, referring to the example we discussed earlier, we have for $s_n = 1/n$

$$\limsup_{n\to\infty} \frac{1}{n} = \lim_{k\to\infty} \frac{1}{k} = 0, \quad \liminf_{n\to\infty} \frac{1}{n} = \lim_{k\to\infty} 0 = 0$$

Similarly, for $s_n = (-1)^n$

$$\limsup_{n\to\infty} (-1)^n = \lim_{k\to\infty} 1 = 1, \quad \liminf_{n\to\infty} (-1)^n = \lim_{k\to\infty} (-1) = -1$$

And for $s_n = 2n$

$$\limsup_{n\to\infty} (2n) = \infty, \quad \liminf_{n\to\infty} (2n) = \lim_{k\to\infty} 2k = \infty$$

Notice that of the above three sequences, only $1/n$ converges to a real number, and it has the property that its limit supremum and limit infimum are equal real numbers. These features characterize all convergent sequences as we see next.

Theorem 64 *A sequence s_n converges to a real number s if and only if*

$$\limsup_{n\to\infty} s_n = \liminf_{n\to\infty} s_n = s \tag{3.26}$$

Proof First, we assume that (3.26) is given and prove that s_n must converge to the number s. By (3.24)

$$l_n \leq s_n \leq u_n$$

Since by (3.26) $\lim_{n\to\infty} l_n = s$ and also $\lim_{n\to\infty} u_n = s$, the squeeze theorem implies that $\lim_{n\to\infty} s_n$ exists and equals s.

Conversely, assume that s_n converges to a number s. Then by the definition of convergence, for every $\varepsilon > 0$, we can find a positive integer N such that

$$|s_n - s| < \varepsilon \quad \text{for all } n \geq N$$

or equivalently,

$$s - \varepsilon < s_n < s + \varepsilon \quad \text{for all } n \geq N$$

In particular, $s_{n+j} < s + \varepsilon$ for all $j = 1, 2, 3, \ldots$ and $n \geq N$, and this implies that

$$\sup\{s_n, s_{n+1}, s_{n+2}, \ldots\} \leq s + \varepsilon \quad \text{for all } n \geq N \tag{3.27}$$

because $s + \varepsilon$ is an upper bound of the set $\{s_n, s_{n+1}, s_{n+2}, \ldots\}$, while the supremum is its least upper bound.

Similarly, because $s_{n+j} > s - \varepsilon$ for all j, we infer that $s - \varepsilon$ is a lower bound for $\{s_n, s_{n+1}, s_{n+2}, \ldots\}$, and therefore, the greatest lower bound of this set satisfies

$$\inf\{s_n, s_{n+1}, s_{n+2}, \ldots\} \geq s - \varepsilon \quad \text{for all } n \geq N \tag{3.28}$$

From (3.27), (3.28), and (3.24), we conclude that

$$s - \varepsilon \leq l_n \leq u_n \leq s + \varepsilon \quad \text{for all } n \geq N$$

These inequalities imply the following:

$$|u_n - s| \leq \varepsilon, \quad |l_n - s| \leq \varepsilon \quad \text{for all } n \geq N$$

Since these inequalities hold for all $\varepsilon > 0$, we conclude that

$$\lim_{n\to\infty} u_n = s \quad \text{and} \quad \lim_{n\to\infty} l_n = s$$

and these limits are none other than (3.26). ∎

An immediate consequence of the above theorem is the following:

Corollary 65 *If* $\lim\sup_{n\to\infty} s_n \neq \lim\inf_{n\to\infty} s_n$ *for a sequence* s_n, *then* s_n *diverges.*

Notice that the above corollary includes the cases where limit supremum or limit infimum are ∞ or $-\infty$. For example, Corollary 65 implies the divergence of both of the sequences $(-1)^n$ and $2n$ that we discussed earlier.

It is worth summarizing the preceding discussion for later reference. A sequence s_n can have a limit s only when the upper and lower bounding sequences meet:

$$l_1 \leq l_2 \leq \cdots \leq l_k \leq \cdots \to s \leftarrow \cdots \leq u_k \leq \cdots \leq u_2 \leq u_1 \qquad (3.29)$$

If the lower sequence does not meet the upper one, then there is a nonempty open interval of numbers (l, u) between them:

$$l_1 \leq l_2 \leq \cdots \to \lim_{n \to \infty} l_n = l < u = \lim_{n \to \infty} u_n \leftarrow \cdots \leq u_2 \leq u_1 \qquad (3.30)$$

The sequence s_n cannot converge to a limit if the interval (l, u) is not empty, i.e., $u - l > 0$ because not matter how large the index N we choose, there are terms $s_k \leq l$ and other terms $s_m \geq u$ with $k, m \geq N$; so if we choose, say, $\varepsilon = (u - l)/2$, then no valid threshold index N can be found to match such values of ε.

We will come across a situation similar to that depicted in (3.29) and (3.30) when defining the Riemann integral in Chapter 7.

Every sequence s_n has a limit supremum u and a limit infimum l (they could be ∞ or $-\infty$ if s_n is unbounded). Although u and l are limits of monotone sequences u_k and l_k that are derived from s_n, these bounding sequences may or may not contain terms of s_n; in fact, there are sequences where $u_k \neq s_n$ and $l_k \neq s_n$ for every k and every n (see Exercise 93). On the other hand, the definitions of the bounding sequences suggest that the numbers u_k and l_k are increasingly aligned with the terms of s_n as k and n get larger. This raises a natural question: are there *subsequences* of s_n that converge to the *limits* u and l?

To answer this question, let s_n be a given sequence and consider its upper bounding sequence

$$u_k = \sup\{s_k, s_{k+1}, s_{k+2}, \ldots\}$$

If u is the limit supremum of s_n, then because u_k is a nonincreasing sequence,

$$u = \inf_{k \geq n} u_k = \lim_{k \to \infty} u_k$$

If we pick any number $\varepsilon > 0$, then there is a positive integer N such that

$$u_k - u = |u_k - u| < \varepsilon \quad \text{for all } k \geq N \qquad (3.31)$$

Further, $u_k \geq s_n$ for all $n \geq k$ so that

$$s_n \leq u_k < u + \varepsilon \quad \text{for all } n \geq N$$

Next, since u_k is the *least* upper bound of the set $\{s_k, s_{k+1}, s_{k+2}, \ldots\}$ for each k and further, $u \leq u_k$ for all k, it follows that $u - \varepsilon$ is not an upper bound of this set. This means that there is an integer $m \geq 0$ such that $u - \varepsilon < s_{k+m}$. If $k \geq N$, then

$$u - \varepsilon < s_{k+m} < u + \varepsilon \quad k \geq N \tag{3.32}$$

These inequalities help us identify a subsequence of s_n that converges to u. Note that due to the well ordering property, we may assume that N is the least positive integer that makes (3.31) true; so if we choose $k = N$, then $N + m$ gives the least integer for which (3.32) holds.

Set $\varepsilon = 1$. There is a (least) positive integer N_1 such that (3.32) holds for integers k_1 and m_1 with $m_1 \geq 0$ and $k_1 \geq N_1$. We set $k_1 = N_1$ and define $n_1 = N_1 + m_1$ to obtain the following:

$$u - 1 < s_{n_1} < u + 1 \tag{3.33}$$

Next, let $\varepsilon = 1/2$ and note that there is a least integer $N_2 > n_1$ and an index $N_2 + m_2$ ($k_2 = N_2$) such that (3.32) holds with $n_2 = N_2 + m_2$, i.e.,

$$u - \frac{1}{2} < s_{n_2} < u + \frac{1}{2} \qquad n_2 > n_1 \tag{3.34}$$

The integer N_2 above exists by (3.31) since n_1 was the least choice for $\varepsilon = 1$, and we may look for an index larger than n_1 that makes (3.31) true for $\varepsilon = 1/2$.

Using this argument repeatedly, for every $j \in \mathbb{N}$, we find indices n_j such that

$$u - \frac{1}{j} < s_{n_j} < u + \frac{1}{j} \qquad n_j > n_{j-1} > \cdots > n_2 > n_1$$

These inequalities imply that for every $j \in \mathbb{N}$, there is a positive integer n_j such that

$$|s_{n_j} - u| < \frac{1}{j}$$

It follows that the subsequence s_{n_j} converges to u as $j \to \infty$.

Using a similar procedure, we may obtain a subsequence of s_n that converges to the limit infimum l. These arguments prove the following:

Theorem 66 *For each bounded sequence s_n there is a subsequence of s_n that converges to $\lim \sup_{n \to \infty} s_n$ and a subsequence that converges to $\lim \inf_{n \to \infty} s_n$.*

For example, the divergent sequence $s_n = (-1)^n + 1/n$ has the following convergent subsequences:

$$s_{2j} = 1 + \frac{1}{2j} \to 1 = \lim_{n \to \infty} \sup s_n \quad (n_j = 2j)$$

$$s_{2j-1} = -1 + \frac{1}{2j-1} \to -1 = \lim_{n \to \infty} \inf s_n \quad (n_j = 2j - 1)$$

Cluster points and subsequences

Theorem 66 states that for every bounded sequence s_n there are subsequences that converge to the numbers $\lim\sup_{n\to\infty} s_n$ and $\lim\inf_{n\to\infty} s_n$. If these two numbers are different, then Corollary 65 states that s_n doesn't converge. We conclude that every bounded sequence (convergent or divergent) has subsequences that converge.

On the other hand, we saw earlier that in the absence of completeness, limits of sequences may not exist (they may correspond to gaps or holes). For example, if s_n is a sequence of rational numbers, but $\lim\sup_{n\to\infty} s_n$ or $\lim\inf_{n\to\infty} s$ are irrational, then the (rational) subsequences of s_n that converge to them don't have limits in $\mathbb{Q}$!

We tackle this conundrum in Chapter 4 where we show that limits of subsequences exist in the set of real numbers; in Theorem 66, it is tacitly assumed that the sequence s_n is in the set of real numbers, which we discuss in detail in the next section, along with a central fact in real analysis: the celebrated Bolzano–Weierstrass theorem. Here we define the following important concept that helps put limits supremum and infimum in a broader perspective.

> **Cluster points** A *cluster point* of a sequence s_n is the limit of a subsequence of s_n. We also say that ∞ or $-\infty$ are cluster points if there are subsequences that diverge to these in the sense defined earlier.

Cluster points are special cases of limit points of sets that we discuss in Chapter 4.

According to Theorem 66, a bounded sequence (of real numbers) has the numbers $u = \lim\sup_{n\to\infty} s_n$ and $l = \lim\inf_{n\to\infty} s_n$ as its cluster points.

If s_n converges to a limit $s = \lim_{n\to\infty} s_n$, then s *is the unique cluster point of* s_n since every subsequence converges to s by Theorem 61. Sequences that don't converge can have any number of cluster points; see Exercise 94. If we arrange all the countably many rational numbers in the interval $[0,1]$ in the form of a sequence q_n, then all real numbers in $[0,1]$ are cluster points of q_n. This follows quickly from the fact (proved in Chapter 4) that rational numbers form a "dense set" in $[0,1]$. Thus, a sequence can have an uncountable infinity of cluster points!

The sequence $s_n = n(-1)^n$ has two cluster points: ∞ and $-\infty$. Since these are not numbers, it is more accurate to say that $n(-1)^n$ clusters around both ∞ and $-\infty$. A strictly accurate description would be that $n(-1)^n$ has a subsequence that diverges to ∞, namely, s_{2j} ($n_j = 2j$); and a subsequence that diverges to $-\infty$, namely, s_{2j-1} ($n_j = 2j-1$). In this example, we may consider ∞ the limit supremum of $n(-1)^n$

by stretching the meaning of the term "limit supremum" a little bit; in the same sense, $-\infty$ is the limit infimum of $n(-1)^n$.

The next result states that $u = \limsup_{n\to\infty} s_n$ and $l = \liminf_{n\to\infty} s_n$ are special cluster points of a sequence.

Theorem 67 *If s_n is a bounded sequence of real numbers, then $u = \limsup_{n\to\infty} s_n$ is its largest cluster point, and $l = \liminf_{n\to\infty} s_n$ is its smallest cluster point.*

Proof By Theorem 66, both u and l are cluster points of s_n. Suppose that c is also a cluster point of s_n, and s_{n_j} is the subsequence converging to it:

$$c = \lim_{j\to\infty} s_{n_j}$$

Let u_k and l_k be the associated bounding sequences for s_n, and recall that for every subindex $j = 1, 2, 3, \ldots$

$$l_{n_j} \leq s_{n_j} \leq u_{n_j}$$

If we take the limit as $j \to \infty$ and apply Theorem 57, we obtain

$$l = \lim_{j\to\infty} l_{n_j} \leq \lim_{j\to\infty} s_{n_j} \leq \lim_{j\to\infty} u_{n_j} = u$$

This shows that $l \leq c \leq u$, which completes the proof. ∎

3.8 Sequences, functions, and infinite direct products of sets

In this section, we take a different look at functions and sequences. We discover that sequences (and functions in general) can be considered mere points in spaces that are formed by taking the direct (or Cartesian) product of infinitely many sets. This study is useful in two ways: it settles questions about cardinal numbers because the resulting product sets often have a larger cardinality than their component sets; it also lets us think of sequences in a way that is useful in the upcoming Chapter 4.

In Chapter 2, we proved that a countable union of sets with cardinality $\aleph_0$ again has cardinality $\aleph_0$. Now we show that a countable product of countable sets does in fact take us from $\aleph_0$ to $2^{\aleph_0}$.

Consider $\mathbb{N}$ and recall that $\mathbb{N} \times \mathbb{N}$ has cardinality $\aleph_0$ and by writing $\mathbb{N} \times \mathbb{N} \times \mathbb{N}$ as $(\mathbb{N} \times \mathbb{N}) \times \mathbb{N}$, we see that $\mathbb{N} \times \mathbb{N} \times \mathbb{N}$ still has cardinality $\aleph_0$. Clearly doing this a *finite* number of times does not change the cardinality, so

$$\# (\underbrace{\mathbb{N} \times \mathbb{N} \times \cdots \times \mathbb{N}}_{n \text{ times}}) = \aleph_0$$

We can rephrase this symbolically as

$$\aleph_0^n = \aleph_0, \quad n \in \mathbb{N}$$

But what if we take the direct product of $\mathbb{N}$ with itself $\aleph_0$ times? To answer this question, it is necessary to first explain what a "direct product of a countable infinity of sets" means.

Let us begin by looking at the definition of a finite direct product again (here the sets A_i are arbitrary and not necessarily countable):

$$\prod_{i=1}^{n} A_i = \{(x_1, x_2, ..., x_n) : x_i \in A_i \text{ for } i = 1, 2, ..., n\}$$

The objects that make up this set are often called "n-tuples" or "n-dimensional vectors."

Now, think of each infinite sequence $x_1, x_2, x_3, \ldots$ as a "vector with $\aleph_0$ components"; then the *set of all sequences* is a natural candidate for the direct product of a countable infinity of sets:

$$\prod_{i \in \mathbb{N}} A_i = A_1 \times A_2 \times A_3 \times \cdots = \{(x_1, x_2, x_3, ...) : x_i \in A_i \text{ for each } i \in \mathbb{N}\}$$

In the special case that $A_i = A$ are all the same set, the common notation is the power notation $A^{\mathbb{N}}$; this is shorthand for the *set of all sequences in the set A*:

$$A^{\mathbb{N}} = A \times A \times A \times \cdots = \{(x_1, x_2, x_3, ...) : x_i \in A \text{ for each } i \in \mathbb{N}\}$$

In particular, if we pick $A = \mathbb{N}$, then we have the following:

The direct product of $\mathbb{N}$ with itself a countable infinity of times is the set $\mathbb{N}^{\mathbb{N}}$ of all sequences of positive integers.

So now it remains to show that $\mathbb{N}^{\mathbb{N}}$ is uncountable. We write the cardinal number of $\mathbb{N}^{\mathbb{N}}$ symbolically as $\aleph_0^{\aleph_0}$.

Cantor's theorem shows that $\aleph_0 < 2^{\aleph_0}$ and $\aleph_0^{\aleph_0}$ looks like it cannot be smaller than $2^{\aleph_0}$. To actually prove this latter fact, consider a *subset* of $\mathbb{N}^{\mathbb{N}}$, specifically, the set of all binary sequences of 0's and 1's. This set is the product of the finite set $\{0, 1\}$

with itself a countable infinity of times, or symbolically, $\{0, 1\}^{\mathbb{N}}$. The fact that this set has cardinality $2^{\aleph_0}$ and $\{0, 1\}$ has cardinality 2 is a convenient coincidence.

Theorem 68 *The set $\{0, 1\}^{\mathbb{N}}$ is equipollent to the power set $\mathcal{P}(\mathbb{N})$.*

The basic idea is simple: to find a bijection $f : \{0, 1\}^{\mathbb{N}} \rightarrow \mathcal{P}(\mathbb{N})$, we assign a unique subset of $\mathbb{N}$ to each and every binary sequence. The natural thing to consider is using the indices of the terms in each sequence. For instance, the sequence $0, 1, 0, 1, \ldots$ has a 1 for every even index 2,4,6, ... and a 0 elsewhere, so we assign to the sequence $0, 1, 0, 1, \ldots$ the set of all even numbers $\{2, 4, 6, \ldots\}$. We may write

$$f(0, 1, 0, 1, 0, 1, \ldots) = \{2, 4, 6, \ldots\}$$

Similarly, we assign the finite set $\{1, 5, 8\}$ to the sequence 1,0,0,0,1,0,0,1,0,... with all zeros after the eighth place. In symbols,

$$f(1, 0, 0, 0, 1, 0, 0, 1, 0, \ldots) = \{1, 5, 8\}$$

This is the idea behind the proof.

Proof Define the map $f : \{0, 1\}^{\mathbb{N}} \rightarrow \mathcal{P}(\mathbb{N})$

$$f(x_1, x_2, x_3, \ldots) = \{n \in \mathbb{N} : x_n = 1\} \tag{3.35}$$

We prove that f is a bijection. To show that it is an injection (one-to-one), consider two distinct sequences $x_1, x_2, x_3, \ldots$ and $y_1, y_2, y_3, \ldots$ in $\{0, 1\}^{\mathbb{N}}$. These are different in at least one entry, say, $x_k \neq y_k$. If, say, $x_k = 1$ and $y_k = 0$, then k is in the set $f(x_1, x_2, x_3, \ldots)$ but not in the set $f(y_1, y_2, y_3, \ldots)$, which leads us to conclude

$$f(x_1, x_2, x_3, \ldots) \neq f(y_1, y_2, y_3, \ldots)$$

We get the same inequality if $x_k = 0$ and $y_k = 1$, so f is injective.

Next, we show that f is surjective (onto). We pick any set S in $\mathcal{P}(\mathbb{N})$ so that S is a set of positive integers

$$S = \{n_1, n_2, n_3 \ldots\}$$

where we may assume that n_1, n_2, etc., are listed in increasing order: $n_1 < n_2 < n_3 < \cdots$ Now consider the sequence $x_1, x_2, x_3, \ldots$ where

$$x_n = 1 \quad \text{if } n = n_i \text{ for some index } i$$

$$x_n = 0 \quad \text{for all other indices } n$$

Then $x_1, x_2, x_3, \ldots$ is a binary sequence in $\{0, 1\}^{\mathbb{N}}$, and $f(x_1, x_2, x_3, \ldots) = S$, so f is onto $\mathcal{P}(\mathbb{N})$. This completes the proof. ∎

For an alternative argument that suggests $\{0, 1\}^{\mathbb{N}}$ has cardinality $2^{\aleph_0}$, see Exercise 97.

Next, note that every countable (finite or countably infinite) set with at least two elements contains a set like $\{a, b\}$ where $a \neq b$. It follows that if A_n is a countable set for every positive integer n, then the product $\prod_{i \in \mathbb{N}} A_i$ must have no fewer elements than the set $\{a, b\}^{\mathbb{N}}$. Therefore, Theorem 15 implies that $\prod_{i \in \mathbb{N}} A_i$ is uncountable, and we have the following corollary:

Corollary 69 *The direct product of a countable infinity of countable sets each containing at least two elements is uncountable. In particular, $\mathbb{N}^{\mathbb{N}}$ is uncountable.*

We can actually say more: $\mathbb{N}^{\mathbb{N}}$ has the same cardinal number as $\{0, 1\}^{\mathbb{N}}$, which shows in effect that $\aleph_0^{\aleph_0} = 2^{\aleph_0}$. This may be proved using the fact that every positive integer has a binary expansion so that all points in $\mathbb{N}^{\mathbb{N}}$ are matched with points in $\{0, 1\}^{\mathbb{N}}$. There are some technical issues to iron out, but this is the basic idea.

Functions as points in an infinite product of sets

I close this section by extending the concept of direct product further to *all* possible index sets. Recall that every sequence in A is actually a *function* $f : \mathbb{N} \to A$. Now think of an arbitrary set $\mathcal{J}$ as our index set and a collection of sets A_i for $i \in \mathcal{J}$. Define the set of all functions $f : \mathcal{J} \to \bigcup_{i \in \mathcal{J}} A_i = A$ to be the direct product of the A_i, or in symbols

$$\prod_{i \in \mathcal{J}} A_i = \{f : \mathcal{J} \to A : f(i) \in A_i \text{ for every } i \in \mathcal{J}\}$$

In particular, if all $A_i = A$ are the same set, then we obtain the following:

The set of all functions from $\mathcal{J}$ into A is

$$A^{\mathcal{J}} = \{f : \mathcal{J} \to A : f(i) \in A \text{ for every } i \in \mathcal{J}\}.$$

For example, $A = \mathbb{N}^{\mathbb{R}}$ is the set of all functions with domain $\mathbb{R}$ taking values in $\mathbb{N}$. An example is the step function $f(x) = \lceil x \rceil$, defined as the least integer that is greater than or equal to x (or the ceiling function).

If $\mathcal{J} = \mathbb{N}$, then we refer to $A^{\mathbb{N}}$ as *the set of all sequences in A*; for example, $\mathbb{Z}^{\mathbb{N}}$ is the set of all sequences of integers. On the other hand, $A^{\mathbb{Z}}$ is the set of all doubly infinite sequences in A; for instance, $\mathbb{R}^{\mathbb{Z}}$ is the set of all doubly infinite sequences of real numbers, like

$$\ldots, -2\pi, -\pi, 0, \pi, 2\pi, \ldots$$

Bear in mind that this *entire* sequence is just *one point* in $\mathbb{R}^{\mathbb{Z}}$.

In the field of *functional analysis*, the A in $A^{\mathcal{J}}$ is often a subset of the real or complex numbers, and $\mathcal{J}$ is an infinite set, like $\mathbb{N}$, an interval $[a, b]$, etc. In these cases, $A^{\mathcal{J}}$ often has additional structure like a metric (distance function) that turns it into a space of functions known as *functionals*.

An example of a functional is the length of a curve: for instance, every curve in the regular three-dimensional space from a point P to a point Q has a real number[6] as its length; the correspondence that assigns lengths to curves is a functional. We briefly discuss this topic in Appendix 9.2 to explain the discontinuity of the length functional that is defined on the space of continuous functions from $\mathcal{J} = [0, 1]$ into $A = [0, \infty)$. The discussion in this appendix is tied to the problem of staircase curves that we discussed in Chapter 1.

If a set A is totally ordered by a relation $\leq$ like the set $\mathbb{R}$ with its ordinary ordering relation, then we can define a partial ordering $\preceq$ on the set of all functions $A^{\mathcal{J}}$ as follows:

For functions $f(x)$ and $g(x)$ in $A^{\mathcal{J}}$, define $f \preceq g$ if $f(x) \leq g(x)$ for all $x \in \mathcal{J}$.

For example, let $A = \mathcal{J} = \mathbb{R}$ a $f(x) = x^2 + 1$ and $g(x) = 2x$. Then $f(x) \leq g(x)$ for all $x \in \mathcal{J}$, so $f \preceq g$. On the other hand, if $f(x) = x^2$, $g(x) = 2x$, then $x^2 \leq 2x$ only for $0 \leq x \leq 2$, while $x^2 > 2x$ for $x < 0$ or $x > 2$. It follows that $f \npreceq g$ and $g \npreceq f$ (f and g are not comparable elements relative to $\preceq$). But if we choose $\mathcal{J} = [0, 2]$, $A = \mathbb{R}$, then $f(x) \leq g(x)$ for all $x \in \mathcal{J}$ and have $f \preceq g$.

3.9 Exercises

Exercise 70 *Consider the sequence $s_n = \sin(\pi n/4)$.*

(a) Find the first 20 terms of this sequence in exact form using special-angle values. Can you tell if this a periodic sequence? If so, of what period?

(b) Verify that $s_{n+8} = s_n$ for all n using the fact that $\sin x$ is a periodic function with period 2π.

(c) Can you figure out the period of $\sin(\pi n/8053)$? You will find the approach in (b) easier than that in (a); and as a bonus, in the process of calculation, you also prove that this sequence is indeed periodic! Start by letting k be a potential period and figure out what k has to be so that $s_{n+k} = s_n$ for all n.

---

[6] Assuming bounded variation.

Exercise 71 *Suppose that a sequence is defined as follows:*

$$s_n = 10 \cos \frac{\pi n}{3} + \max\left(\frac{50}{n^2}, 1\right)$$

where max(x, y) *means the larger of the two numbers* x *and* y *inside the parenthesis. Explain why this sequence is eventually periodic, what the eventual period is, and list the entire transient part.*

Exercise 72 *(a) Prove that* $\cos n$ *is not a periodic sequence. What about the sequence* $\cos(\pi n)$?

(b) Explain why the sequence $\tan n$ *is defined on its entire domain* $\mathbb{Z}$ *(has no vertical asymptotes, unlike the function* $\tan x$, *which is not defined where* $\cos x = 0$). *On the other hand,* $\tan x$ *is periodic on its domain and has period* π. *Is* $\tan n$ *periodic? Why or why not?*

(c) Calculate and plot the first 200 terms of each of $\cos n$ *and* $\tan n$ *versus* n *using a computing device. Do you see any identifiable patterns?*

Next, plot $\cos(n + 1)$ *versus* $\cos n$ *and also* $\tan(n + 1)$ *versus* $\tan n$ *for* $n = 0, 1, 2, \dots 200$. *Can you explain what you see? Think of* $y = \tan(n+1)$ *as a function of* $x = \tan n$ *and use trigonometric identities to expand* $\tan(n + 1)$ *and get a more familiar equation in terms of* x *and* y; *use the same idea for the cosine case.*

Exercise 73 *(a) For the following sequence, show that* $s_{n+1} - s_n > 0$ *for every* n, *thereby establishing that this sequence is increasing:*

$$s_n = \frac{n-1}{n}$$

(b) Use the same idea to determine the monotone nature of the following sequence, and prove your conclusion:

$$s_n = \frac{n+1}{n}$$

Exercise 74 *Consider the sequence of power functions* $f_n(x) = x^n$. *Explain whether this sequence is increasing, decreasing, or neither on each of the following intervals:*

(a) $[0, 1]$ *(b)* $[1, 2]$ *(c)* $[0, 2]$

NOTE: See the definition of ordering of functions in the space A^J; *we are not asking if the functions themselves are increasing or decreasing. It should be helpful to sketch the graphs of a few of these functions in the same coordinate systems over the interval [0,2].*

Exercise 75 *Calculate and plot s_1 through s_8 for each of the sequences below:*

(a) $s_n = \dfrac{(-1)^n}{n}$
(b) $s_n = n + \dfrac{(-1)^n}{n}$
(c) $s_n = \dfrac{n}{2} - \dfrac{(-1)^n}{n}$

Use your plot to tell if each sequence is monotone or eventually monotone.

Exercise 76 *For almost all values of the coefficient a in the logistic equation there is no known formula for the sequence that (3.6) generates. One of the exceptions is the value a = 4, where the generated sequence does in fact have a simple formula*

$$x_n = \sin^2\left(2^n\theta\right)$$

where θ is any fixed positive number. Note that the initial value is $x_0 = \sin^2\theta$ and is determined by the value of θ. Prove that the above sequence x_n satisfies (3.6).

NOTE. *For most values of θ that are not commensurate with π (say, $\theta = 1$), the sequence $\sin^2\left(2^n\theta\right)$ will be chaotic: it shows sensitivity to initial values, it may be aperiodic and dense in the interval [0,1], or it may be one of the (unstable) periodic orbits of all possible periods that may occur depending on the value of θ.*

Exercise 77 *Use the definition of convergence to explain why the constant sequence $c, c, c, \ldots$ converges to the number c. Explain why $N_\varepsilon = 1$ for every $\varepsilon > 0$.*

Exercise 78 *(a) Use an argument involving ε and N to prove that for any (fixed) real number c*

$$\lim_{n\to\infty} \frac{c}{n} = 0$$

(b) Show that the ε-index in this case is $N_\varepsilon = |c|/\varepsilon$.
(c) Generalize: suppose that p and c are real numbers, $p > 0$. Use the definition of convergence (with ε and N) to prove that

$$\lim_{n\to\infty} \frac{c}{n^p} = 0$$

Exercise 79 *Consider the two sequences*

$$s_n = 1 - \frac{1}{n^2} \qquad s_n{}' = 1 + \frac{1}{\sqrt{n}}$$

(a) List the first six terms $s_1, s_2, \ldots, s_6$ of s_n explicitly; round your answers to three decimal places.
(b) For $\varepsilon = 0.003$, find a positive integer N such that $|s_n - 1| < \varepsilon$ for all $n > N$. What is the ε-index $N_{0.003}$?

(c) Find the ε-index N_ε for any (unspecified) $\varepsilon > 0$, whose existence as a finite quantity proves that $\lim_{n \to \infty} s_n = 1$.
(d) Repeat (a)–(c) for $s_n{}'$.

Exercise 80 *Let s_n be a convergent sequence. Explain why its ε-index N_ε is a nondecreasing function of ε, i.e., if $\varepsilon_1 < \varepsilon_2$, then $N_{\varepsilon_1} \leq N_{\varepsilon_2}$.*

Exercise 81 *Suppose that s_n is a sequence of numbers with period $k \geq 2$. Use the definition of convergence to prove that s_n is bounded but does not converge, i.e., it is divergent. (Note. One way to proceed is by contradiction: assume $s_n \to s$ for some number s and reach a contradiction by choosing a value for ε that is much smaller than the spacing between different terms of s_n of which there are only finitely many.)*

Exercise 82 *Which of the following sequences diverges to ∞ or $-\infty$? Explain.*

$$a_n = 2n - 3, \qquad b_n = 3 - n^2, \qquad c_n = n \cos\left(\frac{\pi n}{3}\right)$$

Exercise 83 *Prove that the following sequences diverge to ∞. Is either one monotone (nondecreasing)? Explain.*

$$s_n = 2n + (-1)^n n, \qquad r_n = n^2 + (-1)^n n$$

Exercise 84 *The inequality in (3.22) is not strict in general, even when $x_n < y_n$ is a strict inequality for all n. Give a simple example to illustrate this fact.*

Exercise 85 *The sequence $s_n = 1/n$ converges to 0, so all of its subsequences converge to 0. Explain why in each case below we have a subsequence of s_n and list the first five terms of the subsequence in each case:*
    (a) $n_k = 2k$ (b) $n_k = 2^k$

Exercise 86 *Identify a subsequence of the following sequence that converges and a subsequence that diverges:*

$$s_n = n\left(1 + \sin\frac{n\pi}{4}\right)$$

Exercise 87 *Consider the set of rational numbers $A = \{x \in \mathbb{Q} : x^2 < 2\}$.*
    (a) Show that $\inf A = -\sqrt{2}$ and $\sup A = \sqrt{2}$.
    (b) Considering A as a subset of the real numbers $\mathbb{R}$, explain why both $\inf A$ and $\sup A$ exist although neither one is contained in A.
    (c) Considering A as a subset of the rational numbers $\mathbb{Q}$, explain why neither $\inf A$ nor $\sup A$ exist although A has both upper bounds and lower bounds in $\mathbb{Q}$.

Exercise 88 *Complete the proof of Lemma 63.*

Exercise 89 *Calculate the limit supremum and the limit infimum in each case below:*

(a) $s_n = 1 - \dfrac{1}{n}$ (b) $t_n = 1 + \dfrac{1}{n}$

(c) $u_n = [1 + (-1)^n]n$ (d) $v_n = (-1)^n n$

Exercise 90 *(a) Give an example of a sequence of numbers s_n such that $\liminf_{n \to \infty} s_n = \infty$.*
(b) Give an example of a sequence of numbers s_n such that $\limsup_{n \to \infty} s_n = -\infty$.

Exercise 91 *For every sequence s_n explain why $\liminf_{n \to \infty} s_n$ and $\limsup_{n \to \infty} s_n$ are unique; i.e., s_n cannot have more than one of each (recall Theorem 57).*

Exercise 92 *Let x_n and y_n be bounded sequences of real numbers.*
(a) Prove that

$$\liminf_{n \to \infty} x_n + \limsup_{n \to \infty} y_n \le \limsup_{n \to \infty}(x_n + y_n) \le \limsup_{n \to \infty} x_n + \limsup_{n \to \infty} y_n$$

(b) If $x_n > 0$ and $y_n > 0$ for all n, then

$$\limsup_{n \to \infty}(x_n y_n) \le \left(\limsup_{n \to \infty} x_n \right) \left(\limsup_{n \to \infty} y_n \right)$$

(c) Show that the inequality in (b) can be strict [consider $x_n = 1 + (-1)^n$ and $y_n = 1 - (-1)^n$ for all $n \in \mathbb{N}$].

Exercise 93 *Consider the sequence*

$$s_n = \begin{cases} 2 - 1/n & \text{if } n \text{ is odd} \\ 1/n & \text{if } n \text{ is even} \end{cases}$$

(a) Show that the bounding sequences of s_n are the constant sequences $u_k = 2$ and $l_k = 0$ for all k. Note that $s_n \ne 0, 2$ for all n.
(b) Identify a subsequence of s_n that converges to $\limsup_{n \to \infty} s_n$ and a subsequence that converges to $\liminf_{n \to \infty} s_n$.

Exercise 94 *Explain why for every positive integer k there is a sequence with exactly k cluster points (think about periodic sequences).*

Exercise 95 *Assume that a ≠ b, and specify a bijection between {a, b}$^{\mathbb{N}}$ and {0, 1}$^{\mathbb{N}}$.*

Exercise 96 *Explain in words what the following sets represent, and give examples of some elements in each:*

$$\mathbb{R}^{\mathbb{N}}, \quad \mathbb{R}^{\mathbb{R}}, \quad \mathbb{N}^{\mathbb{Z}}$$

Exercise 97 *Consider all sequences of 0's and 1's that are eventually constant with value 0. For each positive integer N, the eventually constant sequence*

$$s_n = s_1, s_2, ..., s_N, 0, 0, ...$$

has a transient of length N. Each of $s_1, s_2, ..., s_N$ is either 0 or 1.
(a) How many sequences of 0's and 1's with transients of length N are there?
(b) What happens to the answer in (a) as N → ∞? Compare to Theorem 68.

4

The Real Numbers

The real numbers came up often in previous chapters where we thought of them intuitively. Now it is time to answer questions like what exactly are real numbers, and why do we need them?

We define a rational number simply as the quotient of two integers, as long as we don't divide by 0. This feature endows the rational numbers with a useful feature that the integers lack: the reciprocal of every rational number is also a rational number. However, this is not enough to include numbers like $\sqrt{2}$ or π that are also extremely important in all areas of pure and applied mathematics.

The key property that defines and distinguishes real numbers from the rational ones is "completeness." This property is different from all the algebraic properties that are common to both rational and real numbers. Completeness is a collective set property, like *closure*: the sum and product of two rational numbers is rational, so the *set* of all rational numbers is *closed* under addition and multiplication.

In the case of completeness, it is necessary to invoke infinity since the set of real numbers must contain the *limits* of so-called Cauchy infinite sequences. Loosely speaking, if we think of the decimal expansion of, say, $\pi = 3.14159\ldots$, then we can imagine it being constructed progressively using a sequence of rational numbers like $3, 3.1 = 31/10, 3.14 = 314/100$, and so on. The proper description of this process leads to greater technical challenges, like how to define the rational numbers that make up the sequence, how to make sure that adding or multiplying (irrational) real numbers gives a real number again, and of course, how to ensure that limits of Cauchy sequences of real numbers are again real numbers, i.e., the set of all real numbers is *closed under limits*.

It is common practice in analysis textbooks to introduce completeness *as an axiom* that is added to the field axioms that define the rational numbers once the concept of Cauchy sequence has been defined. Logically equivalent versions of completeness, such as the often used "least upper bound axiom," don't even need Cauchy sequences and can be defined as a property that subsets of the set of real numbers must satisfy. Many textbooks introduce more than one version and then prove the equivalence of those versions.

This approach certainly makes it possible to enter the theory faster, but it offers little intuition and insufficient insight into the real numbers, or in particular, what the irrational numbers are. From a pedagogical point of view, the axiomatic approach to introducing completeness seems contrived.

Real Analysis and Infinity. Hassan Sedaghat, Oxford University Press.
© H. Sedaghat (2022). DOI: 10.1093/oso/9780192895622.003.0004

We opt for developing the real numbers more constructively starting from the rational numbers and proceed to Cauchy sequences of these numbers. Although this approach slows things down a little, the deeper insight into real numbers that is gained from it is well worth the effort. Further, this bottom to top approach is less pedagogically contrived and provides an opportunity to see how infinity is fundamentally involved in defining a real number, in line with what we observe in the decimal expansions of numbers where the irrational numbers have non-periodic infinite decimal expansions.

We begin this chapter with a discussion of the rational numbers as algebraic concepts. The rational numbers serve as a form of skeletal structure for the real numbers. There are two well-known approaches to defining the real numbers starting with the rational ones: one based on sets and the other based on sequences. Before discussing the one that we use here, i.e., sequences, I briefly mention the other.

The approach based on sets is due to Dedekind[1] and involves splitting the set of all rational numbers into pairs of disjoint sets called *Dedekind cuts* that flank and encode real numbers. For example, the number $\sqrt{2}$ uniquely corresponds to the set of all the rational numbers that are smaller than it; more precisely, $\sqrt{2}$ may be defined by the inequality $q^2 < 2$ for $q > 0$ plus all $q \leq 0$; such a set (or its complement) is a cut that uniquely corresponds to $\sqrt{2}$. The numbers in this cut include 1.3 because $1.3^2 = 1.69$, which is less than 2. Similarly, the numbers 1.4, 1.41, 1.414, etc., are all in the same cut. For $\sqrt[3]{2}$, the inequality $q^3 < 2$ defines a cut of rational numbers, and so on. Although not all cuts are so easily expressed, Dedekind's theory constructs all real numbers using this basic idea.

In the second approach, due to Cantor, irrational numbers are associated with limits of sequences of rational numbers. We follow Cantor's approach in this book for multiple reasons: first, the important concept of Cauchy sequence that is used throughout analysis is introduced early on in a context that makes its essence clear; second, it is more closely aligned with our goal of using sequences to define limits, so there is greater uniformity in the presentation of later material in this book; and finally, we have an explicit way of illustrating the foundational role of infinity in analysis.

4.1 Rational numbers

The most basic numbers, namely, the natural numbers, are defined axiomatically via set theory. Zero and the negative integers are added via algebraic axioms and properties. We take all this as given because they are not analytic in nature. We use them to define fractions and establish their algebraic properties. Taking fractions,

---

[1] Richard Dedekind (1831–1916), an accomplished German mathematician, was also a doctoral student of Gauss.

or rational numbers, as a starting point may seem arbitrary, but greater exposure to these numbers serves a useful pedagogical purpose since rational numbers play a critical role in proving many results in real analysis.

The rational numbers

The set of all ratios of pairs of integers or integer fractions

$$\mathbb{Q} = \left\{ \frac{m}{n} : m \in \mathbb{Z}, \, n \in \mathbb{N} \right\}$$

constitutes the set of all *rational numbers*. As usual, the number m is the numerator and n is the denominator. We identify rational numbers that have a common integer factor in their numerator and denominator

$$\frac{m}{n} = \frac{mk}{nk}$$

for every nonzero integer k. A fraction having no common factors in its numerator and denominator is in *reduced form*. Note that each rational number has a *unique* reduced form that is obtained by removing the largest common integer factor, which is none other than the greatest common divisor of the numerator and the denominator.

Note that when we fix $n = 1$, we obtain the set $\mathbb{Z}$ of all integers as a proper subset of $\mathbb{Q}$.

Countably many rational numbers

The following basic feature of the rational numbers that we discussed earlier in Chapter 2 is worth repeating here since it will be important in our later work with the real numbers.

The set $\mathbb{Q}$ of rational numbers is countably infinite (equipollent to a subset of $\mathbb{Z} \times \mathbb{N}$).

Rational numbers form a totally ordered field

The two common algebraic operations on rational numbers are the familiar addition and multiplication.

The ordinary addition and multiplication of rational numbers is defined the way we remember them from past algebra experience:

$$\frac{j}{k} + \frac{m}{n} = \frac{km + jn}{kn}, \quad \frac{j}{k}\frac{m}{n} = \frac{jm}{kn}$$

Here $j, m \in \mathbb{Z}$, and $k, n \in \mathbb{N}$. Since sums and products of integers are again integers, the above definitions indicate that sums and products of rational numbers are also rational.

When adding rational numbers, we often use the *least common multiple* of the denominators k and n to speed up the process; if not, then the greatest common divisor of the two denominators is canceled out in the *reduced form* as in

$$\frac{19}{18} - \frac{11}{12} = \frac{6 \times 2 \times 19 - 6 \times 3 \times 11}{6 \times 3 \times 6 \times 2} = \frac{6(38 - 33)}{6(36)} = \frac{5}{36}$$

Notice that the least common multiple of the denominators 12 and 18 is indeed 36.

The set $\mathbb{Q}$ is a more structured set than $\mathbb{N}$ or $\mathbb{Z}$. From an algebraic point of view, $\mathbb{Q}$ is a *field*, but neither $\mathbb{N}$ nor $\mathbb{Z}$ are fields. Specifically, when we divide two numbers in $\mathbb{Q}$, the result is in $\mathbb{Q}$ (as long as we do not divide by 0) because if $p = m/n$ and $q = j/k$ then

$$\frac{p}{q} = p\frac{1}{q} = \frac{m}{n}\frac{k}{j} = \frac{mk}{nj}$$

is again rational; for instance, operations as simple as taking the average of two numbers $(p + q)/2$ are possible in $\mathbb{Q}$ but not in $\mathbb{Z}$.

The operations of addition and multiplication defined above are *field operations* on the set of rational numbers, and the next result lists their main properties. The statements in the next theorem may be considered "field axioms" in order to expedite the journey into real analysis.[2] But here by accepting the integers and their basic properties axiomatically and then using them to prove the properties of rational numbers there is an opportunity to get better acquainted with rational numbers and their algebraic properties.

It is worth a mention that properties (i)–(vii) in Theorem 98 are shared by the set of integers $\mathbb{Z}$, which is not a field. But property (viii) is a defining property of all fields, including $\mathbb{Q}$.

Theorem 98 *(Field of rationals) The operations of addition and multiplication defined above satisfy the following on the set $\mathbb{Q}$:*

---

[2] If pressed for time, then you may accept the statements in Theorem 98 as axioms that define the set of rational numbers and move on without loss of continuity. Whether the starting point is with the integers or directly with the rational numbers doesn't makes a significant difference to the study of real analysis.

(i) *Closure property:* $\mathbb{Q}$ *is closed under addition and multiplication, i.e., for every pair of numbers* $p, q \in \mathbb{Q}$

$$p + q \in \mathbb{Q} \quad and \quad pq \in \mathbb{Q}$$

(ii) *Associative property: addition and multiplication are associative, i.e., for all rational numbers* p, q, r

$$p + (q + r) = (p + q) + r$$
$$p(qr) = (pq)r$$

(iii) *Commutative property: addition and multiplication are commutative, i.e., for all rational numbers* p, q

$$p + q = q + p \quad and \quad pq = qp$$

(iv) *Distributive property: multiplication distributes over addition, i.e., for all rational numbers* p, q, r

$$p(q + r) = pq + pr$$

(v) *Additive identity and zero element: the integer 0 is the unique rational with the properties that for all rational* q

$$q + 0 = q \quad and \quad q0 = 0$$

(vi) *Multiplicative identity (unit element): the integer 1 is the unique rational with the property that for all rational* q

$$q1 = q$$

(vii) *Additive inverse (negatives): for each rational number* q, *there is a unique rational, which we may write as* $-q$, *such that*

$$q + (-q) = 0$$

(viii) *Multiplicative inverse (reciprocals): for each rational number* $q \neq 0$, *there is a unique rational, which we may write as* $1/q$, *such that*

$$q\left(\frac{1}{q}\right) = 1$$

Proof I prove (iv) and (viii) to give a flavor of the arguments needed in this context. The rest are left as exercises.

To prove (iv), let $p = k/n$, $q = j/m$, and $r = i/l$, and note that

$$p(q + r) = \frac{k}{n}\left(\frac{j}{m} + \frac{i}{l}\right) = \frac{k}{n}\left(\frac{jl + im}{ml}\right) = \frac{k(jl) + k(im)}{n(ml)}$$

where in the last step, we used the distributive property for integers in the numerator. Next, we examine the right hand side of the equality in (iv):

$$pq + pr = \frac{k}{n}\left(\frac{j}{m}\right) + \frac{k}{n}\left(\frac{i}{l}\right) = \frac{kj}{nm} + \frac{ki}{nl} = \frac{(kj)(nl) + (ki)(nm)}{(nm)(nl)}$$

Applying the commutative and associative properties to the last fraction gives

$$pq + pr = \frac{(nk)(jl) + (nk)(im)}{(nn)(ml)}$$

Now, again using the distributive property for integers in the numerator and the associative property in the denominator, we get

$$pq + pr = \frac{n[k(jl) + k(im)]}{n[n(ml)]} = \frac{k(jl) + k(im)}{n(ml)}$$

The last fraction equals the fraction for $p(q + r)$ and completes the proof of (iv).

As for (viii), let $q = m/n$ where $m \neq 0$. Then n/m is a rational number with the property

$$q\frac{n}{m} = \frac{m}{n}\frac{n}{m} = \frac{mn}{nm} = 1$$

If we define $1/q = n/m$, then the above calculation shows that $1/q$ has the required properties. Thus, the multiplicative inverse of q exists as stated. If m/n is the reduced form of q, then n/m is the unique reduced form of $1/q$.   ■

The following is a corollary that highlights basic features of $\mathbb{Q}$ (and $\mathbb{R}$ later) that are not shared by more general algebraic structures such as rings of matrices or rings of polynomials. Proving the statements in the corollary illustrate immediate applications of Theorem 98.

Corollary 99 *Assume that p and q are rational numbers.*
(a) *If $pq = 0$, then $p = 0$ or $q = 0$ (or both)*
(b) *If $r \neq 0$ and $pr = qr$, then $p = q$*

Proof (a) Suppose that $q \neq 0$. Then using properties (ii), (v), (vi), and (viii) in Theorem 98 we obtain

$$p = p1 = p\left(q\frac{1}{q}\right) = (pq)\frac{1}{q} = 0\frac{1}{q} = 0$$

(b) If $r \neq 0$, then it has a reciprocal by Theorem 98 (viii), so using various properties,

$$p = p\left(r\frac{1}{r}\right) = (pr)\frac{1}{r} = (qr)\frac{1}{r} = q \qquad \blacksquare$$

Property (b) above is called the *cancellation property*, while (a) ensures that there are no *zero divisors* in $\mathbb{Q}$ (or in $\mathbb{R}$), namely, a nonzero pair of numbers that multiply out to 0.

It is worth mentioning that although Property (viii) of Theorem 98 fails in $\mathbb{Z}$, the above corollary is valid. This is not true of other algebraic structures that satisfy (i)–(vii) in Theorem 98 but not (viii). For example, the set of all 2×2 matrices with rational entries forms an algebraic structure that satisfies all of the properties in Theorem 98 except for (iii) and (viii). Both (a) and (b) in Corollary 99 fail for this set (algebraically a *ring*); for counterexamples, consider

$$\begin{pmatrix} 1 & 0 \\ 0 & 0 \end{pmatrix}\begin{pmatrix} 0 & 0 \\ 0 & 1 \end{pmatrix} = \begin{pmatrix} 0 & 0 \\ 0 & 0 \end{pmatrix}$$

We see that (a) in Corollary 99 fails, as the product of two nonzero metrices above can be the zero matrix; further,

$$\begin{pmatrix} 0 & -1 \\ 1 & 0 \end{pmatrix}\begin{pmatrix} 1 & -2 \\ -1 & 2 \end{pmatrix} = \begin{pmatrix} 1 & -2 \\ 1 & -2 \end{pmatrix}$$

while

$$\begin{pmatrix} 1 & 0 \\ 0 & -1 \end{pmatrix}\begin{pmatrix} 1 & -2 \\ -1 & 2 \end{pmatrix} = \begin{pmatrix} 1 & -2 \\ 1 & -2 \end{pmatrix}$$

so the cancellation property (b) in Corollary 99 fails.[3]

An algebraic structure where all of the properties except (viii) in Theorem 98 are satisfied (including the commutative property) but Corollary 99 fails for it is the set of all continuous functions on an interval of real numbers with the operations of ordinary addition and multiplication (see Exercise 121).

---

[3] The ring of 2×2 matrices with rational entries has a richer structure than $\mathbb{Z}$; in particular, the only elements in $\mathbb{Z}$ that have a reciprocal are -1 and 1, but there are infinitely many matrices that have reciprocals, or inverses. In fact, as long as a matrix has a nonzero determinant, it has an inverse.

We now define the important ordering property, starting with usual ordering $<$ of the set $\mathbb{N}$. This can be readily extended to a total ordering of all the integers in $\mathbb{Z}$ by requiring that

$$m < n \quad \text{if } n - m \in \mathbb{N}$$

For instance, $-3 < -1$ because $-1 - (-3) = 2$ is a natural number; however, $2 \not< 1$ since $1 - 2 = -1 \notin \mathbb{N}$.

A similar thing happens in $\mathbb{Q}$ but less directly; we may say that if m/n and j/k are rational numbers, then

$$\frac{m}{n} < \frac{j}{k} \quad \text{if } km < jn \text{ in } \mathbb{Z}$$

For instance, $1/2 < 2/3$ because $3 < 4$, and $-1/2 < -1/3$ because $-3 < -2$. Notice that if we fix $n = k = 1$, then we recover the ordering of integers.

By adding the equality option to $<$, we obtain its reflexive extension $\leq$.

Proving that the relation $<$ above in fact totally orders $\mathbb{Q}$ is left as an exercise.

Finally, $\mathbb{Q}$ is an *ordered field* not just an ordered set. This is proved in the next often useful theorem.

Theorem 100 *Addition and multiplication preserve the usual ordering of rational numbers, i.e., for all p, q in $\mathbb{Q}$, the following hold:*
   (a) if $p < q$, then $p + r < q + r$ for every rational number r
   (b) if $p < q$ and $r > 0$, then $pr < qr$

Proof (a) Let $p = j/m$ and $q = k/n$. Then by definition, $p < q$ if $jn < km$. As integers, this is equivalent to $km - jn > 0$. Next, let $r = s/t$ where $s \in \mathbb{Z}$ and $t \in \mathbb{N}$ and using the addition of rationals above

$$p + r = \frac{jt + ms}{mt} \qquad p + q = \frac{kt + ns}{nt}$$

Again, by definition of ordering, $p + r < q + r$ if

$$mt(kt + ns) - nt(jt + ms) > 0$$
$$mkt^2 - jnt^2 > 0$$
$$(mk - jn)t^2 > 0$$

The last inequality is true since the product of two positive integers $mk - jn$ and t^2 is positive, and the proof of (a) is complete.

(b) Proceeding similarly to the proof of (a), $pr < qr$ if

$$\frac{js}{mt} < \frac{ks}{nt}$$

and this inequality holds if

$$ksmt - jsnt > 0$$
$$(km - jn)st > 0$$

By hypothesis, $r > 0$ so $s > 0$, and again we have a product of three positive integers that yields a positive integer. This completes the proof. ∎

No consecutive rational numbers but a lot of gaps in $\mathbb{Q}$

We speak of *consecutive* integers because for each n in $\mathbb{N}$ or in $\mathbb{Z}$, there are no integers between n and $n+1$. But we don't speak of "consecutive rationals" because between every pair of rational numbers a and b, no matter how close they are to each other, there is another rational number, say, their average $(a + b)/2$. So if we pick any rational q, we find no rational number that sits immediately next to q with no other rationals in between. Later we see that rationals have a much more important property in that they are *dense* in the vastly larger set of all *real* numbers.

The number $\sqrt{2}$ with the approximate value of 1.4142 is between 1 and 2. We pointed out earlier that this number is not rational, and now we prove it. This proof is one of the best known examples of proof by contradiction.

The number $\sqrt{2}$ is not rational.

Proof By way of contradiction, suppose that $\sqrt{2}$ is rational. Then there are (positive) integers m and n such that $m/n = \sqrt{2}$. We can assume that m/n is in reduced form since the greatest common divisor may be cancelled out without changing the fraction's value. Now, squaring both sides gives

$$\frac{m^2}{n^2} = \left(\frac{m}{n}\right)^2 = \left(\sqrt{2}\right)^2 = 2 \quad \text{so:}$$
$$m^2 = 2n^2$$

The occurrence of 2 on the right hand side of the second equality above means that m^2 is an even number. Since the square of an odd number is again odd, it follows that m is even, say, $m = 2k$ where k is another positive integer.

Then

$$2n^2 = (2k)^2 = 4k^2 \quad \text{so:}$$
$$n^2 = 2k^2$$

Again, this implies that n is even too. But now m and n do have a common factor of 2, contrary to our assumption that m/n is in reduced form. The only way of avoiding this contradiction is if $\sqrt{2}$ is not rational. ∎

Now we construct a sequence of rational numbers that converges to $\sqrt{2}$, which is a number r with the property that $r^2 = 2$. Recall that the square function x^2 is an increasing function when $x > 0$; we use this fact, and the fact that the average of two rational numbers is rational, to obtain rational approximations for $\sqrt{2}$.

We start with the observation that $1^2 = 1$ and $2^2 = 4$. Since $r^2 = 2$, and 2 is between 1 and 4, the increasing nature of the square function implies that $\sqrt{2}$ is between 1 and 2. It is not quite as large as 1.5, since $1.5^2 = 2.25$, but it is larger than 1.4, since $1.4^2 = 1.96$. Since 1.4 is rational (14/10, or 7/5 in reduced form) and closer to $\sqrt{2}$ than 1.5 is, set $q_1 = 1.4$ as a first rational approximation to $\sqrt{2}$.

Next, we compute the average of 1.4 and 1.5, which is 1.45, and notice that $1.45^2 = 2.1025 > 2$, so $\sqrt{2}$ is between 1.4 and 1.45. Now, 1.96 is still closer to 2 than 2.1025 is, so we calculate the average of 1.4 and 1.45. This gives 1.425, whose square is 2.030625, and this is closer to 2 than 1.96 is. Thus, a second approximation in $\mathbb{Q}$ to $\sqrt{2}$ is $q_2 = 1.425$.

We continue in this way[4] to generate a sequence of approximations in $\mathbb{Q}$ by finding average values

$$q_1, q_2, q_3, \ldots = 1.4, 1.425, 1.4125, \ldots \tag{4.1}$$

that gets closer and closer to $\sqrt{2}$, which as we have claimed is *not* in $\mathbb{Q}$.

So there is an infinite sequence of rational numbers between 1 and 2 that accumulate around $\sqrt{2}$ and get very close to it, but they do not actually reach it since $\sqrt{2}$ is not rational. We see that there is a gap in $\mathbb{Q}$ that contains no number but is the *limit* of the sequence in (4.1).

To fill this gap, we must *add* the limit of the sequence to the set $\mathbb{Q}$; this fills in the gap for $\sqrt{2}$ and creates a slightly larger set than the rational numbers. We may use the same averaging process to generate sequences of rational numbers that converge to $\sqrt{3}$ or to $\sqrt[3]{2}$ (using the cubic function x^3 in the latter case), then add

---

[4] This recursive averaging procedure is known as the *bisection algorithm*.

the limits of those sequences to $\mathbb{Q}$ to fill in the corresponding gaps. But it is harder to do this type of calculation for irrational numbers like π or e and for the many more quantities like them.

So how do we fill in *all* the gaps in $\mathbb{Q}$? *How do we even identify all of these gaps* (corresponding to the irrational numbers)?

These questions are not easy to answer; in fact, there is nothing obvious or simple about the real numbers because infinity is so intricately woven into their fabric. But we do have a clue in the above approximation of $\sqrt{2}$; specifically, the terms q_n of the sequence in (4.1) are rational numbers, but they are practically indistinguishable from $\sqrt{2}$ when the index n is very large. It is hard to distinguish the rational numbers in the tail end of this sequence from $\sqrt{2}$ when we go far with the midpoint or averaging process.

What is important is that *there is this sequence of rational numbers that can be associated with $\sqrt{2}$*; instead of looking for irrational numbers, we work with sequences of rational numbers whose limits are *not* rational numbers!

This observation is the starting point of Cantor's construction of real numbers, which we discuss later in this section after defining another important concept in real analysis, namely, Cauchy sequences.

4.2 Cauchy sequences

Our goal is to start from the set $\mathbb{Q}$ of all rational numbers and build up to the set of all real numbers $\mathbb{R}$, including all of the irrationals. We now examine sequences of rational numbers that we can be sure will converge without requiring a knowledge of their limits. This suits our purpose since we don't know all the irrational numbers; *but is it possible?*

We must rule out rational sequences that don't converge, like the sequences $q_n = n^2$ (or $1^2, 2^2, 3^2, \ldots$) and $q_n = (-1)^n$ (or $-1, 1, -1, 1, \ldots$), which don't approach any number, be it rational or not. So what is different about the sequence in (4.1) that does converge? How can we tell if a sequence converges if we don't know a limit for it?

Our next major step on the way to the real numbers is to identify these special sequences of rational numbers.

Earlier we saw how to estimate $\sqrt{2}$ using a sequence of rational numbers. We found that the more terms we used, the closer the approximation became. Now since we don't know the exact digital value of $\sqrt{2}$, consider a *practical* question: *how many terms of a rational sequence should we use to get a desired accuracy, say, 10 decimal places?*

Here is the key observation: *as the terms of the rational sequence approach their limit, they also get closer to each other.*

If we use many terms of the sequence, then the later terms form a dense patch of rationals around the limit; in the language of Chapter 3, the numbers *cluster* around the limit. The distance between any pair of terms q_m and q_n with large enough indices m and n can be as small as we need it to be, like a target threshold for a desired level of accuracy. For the 10-decimal-place accuracy, this threshold is 0.00000000005 or 5×10^{-11}.

More precisely, we must reach an index N that is large enough that the difference between *any pair* of terms q_m and q_n is less than 0.00000000005 as long as m, n both exceed N. Let's illustrate this idea using the recursion (1.3) in Exercise 3 (the *divide and average rule*).[5] Iterating five times generates the first five terms of the approximating sequence as

n	q_n
1	1.5
2	1.41666666667
3	1.41421568627
4	1.41421356238
5	1.41421356237

I have rounded the rational approximations to 11 decimal places for ease of reading.[6] Notice that the difference between q_4 and q_5 is 0.00000000001, which hits our target threshold (it is indeed less than 0.00000000005). If we continue repeating the divide and average process, then we find that after the fifth term listed above ($N = 5$), subtracting any pair of terms q_m and q_n from the sixth on up ($m, n \geq 6$) indeed comes up less than 0.00000000005 (and getting smaller with additional iterations).

If we continue using the recursion (1.3), then the first 10 decimal places are not affected. They stay fixed at 1.4142135623, so up to 10 decimal places, we seem to have reached a limit. As we go on with this process, more decimal places get fixed. But does this mean that this sequence of rational numbers q_n really does approach a limit?

Again, the answer seems to be yes, but it is not obvious why; in fact, we can't be sure yet that the limit is not a rational number. What if it is rational with a decimal expansion whose period is hundreds of decimal places long?

We have made an interesting discovery, but evidently more study is necessary to answer the above questions that strike at the heart of the concept of "real number." We continue our quest by giving special sequences like q_n above a name.[7]

---

[5] We could use the averaging process (bisection algorithm) of the last section instead, but that process generates accurate results much more slowly.
[6] The recursion actually generates *exact* rational terms as quotients of integers: $q_1 = (1/2)/(1 + 2/1) = 3/2$, $q_2 = (1/2)/(3/2 + 2/[3/2]) = 17/12$, etc.
[7] This is named after Augustin-Louis Cauchy (1789–1857), one of the founders of modern analysis. He may be better known to some people for his famous integral formula in complex analysis.

Cauchy sequence A sequence of rational numbers $q_1, q_2, q_3, \ldots$ is called *Cauchy* (pronounced *coashi*) if for every $\varepsilon > 0$, there is a positive integer N such that

$$|q_m - q_n| < \varepsilon \quad \text{for all } m, n \geq N \tag{4.2}$$

The bars as usual give the absolute value of the difference $|q_m - q_n|$, which tells us how far apart the terms q_m and q_n are, or the distance between them, regardless of their signs. The symbol $m, n \to \infty$ may be read "m and n both go to infinity"; they may do so in tandem, but in general, they are assumed to be independent of each other.

It is worth emphasizing that indices m and n must be unrestricted; for example, if we prove that the difference between consecutive terms, i.e., $|q_{n+1} - q_n|$ converges to 0, then we have not proved that the sequence is Cauchy (and it doesn't have to be; see Exercise 124).

For a simple example, consider the sequence

$$q_n = \frac{n-1}{n}, \quad n = 1, 2, 3, \ldots \tag{4.3}$$

It takes just a little bit of algebra to rigorously verify that this is a Cauchy sequence. First, write it as

$$q_n = 1 - \frac{1}{n}$$

and notice that

$$|q_m - q_n| = \left| \left(1 - \frac{1}{m}\right) - \left(1 - \frac{1}{n}\right) \right| = \left| \frac{1}{n} - \frac{1}{m} \right|$$

Given any $\varepsilon > 0$, we look for the index threshold N by requiring that

$$\left| \frac{1}{n} - \frac{1}{m} \right| < \varepsilon \tag{4.4}$$

be true for $m, n \geq N$. We can't solve (4.4) for both m and n, so we eliminate one of them. We may assume that $m \leq n$ since one of them must be smaller than (or the same as) the other. Then $1/n \leq 1/m$ (see Exercise 118), and it follows that

$$\left| \frac{1}{n} - \frac{1}{m} \right| = \frac{1}{m} - \frac{1}{n} < \frac{1}{m}$$

This inequality implies that if we choose m so large that $1/m < \varepsilon$, i.e., $m > 1/\varepsilon$, then (4.4) holds for all $m, n > 1/\varepsilon$ since $n \geq m$. So we choose any integer N larger than $1/\varepsilon$ to conclude that $m, n \geq N$ implies (4.4) and prove that the sequence in (4.3) is Cauchy.

In the above example, we did not use the fact that the sequence q_n actually has a limit, namely, 1. The next result shows that if a sequence converges, then it must be Cauchy.

Theorem 101 *Every convergent sequence is Cauchy.*

Proof Suppose that x_n converges to some number x, and $\varepsilon > 0$ is given. Then there is a positive integer N such that

$$|x_n - x| < \frac{\varepsilon}{2} \quad \text{if } n \geq N$$

Now for all $m, n \geq N$, we can use the triangle inequality to infer

$$|x_m - x_n| = |x_m - x + x - x_n| \leq |x_m - x| + |x - x_n| < \frac{\varepsilon}{2} + \frac{\varepsilon}{2} = \varepsilon$$

This proves that x_n is Cauchy. ∎

The converse of Theorem 101, namely, "every Cauchy sequence converges," is not always true; this is a subtle fact that will be crucial in our upcoming discussion. We saw earlier that there are Cauchy sequences of rational numbers (like the divide and average rule's output) that do not converge to rational numbers. Such a Cauchy sequence "would" converge if the presumed limit existed in the set of interest. In the set of all rational numbers $\mathbb{Q}$, any sequence that converges to an irrational number ends up in a hole, or gap, in $\mathbb{Q}$.

Before moving on to the next step in our journey, let's clarify when a sequence is *not* Cauchy.

A sequence q_n is not Cauchy when there is some value of ε, say, ε_0, such that *no matter how large an integer N we pick, we can always find a pair of indices $n, m \geq N$ such that $|q_n - q_m| \geq \varepsilon_0$.*

For example, consider the periodic sequence $q_n = (-1)^n$. If we pick any positive integer N, no matter how large, then there is a pair of integers, say, $m = N$ and $n = N + 1$, such that

$$|q_n - q_m| = |(-1)^{N+1} - (-1)^N| = 2$$

because one of N or $N+1$ is odd while the other is even, so the difference $(-1)^{N+1} - (-1)^N$ is always ± 2 with absolute value 2. Now, if ε_0 is any number between 0 and 2, say, $\varepsilon_0 = 1$, then (4.2) fails for every (finite) integer N.

4.2.1 Equivalent Cauchy sequences

We discovered that every Cauchy sequence of rational numbers tends to coalesce around a single number. The limit may be rational again, as in the case of (4.3) where the limit is 1; or the limit may be irrational and obtained by approximation, like $\sqrt{2}$ approximated by the divide and average rule.

Suppose that we associate a Cauchy sequence to the number that it either converges to, or it *would* converge to, if it didn't reach a gap. Then we can identify irrational numbers using a Cauchy sequence of rational numbers; we don't need to have the actual limit!

This idea is quite important since it doesn't require conjuring up irrational numbers to serve as limits. But things are not as clear-cut as they may seem at first; the idea needs clarification because many Cauchy sequences approach the same number and therefore, they can all be identified with it; for instance, if we use the divide and average rule with rational values for q_0 that are slightly different from 1, say, 1.5 or 0.5, then we get infinitely many different sequences that also approach the same limit, namely, $\sqrt{2}$. Which of these sequences should represent $\sqrt{2}$?

The short answer is, *all of them!* These sequences are *equivalent* in the sense that any one of them is as good as any other; there is no obvious reason for choosing any one over the others. This naturally leads to equivalence relations, the same concept that we discussed earlier.

We begin by collecting *all* Cauchy sequences of rational numbers *in one set* $\mathcal{C}$ of sequences and define a relation $\sim$ in $\mathcal{C}$ as follows:

Equivalent Cauchy sequences Let $s = q_1, q_2, q_3, \ldots$ and $s' = q_1', q_2', q_3', \ldots$ be sequences in $\mathcal{C}$. We say that $s \sim s'$ if $|q_n - q'_n| \to 0$ as $n \to \infty$.

For example, the sequence $q_n = 1 + 1/n$ is equivalent to the constant sequence $1, 1, 1, \ldots$ since $|(1 + 1/n) - (1)| = |1/n| \to 0$ as $n \to \infty$. Likewise, the sequences $1+2/n, 1+1/(2n), 1-1/n^2$ are all equivalent in the above sense. And as mentioned above, all (Cauchy) sequences that converge to $\sqrt{2}$ are equivalent. On the other hand, $1+1/n$ is not equivalent to $1/n$ because the absolute value of their difference is the sequence that is constantly equal to 1 and therefore doesn't converge to 0.

We now show that the above relation is indeed an equivalence relation.

Theorem 102 *The relation $\sim$ is an equivalence relation in the set $\mathcal{C}$.*

Proof The reflexive and symmetric properties are obvious. As for the transitive case, let $s \sim s'$ and $s' \sim s''$. Then both $|q_n - q'_n| \to 0$ and $|q'_n - q''_n| \to 0$ as $n \to \infty$; so for each $\varepsilon > 0$, we can find positive integers N and N' large enough that $|q_n - q'_n| < \varepsilon/2$ for all $n \geq N$, and $|q'_n - q''_n| < \varepsilon/2$ for all $n \geq N'$. Further, by the triangle inequality,

$$|q_n - q''_n| = |(q_n - q'_n) - (q'_n - q''_n)|$$
$$\leq |q_n - q'_n| + |q'_n - q''_n|$$

Let $N'' = \max(N, N')$ (the larger of N and N') and notice that if $n \geq N''$, then

$$|q_n - q''_n| \leq \frac{\varepsilon}{2} + \frac{\varepsilon}{2} = \varepsilon$$

It follows that $|q_n - q''_n| \to 0$ as $n \to \infty$, which means that $s \sim s''$. ∎

For each sequence s in $\mathcal{C}$, let's write $[s]$ for its equivalence class. For example, the sequence $1 + 1/n$ is in $[1]$ and $1/n$ is in $[0]$. Indeed, $[1/n] = [0]$ since by symmetry, the constant sequence $0, 0, \ldots$ is a member of the equivalence class $[1/n]$. Of course, there are infinitely many sequences in $[0]$; in particular, sequences that are equal to 0 after some point, like $3, 2, 1, 0, 0, \ldots$ are also in $[0]$ as well as Cauchy sequences that are equal to 0 infinitely often, like the sequence $q_n = [1 + (-1)^n]/n$ (i.e., $0, 1, 0, 1/2, 0, 1/3, \ldots$). On the other hand, if a Cauchy sequence is *not* in $[0]$, then its terms cannot accumulate around 0. For example, the constant sequence $1, 1, 1, \ldots$ is not in $[0]$ and neither is the sequence $s_n = 1 + 1/n$ that converges to 1.

Placing the rational numbers in a larger set

Each rational number q generates the constant sequence $q, q, q, \ldots$ which is trivially Cauchy, and hence in $\mathcal{C}$. Every Cauchy sequence of rational numbers that converges to q is in the equivalence class $[q]$. The collection of all equivalence classes $[q]$ of rationals is in a one-to-one correspondence with the set $\mathbb{Q}$ of rationals because if q and p are distinct rationals, then $[q]$ and $[p]$ are disjoint sets (by Theorem 100 the limit of a sequence is unique, so the sequence cannot converge to both q and p).[8]

We use the notation $[\mathcal{C}]$ for *the set of all equivalence classes* of the relation $\sim$. Each rational number q in $\mathbb{Q}$ is uniquely associated with the class $[q]$ in $[\mathcal{C}]$.

---

[8] The collection $[C]$ is formally called the "completion" of the set of rational numbers. This concept is of fundamental importance in analysis, as it extends to general metric spaces. These include Hilbert spaces, specifically, spaces of functions that are important to applications of analysis to scientific theories.

This idea may seem abstract, but it is not hard to grasp. Think of the set $[\mathcal{C}]$ of equivalence classes as a collection of boxes filled with (infinitely long) chains, each chain being a Cauchy sequence of rational numbers. We use the rational number q to mark or label the box or class that contains all Cauchy sequences that are equivalent to the constant sequence $q, q, q, \ldots$ This is how we may think of the ordinary rational numbers as members of $[\mathcal{C}]$.

After using up all of our labels in $\mathbb{Q}$, there are still many boxes in $[\mathcal{C}]$ left without a label; for instance, all Cauchy sequences generated by the divide and average rule are in a single box (equivalence class) that we may label $\sqrt{2}$. Similarly, we may label by π a box that contains rational sequences that converge to π, and so on. These leftover boxes in $[\mathcal{C}]$ are classes that cannot be labeled by rational numbers because the Cauchy sequences contained in them don't converge to rational numbers.

I emphasize that *all* of these leftover classes contain *only rational* sequences. They don't have rational *labels* because they don't contain constant sequences of rational numbers. We can use familiar symbols like $\sqrt{2}$ or π to label a few of these boxes whose contents have special meanings to us (length of the diagonal of a unit square or the area of a unit circle). But as we discover soon, there are far too many leftover boxes for us to be able to label each and every one of them.

4.3 Real numbers

With the preceding discussion in mind, we now define:

A *real number* is the equivalence class of a Cauchy sequence of rational numbers. We generally use the symbol $\mathbb{R}$ for the set of all real numbers rather than the notation $[\mathcal{C}]$.

It may seem strange to think of a number like $\sqrt{2}$ as a *collection* of sequences, not even an individual sequence; but as we soon discover, all of the properties of real numbers that we are familiar with hold for these equivalence classes. Because the distinction between these objects and what we are accustomed to as numbers is semantic, we may as well consider these equivalence classes to be numbers.

I have moved some of the more tedious proofs in the remainder of this chapter to Appendix 9.1 so that the technical nature of these proofs don't deter us from our primary goal: understanding the remarkable way in which real numbers are constructed from the rational ones through the concepts of a Cauchy sequence and equivalence class. Once comfortable with the basic arguments, the proofs will be there for us to explore at leisure. These proofs are not essential for understanding the rest of the book.

Real numbers form a totally ordered field

Recall that the set of all rational numbers $\mathbb{Q}$ is a totally ordered field. If $\mathbb{R}$ is also to be such a field, then it needs the operations of addition and multiplication plus an ordering defined on it that are compatible with the ordinary addition and multiplication and the ordering in the set $\mathbb{Q}$ of rationals that we discussed earlier.

The operations of addition and multiplication in $\mathbb{R}$ are defined naturally using the same operations on the rational numbers. For each pair of real numbers r and r' there are Cauchy sequences of rational numbers such that $r = [q_n]$ and $r' = [q'_n]$. Using these, we define

$$r + r' = [q_n + q'_n] \qquad rr' = [q_n q'_n]$$

The brackets above indicate equivalence classes of sequences of rational numbers that are obtained by adding or multiplying two sequences. We show in Appendix 9.1 (Theorem 4) that these operations are *well-defined* in the sense that any choice of rational sequence in $[q_n]$ and in $[q'_n]$ will produce the same result as any other choice in the same class.

Recall that if both r and r' are *rationals*, then each can be the equivalence class of a constant sequence. So, for example, if $r = [2, 2, 2, ...]$ and $r' = [-1/3, -1/3, -1/3, ...]$, then

$$r + r' = \left[2 - \frac{1}{3}, 2 - \frac{1}{3}, 2 - \frac{1}{3}, ...\right] = \left[\frac{5}{3}, \frac{5}{3}, \frac{5}{3}, ...\right]$$

$$rr' = \left[2\left(-\frac{1}{3}\right), 2\left(-\frac{1}{3}\right), 2\left(-\frac{1}{3}\right), ...\right] = \left[-\frac{2}{3}, -\frac{2}{3}, -\frac{2}{3}, ...\right]$$

The same rules apply to equivalence classes of irrational numbers, although sequences that correspond to the irrationals are not eventually constant. For instance, in Chapter 8 we show that the rational sequence

$$q_n = 1, 1 - \frac{1}{3}, 1 - \frac{1}{3} + \frac{1}{5}, 1 - \frac{1}{3} + \frac{1}{5} - \frac{1}{7}, ... = 1, \frac{2}{3}, \frac{13}{15}, \frac{76}{105}, ...$$

converges to $\pi/4$. This lets us define numbers like the sum $-1/3 + \pi/4$ and the product $(-1/3)(\pi/4) = -\pi/12$ as

$$-\frac{1}{3} + \pi = \left[-\frac{1}{3}, -\frac{1}{3}, -\frac{1}{3}, ...\right] + \left[1, \frac{2}{3}, \frac{13}{15}, \frac{76}{105}, ...\right]$$

Adding the corresponding numbers in the brackets gives

$$\left[-\frac{1}{3} + 1, -\frac{1}{3} + \frac{2}{3}, -\frac{1}{3} + \frac{13}{15}, -\frac{1}{3} + \frac{76}{105}, ...\right] = \left[\frac{2}{3}, \frac{1}{3}, \frac{8}{15}, \frac{123}{315}, ...\right]$$

Likewise,

$$-\frac{\pi}{12} = \left[-\frac{1}{3}, -\frac{1}{3}, -\frac{1}{3}, \ldots\right]\left[1, \frac{2}{3}, \frac{13}{15}, \frac{76}{105}, \ldots\right]$$

$$= \left[-\frac{1}{3}, -\frac{2}{9}, -\frac{13}{45}, -\frac{76}{315}, \ldots\right]$$

Remarkably, the above two operations on the set $[\mathcal{C}] = \mathbb{R}$ of all equivalence classes of Cauchy sequences of rational numbers have the familiar properties of ordinary addition and multiplication of real numbers.

To begin with, the classes $[0]$ and $[1]$ are in fact like the ordinary 0 and 1; if $r = [q_n]$ is any real number, then

$$r + [0] = [q_n] + [0] = [q_n + 0] = [q_n] = r$$
$$r[0] = [q_n][0] = [q_n 0] = [0] = 0$$
$$r[1] = [q_n][1] = [q_n 1] = [q_n] = r$$

We use the equivalence class $[-q_n]$ of the rational sequence of negatives to define $-r$:

$$-r = -[q_n] = [-q_n]$$

This definition implies the expected result

$$r + (-r) = [q_n] + [-q_n] = [q_n + (-q_n)] = [0] = 0$$

If $r \neq 0$, then the reciprocal is defined similarly, but we need to pick an equivalence class $[q_n]$ with $q_n \neq 0$ for all n. The condition $r \neq 0$, i.e., $[q_n] \neq [0]$, does not guarantee this, since the equivalence class $[q_n]$ contains sequences containing many zeros, though not converging to 0.[9] With this caveat in mind,

$$\frac{1}{r} = r^{-1} = [q_n]^{-1} = [q_n^{-1}] = \left[\frac{1}{q_n}\right]$$

For example, if $r = \pi/4$, then as noted above, r is represented by equivalence class $[1, 2/3, 13/15, 76/105, \ldots]$. Because none of the terms of the sequence in this equivalence class is zero, we obtain

$$\frac{1}{r} = \left[1, \frac{3}{2}, \frac{15}{13}, \frac{105}{76}, \ldots\right]$$

Notice that *the class $[0]$ has no reciprocal*. Even if we choose a sequence like $q_n = 1/n$, i.e., $1, 1/2, 1/3, \ldots$ in $[0]$, which has no zero terms, then the sequence

---

[9] This issue is resolved in Appendix 9.1.

$1/q_n$ of reciprocals of q_n, i.e., $1, 2, 3, \ldots$ diverges to infinity and thus is not Cauchy. In particular, the sequence $1/q_n$ isn't in the set $\mathcal{C}$ and cannot serve as a reciprocal sequence for q_n.

In Appendix 9.1, we establish the other field properties mentioned in Theorem 98 like the closure, commutative, associative, and distributive properties, so that $[\mathcal{C}] = \mathbb{R}$ is indeed an algebraic field, just like $\mathbb{Q}$.

We now have a *field* of objects (equivalence classes) $\mathbb{R}$ that contains the rational numbers $\mathbb{Q}$ (or rather, a copy of $\mathbb{Q}$ in the form of classes $[q]$ equivalent to the constant rational sequences).

Next, we introduce an ordering on $\mathbb{R}$ that is compatible with the usual ordering $<$ of $\mathbb{Q}$ as previously defined. This requires that the intended ordering of $\mathbb{R}$ when restricted to the rational numbers must be the same as the integer-based, usual ordering of $\mathbb{Q}$ that we discussed earlier.

In order to motivate the definitions below, let us take another look at the ordering of $\mathbb{Q}$; since $p < q$ is equivalent to $0 < q - p$, we see that to say $p < q$ is to say that "$q - p$ is positive." It is now easy to define what it means to say something is "positive" in $\mathbb{R}$.

The real number $r = [q_n]$ is *positive* if the rational Cauchy sequence $q_1, q_2, q_3, \ldots$ is eventually positive *and* its terms do not approach 0.

For example, consider the sequence $q_n = 1/4 - 1/n$ of rational numbers whose first few terms are

$$-\frac{3}{4}, -\frac{1}{4}, -\frac{1}{12}, 0, \frac{1}{20}, \frac{1}{12} \ldots$$

All terms after the fourth are positive and approach $1/4$ as $n \to \infty$, so the equivalence class of this sequence is a positive real number. Indeed, this sequence is equivalent to the rational number $1/4$, which we deem to be positive. On the other hand, even though *all* terms of the sequence

$$1, \frac{1}{2}, \frac{1}{3}, \frac{1}{4}, \ldots, \frac{1}{n}, \ldots$$

are positive, it does not represent a positive real number *because it converges to 0* and therefore, it is in the equivalence class $[0]$.

Now, in an analogy with the ordinary rational numbers, we define a relation $\prec$ in $\mathbb{R}$.

Let $x = [p_n]$ and $y = [q_n]$. Then $x \prec y$ if $[0] \prec [q_n] - [p_n]$, that is, $x \prec y$ if $[q_n] - [p_n] = [q_n - p_n]$ is positive.

The following properties of the ordering $\prec$ are true; the proofs are given in Appendix 9.1.

Theorem 103 *For all* $x, y, z \in \mathbb{R}$,

      (a) If $x \prec y$, *then* $x + r \prec y + r$ *for every* r *in* $\mathbb{R}$.
      (b) If $x \prec y$, *then* $xr \prec yr$ *for every positive* r *in* $\mathbb{R}$.
      (c) $\prec$ *is a total ordering: if* $x \neq y$, *then either* $x \prec y$ *or* $y \prec x$ *(not both)*
      (d) $\prec$ *is transitive: if* $x \prec y$ *and* $y \prec z$ *then* $x \prec z$

In addition to the properties in Theorem 103, the ordering $\prec$ on $\mathbb{R}$ is compatible with the usual ordering $<$ of the rational numbers, that is,

---

If p, q are rational numbers such that $p < q$ then $[p] \prec [q]$.

Proof Since $p < q$, it follows that $[p] \neq [q]$. Further, $[q-p]$ is positive because the difference between the constant sequences $q, q, q, \dots$ and $p, p, p, \dots$ is positive and does not approach 0. Therefore,

$$[0] \prec [q - p] = [q] - [p]$$

from which we conclude that $[p] \prec [q]$ and complete the proof.   ■

Based on the above result, from now on we follow the customary (and convenient, if not precise) practice of using the symbol $<$ for the ordering of real numbers instead of the symbol $\prec$.

4.4 Completeness and other foundational theorems of real analysis

We just showed that $\mathbb{R}$ is a totally ordered field that contains (a copy of) the field of rational numbers $\mathbb{Q}$. Certainly $\mathbb{R}$ is larger than $\mathbb{Q}$, but by how much? Does $\mathbb{R}$ still have gaps or holes like $\mathbb{Q}$ where limits of Cauchy sequences may fall through?

We answer these questions and discuss important related issues in this section. Our focus is now on $\mathbb{R}$ *as a set* rather than on individual real numbers. We discuss several fundamental theorems of real analysis in this section. Many are logically equivalent to the completeness Theorem 104 below, including the *nested intervals theorem*, the *least upper bound theorem*, and the *Bolzano–Weierstrass theorem*. These are often used to prove fundamental results about functions like the intermediate value theorem, the extreme value theorem, and the mean value theorem that we encounter and prove in Chapter 6.

Further interesting facts about the real numbers are discussed in Chapters 5 and 8 using infinite series.

Having constructed the set $\mathbb{R}$ as a totally ordered field, at this point, it is harmless for our purposes to return to our intuitive view of the real numbers as points on the number line.

4.4.1 Completeness of $\mathbb{R}$ and the Cauchy convergence criterion

The set $\mathbb{R}$ has no gaps or holes where the limit of a Cauchy sequence might end up. This observation leads to the following fundamental result (proved in Appendix 9.1).

Theorem 104 *(Completeness theorem) The set $\mathbb{R}$ of all real numbers is complete, i.e., every Cauchy sequence of real numbers converges to a real number.*

The above theorem states that all *real* Cauchy sequences converge, not just the rational sequences; and since the limits are again real numbers, no new numbers are created by the limit process (unlike sequences in $\mathbb{Q}$). The following corollary, often referred to as the *Cauchy criterion for convergence*, is an immediate consequence of Theorems 101 and 104.

Corollary 105 *(Cauchy Criterion) A sequence of real numbers converges if and only if it is a Cauchy sequence.*

4.4.2 Density of rational numbers in $\mathbb{R}$

If $x = [q_n]$, then all sequences that are equivalent to q_n accumulate around the same limit that is represented by the symbol x. Intuitively we may think of x as a point on the number line (the x-axis). If r is another point on the line, say, $r < x$, then $r = [p_n]$ for some Cauchy sequence of rationals p_n whose limit falls short of the limit of q_n. In particular, q_n will surpass the limit of p_n, which means that some term q_N must exist in between the two limits for sufficiently large N, i.e., $r < [q_N] < x$ where $[q_N]$ is the equivalence class of the constant sequence $q_N, q_N, q_N, \ldots$ Since q_N is a rational number by design, we have proven the next important theorem (see Theorem 5 in Appendix 9.1 for a proof).

Theorem 106 *The set $\mathbb{Q}$ of rational numbers is dense in $\mathbb{R}$; i.e., between every pair of distinct real numbers x and y, there is a rational number q.*

This statement actually implies that there are infinitely many rational numbers between every pair of real numbers; see Exercise 128. Thus, every open interval (a, b) of real numbers contains infinitely many rational numbers. If we think of the real numbers as points on the number line, then the rational numbers are densely and uniformly spread all over the line. However, *they do not form any continuous stretch* (a continuum) because as the next corollary of Theorem 106 implies there are irrational numbers between every pair of rational numbers too. In particular, both the set of all rational numbers and the set of all irrational numbers are incomplete by themselves, i.e., each contains gaps where numbers from the other type of set go. When they are put together, all gaps are filled, and a continuum is obtained.

Corollary 107 *The set of all irrational numbers is dense in $\mathbb{R}$; i.e., between every pair of distinct real numbers x and y, there is an irrational number.*

Proof Note that if q is any positive rational number, then $q/\sqrt{2}$ is irrational (see Exercise 129). Now if $0 < x < y$, then $x\sqrt{2} < y\sqrt{2}$; and by Theorem 106, there is a rational number q such that $x\sqrt{2} < q < y\sqrt{2}$. Dividing by $\sqrt{2}$ thus gives $x < q/\sqrt{2} < y$, and we have found an irrational number between x and y. If $x \leq 0 < y$, then there is a rational q' such that $0 < q' < y$, and we have seen that we can find an irrational between q' and y. Such an irrational is obviously also between x and y. Finally, if $x < y \leq 0$, then $0 \leq -y < -x$, and we can find an irrational number, say, z, where $-y < z < -x$. Now $-z$ is the number we are looking for since $x < -z < y$.   ■

Density is a very useful property; for example, the following result implies that every real number can be approximated by a sequence of rational numbers as well as a sequence of irrational numbers.

Theorem 108 *For every real number r, there is a sequence q_n of rational numbers and a sequence x_n of irrational numbers such that*

$$r = \lim_{n \to \infty} q_n = \lim_{n \to \infty} x_n$$

Proof Here we only know the limit r and want to find the sequences q_n and x_n. By the density of rational numbers, there is a rational number q_1 between $r + 1/2$ and $r + 1$, a rational q_2 between $r + 1/3$ and $r + 1/2$, a rational q_3 between $r + 1/4$ and $r + 1/3$, and so on. This process generates a sequence of distinct numbers

$$r + 1 > q_1 > r + \frac{1}{2} > q_2 > r + \frac{1}{3} > q_3 > r + \frac{1}{4} > \cdots > r$$

Since $r < q_n < r + 1/n$ for every $n = 1, 2, 3, \ldots$ and $1/n \to 0$ as $n \to \infty$, the squeeze theorem implies that q_n converges to r. We do not need to know what each number q_n is specifically, but we do have a recipe for defining it. Repeating this same argument with all q_n changed to x_n using the density of irrational numbers proves the statement about irrationals.   ∎

The Archimedean property

We know that the set of positive integers $1, 2, 3, \ldots$ is unbounded and goes on to infinity. But are there real numbers that are greater than all of the integers? The answer seems to be no; but why not? The answer turns out to be a consequence of the density of rational numbers in $\mathbb{R}$.[10]

Theorem 109 *(Archimedean property) For every positive real number x, there is a positive integer n such that $n > x$.*

Proof   For rational numbers this is easy to see: if $q = k/m > 0$, then since $m \geq 1$, we see that $q \leq k$; so the integer $n = k + 1$ works. If r is any positive real number, then by the density property of rational numbers, there is a (positive) rational number q such that $r < q < r + 1$. If n is a positive integer such that $n > q$, then by transitivity, $n > r$, and we are done!   ∎

Notice that the above theorem works for negative numbers too, i.e., if x is a negative real number, then there is a negative integer $-n$ such that $-n < x$. The proof is simple: if $x < 0$, then $-x > 0$; and by Theorem 109, there is a positive integer such that $n > -x$. Therefore, $-n < x$.

The Archimedean property is often used in the following way when working with limits:

If r is any positive real number, then for every real $\varepsilon > 0$, *no matter how small,* there is a positive integer n such that $r/n < \varepsilon$ (set $x = r/\varepsilon$ in the statement of the Archimedean property).

The fact that the integers stretch as far as the real numbers lets us choose numbers that are arbitrarily close to 0 simply by dividing a given number by a very large integer. For instance, if $r = 1$, and we pick any very small number, say,

$$\varepsilon = 0.00000000000000123 = 1.23 \times 10^{-15}$$

---

[10] This property of real numbers was named after Archimedes by the mathematician O. Stolz in the 1880s. Archimedes in turn credited it to Eudoxus. There are abstract number fields that do not have this property, such as the hyperreal numbers of nonstandard analysis. These are number fields that contain the real numbers as well as infinite numbers and their reciprocals, the infinitesimals.

then dividing r by the integer $n = 10^{15}$ (a quadrillion) gives $r/n = 10^{-15}$, which is less than the above value of ε.

4.4.3 Least upper bounds and nested intervals

The completeness property has deep consequences that are essential for proving other profound statements. A logical foundation on which our mental intuition of the continuum is based on such statements, many of which we encounter in later chapters.

We discuss two more fundamental consequences of completeness in this section to illustrate the intricate nature of the aforementioned logical foundation.

The first property has to do with bounded sets. Consider a nonempty set S of real numbers. Recall from Chapter 3 that an upper bound of S is a real number u such that every number x in S is less than or equal to u. Likewise, a lower bound of S is a real number l that is less than or equal to every number in S. If S has an upper bound, then it is bounded from above; and if it has a lower bound, then it is bounded from below. S is a bounded set if it has both an upper bound and a lower bound.

Recall also that the *least upper bound*, or *supremum* of S (sup S for short) is, as its name indicates, the smallest of all upper bounds. Likewise, the *greatest lower bound* or *infimum* of S (inf S for short) is the largest of all lower bounds of S. If they exist, then by Lemma 62, sup S and inf S are uniquely defined for every set S.

In Exercise 87 in Chapter 3, we find that a bounded set of rational numbers such as

$$S = \{q \in \mathbb{Q} : q^2 \leq 2\}$$

may fail to have a least upper bound or a greatest lower bound in $\mathbb{Q}$. In $\mathbb{R}$ though, both sup $S = \sqrt{2}$ and inf $S = -\sqrt{2}$ exist. This case illustrates that the existence of a supremum or an infimum is implied by the nonexistence of gaps in $\mathbb{R}$, i.e., completeness.

The set S above has the least upper bound property in $\mathbb{R}$ but not in $\mathbb{Q}$ because $\sqrt{2}$ is not rational. The following result establishes the same property for increasing sequences. Its first part is proved in Appendix 9.1 (Theorem 9) as a corollary of the Archimedean property, Theorem 109. The second part follows readily from the first part; see Exercise 131.

Corollary 110 *(a) Every nondecreasing sequence x_n of real numbers that is bounded from above is Cauchy and converges to the supremum of its range, i.e.,*

$$\lim_{n \to \infty} x_n = \sup\{x_n : n \geq 1\}$$

(b) Every nonincreasing sequence r_n of real numbers that is bounded from below is Cauchy and converges to the infimum of its range, i.e.,

$$\lim_{n \to \infty} r_n = \inf\{r_n : n \geq 1\}$$

For example, the increasing sequence $x_n = 1 - 1/n$ converges to its supremum 1. The supremum is 1 because 1 is larger than every number in the sequence, and no other real number is less than 1 and larger than *all* terms of the sequence. Less obvious but more interesting is the sequence

$$q_n = \sum_{k=1}^{n} \frac{1}{k!} = \frac{1}{1!} + \frac{1}{2!} + \frac{1}{3!} + \cdots + \frac{1}{n!}$$

where the *factorial* is defined as $k! = k(k-1)(k-2)\cdots(2)(1)$. So

$$q_1 = 1, \qquad q_2 = 1 + \frac{1}{2} = \frac{3}{2}, \qquad q_3 = 1 + \frac{1}{2} + \frac{1}{6} = \frac{5}{3}, \quad \text{etc.}$$

Notice that q_n is an increasing sequence of rational numbers. In calculus, you may have come across such a sequence in which case you may recall that its limit is $e - 1$, an irrational number approximately equal to 1.718 (we also discuss the number e and related concepts in later chapters, proving its irrationality in Chapter 8). We may write this result as

$$\lim_{n \to \infty} q_n = \sup\{q_n : n \geq 1\} = e - 1$$

This example also shows that Corollary 110 is not true if limited to the set of rational numbers.

The most important application of Corollary 110 is to cases where we *don't know the limit.* For example, the sequence of real numbers

$$r_n = \sum_{k=1}^{n} \frac{\sin^2 k}{k!} = \frac{\sin^2 1}{1!} + \frac{\sin^2 2}{2!} + \frac{\sin^2 3}{3!} + \cdots + \frac{\sin^2 n}{n!}$$

is nondecreasing and it is bounded by the sequence q_n for every index n, i.e., $r_n \leq q_n$. Therefore, by Corollary 110, it is Cauchy and converges to a number less than $e - 1$, which is an upper bound for r_n; you can work out the details in Exercise 132, but keep in mind that you won't find the exact limit by elementary methods.

You can *estimate* the limit using a computer; rounded to three decimal places, the limit is 1.157, which is substantially less than $e - 1$. While this estimate by itself does not prove that r_n converges to anything, using Corollary 110, we can state with confidence that

$$\lim_{n\to\infty} r_n = \sup\{r_n : n \geq 1\} \approx 1.157$$

The nested intervals property

The next result is an important enough fact in real analysis to be given a name.[11]

Theorem 111 *(Nested intervals property) Consider a decreasing sequence of nested closed intervals in $\mathbb{R}$:*

$$[a_1, b_1] \supset [a_2, b_2] \supset [a_3, b_3] \supset \cdots$$

If the lengths of these intervals vanish, i.e.,

$$\lim_{n\to\infty} (b_n - a_n) = 0 \tag{4.5}$$

then there is exactly one real number r that is common to all of these intervals, i.e.,

$$\bigcap_{n=1}^{\infty} [a_n, b_n] = \{r\} \tag{4.6}$$

Proof The nondecreasing sequence of left end points $a_1, a_2, a_3, \ldots$ is bounded by any one of the right endpoints b_n; so by Corollary 110, it has a least upper bound, say α. Similarly, the nonincreasing sequence $b_1, b_2, b_3, \ldots$ of right end points has a greatest lower bound β (Exercise 131). Therefore,

$$a_1 \leq a_2 \leq a_3 \leq \ldots \leq \alpha \leq \beta \leq \ldots \leq b_3 \leq b_2 \leq b_1$$

In particular, $0 \leq \beta - \alpha \leq b_n - a_n$ for every index n. Since $\lim_{n\to\infty}(b_n - a_n) = 0$, the squeeze theorem implies that $\beta - \alpha = 0$. So $\alpha = \beta$, and this common value is the number r in (4.6). ∎

---

[11] The nested intervals property as presented here is a special case of a much more general result in topology; in particular, it can be extended to complete metric spaces. Cantor's name is usually associated with this property, but many other mathematicians among Cantor's contemporaries and even predecessors knew about it in one form or another.

The nested intervals property holds because the completeness of real numbers ensures that there are no gaps or holes in $\mathbb{R}$. Within the set of just the rational numbers (which is incomplete), the intersection in (4.6) can be empty. For example, if a_n is a sequence of rational numbers that is increasing and converges to $\sqrt{2}$ from below, and b_n is a sequence of rational numbers that is decreasing and converges to $\sqrt{2}$ from above[12], then

$$a_1 < a_2 < a_3 < \cdots < \sqrt{2} < \cdots < b_3 < b_2 < b_1$$

and we have a sequence of nested intervals with rational endpoints:

$$[a_1, b_1] \supset [a_2, b_2] \supset [a_3, b_3] \supset \cdots$$

whose intersection is $\sqrt{2}$, an irrational number.[13] If we ignore irrational numbers like $\sqrt{2}$, then the intersection of these closed intervals will be empty.

The other conditions in Theorem 111 are also necessary. If the intervals are not closed, then the intersection in (4.6) may be the empty set; see Exercise 134 where you also show that condition (4.5) is essential, too.

The least upper bound property

The nested intervals property is often useful in proving other properties of sets of real numbers, like the next result that generalizes Corollary 110 to the supremum and infimum of general sets and is a fundamental result in real analysis. In fact, in many textbooks, it is taken as an axiom from which completeness and all other properties (density of rationals, Archimedean property, etc.) are derived.

Theorem 112 (*The least upper bound property*) *Every nonempty set S of real numbers that is bounded from above has a least upper bound* $\sup S \in \mathbb{R}$.

Proof Given that S is bounded from above, there is a real number u_1 such that $u_1 \geq x$ for every x in S. Let x_1 be a number that is not an upper bound of S; e.g., $x_1 = x - 1$, where x is any number in S (x_1 need not be in S). Consider the interval $[x_1, u_1]$; if the midpoint of this interval, i.e., the average $\alpha = (x_1 + u_1)/2$ is an upper bound for S, then define $u_2 = \alpha$ (this takes us to a smaller upper bound) and $x_2 = x_1$. Otherwise, define $x_2 = \alpha$ and $u_2 = u_1$. In either case, the (possibly) new number u_2 is an upper bound for S, and the updated x_2 is not an

---

[12] One way of obtaining these sequences is by using the averaging or midpoint method that we discussed earlier. Alternatively, a faster method would be the divide and average rule that generates a_n and b_n as two of its subsequences.

[13] Notice that $\sqrt{2}$ is both the least upper bound or supremum of the sequence a_n and the greatest lower bound or infimum of the sequence b_n.

upper bound for S; more specifically,

$$x_1 \leq x_2 < u_2 \leq u_1$$

and further, the length of the updated interval $[x_2, u_2]$ is half of the length of the original interval, i.e.,

$$u_2 - x_2 = \frac{u_1 - x_1}{2}$$

This updating and shrinking process may be repeated. The next step starts with $[x_2, u_2]$ and by using its midpoint $(x_2 + u_2)/2$, gets an updated interval $[x_3, u_3]$ such that

$$x_1 \leq x_2 \leq x_3 < u_3 \leq u_2 \leq u_1$$

and with length half of the length of $[x_2, u_2]$

$$u_3 - x_3 = \frac{u_2 - x_2}{2} = \frac{u_1 - x_1}{2^2}$$

Repeating n times gives

$$x_1 \leq x_2 \leq x_3 \leq \cdots \leq x_n < u_n \leq \cdots \leq u_3 \leq u_2 \leq u_1$$

with

$$u_n - x_n = \frac{u_{n-1} - x_{n-1}}{2} = \cdots = \frac{u_1 - x_1}{2^{n-1}}$$

This process generates a nested sequence of closed intervals

$$[x_1, u_1] \supset [x_2, u_2] \supset [x_3, u_3] \supset \cdots \supset [x_n, u_n] \supset \cdots$$

where the length of the nth interval is given by

$$\frac{u_1 - x_1}{2^{n-1}}$$

for every $n = 1, 2, 3, \ldots$ The conditions of Theorem 111 are satisfied, so by the nested intervals property, there is one real number r that is in all of these intervals. We claim that $r = \sup S$ and use two contradiction arguments to prove it.

First notice that r is an upper bound for S; if not, then (by way of contradiction) S contains a number $x > r$ so that $x - r > 0$. Since u_n and x_n both converge to r, there exists an index N large enough that

$$u_N - x_N = \frac{u_1 - x_1}{2^{N-1}} < x - r$$

Since $x_n \leq r$ for all n, this implies that

$$u_N < u_N + (r - x_N) < x$$

This contradicts the fact that u_N is an upper bound for S; to avoid this contradiction, r has to be an upper bound for S. To show that it is in fact the *least* upper bound, suppose (again by way of contradiction) that there is an upper bound u that is smaller than r. Then, since $r \leq u_n$, it follows that $u_n > u$ for all n. Then there is an index K large enough that

$$u_K - x_K = \frac{u_1 - x_1}{2^{K-1}} < r - u \leq u_K - u$$

which implies that $x_K > u$. But this means that x_K is an upper bound of S and this contradicts the definition of x_K. Therefore, r must be the least upper bound, and our claim, and the theorem, is proved. ∎

Using Theorem 112 and a short argument, you can prove the following mirror image version in Exercise 133.

Corollary 113 (*greatest lower bound property of sets*) *Every nonempty set S of real numbers that is bounded from below has a greatest lower bound* inf $S \in \mathbb{R}$.

Notice that completeness is essential in Theorem 112. For example, the set $\{q \in \mathbb{Q} : q^2 \leq 2\}$ has neither a least upper bound nor supremum nor a greatest lower bound nor infimum in the (incomplete) set of rational numbers, but it has both a supremum $\sqrt{2}$ and an infimum $-\sqrt{2}$ in $\mathbb{R}$.

4.4.4 The Bolzano–Weierstrass theorem

In our discussion of cluster points in Chapter 3, it seemed that *every bounded sequence, convergent or not, has cluster points,* and the largest of these is the limit supremum and the smallest is the limit infimum of the sequence (Theorem 66). However, the italicized statement is false without further qualification. If we take a sequence of rational numbers that is generated by the divide and average rule, then such a sequence is in $\mathbb{Q}$; but it converges to $\sqrt{2}$, which isn't in $\mathbb{Q}$. Thus, the would-be cluster point $\sqrt{2}$ doesn't exist in $\mathbb{Q}$.

This subtle point is quite important for abstract structures like infinite-dimensional metric spaces of functions that contain the solutions of important differential equations of science. These solutions are often limits of sequences of simpler functions. The question is, how can we be sure that a bounded sequence in a general metric space has any cluster points?

For the special case of real numbers, completeness is all that we need. If we think of the divide and average rule as a sequence of real numbers, then such a sequence does in fact have a (unique) cluster point, namely, its limit $\sqrt{2}$. But how does completeness ensure the existence of cluster points for all bounded sequences?

Since cluster points are limits of subsequences (by definition), the next fundamental result in real analysis answers our question. It was first proved in 1817 by Bernard Bolzano (1781–1848) as part of his proof of the intermediate value theorem that we discuss in Chapter 6. Fifty years after, Bolzano and Weierstrass independently proved it again as a result that was of interest in its own right.[14]

Theorem 114 (Bolzano–Weierstrass) *Every bounded sequence of real numbers has a convergent subsequence (hence, a cluster point).*

Proof Let r_n be a bounded sequence in $\mathbb{R}$, and let the set $S = \{r_n : n \in \mathbb{N}\}$ be the range of r_n. If S is a finite set, then at least one point r in it must repeat infinitely often; i.e., there is an increasing sequence of indices $n_1 < n_2 < \cdots$ such that $r_{n_j} = r$ for all $j = 1, 2, 3, \ldots$ It follows that the subsequence r_{n_j} converges (trivially) to r, making r a cluster point.

If S is infinite, then given the boundedness of r_n, there are real numbers a, b such that $S \subset [a, b]$. If we partition the closed interval $[a, b]$ into halves

$$[a, b] = \left[a, \frac{a+b}{2}\right] \cup \left[\frac{a+b}{2}, b\right]$$

then at least one of the half intervals must contain infinitely many distinct elements of S. Denote that half interval by $[a_1, b_1]$ where either a_1 or b_1 is $(a+b)/2$ and note that the length of $[a_1, b_1]$ is half the length of $[a, b]$, i.e., $(b-a)/2$.

If we further partition $[a_1, b_1]$ into half intervals, then one of the new half intervals, call it $[a_2, b_2]$, contains infinitely many (distinct) points of S, and its length is $(b-a)/4$. Repeating this partitioning process generates a nested sequence of half intervals $[a_k, b_k]$ such that

(i) $[a, b] \supset [a_1, b_1] \supset [a_2, b_2] \supset [a_3, b_3] \supset \cdots$
(ii) the length of $[a_k, b_k]$ is $(b-a)/2^k$ for every $k = 1, 2, 3, \ldots$
(iii) $[a_k, b_k]$ contains infinitely many points of S.

---

[14] There are different versions of this theorem that are equivalent for metric spaces; the version that we present here based on sequences is also known as the *sequential compactness theorem*.

By Theorem 111 (the nested intervals property, a consequence of completeness),

$$\bigcap_{k=1}^{\infty} [a_k, b_k] = \{r\}$$

This number r is in $[a_k, b_k]$ for every k, which by (iii) above must also contain at least one term of r_n, say, r_{n_k}, where $r_{n_k} \neq r$. It follows that the distance between r_{n_k} and r cannot exceed the length of the interval $[a_k, b_k]$, i.e.,

$$|r_{n_k} - r| \leq \frac{b - a}{2^k}$$

The fraction on the right hand side converges to 0 as $k \to \infty$, so given any $\varepsilon > 0$ there is some positive integer N such that $(b - a)/2^k < \varepsilon$ for all $k \geq N$. Therefore,

$$|r_{n_k} - r| < \varepsilon, \quad k \geq N$$

We conclude that the subsequence r_{n_k} converges to r and therefore, r is a cluster point. ∎

We use Theorem 114 to prove some important results later on. To illustrate the utility of the Bolzano–Weierstrass theorem for now, consider the sequence $x_n = \sin n$. We saw earlier that this sequence isn't periodic, so it is not obvious whether it has any cluster points or not. However, it is bounded with range contained in the interval $[-1, 1]$, so Theorem 114 guarantees that $\sin n$ has a cluster point in $[-1, 1]$.

Proving that any particular point, e.g., 1, is actually a cluster point, although intuitively compelling, is difficult since we must show that $\sin n$ is in intervals of type $(1 - \varepsilon, 1)$ for infinitely many values of n and all $\varepsilon > 0$. But we can do more modest things without a lot of calculation, thanks to the Bolzano–Weierstrass theorem.

For example, suppose we want to prove that $\sin n$ has a cluster point in the interval $(0, 1)$. Since $\sin n$ is not contained in $(0, 1)$ for all n, we need to show that it is in $(0, 1)$ for infinitely many n in order to use Theorem 114. To do this, observe that $\sin n \in (0, 1)$ if

$$2\pi k < n < 2\pi k + \pi, \quad k = 0, 1, 2, \ldots$$

This interval has length $\pi > 3$, which makes it large enough to contain at least one integer (up to three in fact) for every k. Hence, $\sin n$ is in $(0, 1)$ for infinitely many n, and it follows that the sequence has a cluster point in $(0, 1)$; more precisely, if we choose one n_k in the interval $(2\pi k, 2\pi k + \pi)$ for every k, then $\sin(n_k)$ (and thus, $\sin n$) has a convergent subsequence in $(0, 1)$. We continue this discussion in the exercises.

4.5 The set of real numbers is uncountable

In the preceding sections of this chapter, we elevated each rational number q to an equivalence class $[q]$ of the constant sequence $q, q, q, \ldots$ We added the equivalence classes $[q_n]$ of all other Cauchy sequences (not eventually constant) that correspond to the irrational numbers and showed that the enlarged set forms a totally ordered field that contains the (equivalence classes of) rational numbers. We showed that the new, larger set $\mathbb{R}$ of real numbers is complete; that is, it had no gaps or holes. Further, we discovered that the set of all rational numbers $\mathbb{Q}$ is dense in the set $\mathbb{R}$.

Now we answer the question: *how big is the set of all real numbers?*

Since the set of rational numbers is countable, the precise question that we answer is the following: *is the set of real numbers countable or uncountable?*

Cantor's diagonal argument

One of the most famous proofs of the uncountability of real numbers was given by Cantor and has come to be known as the "diagonal argument." This is a simple argument that uses the decimal expansion of real numbers that we prove in Chapter 5 after some exposure to infinite series. At that time, we also see another proof of the uncountability of real numbers that achieves more than the diagonal argument since it actually gives the cardinal number of $\mathbb{R}$ as $2^{\aleph_0}$, not just uncountable.

Theorem 115 *(Uncountability, the diagonal argument) The open interval (0,1) and thus, the set $\mathbb{R}$ of all real numbers, is uncountable.*

Proof We limit our attention to all real numbers that lie between 0 and 1, namely, the open interval $(0, 1)$. As a subset of $\mathbb{R}$, if $(0, 1)$ is uncountable, then so is $\mathbb{R}$. The diagonal argument is a proof by contradiction: we assume that $(0, 1)$ is countable and show that this assumption leads to a contradiction.

Every real number r in $(0, 1)$ has a decimal expansion $0.d_1 d_2 d_3 \ldots$ where each digit d_i is an integer between 0 and 9 (see Chapter 5). If $(0, 1)$ is countable, then all of its elements can be listed in a one-to-one correspondence with positive integers:

$$
\begin{array}{ll}
1 & 0.d_{11}d_{12}d_{13}\ldots d_{1n}\ldots \\
2 & 0.d_{21}d_{22}d_{23}\ldots d_{2n}\ldots \\
3 & 0.d_{31}d_{32}d_{33}\ldots d_{3n}\ldots \\
\vdots & \qquad \vdots \\
n & 0.d_{n1}d_{n2}d_{n3}\ldots d_{nn}\ldots \\
\vdots & \qquad \vdots
\end{array}
$$

Consider the real number $r = 0.d_1 d_2 d_3 \ldots$ with d_n defined as follows using the diagonal entries in the above list:

$$d_n = \begin{cases} 1 & \text{if } d_{nn} = 0 \text{ or } 9 \\ 0 & \text{if } d_{nn} \neq 0 \text{ or } 9 \end{cases} \qquad n = 1, 2, 3, \ldots$$

The number r cannot be in the above list because it is different from every number in the list in at least one digit. Since the existence of the above list was implied by the assumption that $(0, 1)$ was countable, that assumption had to be false. ■

Although $(0, 1)$ and $\mathbb{R}$ are both uncountable, you may wonder whether $\mathbb{R}$ has a larger cardinal number because it is an infinite union of infinite sets:

$$\cdots \cup (-2, -1) \cup (-1, 0) \cup (0, 1) \cup (1, 2) \cup \cdots \cup \mathbb{Z} = \mathbb{R}$$

Nevertheless, both $(0, 1)$ and $\mathbb{R}$ have the same cardinal number. The equipollence is established by a bijection like

$$\frac{1}{\pi} \tan^{-1} x + \frac{1}{2}$$

that maps $\mathbb{R}$ one-to-one onto $(0, 1)$; see Figure 4.1.

A simple modification of the above inverse tangent function in the form

$$\frac{2}{\pi} \tan^{-1} x$$

shows that the set $(0, \infty)$ of all *positive* real numbers is also equipollent to the interval $(0, 1)$. In Exercise 137, we see that every open interval of real numbers is equipollent to $(0, 1)$ and thus also to $\mathbb{R}$.

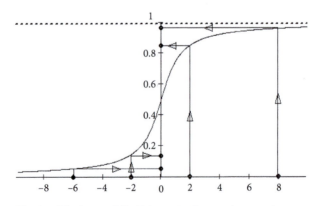

Fig. 4.1 The interval $(0, 1)$ is equipollent to $(-\infty, \infty)$

The uncountability of the set of real numbers gives us a low-hanging fruit to pluck. The set of all real numbers is the disjoint union of all irrational numbers and all rational numbers. We know that $\mathbb{Q}$ is countable, and by Theorem 115, $\mathbb{R}$ is uncountable. If the irrational numbers were countable, then their union with the rationals would be countable by Theorem 12 in Chapter 2. So we conclude the following:[15]

Corollary 116 *The set of all irrational numbers is uncountable.*

This result, together with countability of the rationals, point to a striking fact about real numbers: *the overwhelming majority of real numbers are irrational.* This is all the more significant because the rational numbers are the only ones that we can concretely deal with: all measurements that we take in nature by any type of instrument, any set of data that is ever saved in texts or in the "cloud" or any other type of computer memory, algorithms that our machines can ever process, all of these can only handle *rational* numbers. Further, rational numbers possess algebraic structure that is useful for calculations and solving equations, something that the rest of the real numbers without them lack.

A unique feature of the rational numbers that makes calculus (and the continuum) manageable is their *density* in $\mathbb{R}$. As our construction above shows, each irrational number can be approximated to an arbitrary degree of accuracy by a sequence of rational numbers. We do not need to know all the digits in $\sqrt{2}$, π and the like in order to calculate with them or to record the results involving them, because we can always find rational numbers that are indistinguishable from them for all practical purposes. Rational numbers may be a tiny minority, but fortunately, they are pervasive!

Irrational numbers are unavoidable though. While the rational numbers give algebraic structure to the real numbers and, through density, provide approximation power, the irrational numbers are required to fill the gaps that the rationals leave on the number line. Being limits of sequences of rational numbers, the irrationals are the fillers, the glue that bind the rationals to make a continuous (or complete) whole out of the real numbers. Without the irrational numbers, we can't have limits; and without limits, there can be no calculus or analysis!

Algebraic and transcendental numbers

We close this chapter by discussing a classification of real numbers that enlarges the set of rational numbers by appending some irrational numbers to it. This brings algebraic structure to some irrational numbers, although the newly structured irrationals remain a small minority among all the irrational numbers.

---

[15] A technical term for this type of proof by contradiction is a *cardinality (or countability) argument.*

You may remember *polynomials* from algebra; they are built from numbers and positive-integer powers of a variable x:

$$P(x) = c_0 + c_1 x + c_2 x^2 + \cdots + c_n x^n$$

We say that $P(x)$ is a *polynomial of degree n*, assuming that the number c_n (called the leading coefficient) is not zero. Examples are $2 + 3x$ (with degree 1), $x^2 - x - 1$ (degree 2), x^5 (degree 5), and $7/4$ (degree 0, like all numerical constants).

A *root* of $P(x)$ is a number r that makes $P(x)$ zero if we set $x = r$. For instance, if we set $2 + 3x = 0$ and solve for x, we find the only root $x = -2/3$ of the polynomial $2x + 3$. Though less obvious, you may recall that the roots of $x^2 - x - 1$ are found using the "quadratic formula":

$$x = \frac{1 \pm \sqrt{5}}{2} = \frac{1}{2} \pm \frac{\sqrt{5}}{2} \tag{4.7}$$

The *Fundamental Theorem of Algebra*[16] states that a *polynomial of degree n has at most n distinct roots in the field of complex numbers*. Some, or all, of the roots may be repeated (or multiple) roots; for example, $x^5 = 0$ has a single solution 0, which is also a 5-times repeated root (a root of "multiplicity" 5). Complex (non-real) roots routinely appear in "conjugate pairs"; for example, the quadratic formula gives the roots of $x^2 - x + 1$ as the two numbers

$$\frac{1}{2} \pm \frac{\sqrt{3}}{2} i, \quad i = \sqrt{-1} \tag{4.8}$$

An *algebraic number* is a root of a polynomial with *rational coefficients* (the numbers c_0, c_1, etc. are rationals). For example, the numbers in (4.7) are algebraic, or more precisely, *real* algebraic to distinguish them from the *complex* algebraic numbers like those in (4.8).

The set of real algebraic numbers is considered to be an extension of the set of rational numbers because each rational number m/n is a root of $nx - m$ or $x - m/n$ that are polynomials with rational coefficients; therefore, it is an algebraic number. The set of algebraic numbers contains a lot of irrational numbers, but still the following is true:

The set of all algebraic numbers is countable.

---

[16] This theorem shows that the set of complex numbers has an *algebraic* property that the real numbers lack: all polynomials factor completely in the set of complex numbers. There is a simple proof of this important result using complex analysis. The algebraic proof is more elaborate, but it applies in more general contexts than real or complex numbers. Both of these proofs require some theory beyond our focus here.

This is not surprising; the rationals are countable, so finite sets of them (in the form of polynomial coefficients) are still countable as a countable union of countable sets by Corollary 14 in Chapter 2. Further, each polynomial comes with a finite number of roots, so we are still able to use Corollary 14 to complete the proof. Although not all algebraic numbers are real numbers, it is clear that the set of all real algebraic numbers must be countable as a subset of all algebraic numbers.

This discussion shows that most real numbers are actually *not algebraic*. Real numbers that are not algebraic are called *transcendental*.

All transcendental numbers are irrational, and the set of all transcendental numbers is uncountable.

Although the vast majority of irrational numbers are transcendental, conjuring up transcendental numbers or verifying that any given irrational number is transcendental is typically hard. For example, is π algebraic or transcendental?

It is trivially a root of the polynomial $x - \pi$, so is the coefficient $c_0 = \pi$ rational? It is known that π is not rational,[17] so π is not the root of a polynomial with rational coefficients of degree 1. But are there polynomials of greater degree with rational coefficients that have π as a root? This is a more difficult question to answer.

We didn't know for sure if any known number was transcendental until 1844 when Liouville came up with transcendental numbers *on demand*; we discuss Liouville's numbers briefly in Chapter 5, but familiar numbers like π were not among Liouville's numbers.

That π was transcendental was finally proved in 1882 by the German mathematician Ferdinand Lindemann. The most famous transcendental number (in calculus) that we often come across is $e \simeq 2.71828$, named in honor of Leonhard Euler (1707–1783). It was proved to be transcendental in 1873 by the French mathematician Charles Hermite (1822–1901) of the "Hermitian operators" fame; these operators are important in quantum theory and elsewhere in science and mathematics.

It seems that "transcendental" is an apt name for these abundant, yet highly illusive numbers. It is even hard to tell whether simple algebraic combinations of known transcendental numbers are again transcendental. For example, although we know that each of e and π is transcendental, we don't know yet if $e + \pi$ or $e\pi$ are also transcendental.

---

[17] The Swiss polymath J. H. Lambert seems to have been the first to prove the irrationality of π in 1768. Different proofs have since appeared.

It is usually easier to prove that a number is *not* transcendental by showing that it is algebraic. For instance, if $r = \sqrt{3 + \sqrt{2}}$, then $r^2 = 3 + \sqrt{2}$; so

$$(r^2 - 3)^2 = 2$$
$$r^4 - 6r^2 + 9 = 2$$
$$r^4 - 6r^2 + 7 = 0$$

We see that r is a root of the polynomial $x^4 - 6x^2 + 7$. Since this has rational coefficients, r is algebraic.

4.6 Exercises

Exercise 117 *Prove the rest of Theorem 98 (assume that addition and multiplication have their usual properties in the set of integers $\mathbb{Z}$).*

Exercise 118 *Prove each of the following statements; assume that all letters p and q indicate rational numbers:*
  (a) *If $q < 0$, then $-q > 0$.*
  (b) *If $q \neq 0$, $p > 0$, and $p \leq q$, then $1/q \leq 1/p$. What happens if $p < 0$?*

Exercise 119 *Recalling the definition of a total ordering in Chapter 2, verify that the ordering of $\mathbb{Q}$ by the relation $<$ is total.*

Exercise 120 *Consider the set $\mathbb{Q}_C$ of complex rationals, i.e., the subset of complex numbers consisting of numbers of type $p + qi$ where $i = \sqrt{-1}$ and $p, q \in \mathbb{Q}$. Addition and multiplication are defined as*

$$(p + qi) + (p' + q'i) = p + p' + (q + q')i$$
$$(p + qi)(p' + q'i) = pp' - qq' + (pq' + p'q)i$$

where in the case of multiplication, we multiply all terms in the parentheses and then combine terms.
  (a) *Prove that $\mathbb{Q}_C$ is a field with the reciprocal of $z = p + qi \neq 0$ defined as follows:*

$$\frac{1}{z} = \frac{\bar{z}}{z\bar{z}} = \frac{p - qi}{p^2 + q^2}$$

Here $\bar{z} = p - qi$ is the complex conjugate of z. You need to prove that the version of Theorem 98 for $\mathbb{Q}_C$ is true; you may use that theorem in your proof for the $\mathbb{Q}_C$ version rather than reverting back to the integers.

(b) Define a relation $\leq$ on $\mathbb{Q}_C$ as follows: if $z = p + qi$ and $z' = p' + q'i$ then

$$z \leq z' \quad \text{if} \quad p \leq p' \text{ and } q \leq q'$$

Prove that $\leq$ is an ordering on $\mathbb{Q}_C$ that is not total. Give some examples of $\leq$ non-comparable numbers.

(c) Is $\mathbb{Q}_C$ complete? Explain.

Exercise 121 Consider the set $C[-1, 1]$ of all continuous functions $[-1, 1] \rightarrow \mathbb{R}$.[18]

(a) Define addition and multiplication in the usual way; if $f(x)$ and $g(x)$ are two elements of $C[-1, 1]$, then the sum $f + g$ and fg are

$$(f + g)(x) = f(x) + g(x), \qquad (fg)(x) = f(x)g(x)$$

for every x in $[-1, 1]$. Verify that $C[-1, 1]$ satisfies (i)–(vii) in Theorem 98. Note that the constant functions $f(x) = 0$ and $f(x) = 1$ are continuous.

(b) Show that Property (viii) in Theorem 98 is not satisfied by showing that there are many functions in $C[-1, 1]$ whose reciprocals are not continuous and thus not in $C[-1, 1]$ (consider functions whose graphs cross the x-axis).

(c) Give examples of functions $f(x), g(x)$ with the property

$$f(x)g(x) = 0 \quad \text{for all } x \in [-1, 1], \text{ but neither } f(x) \text{ nor } g(x) \text{ is the constant}$$
zero function.

(d) Give examples of functions $f(x), g(x), h(x)$ with the property

$$f(x)h(x) = g(x)h(x) \quad \text{for all } x \in [-1, 1], \ h(x) \text{ is not the constant zero}$$
function, and $f(x) \neq g(x)$.

Exercise 122 Use the definition of Cauchy sequence to prove that the following sequence is Cauchy:

$$s_n = \frac{2n + 9}{3n}$$

Exercise 123 (a) A sequence q_n of rational numbers is "eventually constant" if there is a rational number q and an index K such that $q_n = q$ for all $n \geq K$. Is this a Cauchy sequence? Explain using the definition of Cauchy sequence.

(b) Consider a periodic sequence q_n of rational numbers with period $k \geq 2$. Can this type of sequence be Cauchy? Explain.

---

[18] Think of continuous in the intuitive sense, with a graph that has no holes. We discuss continuity in a formal way later in this book.

Exercise 124 *Consider the sequence of rational numbers*

$$q_n = \sum_{k=1}^{n} \frac{1}{k} = 1 + \frac{1}{2} + \cdots + \frac{1}{n}$$

Show that $|q_{n+1} - q_n|$ converges to 0. However, as we discover in Chapter 5, q_n is not Cauchy; far from it; as the nth partial sum of the harmonic series, q_n diverges to infinity.

Exercise 125 *Consider the sequence in (4.3); let's call this sequence s. Which of the following sequences are equivalent to s and which are not? Explain, remembering that all equivalent sequences must accumulate around the same number:*

(a) $1, 1, 1, \ldots$ (b) $1 + \frac{1}{n}$ (c) $2 - \frac{1}{n}$ (d) $\frac{1 + (-1)^n}{2}$

Exercise 126 *Let q_n be any sequence of rational numbers that converges to $\sqrt{2}$, i.e., $[q_n] = \sqrt{2}$.*
(a) Find a constant sequence of rational numbers that is equivalent to the sequence of squares q_n^2.
(b) Explain why the sequence q_n^3 is equivalent to $2q_n$.

Exercise 127 *(a) Let $q_1, q_2, q_3, \ldots$ be a Cauchy sequence of rational numbers that is **not** in [0]. Assuming that $q_n \neq 0$ for all indices $n \geq N$, prove that the sequence of reciprocals $1/q_{N+1}, 1/q_{N+2}, 1/q_{N+3}, \ldots$ is Cauchy.*
(b) Suppose that $q_1, q_2, q_3, \ldots$ is a Cauchy sequence in [0] with $q_n \neq 0$ for all indices n. The sequence of reciprocals $1/q_1, 1/q_2, 1/q_3, \ldots$ is well defined, but is it Cauchy?

Exercise 128 *Let x and r be distinct real numbers. Explain why there are infinitely many rational numbers between x and r.*

Exercise 129 *Suppose that x, y are irrational numbers, and q is rational ($q \neq 0$). Explain why $x + q$ and xq are always irrational, but $x + y$ and xy may be irrational or rational. In particular, this shows that irrational numbers do not form a field under ordinary addition and multiplication.*

Exercise 130 *Suppose that x and y are distinct positive real numbers. Why must there be an integer n such that $ny > x$? Note that x/y is a real number.*

Exercise 131 *Prove Corollary 110(b): consider using Corollary 110(a) on the sequence $x_n = -r_n$.*

Exercise 132 *(a) Prove that the sequence of real numbers*

$$r_n = \sum_{k=1}^{n} \frac{\sin^2 k}{k!} = \frac{\sin^2 1}{1!} + \frac{\sin^2 2}{2!} + \frac{\sin^2 3}{3!} + \cdots + \frac{\sin^2 n}{n!}$$

converges to a real number r and that r is strictly less than e − 1.
  (b) Go a little further: show that

$$\frac{3}{4} < r < e - \frac{5}{4}$$

  Note that $r > \sin^2 1$ *and also that 1 radian is just under* $\pi/3$ *radians so that* $\sin^2 1$ *is slightly less than* $\sin^2(\pi/3)$; *both the aforementioned lower bound and the upper bound for r can be improved.*

Exercise 133 *Prove Corollary 113 using the idea in Exercise 131.*

Exercise 134 *(a) Consider the nested sequence of open intervals*

$$(0,1) \supset \left(0, \frac{1}{2}\right) \supset \left(0, \frac{1}{3}\right) \supset \cdots \supset \left(0, \frac{1}{n}\right) \supset \cdots$$

Show that the lengths of these intervals vanish as $n \to \infty$, *but*

$$\bigcap_{n=1}^{\infty} \left(0, \frac{1}{n}\right) = \varnothing$$

  (b) Consider the nested sequence of closed intervals

$$[1, \infty) \supset [2, \infty) \supset [3, \infty) \supset \cdots \supset [n, \infty) \supset \cdots$$

Notice that the lengths of these intervals do not vanish as $n \to \infty$. *What is the intersection of all of these intervals?*

Exercise 135 *Find the limit supremum, the limit infimum, and all other cluster points of the following sequence:*

$$s_n = \cos \frac{2\pi}{n} + \cos \frac{n\pi}{2}$$

Exercise 136 *Prove that the sequence* $\sin n$ *has cluster points in the interval* $(-1/2, 1/2)$.

Exercise 137 (a) Consider an arbitrary open interval of real number (a, b) from a to b where $a < b$. Verify that the function $f : (a, b) \to (0, 1)$ defined as

$$f(x) = \frac{x - a}{b - a}$$

is a bijection, thus confirming that (a, b) is equipollent to $(0, 1)$. Note that $f(x)$ here is a linear function.

(b) Determine the specific form of $f(x)$ for $(a, b) = (-1, 3)$ and $(a, b) = (0, 1/4)$. Consider sketching the simple graph of $f(x)$ to get a better feel for how this function works.

Exercise 138 Show that the following numbers are algebraic by finding the associated polynomials with rational coefficients:

$$1 + \sqrt{3}, \qquad \sqrt[3]{2 - \sqrt{3}}, \qquad \frac{\sqrt{2}}{1 + \sqrt{3}}, \qquad \cos \frac{\pi}{12}$$

NOTE: $\pi/12 = \pi/3 - \pi/4$ is the difference of two special angles. Use trigonometric identities to find a polynomial expression.

5

Infinite Series of Constants

Our aim in this chapter is to explain what is meant by "adding infinitely many numbers" or an infinite series of numbers in terms of limits. In particular, we will see that infinite series are rarely computed exactly, but when they are, it is never done by actually adding infinitely many numbers. Rather, we add finitely many numbers to generate a partial sum of the series, then make a sequence of these partial sums, and finally take the limit of that sequence. The key words here are *limit* and (finite) *partial sums*.

We also use the knowledge that we gain from the study of infinite series to explore the role of infinity in defining transcendental real numbers such as π and e as sums of infinite series of rational numbers.

5.1 On adding infinitely many numbers

Consider the infinite sum

$$\frac{1}{2} + \left(\frac{1}{2}\right)^2 + \left(\frac{1}{2}\right)^3 + \left(\frac{1}{2}\right)^4 + \cdots \tag{5.1}$$

Should this sum have infinite value because we are adding an infinite number of positive terms? We look at this question in the following way: imagine dividing the interval $[0,1]$ infinitely often and writing it as a *disjoint* union:

$$[0, 1] = [0, 1) \cup \{1\}$$

$$= \{1\} \cup \left[0, \frac{1}{2}\right) \cup \left[\frac{1}{2}, \frac{3}{4}\right) \cup \left[\frac{3}{4}, \frac{7}{8}\right) \cup \left[\frac{7}{8}, \frac{15}{16}\right) \cup \cdots$$

$$= \{1\} \cup \left[0, \frac{1}{2}\right) \cup \left[\frac{1}{2}, \frac{1}{2} + \frac{1}{2^2}\right) \cup \left[\frac{3}{2^2}, \frac{3}{2^2} + \frac{1}{2^3}\right) \cup \cdots$$

Notice that the total lengths of all the intervals on the right-hand side is the infinite sum in (5.1) with the length of $\{1\}$ being 0, while the length of the whole interval $[0,1]$ is 1. From this observation, we may reasonably conclude that the total value of the sum in (5.1) is just 1.

Real Analysis and Infinity. Hassan Sedaghat, Oxford University Press.
© H. Sedaghat (2022). DOI: 10.1093/oso/9780192895622.003.0005

The expression in (5.1) is a special case of the following more general expression called a *geometric series*

$$r + r^2 + r^3 + r^4 + \cdots \tag{5.2}$$

where we take *any* nonzero real number r and multiply it repeatedly by itself and add the results as we go. In the special case of (5.1), $r = 1/2$.

Assume that the expression in (5.2) does add up to some real number S; it is then possible to find S by using simple algebra: we multiply S by r to get

$$Sr = (r + r^2 + r^3 + r^4 + \cdots)r = r^2 + r^3 + r^4 + \cdots$$

If we subtract Sr from S, then

$$S - Sr = r + r^2 + r^3 + r^4 + \cdots$$
$$- r^2 - r^3 - r^4 - \cdots$$

On the right-hand side, everything except the first r cancels out and we are left with

$$S - Sr = r$$

Now, we first factor out the common S and then divide the result by $1 - r$ to get:

$$S = \frac{r}{1 - r}$$

This gives the formula

$$r + r^2 + r^3 + r^4 + \cdots = \frac{r}{1 - r} \tag{5.3}$$

In particular, if $r = 1/2$, then the right-hand side of (5.3) works out to 1 as we might expect.

If you examine the above calculation carefully, then you will notice that in addition to assuming that (5.3) adds to a real number, to obtain the equality in (5.3) we divided by the number $1 - r$, which is possible if $r \neq 1$. What happens if $r = 1$? If we set $r = 1$ in (5.3) then we get

$$1 + 1^2 + 1^3 + 1^4 + \cdots = \frac{1}{0} \tag{5.4}$$

Since $1 = 1^2 = 1^3$ etc., the left-hand side is 1 added to itself infinitely often; so this side of (5.4) is infinitely large, which is certainly not a real number S. The right-hand side is also problematic since division by 0 is algebraically meaningless.

This observation suggests that the sum in (5.2) may not result in a real number S for all possible values of r. If we set $r = 2$ in (5.3), then we obtain

$$2 + 2^2 + 2^3 + 2^4 + \cdots = \frac{2}{1-2}$$

$$2 + 4 + 8 + 16 + \cdots = -2 \qquad (5.5)$$

This may be sufficiently absurd to convince us that (5.3) is not valid when $r = 2$, or any other number larger than 1, which gives a negative number on the right-hand side.

If we insert a negative value for r, then the odd powers of r generate negative values so that the terms of the sum alternate in sign. For instance, if $r = -1$ in (5.3), then

$$(-1) + 1 + (-1) + 1 + \cdots = -\frac{1}{2} \qquad (5.6)$$

Notice that there are no infinite magnitudes here, and the answer $-1/2$ does not seem as implausible as those in (5.4) or (5.5).

But $-1/2$ isn't an intuitively clear result; the left-hand side of (5.6) looks more like it should be zero with all the possible cancellations:

$$[(-1) + 1] + [(-1) + 1] + \cdots = 0 + 0 + \cdots = 0 \qquad (5.7)$$

But this is not the end of our problems; here is yet another possible answer: suppose that we set the first -1 aside in (5.6) but pair up the rest of the numbers to get

$$(-1) + [1 + (-1)] + [1 + (-1)] + \cdots = -1 + 0 + 0 + \cdots = -1$$

This answer is no less valid than 0 because it was derived in the same way, only with a slightly different grouping. So far we have come across three possible answers: $-1/2, 0,$ and -1.

Not bad enough? Well, there are still more possibilities! For example, starting after the first number -1, let's add an even number of $+1$ terms followed by adding an odd number of -1 terms as in

$$\underbrace{-1}_{1} + \underbrace{(1 + 1)}_{2} + \underbrace{(-1 - 1 - 1)}_{3} + \underbrace{(1 + 1 + 1 + 1)}_{4} + \cdots = -1 + 2 - 3 + 4 - 5 + \cdots$$

Since we have an *infinite supply* of $+1$ and -1 terms, *if there are no restrictions* on how we add them up, then the series on the right-hand side is a possible outcome. The last series may also be added in a variety of ways. For instance, if we add the numbers one pair at a time, then

$$-1 + 2 - 3 + 4 - 5 + \cdots = (-1 + 2) + (-3 + 4) + (-5 + 6) + \cdots$$

$$= 1 + 1 + 1 + 1 + \cdots$$

Adding 1 to itself an infinite number of times give infinity!

With $r = -1$, we don't divide by 0 on the right-hand side of (5.3); but the preceding discussion makes it clear that the problems are worse in this case than when $r = 1$ because we seem to be getting a variety of possible answers, and we cannot decide which one is right, or even better.

So after all this, the question remains: *for which values of r is the equality in (5.3) valid?*

To answer this type of question, it is necessary to clarify what we mean by adding infinitely many numbers; that is just what we do in the next section of this chapter.

Refuting a conjecture

Before getting into the theory of infinite series, we consider one more issue that will be important to remember later. You may have noticed above that as we keep adding fractions in (5.1), we are adding smaller ones as we proceed: the first fraction is $1/2$, the second is $1/4$, which is half of the first, the third is $1/8$ or half of the second, and so on. As we keep adding new fractions, we accumulate less at each step as we go on.

A conjecture at this point might be that if the terms of the sum decrease to 0 as we add them, then the sum will be a well-defined, finite number.

This conjecture would make it easy to identify infinite sums with finite values; but alas, it is not true!

To illustrate this essential point, let's consider the following:

$$1 + \frac{1}{2} + \frac{1}{3} + \frac{1}{4} + \frac{1}{5} + \frac{1}{6} + \cdots \tag{5.8}$$

This is an important enough series that like (5.3) it has been given a name: the *harmonic series*. You can see that in (5.8) we are also adding numbers that get smaller at each step and approach 0 as $n \to \infty$. But this series does not add up to a real number, a fact that has been known since the 1300s.[1]

Here is the proof: arrange the numbers in (5.8) in blocks of size a power of 2 as follows:

$1 + \frac{1}{2} +$	$> \frac{1}{2}$	(2^0 terms)
$\frac{1}{3} + \frac{1}{4}$	$> \frac{1}{4} + \frac{1}{4} = \frac{1}{2}$	(2^1 terms)
$\frac{1}{5} + \frac{1}{6} + \frac{1}{7} + \frac{1}{8} +$	$> \frac{1}{8} + \frac{1}{8} + \frac{1}{8} + \frac{1}{8} = \frac{1}{2}$	(2^2 terms)
$\vdots$	$\vdots$	

---

[1] This argument predates calculus, dating back to around 1350 and the French bishop Nicholas (or Nicole) Oresme.

The table has an infinite number of rows, and the sum of the numbers in each row is greater than 1/2, so the sum in (5.8) is greater than

$$\frac{1}{2} + \frac{1}{2} + \frac{1}{2} + \cdots$$

which is 1/2 added to itself repeatedly without end. Excluding the first 1/2, each successive 1/2 is less than the sum of 2^{n-1} terms of the harmonic series; for instance, the third 1/2 is less than the sum of $2^2 = 4$ terms of the harmonic series as shown in the third row of the table. It follows that the sum in (5.8) can exceed any real number if enough fractions are added in; the harmonic series cannot add up to a real number.

Therefore, our conjecture above is false; just because the numbers that we add shrink in size all the way down to 0 does not mean that the total is a real number.

5.2 Infinite series as limits of sequences of finite sums

The issues that we discussed in the last section are among problems that led mathematicians in the nineteenth century to clarify the concept of infinite summation. They used a concept that we discussed earlier: sequences. In this section, we study this idea and discover that it resolves the aforementioned issues.

Suppose that we have an infinite sequence of real numbers

$$a_1, a_2, a_3, \ldots, a_n, \ldots \tag{5.9}$$

and we want to add all of these numbers to see if we get a real number. If we do it one step at a time, then we create *another* sequence of real numbers as follows:

$$s_1 = a_1$$
$$s_2 = a_1 + a_2$$
$$s_3 = a_1 + a_2 + a_3$$
$$\vdots$$
$$s_n = a_1 + a_2 + \cdots + a_n$$

Each of the finite sums above is a part of the total, or the sum of all the infinite number of terms. As the value of the index n increases, the finite sums will get closer to the value of the infinite series, if such a value exists. We formalize this promising idea.

Partial sums and convergence of series. The finite sum in each of the above rows is called a *partial sum* of the terms in (5.9). The *n*th term s_n is called the *n*th partial sum; the sequence

$$s_1, s_2, s_3, \ldots, s_n, \ldots$$

is *the sequence of partial sums* of the original sequence in (5.9). If as $n \to \infty$ the sequence s_n converges to a real number s, then we say that the *infinite series*

$$a_1 + a_2 + \cdots + a_n + \cdots$$

converges and call the limit s *the sum of the infinite series:*

$$a_1 + a_2 + \cdots + a_n + \cdots = s = \lim_{n \to \infty} s_n$$

The numbers $a_1, a_2, \ldots$ are called the *terms of the series*, and the generic term a_n that typically gives the formula for the sequence is called the *n*th *term*.

The sigma notation $\sum$ is the standard shorthand for summation, finite or infinite, so we may write the finite sum as

$$a_1 + a_2 + \cdots + a_n = \sum_{k=1}^{n} a_k$$

The infinite series is written as

$$a_1 + a_2 + \cdots + a_n + \cdots = \lim_{n \to \infty} \sum_{k=1}^{n} a_k = \sum_{k=1}^{\infty} a_k$$

Because sequences of partial sums may converge or diverge as discussed in Chapter 3, we use the same terminology for series.

Convergence *The infinite series* $a_1 + a_2 + \cdots + a_n + \cdots$ **converges** *to a real number s if the infinite sequence of partial sums* $s_1, s_2, \ldots, s_n, \ldots$ *converges to s.*
  Divergence *The infinite series* $a_1 + a_2 + \cdots + a_n + \cdots$ **diverges** *if the partial sums sequence* $s_1, s_2, \ldots, s_n, \ldots$ *does not converge to a real number.*

Let's use the above definitions to re-examine the series on the left-hand side of (5.6). The partial sums sequence of this series is

$$s_1 = -1,$$

$$s_2 = -1 + 1 = 0$$

$$s_3 = -1 + 1 - 1 = -1$$

$$s_4 = -1 + 1 - 1 + 1 = 0$$

and so on. The numbers $s_1, s_2, s_3, s_4, \ldots$ oscillate with period 2 (repeating every other term). This pattern of numbers does not approach any real number s; so the sequence, and thus, the series, diverges.

There is also the type of divergence that involves infinity directly.

Infinite values *The infinite series $a_1 + a_2 + \cdots + a_n + \cdots$ **diverges to ∞** (or to $-\infty$) or has infinite value if its sequence of partial sums $s_1, s_2, \ldots, s_n, \ldots$ diverges to ∞ (respectively, $-\infty$).*

Recall that a sequence $r_1, r_2, \ldots, r_n, \ldots$ of real numbers diverges to ∞ if for every positive real number M, the terms of the sequence eventually exceed M; that is, there is an index N such that $r_n > M$ for all $n \geq N$. Also $r_1, r_2, \ldots, r_n, \ldots$ approaches $-\infty$ if $-r_1, -r_2, \ldots, -r_n, \ldots$ approaches ∞.

We often write $\sum_{k=1}^{\infty} a_k = \infty$ or $\sum_{k=1}^{\infty} a_k = -\infty$ as appropriate.[2] For example,

$$1 + 1 + 1 + \cdots = \infty$$

because the sequence $s_n = n$ diverges to infinity. The proof goes like this: if M is any positive real number, then (by the Archimedean property!) there is an integer $N > M$. So $n > M$ for every $n \geq N$, that is, $s_n > M$ for all $n > N$.

Defining infinite sums using *the sequence* of partial sums avoids the difficulties mentioned earlier without compromising the usefulness of the concept of infinite series or their potential applicability in scientific or engineering contexts. *We are not free to add the terms of a series any way we like*; that freedom exists only for finite sums.

There is *just one way of adding infinitely many numbers*: we start from the first term and add the other terms successively in the order given by the sequence in (5.9).

Using the limit theorems for sequences of partial sums, the following useful result is obtained.

---

[2] Keep in mind that the symbols $\pm\infty$ are abbreviations and not actual numerical magnitudes.

Theorem 139 *(a) If the series $\sum_{k=1}^{\infty} a_k$ and $\sum_{k=1}^{\infty} b_k$ both converge, then $\sum_{k=1}^{\infty}(a_k + b_k)$ converges and is equal to $\sum_{k=1}^{\infty} a_k + \sum_{k=1}^{\infty} b_k$.*

(b) If $\sum_{k=1}^{\infty} a_k$ converges, but $\sum_{k=1}^{\infty} b_k$ diverges, then $\sum_{k=1}^{\infty}(a_k + b_k)$ diverges too.

(c) If $\sum_{k=1}^{\infty} a_k$ converges, then for every constant $c \in \mathbb{R}$ the series $\sum_{k=1}^{\infty} ca_k$ converges and is equal to $c \sum_{k=1}^{\infty} a_k$.

(d) If $\sum_{k=1}^{\infty} a_k$ diverges and $c \neq 0$, then the series $\sum_{k=1}^{\infty} ca_k$ diverges too.

Proof I prove (b) and leave the proofs of the remaining statements as an exercise.

(b) For each n, the nth partial sum of the series $\sum_{k=1}^{\infty}(a_k + b_k)$ is

$$\sum_{k=1}^{n}(a_k + b_k) = a_1 + b_1 + a_2 + b_2 + \cdots + a_n + b_n = s_n + t_n$$

where s_n and t_n are partial sums of $\sum_{k=1}^{\infty} a_k$ and $\sum_{k=1}^{\infty} b_k$, respectively. By assumption, $\lim_{n \to \infty} s_n$ exists, and we may call it s. Now if the series $\sum_{k=1}^{\infty}(a_k + b_k)$ converges, then it is equal to $L = \lim_{n \to \infty}(s_n + t_n)$. This implies that the sequence t_n converges too since

$$t_n = \sum_{k=1}^{n}(a_k + b_k) - s_n$$

$$\lim_{n \to \infty} t_n = L - s$$

Therefore, $\sum_{k=1}^{\infty} b_k$ converges, which contradicts the hypothesis. It follows that $\sum_{k=1}^{\infty}(a_k + b_k)$ diverges. ∎

Theorem 139 can be used to prove that the series

$$\sum_{k=1}^{\infty} \frac{2^{k-1} - 3}{3^k}$$

converges. We note that its partial sums can be written as

$$\sum_{k=1}^{n} \frac{2^k 2^{-1}}{3^k} + \sum_{n=1}^{n} \frac{-3}{3^k} = \frac{1}{2} \sum_{n=1}^{n} \left(\frac{2}{3}\right)^k - 3 \sum_{n=1}^{n} \left(\frac{1}{3}\right)^k \tag{5.10}$$

By (5.3) with $r = 2/3$, we have

$$\sum_{k=1}^{\infty} \left(\frac{2}{3}\right)^k = \frac{2/3}{1 - 2/3} = 2, \quad \sum_{n=1}^{\infty} \left(\frac{1}{3}\right)^k = \frac{1/3}{1 - 1/3} = \frac{1}{2}$$

So by (5.10) and Theorem 139,

$$\sum_{k=1}^{\infty} \frac{2^{k-1}+3}{3^k} = \frac{1}{2}(2) - 3\left(\frac{1}{2}\right) = -\frac{1}{2}$$

How about the following series?

$$\sum_{k=1}^{\infty} \frac{3^{k-1}-2}{3^k} \tag{5.11}$$

The nth partial sum of this series is

$$\sum_{k=1}^{n} \frac{3^{k-1}-2}{3^k} = \sum_{k=1}^{n} \frac{1}{3}\frac{3^k}{3^k} + \sum_{k=1}^{n} \frac{-2}{3^k} = \sum_{k=1}^{n} \frac{1}{3} - 2\sum_{k=1}^{n}\left(\frac{1}{3}\right)^k$$

The last term on the right-hand side converges, but since $\sum_{k=1}^{n}(1/3) = n/3 \to$ ∞ as $n \to \infty$, Theorem 139(b) implies that the series in (5.11) diverges.

Later in this section, we discuss several tests for convergence and divergence that simplify calculations involving infinite series. Many of these results do not require using partial sums in calculations.

The telescoping series

To illustrate how the partial sum sequence is used directly to find the *actual* sum (not just an estimate) of an infinite series, consider the following:

$$\sum_{n=1}^{\infty} \frac{1}{n(n+1)} = \frac{1}{(1)(2)} + \frac{1}{(2)(3)} + \frac{1}{(3)(4)} + \frac{1}{(4)(5)} + \cdots \tag{5.12}$$

We interpret this series as the sum of the terms of the sequence of fractions

$$\frac{1}{(1)(2)}, \frac{1}{(2)(3)}, \frac{1}{(3)(4)}, \frac{1}{(4)(5)}, \cdots$$

The partial sums of this sequence are

$$s_1 = \frac{1}{(1)(2)}$$

$$s_2 = \frac{1}{(1)(2)} + \frac{1}{(2)(3)}$$

$$s_3 = \frac{1}{(1)(2)} + \frac{1}{(2)(3)} + \frac{1}{(3)(4)}$$

and so on. The nth partial sum is the sum of the first n numbers:

$$s_n = \frac{1}{(1)(2)} + \frac{1}{(2)(3)} + \frac{1}{(3)(4)} + \cdots + \frac{1}{(n)(n+1)} \tag{5.13}$$

It is unclear from the right-hand side above whether s_n converges to any real number. But notice that each of the fractions breaks down as

$$\frac{1}{(n)(n+1)} = \frac{1}{n} - \frac{1}{n+1}$$

If we use this split form in (5.13), then we discover a set of cancellations:

$$s_n = \left(\frac{1}{1} - \frac{1}{2}\right) + \left(\frac{1}{2} - \frac{1}{3}\right) + \left(\frac{1}{3} - \frac{1}{4}\right) + \cdots + \left(\frac{1}{n} - \frac{1}{n+1}\right) = 1 - \frac{1}{n+1} \tag{5.14}$$

The in-between terms cancel out, leaving the first and last terms only. Now it is evident that $s_n \to 1$ as $n \to \infty$ because $1/(n+1) \to 0$. By the definition of convergent series, we declare that

$$\sum_{n=1}^{\infty} \frac{1}{n(n+1)} = 1$$

You may check this result by adding the series up to some large number (that is, calculate a big partial sum) using a calculator or a computer.

Because of the cancellations that caused the expression (5.14) for s_n to *fold in*, the series in (5.13) is called a *telescoping series*.[3]

Finitely equivalent infinite series

If we rearrange *finitely* many numbers of the sequence in (5.9) and then sum the new sequence, we obtain the same answer for the series because a finite number of rearrangements in the sequence does not affect the partial sums beyond a certain index. For example, suppose that the rearrangements are confined to at most N numbers in (5.9). Then a_{N+1}, a_{N+2}, etc. are the same and not changed, so the partial sums s_{N+1}, s_{N+2}, etc. are the same for both the old sequence and the rearranged sequence.

Whatever one partial sums sequence converges to, the other must converge to the same number.

---

[3] The numbers or terms of this telescoping series are related to the so-called *triangular numbers*: 1, 3, 6, 10, 15, etc. An easy way to visualize the triangular numbers is to think of the way bowling pins are set up for a game: the front row contains 1 pin, the next row 2 pins, the third row has 3 pins, and so on that when set up, they form a triangle, hence the name. The nth triangular number is the sum of the first n rows: $n(n+1)/2$. The reciprocal of this is $2/n(n+1)$, which is twice a term in the telescoping series. From the formula for the telescoping series we infer that *the sum of the reciprocals of all triangular numbers is just 2.*

For example, suppose that we rearrange the first three terms of the telescoping series in (5.12) to get the series

$$\frac{1}{(3)(4)} + \frac{1}{(1)(2)} + \frac{1}{(2)(3)} + \frac{1}{(4)(5)} + \cdots$$

Then for this series, the partial sums s_1 and s_2 are different from the first and second partial sums of (5.12), but

$$s_3 = \frac{1}{(3)(4)} + \frac{1}{(1)(2)} + \frac{1}{(2)(3)} = \frac{1}{(1)(2)} + \frac{1}{(2)(3)} + \frac{1}{(3)(4)}$$

So s_3 is the same as the third partial sum of (5.12). In fact, we can see that s_n is the same as the nth partial sum of (5.12) for all $n \geq 3$.

Reasoning similarly, if we *change* finitely many numbers in (5.9), then the numerical value of the series may change but *not whether it converges or not*: if the old series converges (or not), then the new series does the same thing because the partial sums for the two sequences beyond some index N are different by a finite amount, namely, the difference between the old s_N and the new one. I summarize this useful fact for future reference:

Finite equivalence *If two infinite series are different in finitely many terms, and one series converges (or diverges), then the other series converges (respectively, diverges).*

Formally, finite equivalence does in fact define an equivalence relation on the set of all infinite series of real numbers. Define a relation $\sim$ as

$$\sum_{n=1}^{\infty} a_n \sim \sum_{n=1}^{\infty} b_n$$

whenever $a_n = b_n$ for all but finitely many indices n. Then $\sim$ is a reflexive relation because every series is different from itself in zero (which is finite) number of indices. That $\sim$ is symmetric is obvious; as for transitivity, if $\sum_{n=1}^{\infty} a_n \sim \sum_{n=1}^{\infty} b_n$, then there is a finite index N_1 such that $a_n = b_n$ for $n \geq N_1$; and if $\sum_{n=1}^{\infty} b_n \sim \sum_{n=1}^{\infty} c_n$, then there is an index N_2 such that $b_n = c_n$ for $n \geq N_2$. So if N is the larger of N_1 and N_2, then $a_n = c_n$ for $n \geq N$, which means that $\sum_{n=1}^{\infty} a_n \sim \sum_{n=1}^{\infty} c_n$ and shows that $\sim$ is transitive.

Two series are finitely equivalent if the relation $\sim$ holds between them.

Now finite equivalence says that *if two infinite series of real numbers are finitely equivalent then they both converge or they both diverge.*

Finally, it is worth remembering that we can change infinitely many terms of a series while leaving infinitely many terms intact. The resulting series is not finitely equivalent to the original series, and it may or may not converge or diverge. For example, if we drop all of the odd-indexed terms, then the resulting series

$$\frac{1}{2} + \frac{1}{4} + \frac{1}{6} + \frac{1}{8} + \frac{1}{10} + \cdots + \frac{1}{2n} + \cdots$$

still diverges (see Exercise 164). However, if we drop all terms except those with denominator a power of 2, then the resulting series

$$\frac{1}{2} + \frac{1}{4} + \frac{1}{8} + \frac{1}{16} + \frac{1}{32} + \cdots + \frac{1}{2^n} + \cdots$$

is the same as (5.1), which converges. In this case, we dropped enough terms of the harmonic series to get convergence. Another example is the telescoping series (5.12)

$$\frac{1}{2} + \frac{1}{6} + \frac{1}{12} + \frac{1}{20} + \frac{1}{30} + \cdots + \frac{1}{n(n+1)} + \cdots$$

that converges, as we proved earlier. It is not easy to find a *smallest set* of terms dropped from the harmonic series to produce a convergent series. For example, Euler posited that the sum of reciprocals of primes

$$\sum_{p \text{ a prime}} \frac{1}{p} = \frac{1}{2} + \frac{1}{3} + \frac{1}{5} + \frac{1}{7} + \frac{1}{11} + \cdots$$

diverges (this fact was rigorously proved later in different ways); this series is obtained by dropping 1 and the reciprocals of all composite integers (i.e., most of the integers) from the harmonic series. Removal of this rather large set is not enough to make the remaining fractions add up to a finite number.

5.3 The geometric series

The series in (5.3) is a special case of the important *geometric series*:

$$a + ar + ar^2 + \cdots \qquad a \neq 0 \qquad\qquad (5.15)$$

Two numbers completely characterize this series. They are the *first term a* and the number r, called the *common ratio* because it is the fixed ratio of each term over the one right before it: $ar/a = r$ and $ar^2/ar = r$ and so on. The series in (5.3) has $a = r$, and that in (5.1) has $a = r = 1/2$.

The geometric series is important because it occurs frequently in calculus; it is "nice" because there is a formula for its exact sum, a rarity for infinite series. The derivation of the formula is essentially the same as that for (5.3), but this time we don't need to *assume* that the series converges. The nth partial sum of the series (5.15) is

$$s_n = a + ar + ar^2 + \cdots + ar^{n-1} \tag{5.16}$$

It ends with power $n - 1$ because s_n is the sum of n terms, starting with a constant a. Now we apply the procedure used for (5.3) to (5.16); the important difference is that this time we are adding a finite number of terms, so we need not worry about infinity. Multiply by r and subtract to obtain

$$s_n - rs_n = a + ar + ar^2 + \cdots + ar^{n-1}$$
$$- ar - ar^2 - ar^3 - \cdots - ar^n$$

After cancelling all the like terms in the middle, we are left with

$$s_n - rs_n = a - ar^n$$
$$(1 - r)s_n = a(1 - r^n)$$

At this point, if $r \neq 1$, we divide by $1 - r$ to obtain a useful formula for s_n in the following.

Lemma 140 *(Formula for partial sums of the geometric series) The sum in (5.16) is known as a "geometric progression" and its value is*

$$s_n = \frac{a(1 - r^n)}{1 - r} \quad \text{if } r \neq 1 \tag{5.17}$$

If $r = 1$ in (5.16), then (trivially by substitution) $s_n = na$.

The sum of a geometric series

Now that we have a formula for s_n, it is easy to see what happens as $n \to \infty$. The situation here is similar to that of the telescoping series that we met earlier, although the formula for s_n now is a little more complex. The only part of (5.17) that changes with n is r^n. It is necessary to determine what happens to r^n as $n \to \infty$. The answer depends on the value of r, so we check the five possible cases:

$$r < -1, \quad r = -1, \quad -1 < r < 1, \quad r = 1, \quad r > 1$$

We dispense with the case $r = 1$ first; note that $s_n = an$ goes to ∞ if $a > 0$ and to $-\infty$ if $a < 0$, so (5.15) diverges if $r = 1$.

To simplify the calculations for the other values of r, we use the absolute value to write the remaining cases above as

$$|r| < 1, \quad r = -1, \quad |r| > 1$$

It is easier to examine what happens to the size or magnitude of r^n using $|r|^n$, which is never negative. If $|r| < 1$, then repeated multiplication gives a decreasing sequence

$$|r| > |r| \cdot |r| = |r|^2 > |r|^2 |r| = |r|^3 > \cdots$$

For instance, if $r = -1/2$, then each time we multiply by $|r| = 1/2$ the previous value is reduced by a half; repeated multiplication gives

$$\left|-\frac{1}{2}\right| = \frac{1}{2}, \quad \left|-\frac{1}{2}\right|^2 = \frac{1}{4}, \quad \left|-\frac{1}{2}\right|^3 = \frac{1}{8} \quad \text{and so on}$$

This decreasing sequence has a limit, namely, its infimum (Corollary 110 in Chapter 4)

$$\lim_{n \to \infty} |r|^n = \inf \{|r|^n : n \geq 1\} = b \geq 0$$

We claim that actually $b = 0$.

If $r = 0$, then this is obvious. If $r \neq 0$, then $0 < |r| < 1$ and by the properties of limits (Chapter 3)

$$|r|b = |r| \lim_{n \to \infty} |r|^n = \lim_{n \to \infty} |r|^{n+1} = b$$

If $b > 0$, then dividing by b gives $|r| = 1$, which contradicts our assumption that $|r| < 1$. So $b = 0$, as claimed.

It follows that r^n approaches 0 in the case $|r| < 1$; so from (5.17), we infer that the partial sums s_n converge to the ratio $a/(1 - r)$.

Next, consider $|r| > 1$; this is the opposite case since repeated multiplication gives an increasing sequence:

$$|r| < |r| \cdot |r| = |r|^2 < |r|^2 |r| = |r|^3 < \cdots$$

In this case, $|r|^n$ diverges to infinity because each time that we multiply by $|r|$, we bump the value up by the fixed amount $|r|$. The formal proof is by contradiction and similar to what we discussed for the previous case; see Exercise 167, a lemma to the next theorem. Thus, from (5.17), we conclude that s_n diverges.[4]

Finally, let $r = -1$. Then $r^n = (-1)^n$ is either 1 when n is even or -1 when n is odd; the sequence r^n is periodic with period 2, so it has no limit. From (5.17), we conclude that s_n diverges.

These conclusions prove the following useful theorem about the geometric series.

---

[4] s_n diverges to infinity if $r > 1$; if $r < -1$, then it has no limit since its limit supremum is ∞ and its limit infimum is $-\infty$.

Theorem 141 *(a) If* $|r| < 1$, *then the geometric series in (5.15) converges, and its value is given by*

$$a + ar + ar^2 + \cdots = \frac{a}{1 - r} \tag{5.18}$$

(b) If $|r| \geq 1$, *then the geometric series in (5.15) diverges.*

As an example, let us calculate the value of

$$1 + 0.9 + 0.9^2 + 0.9^3 + \cdots \tag{5.19}$$

To use the formula (5.18), we must match this series up with the left-hand side in (5.18). Let $a = 1$; then $r = 0.9$, which is the common ratio when dividing successive pairs of terms. Because $|r| = 0.9 < 1$, in this case (5.18) yields

$$1 + 0.9 + 0.9^2 + 0.9^3 + \cdots = \frac{1}{1 - 0.9} = \frac{1}{0.1} = 10$$

You may want to find the sum if 0.9 is changed successively to 0.99 then 0.999 to explore what happens to the sum (or limit) of the series as r approaches 1.

Next, consider the series

$$2 - 0.9 + 0.9^2 - 0.9^3 + \cdots \tag{5.20}$$

Here $r = -0.9$, the ratio of every successive pair except for the first. This indicates that $a \neq 2$. But we may write the series in (5.20) as

$$1 + (1 - 0.9 + 0.9^2 - 0.9^3 + \cdots)$$

Now with $a = 1$ and $r = -0.9$ in (5.18), we obtain

$$2 - 0.9 + 0.9^2 - 0.9^3 + \cdots = 1 + \frac{1}{1 - (-0.9)} = 1 + \frac{10}{19} = \frac{29}{19}$$

5.4 Cauchy criterion and convergence tests

Most infinite series of interest in pure and applied mathematics are not geometric or telescoping in that we can't tell what their *exact* sums are. But *if we somehow know that* $\sum_{n=1}^{\infty} a_n$ *converges*, then we can estimate its value to a desired degree of accuracy, usually by adding enough terms a_n of the series. Knowing the precise limit is often not necessary in practice; we just need to be certain that we are not trying to estimate a non-existent quantity by adding terms of a divergent series.

In this section, we discuss a few methods that are often used to tell whether a series converges or not without finding an exact sum (i.e., the limit). Many of these are also discussed in most calculus textbooks but often without proofs. The most basic idea, however, is usually not discussed in typical calculus texts.

Recall (Corollary 105) that the completeness of the set of real numbers implies that a sequence converges if and only if it is Cauchy. *The important point here is that we can prove a sequence is Cauchy without knowing its limit.*

5.4.1 The Cauchy criterion

If an infinite series $\sum_{k=1}^{\infty} a_k$ converges to a real number s, then the partial sums s_n converge to s as $n \to \infty$. Since every convergent sequence is Cauchy, this means that the sequence of partial sums is Cauchy; so $|s_n - s_m| \to 0$ as $m, n \to \infty$. Without loss of generality, let $m < n$ and notice that

$$s_n = \sum_{k=1}^{n} a_k = \underbrace{a_1 + a_2 + \cdots + a_m}_{s_m} + a_{m+1} + \cdots + a_n = s_m + a_{m+1} + \cdots + a_n$$

Therefore,

$$|s_n - s_m| = |a_{m+1} + a_{m+2} + \cdots + a_n| = \left| \sum_{k=m+1}^{n} a_k \right| \tag{5.21}$$

This observation implies the following result.

Theorem 142 *(Cauchy convergence criterion for infinite series) An infinite series $\sum_{k=1}^{\infty} a_k$ converges if and only if for every $\varepsilon > 0$, there is a positive integer N such that for all integers m, n with $n > m \geq N$*

$$|a_{m+1} + a_{m+2} + \cdots + a_n| < \varepsilon \tag{5.22}$$

Theorem 142 provides a necessary and sufficient condition for a series to converge while most convergence criteria or tests give sufficient conditions (that are easier to apply). Thus, with Theorem 142, we can tell whether a series converges or diverges, something that sufficient conditions cannot provide. On the other hand, Theorem 142 is not easy to apply to specific problems, whereas the sufficient conditions apply more naturally to the type of problem for which they are designed.

A test for divergence

A simple application of Theorem 142 uses its "necessary" aspect to prove that if a_k doesn't converge to 0, then the series diverges. This gives the following corollary that is sometimes called the *nth term test* to emphasize that the test is applied to the individual or generic term of the series (for which a formula is given) rather than to the partial sums (for which we don't have a formula).

Corollary 143 (*A test for divergence*) *The infinite series $\sum_{k=1}^{\infty} a_k$ diverges if the sequence a_k does not converge to 0.*

Proof To avoid extra writing, we prove the equivalent contrapositive version of the statement, namely, if the series $\sum_{k=1}^{\infty} a_k$ converges, then $\lim_{k\to\infty} a_k = 0$. Note that the series itself need not converge to 0, and its actual value is not relevant to our discussion.

Assume that the series converges. Then by Theorem 142 with the special choice $n = m + 1$ in (5.22), for every $\varepsilon > 0$ there is a positive integer N such that

$$|a_n| < \varepsilon \quad \text{for all } n > N$$

This statement is exactly what we mean by saying that the sequence a_k converges to 0, i.e., $\lim_{k\to\infty} a_k = 0$. ∎

As a quick application, consider the series $1 + 1 + 1 + \cdots$ (i.e., the geometric series with $r = 1$). Here $a_n = 1$, a sequence that converges to 1 as $n \to \infty$, not to 0. So by Corollary 143, this series diverges. The same corollary also implies that the series

$$\sum_{n=1}^{\infty} (-1)^{n-1} = 1 - 1 + 1 - 1 + \cdots$$

diverges because the nth term $a_n = (-1)^{n-1}$ does not converge to any number, let alone to 0. Here is another example:

$$\sum_{n=1}^{\infty} n \sin \frac{\pi}{n} = 2 + \frac{3\sqrt{3}}{2} + 2\sqrt{2} + 5 \sin \frac{\pi}{5} + 3 + \cdots \qquad (5.23)$$

Taking the limit of the nth term in (5.23) gives

$$\lim_{n\to\infty} a_n = \lim_{n\to\infty} \left(n \sin \frac{\pi}{n} \right)$$

The last limit may be calculated as follows: let $x_n = \pi/n$ so that $n = \pi/x_n$, and notice that $x_n \to 0$ as $n \to \infty$. Since $\lim_{x\to 0}(\sin x)/x = 1$ (proved in Chapter 6), we obtain

$$\lim_{n\to\infty} a_n = \pi \lim_{n\to\infty} \left(\frac{\sin x_n}{x_n} \right) = \pi$$

Since this is not 0, the series in (5.23) diverges by the divergence test (Corollary 143) even though the sequence of numbers a_n itself converges (to π rather than 0).

I emphasize that to get the divergence test (Corollary 143), it was enough to use only one possibility in Theorem 142, namely, $n = m + 1$. But if we want to use this theorem to prove the convergence of a series, then we must take into account all other possible values of m. This is usually hard to do for most series of interest, so we use alternative (and weaker) convergence tests that we discuss shortly.

Tail end of a convergent series

A related question about Corollary 143 is whether its *converse* (not contrapositive) is true; the converse, namely, of Corollary 143, i.e., "if a_n converges to 0, then the series $\sum_{k=1}^{\infty} a_k$ converges to a real number" is *false* as the example of the harmonic series shows:

$$\sum_{n=1}^{\infty} \frac{1}{n} = 1 + \frac{1}{2} + \frac{1}{3} + \frac{1}{4} + \frac{1}{5} + \cdots \tag{5.24}$$

Note that the sequence of individual terms $1, 1/2, 1/3, ..., 1/n, ...$ (not added) converges to zero but, as we saw earlier, the series as a whole diverges.

Before we discuss tests for the *convergence* of series, let us clarify what is missing in the converse of Corollary 143. If we let $n \to \infty$ in (5.21) while keeping m fixed, then $s_n \to s$ because by assumption the series converges, so we get

$$|s - s_m| = \left| \sum_{k=m+1}^{\infty} a_k \right|$$

The sum inside the absolute value is the *tail end* of the original infinite series $\sum_{k=1}^{\infty} a_k$. If we now let $m \to \infty$ too, then $|s - s_m| \to 0$ and we see that the series' tail end converges to zero. This proves the following version of the Cauchy criterion, which is a necessary and sufficient condition for convergence of series and supplies what is missing in Corollary 143.

Corollary 144 (*Vanishing tail of a convergent series*) *The infinite series* $\sum_{k=1}^{\infty} a_k$ *converges if and only if its tail end converges to 0, i.e.,*

$$\lim_{m \to \infty} \sum_{k=m+1}^{\infty} a_k = 0$$

Intuitively, this states that if the whole series $\sum_{k=1}^{\infty} a_k$ approaches a real number s, then its tail shrinks to zero as we subtract more and more of the numbers a_k from s until nothing is left of s. The divergence test references only one term, a_{m+1} rather than the entire tail end, so we now see why the converse of Corollary 143 is false.

5.4.2 The comparison test

Suppose that $\sum_{k=1}^{\infty} a_k$ is an infinite series where every number a_n is non-negative, i.e., $a_n \geq 0$. If $s_n = \sum_{k=1}^{n} a_k$ is the nth partial sum of the series, then

$$s_n = a_1 + a_2 + \cdots + a_{n-1} + a_n = s_{n-1} + a_n \geq s_{n-1}$$

for all n. It follows that the partial sum sequence s_n is nondecreasing.

This simple observation when used together with the least upper bound property implies the next basic result.

Lemma 145 *(The positive terms criterion) If $a_k \geq 0$ for all indices k, then the infinite series $\sum_{k=1}^{\infty} a_k$ converges if and only if its sequence of partial sums is bounded from above. In this case*

$$\sum_{k=1}^{\infty} a_k = \sup\{s_n : n \geq N\}$$

Otherwise, we write $\sum_{k=1}^{\infty} a_k = \infty$ to indicate that the series diverges to infinity.

As stated, the positive terms criterion is quite general but not easy to use in specific cases because it requires checking the partial sums. There being no formula generally for the partial sums, it is more expedient to translate the criterion into conditions directly on the numbers a_n, the terms of the series. I discuss some of these conditions in the rest of this section.

The first convergence criterion, as its name suggests, compares two infinite series and uses what is known about one series to conclude the same about the other.

Theorem 146 *(Comparison Test) Consider two infinite series $\sum_{k=1}^{\infty} a_k$ and $\sum_{k=1}^{\infty} b_k$ and assume that there is a positive integer N such that*

$$0 \leq a_k \leq b_k \quad \text{for all } k \geq N \tag{5.25}$$

(a) If $\sum_{k=1}^{\infty} b_k$ converges, then $\sum_{k=1}^{\infty} a_k$ also converges; i.e., $\sum_{k=1}^{\infty} b_k < \infty$ implies $\sum_{k=1}^{\infty} a_k < \infty$

(b) If $\sum_{k=1}^{\infty} a_k$ diverges, then $\sum_{k=1}^{\infty} b_k$ also diverges; i.e., $\sum_{k=1}^{\infty} a_k = \infty$ implies $\sum_{k=1}^{\infty} b_k = \infty$

Proof To simplify the writing, we invoke finite equivalence to assume, without loss of generality, that the inequalities in (5.25) hold for all indices k (i.e., $N = 1$); recall that changing the first N terms of both series to zeros (if necessary) does not affect their convergence or divergence.

(a) If $\sum_{k=1}^{\infty} b_k$ converges, then there is a real number B such that $\sum_{k=1}^{\infty} b_k = B$. Since $a_k \leq b_k$ by (5.25) for all k we can write

$$a_1 + a_2 + \cdots + a_n \leq b_1 + b_2 + \cdots + b_n \leq B \quad \text{for every } n = 1, 2, 3, \ldots$$

This shows that the sequence of partial sums of $\sum_{k=1}^{\infty} a_k$ is bounded from above; so by Lemma 145, the series must converge to some number that is less than or equal to B.

(b) Suppose that $\sum_{k=1}^{\infty} a_k$ diverges. Then the sequence of partial sums of $\sum_{k=1}^{\infty} a_k$ is not bounded from above. Since by (5.25)

$$b_1 + b_2 + \cdots + b_n \geq a_1 + a_2 + \cdots + a_n \quad \text{for every } n = 1, 2, 3, \ldots$$

it follows that the sequence of partial sums of $\sum_{k=1}^{\infty} b_k$ is also not bounded from above. Lemma 145 now implies that $\sum_{k=1}^{\infty} b_k$ cannot converge.   ∎

Notice two things about the above statement:

First, it is both a convergence and a divergence criterion; intuitively, it says that if the larger series converges, then so does the smaller one; and if the smaller series diverges, then so does the larger one.

Second, there is no explicit mention of partial sums; the "comparison" is based on the inequalities $a_k \leq b_k$ between the corresponding numbers, or terms, of each series.

Let's use the comparison test to prove the convergence of the following series:

$$\sum_{n=1}^{\infty} \frac{1}{n^2} = \frac{1}{1^2} + \frac{1}{2^2} + \frac{1}{3^2} + \cdots \tag{5.26}$$

Let this be the series $\sum_{k=1}^{\infty} a_k$ in Theorem 146, and let $\sum_{k=1}^{\infty} b_k$ be the telescoping series (5.12). If we add a 1 to that series to get

$$1 + \sum_{n=1}^{\infty} \frac{1}{n(n+1)} = 1 + \frac{1}{(1)(2)} + \frac{1}{(2)(3)} + \frac{1}{(3)(4)} + \cdots$$

then we can make the following comparisons:

$$\underbrace{\frac{1}{1^2} = 1}_{a_1} \leq \underbrace{1}_{b_1}, \quad \underbrace{\frac{1}{2^2}}_{a_2} \leq \underbrace{\frac{1}{(1)(2)}}_{b_2}, \quad \underbrace{\frac{1}{3^2}}_{a_3} \leq \underbrace{\frac{1}{(2)(3)}}_{b_3}, \quad \underbrace{\frac{1}{4^2}}_{a_4} \leq \underbrace{\frac{1}{(3)(4)}}_{b_4}, \text{ and so on}$$

From earlier discussion we know that $\sum_{k=1}^{\infty} b_k$ converges; in fact,

$$1 + \sum_{n=1}^{\infty} \frac{1}{n(n+1)} = 1 + 1 = 2$$

So by Theorem 146, the series (5.26) also converges, and moreover, its value cannot exceed 2.[5]

Here is an example in which the geometric series is used for comparison:

$$\sum_{n=0}^{\infty} \frac{\cos^2 n}{2^n} = \frac{\cos^2 0}{2^0} + \frac{\cos^2 1}{2^1} + \frac{\cos^2 2}{2^2} + \cdots \tag{5.27}$$

Here $2^0 = 1$ when $n = 0$; and unless otherwise stated, the cosine values are calculated in radians. Since

$$0 \le \cos^2 x \le 1$$

we can compare the series in (5.27) with the geometric series

$$\sum_{n=0}^{\infty} \frac{1}{2^n} = \frac{1}{2^0} + \frac{1}{2^1} + \frac{1}{2^2} + \cdots . \tag{5.28}$$

and notice that for every index n,

$$\frac{\cos^2 n}{2^n} \le \frac{1}{2^n}$$

In (5.28), we have the first term $a = 1$ and common ratio $r = 1/2$, so

$$\sum_{n=0}^{\infty} \frac{1}{2^n} = \frac{1}{1 - (1/2)} = 2$$

Theorem 146 now implies that the series in (5.27) converges to a real number less than 2. Furthermore, since the first term of the series $(\cos^2 0)/2^0 = 1$, and all the other terms are non-negative, the sum of the series is a number between 1 and 2; you can estimate this number by adding the first few terms of the series in (5.27).

Next, consider the series

$$\sum_{n=1}^{\infty} \frac{2n - 9}{n^2} = -\frac{7}{1} - \frac{5}{4} - \frac{3}{9} - \frac{1}{16} + \frac{1}{25} + \cdots \tag{5.29}$$

First, we notice that all terms after the fourth are positive since $2n - 9 > 0$ if $n \ge 5$. So Theorem 146 may be applied once we decide on an appropriate series for comparison. If we ignore the constants 2 and -9 for the time being to concentrate on the variable n, then we see that

$$\frac{n}{n^2} = \frac{1}{n}$$

[5] The series in (5.26) is the value of the Riemann "zeta function" $\zeta(s)$ at $s = 2$; this is known to be $\pi^2/6 \simeq 1.645$. This was proved by Euler in 1734 in his solution to the "Basel Problem." We solve this problem in a different way than Euler's in Chapter 8 using Fourier series.

This observation suggests that the series in (5.29) diverges, and we should choose the harmonic series for comparison if we can find a positive integer N such that $2n - 9 \geq n$ for all $n \geq N$. Solving $2n - 9 \geq n$ for n gives $n \geq 9$, so

$$\frac{2n - 9}{n^2} \geq \frac{1}{n} \quad \text{for all } n \geq 9$$

Now by Theorem 146, the series in (5.29) diverges to infinity.

The following is a corollary of Theorem 146 that is sometimes easier to apply than the comparison test.

Corollary 147 *(Limit comparison test) Let $\sum_{k=1}^{\infty} a_k$ and $\sum_{k=1}^{\infty} b_k$ be infinite series of positive real numbers, i.e., $a_k, b_k > 0$ for all k.*

(a) If $\lim_{k \to \infty}(a_k/b_k) = L > 0$, then $\sum_{k=1}^{\infty} a_k$ converges if and only if $\sum_{k=1}^{\infty} b_k$ converges.

(b) If $\lim_{k \to \infty}(a_k/b_k) = 0$ and $\sum_{k=1}^{\infty} b_k$ converges, then $\sum_{k=1}^{\infty} a_k$ converges.

Proof I give the outline of the proof and ask you to fill in the details in Exercise 173.

(a) By the definition of limit, for $\varepsilon = L/2 > 0$, there is a positive integer N such that for all $k \geq N$

$$\left| \frac{a_k}{b_k} - L \right| < \frac{L}{2} \tag{5.30}$$

If $\sum_{k=1}^{\infty} b_k$ converges, then we use (5.30) to obtain $a_k < (3L/2)b_k$ for $k \geq N$. Let $c_k = (3L/2)b_k$ and apply Theorem 146 to conclude that $\sum_{k=1}^{\infty} a_k$ must converge.

If $\sum_{k=1}^{\infty} a_k$ converges, then we can repeat essentially the same argument but with $b_k < (2/L)a_k = d_k$.

(b) In this case, there is a positive integer N such that for $k \geq N$

$$\frac{a_k}{b_k} = \left| \frac{a_k}{b_k} \right| < 1$$

so the proof is completed by applying Theorem 146. ∎

Consider the series in (5.29) again with

$$a_k = \frac{2k - 9}{k^2}, \qquad b_k = \frac{1}{k}$$

Then notice that

$$\lim_{k \to \infty} \frac{a_k}{b_k} = \lim_{k \to \infty} \left(2 - \frac{9}{k} \right) = 2$$

Since $\sum_{k=1}^{\infty} 1/k$ diverges, Corollary 147 implies that the series in (5.29) diverges too.

As this example shows, the limit comparison test can simplify calculations in some problems. But also notice that we cannot apply Corollary 147 to the series in (5.27) using the convergent geometric series $\sum_{k=1}^{\infty} 1/2^k$ for comparison because $\cos^2 k$ doesn't converge as $k \to \infty$.

Also notice that if $\sum_{k=1}^{\infty} a_k$ converges in Part (b) of Corollary 147, then no general conclusions can be drawn about the convergence of $\sum_{k=1}^{\infty} b_k$. For instance, if $a_k = 1/k^2$ and $b_k = 1/k$, then

$$\lim_{k \to \infty} \frac{a_k}{b_k} = \lim_{k \to \infty} \frac{1/k^2}{1/k} = \lim_{k \to \infty} \frac{1}{k} = 0$$

In this case, $\sum_{k=1}^{\infty} a_k$ converges but $\sum_{k=1}^{\infty} b_k$ diverges. On the other hand, in Exercise 172, you show that $\sum_{k=1}^{\infty} 1/k^3$ converges, so with $a_k = 1/k^3$ and $b_k = 1/k^2$ we again obtain

$$\lim_{k \to \infty} \frac{a_k}{b_k} = \lim_{k \to \infty} \frac{1/k^3}{1/k^2} = \lim_{k \to \infty} \frac{1}{k} = 0$$

In this case, both $\sum_{k=1}^{\infty} a_k$ and $\sum_{k=1}^{\infty} b_k$ converge.

5.4.3 The ratio and root tests: Extending the geometric series method

As we explore the comparison test, we come across series like the following for which it is hard to find a closely related sister series with familiar properties for comparison:

$$\sum_{n=1}^{\infty} \frac{n^2}{2^n}, \qquad \sum_{n=1}^{\infty} \frac{10^n}{n!}$$

To see what may work, consider the first series above

$$\sum_{n=1}^{\infty} \frac{n^2}{2^n} = \frac{1^2}{2^1} + \frac{2^2}{2^2} + \frac{3^2}{2^3} + \frac{4^2}{2^4} + \frac{5^2}{2^5} + \frac{6^2}{2^6} + \cdots$$

with the generic term $a_n = n^2/2^n$. Without the n^2, we would have a geometric series that we already discussed.

Inspired by this observation, we consider the ratios of consecutive terms for the series, by dividing the terms a pair at a time:

$$\frac{a_2}{a_1} = 2, \quad \frac{a_3}{a_2} = \frac{9}{8}, \quad \frac{a_4}{a_3} = \frac{8}{9}, \quad \frac{a_5}{a_4} = \frac{25}{32}, \quad \frac{a_6}{a_5} = \frac{36}{50}, \ldots$$

Unlike the geometric series, these ratios are not equal. However, notice that whatever their actual values, *all of the above ratios after the second are less than* 1 and seem to be getting smaller as we keep dividing pairs. Recall that what resulted in the convergence of a geometric series was the magnitude (or absolute value) of the common ratio being less than 1, not that the ratio was constant.

With this fact in mind, consider a generic ratio with unspecified n and do a little algebra to get

$$\frac{a_{n+1}}{a_n} = a_{n+1} \frac{1}{a_n} = \frac{(n+1)^2}{2^{n+1}} \frac{2^n}{n^2} = \left(\frac{n+1}{n}\right)^2 \frac{1}{2}$$

We can rewrite the last quantity to obtain the following equality that is easier to take the limit of:

$$\frac{a_{n+1}}{a_n} = \frac{1}{2}\left(1 + \frac{1}{n}\right)^2$$

Since $1/n$ decreases (to 0), we see that the ratios a_{n+1}/a_n decrease as n moves up through $1, 2, 3, \ldots$ Therefore, once the value of a_{n+1}/a_n drops below 1, it stays below 1 and goes down towards the limit of $1/2$. So if we pick a number r between $1/2$ and 1, say, $r = 8/9$ (the third ratio above), then

$$\frac{a_{n+1}}{a_n} \leq r < 1 \quad \text{for } n = 3, 4, 5, \ldots$$

Next, we remove the fraction by multiplying by a_n to obtain

$$a_{n+1} \leq r a_n \quad \text{for } n = 3, 4, 5, \ldots$$

This implies that

$$a_4 \leq r a_3, \quad a_5 \leq r a_4 \leq r(r a_3) = r^2 a_3, \quad a_6 \leq r a_5 \leq r^3 a_3, \ldots$$

The numbers $r a_3$, $r^2 a_3$, $r^3 a_3$, etc. now look familiar as numbers in a geometric series. In fact, the above calculations let us compare the series that we started with to a geometric one as follows:

$$a_1 + a_2 + a_3 + a_4 + a_5 + a_6 + \cdots \leq a_1 + a_2 + (a_3 + r a_3 + r^2 a_3 + r^3 a_3 + \cdots)$$

The series inside the parenthesis is a convergent geometric series because its common ratio is $r < 1$. Therefore, the series on the left (the series that we started with) converges too!

We see that this argument did not require anything substantial beyond geometric series and the comparison test as long as the ratio of consecutive numbers in the series fell below 1 after some index n and stayed below 1 thereafter (and not approaching 1).

The ratio test

This argument (generalizing the idea behind geometric series) is essentially how the ratio test below is proved.[6] Despite its simplicity, this test for convergence and

---

[6] The ratio test first appeared in some form around 1750 in a work of the French mathematician and physicist Jean-Baptiste le Rond d'Alembert (1717–1783).

divergence is a powerful one, as we discover later when discussing infinite series of functions.

Theorem 148 *(ratio test) Consider an infinite series $\sum_{k=1}^{\infty} a_k$ where $a_k > 0$ for all k greater than or equal to some index m and define the numbers*

$$L = \liminf_{k \to \infty} \left(\frac{a_{k+1}}{a_k} \right), \qquad R = \limsup_{k \to \infty} \left(\frac{a_{k+1}}{a_k} \right) \qquad (5.31)$$

It is possible that $R = \infty$ in (5.31).
(a) If $R < 1$, then the series converges;
(b) If $L > 1$, then the series diverges.

Proof Invoking the finite equivalence property, we may assume that $a_k > 0$ for all k ($m = 1$).
(a) Recall that

$$R = \lim_{k \to \infty} u_k$$

where

$$u_k = \sup \left\{ \frac{a_{n+1}}{a_n} : n \geq k \right\}$$

is a nonincreasing sequence. Since $R < 1$, the terms u_k with sufficiently large index k drop below 1 on their way down toward their limit R. Choose r such that $R < r < 1$ and an index N such that $u_k \leq r$ for all $k \geq N$. Then

$$\frac{a_{k+1}}{a_k} \leq u_k \leq r$$

Thus, for all $k \geq N$, we can write

$$a_{k+1} \leq r a_k$$

This implies that in particular, $a_{N+1} \leq r a_N$ and

$$a_{N+2} \leq r a_{N+1} \leq r^2 a_N$$

and in the same manner, for every $j \geq 1$

$$a_{N+j} \leq r^j a_N$$

The nth partial sum of $\sum_{k=1}^{\infty} a_k$ for $n > N$ is thus bounded above by a geometric series and some extra terms:

$$s_n < (a_1 + \cdots + a_{N-1}) + (a_N + r a_N + r^2 a_N + \cdots)$$

The geometric series with $r < 1$ converges so that

$$\sum_{k=1}^{\infty} a_k = \lim_{n \to \infty} s_n < a_1 + \cdots + a_{N-1} + \frac{a_N}{1-r} < \infty$$

and the series converges.

(b) The argument is similar to the one above for (a): $L = \lim_{k \to \infty} l_k$ where

$$l_k = \inf\left\{\frac{a_{n+1}}{a_n} : n \geq k\right\}$$

is a nondecreasing sequence that converges to L. Choosing r such that $1 < r < L$, there is a positive integer N such that for all $k \geq N$

$$\frac{a_{k+1}}{a_k} \geq l_k \geq r$$

and this implies that the partial sums of $\sum_{k=1}^{\infty} a_k$ for $n > N$ are bounded below by a geometric progression and some extra terms

$$s_n \geq (a_1 + \cdots + a_{N-1}) + (a_N + ra_N + r^2 a_N + \cdots + r^{n-N} a_N)$$

Since $r > 1$ in this case, the geometric progression diverges as $n \to \infty$. It follows that $s_n \to \infty$, i.e., $\sum_{k=1}^{\infty} a_k = \infty$, and the series diverges. ■

We see later in this chapter that the ratio test can be extended to infinite series containing infinitely many *negative* terms.

For now consider a *special case* of the ratio test. The next corollary presents the familiar form of the ratio test that we find in the typical calculus textbook. Recall that when the limit infimum and limit supremum are equal for a sequence, then their common value is in fact the limit of the sequence.

Corollary 149 (*limit ratio test*) *Consider an infinite series $\sum_{k=1}^{\infty} a_k$ where $a_k > 0$ for all k greater than or equal to some index m, and assume that the following limit exists:*

$$L = \lim_{k \to \infty} \frac{a_{k+1}}{a_k} \tag{5.32}$$

If $L < 1$, then the series converges; and if $L > 1$ (including $L = \infty$), then the series diverges.

Theorem 148 and Corollary 149 usually work best when the terms a_k contain exponentials like 2^n and especially *factorials* $n!$. Recall from calculus that

$$n! = n(n-1)(n-2) \cdots (3)(2)(1)$$

In particular,

$$1! = 1, \quad 2! = (2)(1) = 2, \quad 3! = (3)(2) = 6, \quad 4! = (4)(3)(2) = 24$$

and so on. We will encounter factorials later in our discussion of infinite series of functions where we also define

$$0! = 1.$$

For example, consider the series

$$\sum_{n=1}^{\infty} \frac{10^n}{n!} = \frac{10}{1!} + \frac{10^2}{2!} + \frac{10^3}{3!} + \frac{10^4}{4!} + \cdots \qquad (5.33)$$

Before we figure out whether the series in (5.33) converges or diverges, let's check its first few terms:

$$\sum_{n=1}^{\infty} \frac{10^n}{n!} = 10 + 50 + 166.67 + 416.67 + 833.33 + 1388.89 + \cdots$$

Based on what we see above, it is tempting to think that this series diverges. But the first few terms are not indicative of convergence or divergence, so we try the ratio test with

$$a_n = \frac{10^n}{n!}$$

Then

$$\frac{a_{n+1}}{a_n} = \frac{10^{n+1}}{(n+1)!} \frac{n!}{10^n} = \frac{10^n 10}{(n+1)n!} \frac{n!}{10^n} = \frac{10}{n+1}$$

The fraction $10/(n+1)$ approaches 0; so by Corollary 149, the series in (5.33) converges.

Theorem 148 applies to problems that Corollary 149 doesn't because the limits in (5.31) always exist, but the limit in (5.32) may not exist; see Exercise 174.

For some series, it may be that $L \leq 1 \leq R$ in Theorem 148, or $L = 1$ in Corollary 149. In these cases, no conclusions can be drawn from the ratio test; the series may converge or diverge, and another method is needed to determine what the series does.

For example, we discovered earlier that $\sum_{k=1}^{\infty} 1/k$ diverges, while $\sum_{k=1}^{\infty} 1/k^2$ converges. In the latter case, using the limit theorems, we find that

$$\lim_{k\to\infty} \frac{a_{k+1}}{a_k} = \lim_{k\to\infty} \frac{(k+1)^2}{k^2} = \lim_{k\to\infty} \left(1 + \frac{1}{k}\right)^2 = \left[\lim_{k\to\infty} \left(1 + \frac{1}{k}\right)\right]^2 = 1$$

For the harmonic series,

$$\lim_{k\to\infty} \frac{a_{k+1}}{a_k} = \lim_{k\to\infty} \frac{k+1}{k} = \lim_{k\to\infty} \left(1 + \frac{1}{k}\right) = 1$$

Therefore, if $L = 1$ in Corollary 149, then the series in question may converge or diverge.

Note that *the ratio test cannot be directly applied to a series that have infinitely many 0 terms.* For instance, consider

$$\sum_{n=1}^{\infty} \frac{1 - (-1)^n}{n!} = \frac{2}{1!} + \frac{2}{3!} + \frac{2}{5!} + \cdots$$

Here $a_n = [1 - (-1)^n]/n!$ is 0 if n is even, i.e., $a_{2k} = 0$ for all $k = 1, 2, 3, \ldots$ Hence, the ratio a_{2k+1}/a_{2k} is not defined. However, the zeros don't contribute to the sum of a series; in fact, the series as written on the right-hand side contains no zeros, and it can be recast as a positive terms series in the form

$$\sum_{n=1}^{\infty} \frac{2}{(2n-1)!}$$

with the new terms $b_n = 2/(2n-1)!$ that are all positive. This series can now be easily shown to converge using Corollary 149.

The next test, a refinement of the ratio test, applies directly even when an infinite number of terms of a series are zeros.

The root test

In this section, we discuss a test attributed to Cauchy that is similar to the ratio test but more refined and is more easily applicable to some series where the ratio test either does not apply or it is more difficult to apply.

Theorem 150 *(root test) Consider an infinite series $\sum_{k=1}^{\infty} a_k$ where $a_k \geq 0$ for all k greater than or equal to some index m, and define the number*

$$\rho = \limsup_{k \to \infty} \sqrt[k]{a_k} = \limsup_{k \to \infty} a_k^{1/k}$$

If $\rho < 1$, then the series converges; while if $\rho > 1$, the series diverges.

Proof Due to the finite equivalence property of series, we may assume that $m = 1$.

Assume first that $\rho < 1$. Like the proof of the ratio test, the argument here is also based on using geometric series. The nonincreasing sequence

$$u_k = \sup\{a_n^{1/n} : n \geq k\}$$

converges to ρ as $k \to \infty$; so if $\rho < r < 1$, then there is a positive integer N such that

$$a_k^{1/k} \leq u_k \leq r \quad \text{for } k \geq N$$

The power function x^k is increasing if $x \geq 0$ for all integers k so

$$a_k = \left[a_k^{1/k}\right]^k \leq r^k \quad \text{for } k \geq N$$

It follows that the nth partial sum of $\sum_{k=1}^{\infty} a_k$ is bounded above by sum of a convergent geometric series plus a finite sum

$$s_n < (a_1 + \cdots + a_{N-1}) + (a_N + ra_N + r^2 a_N + \cdots)$$
$$= a_1 + \cdots + a_{N-1} + \frac{a_N}{1 - r}$$

for all $n > N$. Since s_n is nondecreasing, it follows that it has a finite limit, which is the sum of the series $\sum_{k=1}^{\infty} a_k$.

(b) If $\rho > 1$, then since u_k converges to ρ in a nonincreasing fashion (goes down toward ρ)

$$\sup\{a_n^{1/n} : n \geq k\} = u_k \geq \rho > 1$$

If for some index k we have $a_n^{1/n} < 1$ for all $n \geq k$, then the supremum u_k will not exceed 1, which means that $u_k \leq 1 < \rho$. Due to the nonincreasing nature of u_k, it follows that $u_j \leq u_k < \rho$ for all $j \geq k$, which contradicts the convergence of u_k to ρ. Therefore, for every k, there is an index n_k such that

$$a_{n_k}^{1/n_k} \geq \rho > 1 \quad \text{for all } k = 1, 2, 3, \ldots$$

or equivalently,

$$a_{n_k} > 1 \quad \text{for all } k = 1, 2, 3, \ldots$$

This implies that the sequence of coefficients a_k in the series doesn't converge to 0; hence, by the divergence test, the series $\sum_{k=1}^{\infty} a_k$ diverges. ∎

Note that like the ratio test, the root test is inconclusive if $\rho = 1$; see Exercise 176.

The root test works best on series $\sum_{n=1}^{\infty} a_n$ where a_n contains variable powers, i.e., n appears in the exponent.

For example, consider the series

$$\sum_{n=1}^{\infty} \frac{2^{(n^2)}}{n^n} = \frac{2}{1} + \frac{2^4}{2^2} + \frac{2^9}{3^3} + \cdots \tag{5.34}$$

Then

$$a_n^{1/n} = \left[\frac{2^{(n^2)}}{n^n}\right]^{1/n} = \frac{2^{(n^2/n)}}{n^{n/n}} = \frac{2^n}{n}$$

We can show that

$$\lim_{n \to \infty} \frac{2^n}{n} = \infty \tag{5.35}$$

so that $\rho = \infty$, and the series in (5.34) diverges by Theorem 150. One way of proving (5.35) is to use the binomial theorem

$$2^n = (1+1)^n = 1 + n + \binom{n}{2} + \binom{n}{3} + \cdots + n + 1$$

to infer

$$2^n > \binom{n}{2} = \frac{n(n+1)}{2}$$

Therefore,

$$\frac{2^n}{n} > \frac{n(n+1)}{2n} = \frac{n+1}{2}$$

and the limit of the last quantity as $n \to \infty$ is infinite; now (5.35) follows.

As the preceding example shows, it is helpful to have limits like the one in (5.35) at our disposal to speed up calculations involving the root test; for instance, (5.35) follows from (5.39) below. We present such a result next; you may first want to try proving that the series in (5.34) diverges using the ratio test in order to convince yourself that the root test is indeed easier to apply.

Theorem 151 *The following equalities are true:*

(a) *(binomial theorem) For every real number r and positive integer n*

$$(1+r)^n = \sum_{i=0}^{n} \binom{n}{i} r^i = \binom{n}{0} + \binom{n}{1} r + \binom{n}{2} r^2 + \cdots + \binom{n}{n} r^n \tag{5.36}$$

where the binomial coefficients are given as

$$\binom{n}{i} = \frac{n!}{i!(n-i)!} \quad i = 1, 2, \ldots, n$$

(b) *For every positive real number x*

$$\lim_{n \to \infty} \sqrt[n]{x} = \lim_{n \to \infty} x^{1/n} = 1 \tag{5.37}$$

(c)

$$\lim_{n \to \infty} \sqrt[n]{n} = \lim_{n \to \infty} n^{1/n} = 1 \tag{5.38}$$

(d) *If x and p are real numbers, and p > 1, then*

$$\lim_{n \to \infty} \frac{n^x}{p^n} = 0 \tag{5.39}$$

(e) For every real number r

$$\lim_{n \to \infty} \frac{p^n}{n!} = 0 \tag{5.40}$$

Proof (a) We use mathematical induction. If $n = 1$, then

$$(1 + r)^1 = \binom{1}{0} + \binom{1}{1} r = 1 + r$$

and (5.36) is true for $n = 1$. So assume it to be true for $n = k$ and note that

$$(1 + r)^{k+1} = (1 + r)^k (1 + r)$$

$$= \left[\binom{k}{0} + \binom{k}{1} r + \binom{k}{2} r^2 + \cdots \binom{k}{k} r^k \right] (1 + r)$$

$$= \binom{k}{0} + \left[\binom{k}{0} + \binom{k}{1} \right] r + \left[\binom{k}{1} + \binom{k}{2} \right] r^2 + \cdots$$

$$+ \left[\binom{k}{k-1} + \binom{k}{k} \right] r^k + \binom{k}{k} r^{k+1}$$

Next, if $1 \le i \le k$, then

$$\binom{k}{i-1} + \binom{k}{i} = \frac{k!}{(i-1)!(k-i+1)!} + \frac{k!}{i!(k-i)!}$$

$$= \frac{k!i + k!(k-i+1)}{i!(k-i+1)!}$$

$$= \frac{k!(k+1)}{i!(k+1-i)}$$

$$= \binom{k+1}{i}$$

Therefore,

$$(1 + r)^{k+1} = \binom{k}{0} + \binom{k+1}{1} r + \cdots \binom{k+1}{k} r^k + \binom{k}{k} r^{k+1}$$

$$= \binom{k+1}{0} + \binom{k+1}{1} r + \cdots \binom{k+1}{k} r^k + \binom{k+1}{k+1} r^{k+1}$$

where the last equality is true because for all positive integers n

$$\binom{n}{0} = 1 = \binom{n}{n}$$

This completes the induction proof of (a).

(b) We use the binomial theorem to prove that (5.37) is true. We define $r_n = x^{1/n} - 1$ and show that r_n converges to 0 as $n \to \infty$. By (5.36)

$$x = (1 + r_n)^n \geq \binom{n}{1} r_n = n r_n$$

so that $r_n \leq x/n$. Since x has a fixed value, it follows that $r_n \to 0$ as $n \to \infty$ as claimed.

(c) We repeat the argument in (b), but this time we define $r_n = n^{1/n} - 1$ and choose a different term of the binomial expansion in the right-hand side of (5.36). Specifically,

$$n = (1 + r_n)^n \geq \binom{n}{2} r_n = \frac{n(n+1)}{2} r_n$$

Solving for r_n gives $r_n \leq 2/(n+1)$, which converges to 0 as $n \to \infty$.

(d) Let m be a positive integer and $x \leq m$ (recall the Archimedean property). If $p = 1 + r$ with $r > 0$, then for $n > 2m$ the binomial theorem implies that

$$p^n = (1 + r)^n > \binom{n}{m+1} r^{m+1} = n(n-1)\cdots(n-m)\frac{r^{m+1}}{(m+1)!}$$

Notice that $n - m > n - n/2 = n/2$, so every term of the product on the right-hand side is less than $n/2$. It follows that

$$p^n > \left(\frac{n}{2}\right)^{m+1} \frac{r^{m+1}}{(m+1)!} = c n^{m+1}, \quad c = \frac{r^{m+1}}{2^{m+1}(m+1)!}$$

Therefore,

$$0 \leq \frac{n^x}{p^n} \leq \frac{n^m}{p^n} < \frac{n^m}{c n^{m+1}} = \frac{1}{cn}$$

Now applying the squeeze theorem completes the proof of (5.39).

(e) Let m be a positive integer and $|p| \leq m$. Note that for $n > m$

$$\left|\frac{p^n}{n!}\right| = \frac{|p|^n}{n!} \leq \frac{m^n}{n!} = \frac{m^{n-m}m^m}{n(n-1)\cdots(m+1)m!} < \frac{m^m}{m!}\frac{m^{n-m}}{(m+1)^{n-m}}$$

If $r = m/(m+1)$, then $r < 1$ and we have

$$0 \leq \left|\frac{p^n}{n!}\right| < cr^n, \quad c = \frac{m^m}{m!r^m} = \frac{(m+1)^m}{m!}$$

Now applying the squeeze theorem completes the proof of (5.40). ∎

For our next example, we use the root test to see if the following series converges or diverges:

$$\sum_{n=1}^{\infty} \frac{[2 + (-1)^n]n}{2^n} = \frac{1}{2} + \frac{3}{2} + \frac{3}{8} + \frac{3}{4} + \cdots \tag{5.41}$$

Consider

$$a_n^{1/n} = \frac{n^{1/n}}{2^{n/n}}[2 + (-1)^n]^{1/n} = \frac{n^{1/n}}{2}[2 + (-1)^n]^{1/n}$$

The term inside the brackets is either 1 or 3 depending on whether n is odd or even. Therefore, for every positive integer k

$$\sup\left\{a_n^{1/n} : n \geq k\right\} \leq \sup\left\{\frac{n^{1/n}3^{1/n}}{2} : n \geq k\right\}$$

Since limits preserve inequalities, taking limits and using (5.37) and (5.38) gives

$$\rho \leq \limsup_{n\to\infty} \frac{n^{1/n}3^{1/n}}{2} = \lim_{n\to\infty} \frac{n^{1/n}3^{1/n}}{2} = \frac{1}{2}$$

Since $\rho < 1$, it follows that the series in (5.41) converges. The series in (5.41) is similar to the series

$$\sum_{n=1}^{\infty} \frac{[2 + (-1)^n]n}{4^n} \tag{5.42}$$

that you may prove converge in Exercise 174 via the ratio test. However, if we try using the ratio test to prove that the series in (5.41) converges, then we fall in the region where the ratio test is inconclusive, i.e.,

$$L \leq 1 \leq R$$

in Theorem 148.

To illustrate this fact, we apply the ratio test to the series in (5.41):

$$\frac{a_{n+1}}{a_n} = \frac{[2 + (-1)^{n+1}](n+1)}{2^{n+1}} \frac{2^n}{[2 + (-1)^n]n} = \frac{2 + (-1)^{n+1}}{2 + (-1)^n}\left(\frac{n+1}{2n}\right)$$

Notice that if n is odd, then

$$\frac{a_{n+1}}{a_n} = \frac{3}{1}\left(\frac{n+1}{2n}\right)$$

while if n is even, then

$$\frac{a_{n+1}}{a_n} = \frac{1}{3}\left(\frac{n+1}{2n}\right)$$

Hence,

$$\liminf_{k\to\infty}\left\{\frac{a_{n+1}}{a_n} : n \geq k\right\} = \lim_{k\to\infty} \frac{1}{3}\left(\frac{k+1}{2k}\right) = \frac{1}{6}$$

while

$$\limsup_{k\to\infty}\left\{\frac{a_{n+1}}{a_n}: n \geq k\right\} = \lim_{k\to\infty} 3\left(\frac{k+1}{2k}\right) = \frac{3}{2}$$

We conclude that

$$L = \frac{1}{6} \leq 1 \leq \frac{3}{2} = R$$

The ratio test is inconclusive. But when we applied the root test, we found $\rho = 1/2$; so for the series in (5.41), we discover that

$$L \leq \rho \leq 1 \leq R$$

The appearance of the parameter ρ is on the correct side of 1, and that made the root test work.

The above example provides an illustration of the next theorem, which explains why the root test is a *refinement* of the ratio test.

Theorem 152 *For every sequence a_n of positive real numbers, the following inequalities hold:*

$$\liminf_{n\to\infty}\frac{a_{n+1}}{a_n} \leq \liminf_{n\to\infty} a_n^{1/n} \leq \limsup_{n\to\infty} a_n^{1/n} \leq \limsup_{n\to\infty}\frac{a_{n+1}}{a_n}$$

Proof Let

$$R = \limsup_{n\to\infty}\frac{a_{n+1}}{a_n} \tag{5.43}$$

If $R = \infty$, then the last inequality in (5.43) is clearly true. So suppose that $R < \infty$, and recall that R is a limit of a nonincreasing sequence:

$$R = \lim_{n\to\infty} u_n, \quad u_n = \sup\left\{\frac{a_{m+1}}{a_m}: m \geq n\right\}$$

Thus, for each $r \geq R$, there is a positive integer N such that for all $n \geq N$

$$R \leq u_n \leq r$$

In particular,

$$\frac{a_{N+1}}{a_N} \leq r \quad \text{i.e., } a_{N+1} \leq ra_N$$

Now arguing as in the proof of Theorem 148, we have

$$a_{N+j} \leq r^j a_N \quad j = 1, 2, 3, \ldots$$

or equivalently, with $n = N + j$,

$$a_n \leq ar^n, \quad a = \frac{a_N}{r^N}$$

Now, taking the nth root, we obtain

$$a_n^{1/n} \le a^{1/n} r \quad \text{for } n \ge N$$

By (5.37) in Theorem 151, $\lim_{n\to\infty} a^{1/n} = 1$ so that

$$\limsup_{n\to\infty} a_n^{1/n} \le r$$

This inequality holds for all $r \ge R$; so to avoid a contradiction, we conclude that

$$\limsup_{n\to\infty} a_n^{1/n} \le R$$

The inequality for the limit infimums is proved similarly, essentially by reversing $\le$ to complete the proof. ∎

As we saw earlier, the inequalities in Theorem 152 are generally strict, but they may be equal for some sequences. For instance, if $b > 0$ and $a_n = b^n$, then

$$\lim_{n\to\infty} \frac{a_{n+1}}{a_n} = \lim_{n\to\infty} \frac{b^{n+1}}{b^n} = b$$

so that $R = L = b$, and thus $\rho = b$ as well. We can verify this directly:

$$a_n^{1/n} = (b^n)^{1/n} = b$$

Theorem 152 can also be used to prove some results that may not be obvious any other way. For example, take $a_n = n!$ and note that

$$\lim_{n\to\infty} \frac{a_{n+1}}{a_n} = \lim_{n\to\infty} \frac{(n+1)!}{n!} = \lim_{n\to\infty} (n+1) = \infty$$

This means that $L = \infty$; and since by Theorem 152 $\rho \ge L$, it follows that $\rho = \infty$, i.e.,

$$\lim_{n\to\infty} \sqrt[n]{n!} = \lim_{n\to\infty} (n!)^{1/n} = \infty$$

This inequality is harder to prove directly using just the definitions. In Exercise 179, you can show that $\sqrt[n]{n!} < n$ so that $\sqrt[n]{n!}$ becomes infinite but no faster than n does.

A paradoxical Stack of Boxes

The fact that the harmonic series $\sum_{n=1}^{\infty} 1/n$ diverges while $\sum_{n=1}^{\infty} 1/n^2$ converges leads to a peculiar construction that is worth discussing here. The construction may seem paradoxical due to the finite nature of our physical intuition while infinite series belong to the realm of infinity. In this instance, calculus reaches beyond the grasp of intuition, and we must use the established theory to understand and manipulate objects and constructions that involve the infinite.[7]

Paradoxes involving infinity must be distinguished from *conceptual* or *logical* paradoxes like "This statement is false." Self-referential statements of this type signal that mathematics is fundamentally incomplete if we insist that it be consistent (which all mathematicians do). After Godel's famous incompleteness theorems, mathematicians gave up the search for an axiom scheme with which to prove all mathematical statements (completeness) without leading to contradictions (consistency).

Infinity paradoxes provide great reasons why mathematicians require consistency in their work. In the absence of experimental verification or falsification and without any physical intuition, it is absolutely essential to have assurance that statements and constructions involving infinity are free of contradictions.

Returning to infinite series, consider a collection of hypothetical boxes, the largest of which has all sides one unit long; the second has a length 1 unit long, but its height and width are $1/2$ unit, etc. as illustrated in Figure 5.1.

The nth box has dimensions $1 \times 1/n \times 1/n$, so its volume is $1/n^2$, which means that the *total volume* of all the boxes is finite and given by the sum S of the series $\sum_{n=1}^{\infty} 1/n^2$; recall that $S < 2$ cubic units.

Now suppose that we stack the boxes up, each on top of the other as shown in Figure 5.1. The *total height* of the tower of boxes is given by the harmonic series, which has infinite value despite its finite volume.

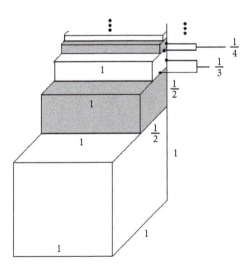

Fig. 5.1 A paradoxical Stack of Boxes

---

[7] The construction here is a discrete version of the well-known "Gabriel's Horn" (or "Torricelli's Trumpet") paradox. The tower of boxes achieves the same results without requiring improper Riemann integration or a formula for the surface area of the trumpet-shaped surface of revolution of the curve $1/x$.

Next, consider the *total surface area* of the tower of boxes: because some of the faces (four out of six) of the nth box have area $(1)(1/n) = 1/n$, the total surface area of all the boxes exceeds the sum of the harmonic series, which is infinite. This leads to a version of the old "painter's paradox": namely, if we fill all of the boxes with paint, then a finite amount of paint (less than 2 cubic units) is enough to cover the infinite surface area of the interiors of all the boxes!

The paradox stems from the fact that our hypothetical Stack of Boxes cannot be physically constructed, let alone be painted. It only exists in the realm of infinity where it is not paradoxical. We can of course consider "hypothetical paint" that has zero thickness (so that it can coat inside infinitesimally thin boxes high up in the Stack). In that case, a finite amount of paint will not fill a finite volume since even a three-dimensional region with finite volume consists of infinitely many two-dimensional cross sections. So we don't have a painter's paradox in the realm of infinity.

Can the infinite surface area be attributed to the infinite height of the tower? To see why it doesn't, let's modify our tower a little bit.

Suppose that the height of box n is $1/n^2$ for every n rather than $1/n$, with all other dimensions kept the same as those shown in Figure 5.1. Then the total height is finite (the sum of the series $\sum_{n=1}^{\infty} 1/n^2$) as is its volume; in this case, the entire stack of boxes fits within a box with dimensions $1 \times 1 \times S$, where $S = \sum_{n=1}^{\infty} 1/n^2 < 2$. It follows that the total volume of all the boxes in the Stack is less than 2. Alternatively, note that since the nth box has volume $(1)(1/n)(1/n^2) = (1/n^3)$, the actual volume of the Stack is given by the series

$$\sum_{n=1}^{\infty} \frac{1}{n^3}$$

which converges to a finite number (see Exercise 172). However, the total surface area of all the boxes is still infinite because the total of all the horizontal sides (top or bottom of each box) exceeds the sum of the divergent harmonic series.

Finally, the Stack of Boxes is a three-dimensional version of the area under the staircase curves that we discussed earlier. In that case, we could construct a triangle-like region (with two straight sides and one zigzag) that has a fixed finite area but an arbitrarily large perimeter.[8]

5.5 Alternating series, conditional and absolute convergence

The infinite in "infinite series" is a source of many non-intuitive results, and the preceding discussion of the Stack of Boxes illustrates this fact. In this section,

---

[8] If you are interested in infinity paradoxes like these, then take a look at my self-study book *Achieving Infinite Resolution*.

we turn to series having infinitely many positive and negative numbers, or more precisely, series with infinitely many sign changes.

Another paradox?

This feature introduces more complexity and produces more (intuitively) paradoxical results; in some cases, series with this feature may converge to different numbers, or not at all, depending on how their terms are rearranged. For example, consider the following list of fractions

$$1, -\frac{1}{2}, \frac{1}{3}, -\frac{1}{4}, \frac{1}{5}, -\frac{1}{6}, \frac{1}{7}, \cdots$$

whose signs alternate between positive and negative. We may add these numbers in the order listed above and call their sum s:

$$s = 1 - \frac{1}{2} + \frac{1}{3} - \frac{1}{4} + \frac{1}{5} - \frac{1}{6} + \cdots \qquad (5.44)$$

We won't worry about what the total is here.[9] But what if we add the numbers in a different order, say,

$$1 + \frac{1}{3} - \frac{1}{2} + \frac{1}{5} + \frac{1}{7} - \frac{1}{4} + \cdots \qquad (5.45)$$

where we add two fractions with odd denominators and then subtract an even one. We are adding exactly the same set of numbers, so it is reasonable to expect that the result also equals s. But this isn't the case!

Suppose that we divide all of the numbers in (5.44) by 2 to get

$$\frac{s}{2} = \frac{1}{2} - \frac{1}{4} + \frac{1}{6} - \frac{1}{8} + \frac{1}{10} - \frac{1}{12} + \cdots$$

When we add the two sides of this equality to those of above (5.44) and cancel out the equal fractions with opposite signs, we get

$$\frac{3s}{2} = 1 + \frac{1}{3} - \frac{2}{4} + \frac{1}{5} + \frac{1}{7} - \frac{2}{8} + \cdots$$
$$= 1 + \frac{1}{3} - \frac{1}{2} + \frac{1}{5} + \frac{1}{7} - \frac{1}{4} + \cdots$$

Notice that this is the way that we added the numbers in (5.45); however, now we get a different answer ($3s/2$) even though we are adding the *same* set of numbers.

Does the above example point to a conceptual inconsistency in the theory of infinite series, or maybe some mistake that we might have made? The answer is

---

[9] We will see in Chapter 8 that $s = \ln 2 \simeq 0.693$, the natural logarithm of 2.

no: *the order in which we add the terms of an infinite sequence with infinitely many sign changes may well affect its sum.*

This is just the tip of a big iceberg. Choose any number such as 0, 1/2, −4/3, $\sqrt{2}$, π, or whatever, and call it r. The *Rearrangement Theorem* of Riemann that we discuss below says that by simply rearranging the fractions in (5.44), it is possible to make them add up to the r that you picked, not just to s or $3s/2$.

An important ingredient of the rearrangement theorem is the fact that if we change all of the negative signs in the series in (5.44) to positive (i.e., take their absolute values), then we obtain the harmonic series that has an infinite sum. The fact that we introduced sign changes into a series known to be divergent is key; if we do the same to a convergent series of positive numbers, then we discover that its sum does not change. In this section, I clarify some important issues involving two types of convergence that characterize this particular facet of infinite series.

5.5.1 The alternating series

We start with the simplest type of series that displays infinitely many sign changes. The *alternating harmonic series* in which the positive and negative terms alternate is an example of such a series:

$$\sum_{n=1}^{\infty} \frac{(-1)^{n-1}}{n} = 1 - \frac{1}{2} + \frac{1}{3} - \frac{1}{4} + \cdots \tag{5.46}$$

Does this series diverge because the harmonic series does, or do the negative numbers reduce the partial sums enough for it to converge?

If you review the criteria and tests that we discussed earlier, you quickly realize that none of them apply to this series! So we begin by checking the partial sums, and it is helpful to do this pictorially. Figure 5.2 shows how it starts:

In Figure 5.2, think of a_1, a_2, etc. as the absolute values 1, 1/2, etc. from the series (5.46). We start with a_1, which is also the first partial sum s_1. Then

$$s_2 = 1 - \frac{1}{2} = a_1 - a_2 = s_1 - a_2$$

$$s_3 = 1 - \frac{1}{2} + \frac{1}{3} = a_1 - a_2 + a_3 = s_2 + a_3$$

$$s_4 = 1 - \frac{1}{2} + \frac{1}{3} - \frac{1}{4} = a_1 - a_2 + a_3 - a_4 = s_3 - a_4$$

and so on. Looking at the numbers on the right ends of the above equalities, we see how the jumps back and forth in Figure 5.2 come about (follow the arrows). Further, because a_{n+1} is less than a_n, for every n the jumps get smaller in length and result in a sequence of numbers (dots in Figure 5.2) that move closer to some

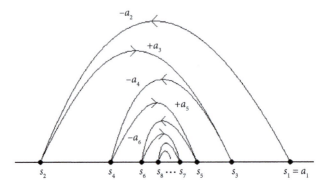

Fig. 5.2 A convergent alternating series

number in the middle. The sequence of partial sums converges to this number, so it must be the sum of the series!

Let's call the sum s. In general, it is not easy to find a formula for s (its exact value), although for the specific series (5.46) $s = \ln 2$ (the natural logarithm of 2); we demonstrate this later when discussing series of functions. The following theorem[10] clarifies and formalizes our discussion above regarding Figure 5.2.

Theorem 153 *(alternating series test for convergence) Let a_k be a sequence of non-negative real numbers such that*
  (a) a_k is nonincreasing,
  (b) $\lim_{k\to\infty} a_k = 0$,
  then the alternating series below converges:

$$\sum_{k=1}^{\infty} (-1)^{k-1} a_k \qquad (5.47)$$

Proof As we saw in the discussion preceding the theorem, for $m < n$

$$|s_n - s_m| = |a_n - a_{n-1} + a_{n-2} - a_{n-3} + \cdots + a_{m+2} - a_{m+1}|$$

By (a) the sequence a_k is nonincreasing (due to finite equivalence, we may assume that $N = 1$), so we can remove the absolute value from the right-hand side and conclude that

$$|s_n - s_m| = a_{m+1} - a_{m+2} + \cdots + a_{n-2} - a_{n-3} + a_n - a_{n-1} \le a_{m+1}$$

Since by (b) a_k converges to 0, the Cauchy convergence criterion (Theorem 142) implies that the series in (5.47) converges. ∎

---

[10] Leibniz is thought to have been the first to have come up with the alternating series test.

To illustrate Theorem 153, consider the following infinite series:

$$\sum_{n=1}^{\infty} \frac{(-1)^{n-1}}{\sqrt{n^2 - n + 1}} = 1 - \frac{1}{\sqrt{3}} + \frac{1}{\sqrt{7}} - \frac{1}{\sqrt{13}} + \cdots \qquad (5.48)$$

In the series (5.48), $a_n = 1/\sqrt{n^2 - n + 1}$. Since $n^2 - n + 1 = n(n - 1) + 1$, the denominator increases with increasing n while the numerator is fixed, so the fraction a_n decreases. In fact, as $n \to \infty$; the numbers a_n decrease all the way to 0. This is all that we need to conclude that the series in (5.48) converges; Theorem 153 doesn't give the sum any more than Theorems 148 or 150 did earlier.

However, there is a way to *estimate* how far from the (unknown) exact value of the sum we are if we cut off the series after a finite number of terms.

Error estimation in alternating series. *Note that the difference between consecutive partial sums is given by the numbers in the alternating series. Specifically,*

$$|s_n - s_{n-1}| = a_n$$

If the series converges to a number s, then as Figure 5.2 shows s is tucked in-between the consecutive jumps in the partial sum values, so for every n

$$|s_n - s| < |s_{n+1} - s_n| = a_{n+1}$$

So if we want to get an estimate of s that is accurate to d decimal places, then we want the smallest index N large enough that $|a_{N+1}| < 0.5 \times 10^{-d}$.

Suppose that we want to estimate the sum of the series in (5.48) to two decimal places; we need to find the least n that satisfies the inequality

$$\frac{1}{\sqrt{(n + 1)^2 - (n + 1) + 1}} < 0.005 = \frac{1}{200}$$

From this we get inequality

$$n^2 + n + 1 > 40000 \quad \text{or} \quad n(n + 1) > 39999$$

The quick way to find the least value of n that satisfies this inequality is to notice that $(199)(200) = 39800$, while $(200)(201) = 40200$; therefore, $N = 200$ will do the job, i.e., adding 200 terms of the series guarantees that the partial sum s_{200} gives the actual sum s of the series in (5.48) accurate to two decimal places whatever the exact value of s may be.

Note that adding 200 terms is sufficient for two-decimal accuracy, but it isn't necessary; fewer terms may in fact be enough.

What can happen if either (a) or (b) in Theorem 153 does not hold?

If (b) doesn't hold, i.e., the terms of the series do not converge to zero, then regardless of (a) the series diverges according to the divergence test discussed earlier. If (b) holds but (a) doesn't, then the series may still diverge; here is an example:

$$\frac{1}{\sqrt{2}-1} - \frac{1}{\sqrt{2}+1} + \frac{1}{\sqrt{3}-1} - \frac{1}{\sqrt{3}+1} + \cdots$$

This series diverges because the subsequence of partial sums with even indices

$$s_{2n} = \frac{2}{1} + \frac{2}{2} + \frac{2}{3} + \cdots + \frac{2}{n-1}$$

diverges to ∞ (Exercise 182).

5.5.2 Absolute convergence

Alternating series display infinitely many sign changes in a very special way. In general, the sign changes may not occur in alternate fashion one after the other. Here is an example:

$$\sum_{n=1}^{\infty} \frac{1}{n^2} \cos \frac{n\pi}{6} = \frac{\sqrt{3}}{2} + \frac{1/2}{2^2} - \frac{1/2}{4^2} + \cdots \tag{5.49}$$

The cosine factor in the numerator changes sign infinitely often in a periodic way (period 12) but not in an alternating fashion. Here is a list of its first 14 values:

n	1	2	3	4	5	6	7
$\cos(n\pi/6)$	$\sqrt{3}/2$	$1/2$	0	$-1/2$	$-\sqrt{3}/2$	-1	$-\sqrt{3}/2$
n	8	9	10	11	12	13	14
$\cos(n\pi/6)$	$-1/2$	0	$-1/2$	$-\sqrt{3}/2$	$\sqrt{3}/2$	$1/2$	0

The alternating series test does not apply to the series in (5.49). How can we tell if it converges or not?

The n^2 in the denominator is reminiscent of a convergent series, and the cosine is always between -1 and 1, so the comparison test comes to mind; however, this test does not apply if there are infinitely many sign changes.

The comparison test *would* work if we replaced $\cos(n\pi/6)$ with its absolute value because then we end up with a series of non-negative numbers:

$$\sum_{n=1}^{\infty} \frac{1}{n^2} \left| \cos \frac{n\pi}{6} \right| \tag{5.50}$$

Since

$$\frac{1}{n^2}\left|\cos\frac{n\pi}{6}\right| \le \frac{1}{n^2} \quad \text{for } n = 1, 2, 3, \ldots$$

and we have already seen that the series $\sum_{n=1}^{\infty} 1/n^2$ converges, the comparison test implies that the series in (5.50) also converges.

We have yet to show that the series in (5.49) converges. We might argue that it must, since the negative terms can only reduce the sum of the infinite series. But notice that there are infinitely many negative terms, so there is a possible issue of divergence to $-\infty$ to think about.

To prove that the series in (5.49) converges, we must determine how it relates to its absolute value series in (5.50). The next result, a consequence of Cauchy's criterion for series (Theorem 142), provides the link.

Theorem 154 (*Absolute-convergence test*) *An infinite series $\sum_{k=1}^{\infty} a_k$ of real numbers converges if the absolute values series $\sum_{k=1}^{\infty} |a_k|$ converges.*

Proof We are given that $\sum_{k=1}^{\infty} |a_k|$ converges. Consider the partial sum sequences for each of the two series:

$$s_n = \sum_{k=1}^{n} a_k = a_1 + a_2 + \cdots + a_n \quad \text{and} \quad s'_n = \sum_{k=1}^{n} |a_k| = |a_1| + |a_2| + \cdots + |a_n|$$

Because s'_n converges, the Cauchy criterion Theorem 142 says that s'_n is a Cauchy sequence, which means that $|s'_n - s'_m| \to 0$ as $m, n \to \infty$. Assuming without loss of generality that $m < n$, we have

$$|s'_n - s'_m| = (|a_1| + |a_2| + \cdots + |a_n|) - (|a_1| + |a_2| + \cdots + |a_m|)$$
$$= |a_{m+1}| + |a_{m+2}| + \cdots + |a_n|$$

Next, let's look at the partial sums of the original series $\sum_{k=1}^{\infty} a_k$ and notice that by the triangle inequality

$$|s_n - s_m| = |a_{m+1} + a_{m+2} + \cdots + a_n| \le |a_{m+1}| + |a_{m+2}| + \cdots + |a_n|$$

Therefore,

$$|s_n - s_m| \le |s'_n - s'_m|$$

Since the right-hand side of the above inequality converges to 0 as $m, n \to \infty$, the even smaller left-hand side does the same, which means that the partial sum sequence of $\sum_{k=1}^{\infty} a_k$ is Cauchy too. Now, once again Theorem 142 implies that the partial sums of $\sum_{k=1}^{\infty} a_k$ converge to a real number and completes the proof. ∎

We have shown the useful fact that if the series of absolute values converges, then so does the original series. We define

---

The series $\sum_{k=1}^{\infty} a_k$ converges absolutely if the series of absolute values $\sum_{k=1}^{\infty} |a_k|$ converges.

Now our task regarding the series in (5.49) is complete: this series converges because it converges absolutely. Along the way, we also found a new and general criterion that can be used for many different types of series that contain infinitely many sign changes. As a corollary of Theorem 154, we have the following useful extension of the earlier ratio and root tests (for convergence) to infinite series containing infinitely many sign changes.

Corollary 155 (a) The infinite series $\sum_{k=1}^{\infty} a_k$ converges absolutely if

$$\limsup_{k\to\infty} \sqrt[k]{|a_k|} = \limsup_{k\to\infty} |a_k|^{1/k} < 1$$

(b) Suppose that $a_k \neq 0$ for all k greater than or equal to some index m. Then the series $\sum_{k=1}^{\infty} a_k$ converges absolutely if

$$\limsup_{k\to\infty} \left| \frac{a_{k+1}}{a_k} \right| < 1$$

To illustrate the application of Corollary 155, consider the series

$$\sum_{n=1}^{\infty} \frac{n^2 \cos n}{2^{n/2}} = \frac{1^2 \cos 1}{\sqrt{2}} + \frac{2^2 \cos 2}{2} + \frac{3^2 \cos 3}{2\sqrt{2}} + \cdots \tag{5.51}$$

We know that $|\cos n| \leq 1$, which we can use to simplify the calculation. We apply the root test in the form of Corollary 155(a), starting with the observation

$$\left| \frac{n^2 \cos n}{2^{n/2}} \right|^{1/n} = \left(\frac{n^2}{2^{n/2}} \right)^{1/n} |\cos n|^{1/n} \leq \frac{n^{2/n}}{\sqrt{2}}$$

Recalling Theorem 151,

$$\lim_{n\to\infty} n^{2/n} = 1^2 = 1$$

we conclude that

$$\lim_{n\to\infty} \left| \frac{n^2 \cos n}{2^{n/2}} \right|^{1/n} = \frac{1}{\sqrt{2}} < 1$$

Therefore the series in (5.51) converges absolutely by Corollary 155, and therefore it converges by Theorem 154.

We could also apply Corollary 155(b) in this case; see Exercise 184.

5.5.3 Conditional convergence and rearrangements of series

It so happens that *the converse of Theorem 154 is false*: if a series converges, then it may not do so absolutely. The alternating harmonic series (5.46) is a prime example of a series that converges but not absolutely. So the following definition is meaningful:

Conditional convergence An infinite series *converges conditionally* if it converges but not absolutely; or more precisely, the infinite series $\sum_{k=1}^{\infty} a_k$ converges conditionally if it converges, but the series of absolute values $\sum_{k=1}^{\infty} |a_k|$ diverges.

Note that the occurrence of infinitely many sign changes in a series is not the same thing as conditional convergence; for example, an alternating series diverges if the magnitudes of its terms do not converge to 0.

The fact that a conditionally convergent series will diverge if the negative signs are all changed to positive is related to some of the most startling effects of infinity that sequences and series allow us to explore at a deep level.

We saw earlier that the value of the alternating harmonic series (5.46):

$$1 - \frac{1}{2} + \frac{1}{3} - \frac{1}{4} + \cdots$$

changes if we add its terms in a different order where we rearrange infinitely many of its terms. Different rearrangements generally result in different values; in fact, *every real number* is a possible value for some rearrangement of the terms of (5.46). To illustrate the essense of the process, we rearrange the fractions in the alternating harmonic series so that the rearranged series converges to, say, 0.

The series starts with a 1, so we go to $-1/2$ or the first negative number and add that to 1. The result is $1/2$, which is still greater than 0, so we move on to the next negative number in (5.46) and continue until we reach 0, or just pass it:

$$1 - \frac{1}{2} - \frac{1}{4} - \frac{1}{6} - \frac{1}{8} \simeq -0.042 < 0$$

Now if we add the first positive number after 1, which is 1/3, then we get

$$1 - \frac{1}{2} - \frac{1}{4} - \frac{1}{6} - \frac{1}{8} + \frac{1}{3} \simeq 0.292 > 0$$

This is back to positive (past 0 again); but notice that it is closer to 0 than 1 was, and this is a small, but important, gain.

Next, we go the next available negative number $-1/10$, which is too small when added to the above sum to reach 0. So continue adding negative numbers till we again reach or pass 0:

$$1 - \frac{1}{2} - \frac{1}{4} - \frac{1}{6} - \frac{1}{8} + \frac{1}{3} - \frac{1}{10} - \frac{1}{12} - \frac{1}{14} - \frac{1}{16} \simeq -0.026 < 0$$

We have overshot 0, but notice that this time the total is closer to zero than the first time we overshot 0 above. We have made more progress.

To get back toward 0, we add positive fractions like before:

$$1 - \frac{1}{2} - \frac{1}{4} - \frac{1}{6} - \frac{1}{8} + \frac{1}{3} - \frac{1}{10} - \frac{1}{12} - \frac{1}{14} - \frac{1}{16} + \frac{1}{5} \simeq 0.174 > 0$$

Again, though we have overshot 0, the total is closer than it was in the previous step where we had a value of about 0.292.

Figure 5.3 captures the above steps in the process; as you can see, it looks like Figure 5.2 though a bit more complex since the back and forth jumps may take several steps in this case. The question now is, if the above process is continued indefinitely, will the total approach 0?

Recall that at each step of the process when a change in direction (or sign) occurs, the numbers to the right of zero and those to the left get smaller:

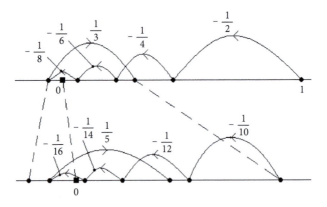

Fig. 5.3 Illustrating Riemann's rearrangement argument

$$-0.042 < -0.026 < -0.018 < \cdots < 0 < \cdots < 0.174 < 0.292 < 1$$

This observation shows that the overshoot sequence to the left of 0 and the one to its right *both* converge to 0, and the key reason for this is that the individual numbers in the series ($1/n$ or $-1/n$) approach 0. For example, in the first step, overshooting 0 occurred by adding $1/3$; but in the second step, we added the smaller number $1/5$. Later steps involve even smaller numbers. Going in the other direction, the negative number that just gets the sum past 0 is smaller in each step: first, $-1/8$; then, $-1/16$; and so on.

Another important question that is not very transparent in the above discussion is this: *how do we know that we will in fact overshoot 0 in each direction?*

To see why this is a relevant question, consider a number that is farther from 1 than 0 was; say $\pi \simeq 3.1416$. Although this is not much farther away, starting from 1, we must add enough positive fractions to reach (or pass) π. This is accomplished with no fewer than 76 fractions:

$$1 + \frac{1}{3} + \frac{1}{5} + \cdots + \frac{1}{151} \simeq 3.1471$$

The next step requires just one negative term to pass π again:

$$1 + \frac{1}{3} + \frac{1}{5} + \cdots + \frac{1}{151} - \frac{1}{2} \simeq 2.6471$$

Although the number 2.6471 is closer to π than 1 was, we must add smaller available positive fractions to get back to the neighborhood of π; in fact, 128 *more fractions* are needed to pass π again:

$$1 + \frac{1}{3} + \frac{1}{5} + \cdots + \frac{1}{151} - \frac{1}{2} + \frac{1}{153} + \frac{1}{155} + \cdots + \frac{1}{409} \simeq 3.1433$$

and so on. The farther the chosen number r is from 1, the more terms (positive or negative) are needed to reach or pass r. For example, if $r = 6$ (a number with modest size), then starting from 1, nearly $23,000$ odd fractions are needed to pass it. So, again, how can we be sure that we will reach or pass r in each direction? The answer is *because the series does not converge absolutely.*

The rearrangement theorems

These above observations constitute the essense of the proof of the rearrangement theorem below.[11] Before discussing it, we must formally define what we mean by "rearrangements."

---

[11] This result appeared in a 1866 paper of the German mathematician Bernhard Riemann (1826–1866) published after his death; but based on his correspondences, he likely proved it around 1853.

Rearrangement of terms of a sequence *Given a sequence a_k of real numbers $a_1, a_2, a_3, \ldots$ a rearrangement of this sequence is a sequence $a_{j(k)}$, i.e., $a_{j(1)}, a_{j(2)}, a_{j(3)}, \ldots$ where $j : \mathbb{N} \to \mathbb{N}$ is a bijection. The corresponding rearrangement of the infinite series $\sum_{k=1}^{\infty} a_k$ is the series $\sum_{j=1}^{\infty} a_{j(k)}$.*

The rearrangement of (5.46) that we used earlier to get $3s/2$ is the following:

$$a_{j(1)}, a_{j(2)}, a_{j(3)}, a_{j(4)}, a_{j(5)}, a_{j(6)}, \ldots = a_1, a_3, a_2, a_5, a_7, a_4, \ldots$$

Now we state and formally prove the rearrangement theorem.

Theorem 156 *(The rearrangement theorem, conditional convergence) Assume that an infinite series $\sum_{k=1}^{\infty} a_k$ converges conditionally. If r is any given real number, then there is a rearrangement $a_{j(k)}$ of the sequence of terms of the series such that*

$$\sum_{k=1}^{\infty} a_{j(k)} = r.$$

Proof Let $a_1^+, a_2^+, a_3^+, \ldots$ be all the positive terms of the series $\sum_{k=1}^{\infty} a_k$ in their originally given order, and let $a_1^-, a_2^-, a_3^-, \ldots$ be the absolute values of all the negative terms, also in the original order. We claim that both of the positive term series $\sum_{k=1}^{\infty} a_k^+$ and $\sum_{k=1}^{\infty} a_k^-$ diverge (we ignore all the zero terms since they do not contribute to the sum).

To prove this claim, define

$$\alpha_k = \frac{|a_k| + a_k}{2} \qquad \beta_k = \frac{|a_k| - a_k}{2}$$

and observe that $\alpha_k - \beta_k = a_k$ and $\alpha_k + \beta_k = |a_k|$ for all indices k. Also notice that if $a_k > 0$, then $\alpha_k = a_k$ and $\beta_k = 0$; whereas if $a_k < 0$, then $\alpha_k = 0$ and $\beta_k = |a_k|$.

Since $\sum_{k=1}^{\infty} |a_k|$ diverges, and for every n

$$\sum_{k=1}^{n} \alpha_k + \sum_{k=1}^{n} \beta_k = \sum_{k=1}^{n} (\alpha_k + \beta_k) = \sum_{k=1}^{n} |a_k| = \infty$$

it follows that at least one of the two series $\sum_{k=1}^{\infty} \alpha_k$ or $\sum_{k=1}^{\infty} \beta_k$ diverges. Further, since $\sum_{k=1}^{\infty} a_k$ converges, and for every n

$$\sum_{k=1}^{n} \alpha_k - \sum_{k=1}^{n} \beta_k = \sum_{k=1}^{n} (\alpha_k - \beta_k) = \sum_{k=1}^{n} a_k$$

both series $\sum_{k=1}^{\infty} \alpha_k$ *and* $\sum_{k=1}^{\infty} \beta_k$ *must diverge as* $n \to \infty$ (otherwise, the right-hand side cannot converge).

Next, we observe that for all n

$$\sum_{k=1}^{n} a_k^+ = \sum_{k=1}^{n} \alpha_k \quad \text{and} \quad \sum_{k=1}^{n} a_k^- = \sum_{k=1}^{n} \beta_k$$

so that as $n \to \infty$, both $\sum_{k=1}^{\infty} a_k^+ = \infty$ and $\sum_{k=1}^{\infty} a_k^- = \infty$, and the above claim is proved.

Now we proceed to construct sequences of indices m_j and n_j such that the rearranged series

$$a_1^+ + \cdots + a_{m_1}^+ - a_1^- - \cdots - a_{n_1}^- + a_{m_1+1}^+ + \cdots + a_{m_2}^+ - a_{n_1+1}^- - \cdots - a_{n_2}^- + \cdots \quad (5.52)$$

adds up to the selected number r. The construction employs the idea that we discussed before the theorem.

We define m_1 to be the smallest integer such that

$$A_1 = a_1^+ + \cdots + a_{m_1}^+ > r$$

Such m_1 exists since $\sum_{k=1}^{\infty} a_k^+ = \infty$. Similarly, because $\sum_{k=1}^{\infty} a_k^- = \infty$, there is a least index n_1 such that

$$B_1 = A_1 - a_1^- - \cdots - a_{n_1}^- < r$$

At this point, we add enough of the a_k^+ terms to *just* pass to the right of r

$$A_2 = B_1 + a_{m_1+1}^+ + \cdots + a_{m_2}^+$$

where m_2 is the least integer that does the job. If we repeat the above index selection process, then after j steps we have

$$A_j = B_{j-1} + a_{m_j+1}^+ + \cdots + a_{m_{j+1}}^+ > r \qquad (5.53)$$

$$B_j = A_{j-1} - a_{n_j+1}^- - \cdots - a_{n_{j+1}}^- < r \qquad (5.54)$$

Since m_{j+1} is the least index for which (5.53) holds, it follows that

$$A_j - a_{m_{j+1}}^+ < r$$

Therefore,

$$0 < A_j - r < a_{m_{j+1}}^+$$

Similarly, from the least nature of n_{j+1} in (5.54), it follows that

$$B_j + a_{n_{j+1}}^- > r$$

Therefore,

$$0 < r - B_j < a_{n_{j+1}}^-$$

By hypothesis, $\sum_{k=1}^{\infty} a_k$ converges, so the (contrapositive of) the divergence test gives

$$\lim_{k\to\infty} a_k^+ = \lim_{k\to\infty} a_k^- = 0$$

This fact, and the squeeze theorem, then imply that

$$\lim_{j\to\infty} A_j = r = \lim_{j\to\infty} B_j \tag{5.55}$$

Finally, let S_n be the nth partial sum of (5.52). If the last term of S_n is of type a_k^+, then

$$B_k < S_n \le A_{k+1}$$

Otherwise, the last term of S_n is of type $-a_k^-$ so that

$$B_{k+1} \le S_n < A_{k+1}$$

From these observations and (5.55), we use the squeeze theorem to conclude that $\lim_{n\to\infty} S_n = r$ and complete the proof of the theorem. ∎

Now a natural question to consider is, - what about rearranging the terms of an *absolutely* convergent series? The next result shows that the sum of such a series is not affected by any rearrangement of its terms.[12]

Theorem 157 (*The rearrangement theorem, absolute convergence*) *If an infinite series $\sum_{k=1}^{\infty} a_k$ converges absolutely to some number s, then every rearrangement of $\sum_{k=1}^{\infty} a_k$ converges to s, too.*

Proof Given that $\sum_{k=1}^{\infty} |a_k| < \infty$ by assumption, Theorem 144 implies that its tail end converges to 0, i.e.,

$$\lim_{m\to\infty} \sum_{k=m}^{\infty} |a_k| = 0$$

Thus, for each given $\varepsilon > 0$, there is a positive integer N such that for all $n \ge m \ge N$

$$\sum_{k=m}^{n} |a_k| < \frac{\varepsilon}{2} \tag{5.56}$$

---

[12] This result first appeared in 1837 in a paper by the German mathematician Peter Gustav Lejeune Dirichlet (1805–1859). It was likely proved by Riemann also.

Let $\sum_{k=1}^{\infty} a_{j(k)}$ be a rearrangement of $\sum_{k=1}^{\infty} a_k$. Since $j(k)$ is a bijection, there is an integer $\ell \geq N$ such that

$$\{1, 2, ..., N\} \subset \{j(1), j(2), ..., j(\ell)\}$$

Now consider the partial sums

$$s_n = \sum_{k=1}^{n} a_k \qquad s'_n = \sum_{k=1}^{n} a_{j(k)}$$

For $n \geq \ell$, the difference $s_n - s'_n$ contains only terms with indices a_k with $k \geq N$ because by our choice of ℓ, the terms $a_1, a_2, ..., a_N$ appear in both s_n and s'_n and thus they cancel out. Therefore, by (5.56) and the triangle inequality

$$|s_n - s_n'| \leq \left| \sum_{k=N}^{n} [a_k - a_{j(k)}] \right| \leq \sum_{k=N}^{n} |a_k| + \sum_{k=N}^{n} |a_{j(k)}| < \frac{\varepsilon}{2} + \frac{\varepsilon}{2} = \varepsilon$$

It follows that $s_n - s'_n$ converges to 0, i.e.,

$$\lim_{n \to \infty} (s_n - s'_n) = 0$$

Since $\sum_{k=1}^{\infty} a_k$ converges by Theorem 154, $\lim_{n \to \infty} s_n$ exists so that $\lim_{n \to \infty} s'_n$ must exist too and

$$\lim_{n \to \infty} s'_n = \lim_{n \to \infty} s_n$$

Thus, $\sum_{k=1}^{\infty} a_{j(k)}$ converges to the same value as $\sum_{k=1}^{\infty} a_k$. ∎

5.6 Real numbers as infinite series, Liouville numbers

Rational numbers are rather easy to understand because each one is just a ratio of two integers. When we divide these integers using long division, we inevitably reach a stage where digits begin to repeat in a periodic fashion. But irrational numbers are more opaque; there is no analog of the long division method that generates them. We can, however, represent all real numbers using infinite series.

In Chapter 4, we showed real numbers can be defined basically as Cauchy sequences of rational numbers. With series we have a more direct representation in terms of numbers (rather than equivalence classes) as well as a greater flexibility, in the sense that every real number can be represented not only by its decimal expansion but also by an infinite series of only 0's and 1's (binary expansion).

In this section, we show that every real number r can be represented as an infinite series, the familiar decimal expansion of r. We also introduce a class of irrational numbers known as "Liouville numbers." The decimal expansions of these numbers are entirely predictable, unlike $\sqrt{2}$ or π.

Series expansions of real numbers

When we write down some of the digits of the decimal expansion of π, we are in fact writing the first few terms of the infinite series

$$\pi = \frac{3}{10^0} + \frac{1}{10^1} + \frac{4}{10^2} + \frac{1}{10^3} + \frac{5}{10^4} + \frac{9}{10^5} + \frac{2}{10^6} + \cdots \tag{5.57}$$

Although we don't have a formula that gives the numerator of 10^n for every value of n, we know that the infinite series on the right converges because the number in the numerator of each fraction is just an integer between 0 and 9, while the number in the denominator is an increasing power of 10. A simple comparison with the geometric series ensures convergence.

The pertinent question at this stage is whether it is possible to write *all* real numbers as an infinite series like the one in (5.57). Given the familiarity of decimal expansions, we expect that the answer is yes. Now we explain how.

Before stating the next result, recall that every real number has an integer part and a fractional part. Since the integer part is just an integer that can simply be added to the fractional part, we need only consider the fractional part; that is, we assume that our real number is between 0 and 1.

Theorem 158 *(Representation of real numbers by infinite series) Let r be a real number in the interval $[0, 1]$ and $b \geq 2$ a fixed positive integer. Then there is a sequence of integers d_n in the finite set $D = \{0, 1, 2, ..., b-1\}$ such that*

$$\sum_{n=1}^{\infty} \frac{d_n}{b^n} = \frac{d_1}{b^1} + \frac{d_2}{b^2} + \frac{d_3}{b^3} + \cdots \tag{5.58}$$

This series, typically abbreviated $0. d_1 d_2 d_3...$, is the "base-b expansion" (decimal expansion if $b = 10$) of r, and the integers d_n are its "digits." The series in (5.58) is uniquely associated with r unless r is a rational number of type m/b^k where $k, m \in \mathbb{N}$. Each of these rationals has two expansions, one ending in 0's and the other ending in $b - 1$'s.

Proof First, note that for every real number x

$$[x] \leq x < [x] + 1$$

where $[x]$ is the greatest integer less than or equal to x.

If $x = br$, and $d_1 = [br]$, then $x \in [0, b]$, $d_1 \in D$ and

$$d_1 \leq br < [br] + 1 = d_1 + 1$$

so dividing by b, we get

$$\frac{d_1}{b} \leq r < \frac{d_1}{b} + \frac{1}{b} \tag{5.59}$$

Therefore, d_1 is the first digit of the expansion of r. Next, define $r_1 = br - d_1$ and $d_2 = [br_1]$. Then $d_2 = [b^2 r - bd_1]$, so we have

$$d_2 \leq b^2 r - bd_1 < [b^2 r - bd_1] + 1 = d_2 + 1$$

It follows that

$$bd_1 + d_2 \leq b^2 r < bd_1 + d_2 + 1$$

and if we divide by b^2, then

$$\frac{d_1}{b} + \frac{d_2}{b^2} \leq r < \frac{d_1}{b} + \frac{d_2}{b^2} + \frac{1}{b^2} \tag{5.60}$$

Now d_2 is the second digit of the expansion of r. Notice that in going from (5.59) to (5.60), we came closer to r from both left and right, and the approximation error was lowered ten times, from $1/b$ to $1/b^2$. If we continue the multiplication and truncation process above, we obtain a sequence of digits $d_1, d_2, \ldots$ recursively. After n steps

$$\frac{d_1}{b} + \frac{d_2}{b^2} + \cdots + \frac{d_n}{b^n} \leq r < \frac{d_1}{b} + \frac{d_2}{b^2} + \cdots + \frac{d_n}{b^n} + \frac{1}{b^n}$$

The sum on the left-hand side above is the nth partial sum of the series in (5.58); call it s_n and subtract it from r to get

$$0 \leq r - s_n < \frac{1}{b^n}$$

Since $1/b^n \to 0$ as $n \to \infty$, by the squeeze theorem, the sequence s_n converges to r. Therefore, r is given by the infinite series in (5.58) as claimed in the theorem, i.e.,

$$r = \frac{d_1}{b^1} + \frac{d_2}{b^2} + \frac{d_3}{b^3} + \cdots \tag{5.61}$$

Next, we prove the (almost) uniqueness of the decimal expansion as the infinite series in (5.58). Suppose that there is another decimal expansion, say, $r = 0.\, d'_1 d'_2 d'_3 \cdots$, that is, the infinite series

$$r = \frac{d'_1}{b^1} + \frac{d'_2}{b^2} + \frac{d'_3}{b^3} + \cdots \tag{5.62}$$

where $d'_n \neq d_n$ for some values of n. If k is the first index where the digits differ, i.e., $d'_k - d_k \neq 0$ but $d'_j = d_j$ for $j < k$, then subtracting the series in (5.61) from that in (5.62) and combining like terms gives

$$0 = \frac{d'_k - d_k}{b^k} + \frac{d'_{k+1} - d_{k+1}}{b^{k+1}} + \frac{d'_{k+2} - d_{k+2}}{b^{k+2}} + \cdots$$

Multiplying the last equality by b^k and simplifying gives

$$0 = d'_k - d_k + \frac{d'_{k+1} - d_{k+1}}{b^1} + \frac{d'_{k+2} - d_{k+2}}{b^2} + \cdots \tag{5.63}$$

Note that since

$$\left| d'_{k+1} - d_{k+1} \right|, \left| d'_{k+2} - d_{k+2} \right|, \ldots \leq b - 1$$

it follows that

$$\frac{\left| d'_{k+1} - d_{k+1} \right|}{b^1} + \frac{\left| d'_{k+2} - d_{k+2} \right|}{b^2} + \cdots \leq \frac{b-1}{b^1} + \frac{b-1}{b^2} + \cdots$$

$$= \frac{(b-1)/b}{1 - 1/b} = 1$$

On the other hand, $\left| d'_k - d_k \right| \geq 1$ since d'_k and d_k are integers. So the only way that (5.63) can hold is if every one of the numerators $d'_{k+1} - d_{k+1}, d'_{k+2} - d_{k+2}, \ldots$ is $b - 1$ if $d'_k - d_k = -1$, or $-(b - 1)$ if $d'_k - d_k = 1$. Therefore, the decimal expansion in (5.58) is unique unless r has one of the following forms:

$$r = \frac{d_1}{b^1} + \frac{d_2}{b^2} + \cdots + \frac{d_k}{b^k} \tag{5.64}$$

or

$$r = \frac{d_1}{b^1} + \cdots + \frac{d_k}{b^k} + \frac{9}{b^{k+1}} + \frac{9}{b^{k+2}} + \cdots \tag{5.65}$$

In either case, (5.64) or (5.65), r is a rational number of type m/b^k as stated in the statement of the theorem. In (5.64) after combining all the fractions, the numerator is found to be:

$$m = b^{k-1} d_1 + b^{k-2} d_2 + \cdots + d_k$$

In (5.65), first we note that

$$\frac{9}{b^{k+1}} + \frac{9}{b^{k+2}} + \cdots = \frac{1}{b^k} \left(\frac{9}{b} + \frac{9}{b^2} + \cdots \right)$$

$$= \frac{1}{b^k} \frac{9/b}{1 - 1/b} = \frac{1}{b^k}$$

so that combining terms gives

$$m = b^{k-1}d_1 + b^{k-2}d_2 + \cdots + d_k + 1$$

This completes the proof.   ∎

If $b = 2$, then the above theorem gives the *binary expansion* of r with digits being 0 or 1. If $b = 3$, then we get the *ternary expansion* and so on.

The infinite series representation theorem above is not a tool for calculating or approximating irrational numbers, but it does confirm that every real number has a decimal expansion. We discuss an important consequence of the existence of this expansion next.

The cardinality of the set of all real numbers

Earlier we used Cantor's diagonal argument to prove that $\mathbb{R}$ is uncountable. This differentiates $\mathbb{R}$ from smaller sets like $\mathbb{N}$ or $\mathbb{Q}$, but it does not say *how large* $\mathbb{R}$ is; we are now ready to show that $\mathbb{R}$ has the same cardinal number as the power set $\mathcal{P}(\mathbb{N})$, i.e., $2^{\aleph_0}$.

By Theorem 158, every real number r where $0 \leq r \leq 1$ has a unique expansion as the infinite series in (5.58); for the exceptional rational numbers, we just take one of the two possibilities, say, the one that ends in all $(b-1)$'s, and discard the other. This gives a bijection between infinite series of type (5.58) and all sequences $d_1, d_2, d_3, \cdots$ of integers between 0 and $b-1$, excluding those sequences that end in zeros (eventually 0 sequences).

In particular, with $b = 2$, we showed earlier in our discussion of infinite products of sets that the set of all such sequences of 0's and 1's has cardinality $2^{\aleph_0}$. We state this important fact as a corollary to Theorem 158.

Corollary 159 *The set $\mathbb{R}$ of all real numbers has cardinality $2^{\aleph_0}$.*

An interesting by-product of this fact is that there is a bijection between $\mathbb{R}$ and the power set $\mathcal{P}(\mathbb{N})$ so every subset of $\mathbb{N}$ uniquely corresponds to a real number. Further, the set of all *transcendental* numbers (numbers that are not algebraic, see Chapter 4) has cardinal number $2^{\aleph_0}$.

The Liouville numbers

We have seen that the decimal expansions of irrational numbers can be unpredictable, like those of π or $\sqrt{2}$. There is no known formula that gives the digit d_n for every index n in the decimal expansions of these numbers (or expansions in any base). But is this true of all irrational numbers? Is it a characteristic property?

The rational numbers have expansions that are periodic (the digits d_n repeat in a fixed pattern). The expansion of an irrational number as an infinite series does not have this property; the digits occur in a non-periodic or non-repetitive way.

But this is not the same thing as disorderly or unpredictable. For instance, here is an irrational number whose decimal expansion is quite orderly and predictable:

$$0.123..89101112..1819202122..2829303132 \cdots$$

This number, which is simply a concatenation of all non-negative integers, is obviously irrational since its digits do not appear in a periodic fashion.[13]

With the help of infinite series, we can discover more: the set of all irrational numbers that have orderly or predictable decimal expansions also has cardinal number $2^{\aleph_0}$, the same as the set of all real numbers. And in a historically remarkable twist, this fact was known three decades before Cantor's work on infinite sets (not in terms of cardinalities, of course).

Back in 1844, the French mathematician Joseph Liouville (1809–1882) discovered certain irrational numbers with interesting features. These *Liouville numbers*, as they are now known, are defined (in base 10) by infinite series as

$$\sum_{n=1}^{\infty} \frac{d_n}{10^{n!}} = \frac{d_1}{10^1} + \frac{d_2}{10^2} + \frac{d_3}{10^6} + \frac{d_4}{10^{24}} + \frac{d_5}{10^{120}} + \cdots \qquad (5.66)$$

In particular, when $d_n = 1$ for every n, we get the so-called *Liouville's constant*:

$$L = 0.11000100000000000000000100 \cdots \qquad (5.67)$$

where we see the in-between zeros grow in number quite rapidly due to the fast growth of the factorial. Since the decimal expansions of Liouville numbers are not eventually periodic, all of them are irrational. Furthermore, by choosing the numbers $d_1, d_2, d_3, \ldots$ to be either 0 or 1, the same argument above that showed the cardinality of $\mathbb{R}$ is $2^{\aleph_0}$ also proves the following:

The set of all Liouville numbers is uncountable with cardinality $2^{\aleph_0}$.

I emphasize that this does not mean the irrational numbers like $\sqrt{2}$ or π with *unpredictable* decimal expansions are few. Recall that proper subsets of infinite sets can be in bijective correspondence with the entire set, so it is entirely possible to have two *disjoint* subsets of $\mathbb{R}$ each having the same cardinality $2^{\aleph_0}$.

We can easily locate Liouville's constant in (5.67) on the number line; writing it as an infinite series.

$$L = \frac{1}{10^1} + \frac{1}{10^2} + \frac{1}{10^6} + \frac{1}{10^{24}} + \frac{1}{10^{120}} + \cdots$$

[13] This number is named after Champernowne after it was introduced by him in 1933. In 1937, K. Mahler showed that this number is in fact transcendental.

we can see that L is greater than the rational number $1/10 = 0.1$ but less than $0.\bar{1}$, which is a convergent geometric series

$$0.\bar{1} = \frac{1}{10^1} + \frac{1}{10^2} + \frac{1}{10^3} + \cdots = \frac{1/10}{1 - 1/10} = \frac{1}{9}$$

So L is sandwiched between two rational numbers

$$0.1 < L < \frac{1}{9} \approx 0.1111$$

These bounds for L can be improved; see the exercises.

When we look at the Liouville constant, we see that there is nothing unpredictable about its decimal expansion: if a digit is indexed by a factorial, then it is 1; otherwise, it is 0, and that is all there is to it!

Given that the set of all Liouville numbers is equipollent to the set of all real numbers, we see that there are lots of irrational numbers with highly predictable decimal expansions.

Additional properties of Liouville numbers. Liouville numbers have many other interesting properties; I list two without proof below because they relate to what we have been doing in this book without going into the (substantial) technical details. Recalling that well-known transcendental numbers like π have unpredictable decimal expansions, it seems natural to conclude that irrational numbers with well-behaved or predictable decimal expansions aren't transcendental. But Liouville proved otherwise:

All Liouville numbers are transcendental.

In other words, Liouville numbers are not roots of polynomials with rational coefficients. Historically, Liouville numbers were the first transcendental numbers to be discovered, before e was shown to be transcendental in 1873 by Hermite. Prior to Liouville numbers, the existence of transcendental numbers was suspected (prominent candidates were π and e) but not yet proved.

Another interesting property of Liouville numbers is the following:

The set of all Liouville numbers is dense in $\mathbb{R}$.

In other words, between any pair of real numbers there is a Liouville number. Thus, Liouville numbers form an uncountable dense subset of the real numbers, like the irrational numbers. Of course, when it comes to density, the smaller the set the more useful it is; and the set of Liouville numbers is a proper subset of all irrational numbers, all of which are transcendental. But the the set of rational numbers is much smaller, and they are much simpler numbers; for these reasons, they are the most important dense subset of $\mathbb{R}$.

5.7 Exercises

Exercise 160 *Consider the infinite series*

$$1 + 1 + 1 - 1 - 1 + 1 + 1 + 1 - 1 - 1 + \cdots$$

in which the pattern $1 + 1 + 1 - 1 - 1$ *keeps repeating. Since this pattern has a net positive value, we conjecture that this series diverges to infinity. To prove it we proceed as follows:*

(a) Calculate and list the values of the 15 partial sums $s_1, \ldots, s_{15}$. *Can you guess what* s_{20} *is without any further additions? How about* s_{100}? s_{1000}?

(b) What is special about every fifth partial sum $s_5, s_{10}, s_{15}, \ldots$? *How do* s_n *compare with these if* $n \neq 5, 10, 15, \ldots$?

(c) Use your observations in (a) and (b) to prove that the above series diverges to ∞.

Exercise 161 *Consider the series:*

$$\sum_{n=1}^{\infty} \frac{2}{n(n+2)}$$

(a) Use an argument similar to the one for telescoping series to show that this series adds up to $3/2$. *To see the cancellations in this case, consider listing a partial sum that has enough numbers in it, say,* s_{10}, *in a form that is similar to (5.14).*

(b) Consider extending the line of reasoning in (a) to the series

$$\sum_{n=1}^{\infty} \frac{k}{n(n+k)}$$

where k *is any fixed (but unspecified) natural number. What do you think the above series adds up to? Examining* $k = 3$ *can be helpful if you get stuck.*

Exercise 162 *Suppose we change the first few terms of the telescoping series (5.12) to get the new equivalent series*

$$\frac{3}{2} - \frac{8}{3} + \frac{1}{(4)(5)} + \frac{1}{(5)(6)} + \frac{1}{(6)(7)} + \cdots$$

Explain why this series converges and what its value is.

Exercise 163 *The tail end of the telescoping series in (5.12) is*

$$\sum_{k=m+1}^{\infty} \frac{1}{k(k+1)} = \frac{1}{(m+1)(m+2)} + \frac{1}{(m+2)(m+3)} + \frac{1}{(m+3)(m+4)} + \cdots$$

(a) Verify that

$$\frac{1}{(m+1)(m+2)} = \frac{1}{m+1} - \frac{1}{m+2}, \quad \frac{1}{(m+2)(m+3)} = \frac{1}{m+2} - \frac{1}{m+3}, \quad etc.$$

Conclude that

$$\sum_{k=m+1}^{n} a_k = \frac{1}{m+1} - \frac{1}{n+1}$$

(b) By first letting $n \to \infty$, find a formula for the tail $\sum_{k=m+1}^{\infty} a_k$. Then use it to verify that $\sum_{k=m+1}^{\infty} a_k$ converges to 0 as $m \to \infty$.

Exercise 164 *Prove that the infinite series*

$$\frac{1}{2} + \frac{1}{4} + \frac{1}{6} + \cdots$$

obtained by dropping all terms of the harmonic series with odd denominators is divergent (what is twice this series?).

Exercise 165 *Let $a_k \geq 0$ for all indices k. Prove that the series $\sum_{k=1}^{\infty} a_k$ converges if and only if the sequence of partial sums s_n has a convergent subsequence s_{n_k}.*

Exercise 166 *Consider the following finite sum where a and d are real numbers:*

$$s_n = a + (a+d) + (a+2d) + \cdots + [a + (n-1)d]$$

Derive a formula for this finite sum, known as an "arithmetic progression." Rewrite s_n backward right under the above sum to make two rows. Then add the two rows to get $2s_n$ on the left and a simplified expression on the right. Solve this equation to get the formula[14]

$$s_n = \frac{n}{2}[2a + (n-1)d] \tag{5.68}$$

Use (5.68) to find a formula for each of the following:

(a) $1 + 2 + 3 + \cdots + n$ *(b) $1 + 3 + 5 + \cdots + 2n - 1$*

Exercise 167 *Prove that $\lim_{n\to\infty} a^n = 0$ if $0 < a < 1$, and $\lim_{n\to\infty} a^n = \infty$ if $a > 1$. By way of contradiction, suppose that $\lim_{n\to\infty} a^n = b$ where b is a positive real number. Use the limit properties to show that this assumption implies $a = 1$ and reach a contradiction; see the discussion before Theorem 141. This exercise can be considered a lemma to Theorem 141.*

---

[14] This derivation is called the "accountant's method" since it is similar to the way accountants used to double-check the sum of entries in a list in the pre-spreadsheet days.

Exercise 168 *Calculate the sum of the first 10 terms of each of the geometric series below using the formula in (5.17). Determine if the infinite series converges and if so, then find its total. Note that the series are not written in standard form, so the number a in (5.15) is not necessarily the first number listed below; you need do a little detecting to find a.*

(a)
$$\frac{1}{4} + \frac{9}{16} - \frac{27}{64} + \cdots$$

(b)
$$-\frac{1}{3} + \frac{16}{9} - \frac{64}{27} + \cdots$$

(c)
$$\sqrt{2} + 1 + \frac{1}{\sqrt{2}} + \frac{1}{2} + \cdots$$

Exercise 169 *Prove the rest of Theorem 139.*

Exercise 170 *Using appropriate series for comparison, prove that each of the following series converges:*

(a) $\displaystyle\sum_{n=1}^{\infty}\left(\frac{\cos n}{n}\right)^2$ (b) $\displaystyle\sum_{n=0}^{\infty}\frac{1000}{2^n + 1}$ (c) $\displaystyle\sum_{n=1}^{\infty}\frac{1}{n}\sin\left(\frac{1}{n}\right)$

In both (a) and (b) specify their upper bounds, i.e., numbers that the values of each of the given series cannot exceed. In (c) recall that $\sin x < x$ if $x > 0$.

Exercise 171 *Prove that each of the following series diverges using any efficient test for divergence:*

(a) $\displaystyle\sum_{n=1}^{\infty}\frac{2 - \cos n}{n}$ (b) $\displaystyle\sum_{n=1}^{\infty}\frac{2n - 1}{6n}$ (c) $\displaystyle\sum_{n=1}^{\infty}\frac{1}{n}\cos\left(\frac{1}{n}\right)$

Exercise 172 *Let m be a fixed integer in $\mathbb{Z}$.*
   (a) Use the comparison test with $\sum_{n=1}^{\infty} 1/n^2$ and mathematical induction to prove that the following series converges if $m > 1$ (i.e., $m \geq 2$)

$$\sum_{n=1}^{\infty}\frac{1}{n^m}$$

(b) Explain why the above series diverges if $m \leq 1$.

Exercise 173 *Fill in the details of the proof of Corollary 147 (the limit comparison test).*

Exercise 174 *Prove that the following infinite series converges using the ratio test:*

$$\sum_{n=1}^{\infty} \frac{[2+(-1)^n]n}{4^n} = \frac{1}{4} + \frac{3}{8} + \frac{3}{64} + \frac{3}{64} + \cdots$$

Exercise 175 *Determine whether each series converges or diverges and prove your conclusion using the ratio test:*

$$(a) \quad \sum_{n=1}^{\infty} \frac{2^n}{n^{10}} \qquad\qquad (b) \quad \sum_{n=1}^{\infty} \frac{2^n n^2}{n!}$$

Exercise 176 *(a) Can the ratio test be applied to the series in Exercise 171(b)? Explain.*

(b) Can the root test be applied to either of the following series? Explain.

$$\sum_{n=1}^{\infty} \frac{1}{n}, \qquad \sum_{n=1}^{\infty} \frac{1}{n^2}$$

Exercise 177 *Determine whether each series converges or diverges, and prove your conclusion using the root test:*

$$(a) \quad \sum_{n=1}^{\infty} \frac{n^n}{2^n n^2} \qquad\qquad (b) \quad \sum_{n=1}^{\infty} \frac{n^{(2n)}}{2^{(n^2)}}$$

Exercise 178 *Determine whether each series converges or diverges, and prove your conclusion using appropriate tests:*

$$(a) \quad \sum_{n=1}^{\infty} \frac{(n!)^2}{(2n)!} \qquad (b) \quad \sum_{n=1}^{\infty} \frac{2^{(n^2)}}{(2n)!} \qquad (c) \quad \sum_{n=1}^{\infty} (\sqrt[n]{n} - 1)^n$$

Exercise 179 *Prove that $\sqrt[n]{n!} < n$ for all $n = 1, 2, 3, \ldots$ (consider the nth power of each side of the inequality). Conclude that for every real number $p > 1$*

$$\lim_{n\to\infty} \frac{\sqrt[n]{n!}}{n^p} = 0$$

Note. It can be shown that

$$\lim_{n\to\infty} \frac{\sqrt[n]{n!}}{n} = \frac{1}{e}$$

so $\sqrt[n]{n!}$ is nearly equal to n/e for large values of n. We discuss the exponential function and related facts in later chapters.

Exercise 180 *Assume that the series $\sum_{k=1}^{\infty} a_k$ converges either by the ratio test or by the root test. Prove that for each positive integer m the series $\sum_{k=1}^{\infty} k^m a_k$ also converges.*

Exercise 181 *Prove that the "alternating p-series" converges for all $p > 0$:*

$$\sum_{n=1}^{\infty} \frac{(-1)^{n-1}}{n^p} = 1 - \frac{1}{2^p} + \frac{1}{3^p} - \frac{1}{4^p} + \cdots$$

Exercise 182 *Verify that the following alternating series satisfies (b) but not (a) in Theorem 153, and prove that it diverges (see the text):*

$$\sum_{k=1}^{\infty} \frac{(-1)^{k-1}}{\sqrt{k+1} + (-1)^k} = \frac{1}{\sqrt{2}-1} - \frac{1}{\sqrt{2}+1} + \frac{1}{\sqrt{3}-1} - \cdots$$

Exercise 183 *(a) Estimate the sum s of the alternating harmonic series in (5.46) that is accurate to three decimal places. How many terms of the series do you need for this accuracy? Use a calculator or computer to compute this approximation and check it against your calculator's value for $\ln 2$.*

(b) The index N that we get in (a) "guarantees" accuracy to three decimal places, but it is not necessarily the smallest index to do the job. Use a calculator, and the value of $\ln 2$ as a check, to show that smaller values of N than that in (a) also give the same level of accuracy.

Exercise 184 *Prove that the following series converges absolutely using the ratio test part of Corollary 155:*

$$\sum_{n=1}^{\infty} \frac{n^2 \cos n}{2^n} = \frac{1^2 \cos 1}{2^1} + \frac{2^2 \cos 2}{2^2} + \frac{3^2 \cos 3}{2^3} + \cdots$$

Exercise 185 *Show that each series below converges absolutely, and therefore, it converges:*

(a) $\displaystyle\sum_{n=1}^{\infty} \frac{(-1)^{n-1} n^3}{2^{n/2} + 1}$ (b) $\displaystyle\sum_{n=1}^{\infty} \frac{3^n (1 + \cos n)}{n!(2 - \sin n)}$

Exercise 186 *Prove the following inequalities for Liouville's constant L*

$$0.11 < L < \frac{109}{990} \approx 0.1101$$

Note. For the upper bound, use the fact that n! is even for all n.

Exercise 187 *Consider the number*

$$0.10100100010000100000100\cdots \qquad (5.69)$$

Here we start with a 1 to the right of the decimal point and every occurrence of 1 is followed by a corresponding number of zeros: the first appearance of 1 is followed by one zero, the second appearance by two zeros, the third by three zeros, and so on.

(a) Explain why this number is irrational, and show that is it smaller than Liouville's constant but still between 1/10 and 1/9.

(b) Prove that there are $2^{\aleph_0}$ real numbers of type

$$0.d_1 0 d_2 00 d_3 000 d_4 0000 d_5 00000 d_6 00\cdots$$

where $d_1, d_2, d_3, \ldots$ are integers ranging from 0 to 9 that replace the 1's in the constant in (5.69);

(c) Write the numbers in (b) as infinite series.

6

Differentiation and Continuity

Velocity, acceleration, slope, rate of change, ... these are concepts associated with a single idea in mathematics, namely, the *derivative*. Since its birth in the seventeenth century, this concept has been associated with infinitely small (or infinitesimal) magnitudes. Despite its intuitive appeal, a proper formulation of the infinitesimal requires substantial abstraction and a significant use of mathematical logic. Its use in calculus historically occurred in a naïve way and was vulnerable to contradictions and inconsistencies.

The standard definition of derivative in analysis does not refer to the infinitesimal and instead uses function limits in the form of the "ε and δ approach." This approach is precise and general, but it is somewhat of an acquired taste that requires patience and diligence to master.

An alternative approach is based on convergent sequences. This approach is equivalent to the standard approach in very general contexts (e.g., metric spaces); and in some ways, it is even simpler and more natural than the more common ε, δ approach (e.g., limits in all dimensions, including infinite, are defined the same way).

I use sequences in this book because they clarify the essential role of infinity in the definition of the derivative in the sense that taking a limit always means a parameter going to infinity. On average, limit calculations using sequences are no more complicated than with the traditional approach. I discuss the traditional approach in a separate section below to give a flavor of it, and to also show its equivalence to the sequence-based approach preferred here. We do use the ε, δ approach when it is the natural way to go or when it simplifies calculations.

6.1 Velocity, slope, and the derivative

Motivated by problems in mechanics, Newton (1643–1727) developed calculus between 1664 and 1666, 2 years after earning his degree from Cambridge. Ten years later, Leibniz did the same (without knowing the content of Newton's work) while studying tangents to curves and the problem of areas of regions with curved

Real Analysis and Infinity. Hassan Sedaghat, Oxford University Press.
© H. Sedaghat, (2022). DOI: 10.1093/oso/9780192895622.003.0006

boundaries. Both of them extended an already existing body of results; and further, they were the first to discover the inverse relation between the derivative and the integral that is now called the "fundamental theorem of calculus."

6.1.1 Velocity and slope

We begin with a discussion of velocity followed by slopes with the aim of discovering what these apparently unrelated concepts have in common, namely, they are different interpretations of the derivative.

As mentioned above, our approach here is based on sequences; so even if you are familiar with the material in this section, I recommend reading it if you are unfamiliar with the sequence definition of derivative.

Defining velocity

The *average velocity* $\bar{v}$ of any moving object is the distance it travels, or the amount of change in its position over a given interval of time:

$$\bar{v} = \frac{\text{change in position}}{\text{change in time}} = \frac{\Delta x}{\Delta t}$$

In this notation, x denotes the *position* or location of the object relative to a point of reference, and t denotes the time as measured from a desired starting point. Velocity may be a negative quantity when the object is moving in the opposite direction. The absolute value, or more generally, the magnitude of velocity, is often called *speed*.

If the velocity of an object is not constant, then the average velocity is not its actual velocity. How do we define velocity that is changing?

We think, as is commonly done, that the position x is a function of time t. To define the actual (or "instantaneous") velocity at a given time τ, we record a sequence of changes in position $x_n = x(t_n)$ over diminishing time intervals $\Delta t_n = t_n - \tau$ to obtain a sequence of average velocity numbers:

$$v_n = \frac{\Delta x_n}{\Delta t_n} = \frac{x(t_n) - x(\tau)}{t_n - \tau}$$

By taking Δt_n smaller with each recording of position $x(t_n)$, we ensure that the changes in position are less substantial. After a large number n of recordings, the time interval Δt_n is so small that only a correspondingly small change in position Δx_n occurs, and the average velocity v_n is thus a more accurate indication of true velocity at time τ.

Ultimately, as Δt_n converges to 0 (i.e., t_n converges to τ), the position $x(t_n)$ approaches the initial position $x(\tau)$, i.e., $\Delta x_n \to 0$. If this happens in such a way

REAL ANALYSIS AND INFINITY 223

that

$$\lim_{n\to\infty} \frac{\Delta x_n}{\Delta t_n} = \lim_{n\to\infty} \frac{x(t_n) - x(\tau)}{t_n - \tau}$$

exists, then the sequence v_n approaches the above limit; this is the most accurate indicator of actual velocity at the instant τ.[1] So we define the velocity at time τ as the quantity

$$v(\tau) = \lim_{n\to\infty} \frac{x(t_n) - x(\tau)}{t_n - \tau} \tag{6.1}$$

An underlying assumption here is that *the specific choice of the sequence t_n does not affect the limit*; for instance, the limit is the same in all of the following cases:

$$t_n = \tau + \frac{1}{n} \qquad t_n' = \tau - \frac{3}{n^2} \qquad t_n'' = \tau + \frac{(-1)^n}{10^n}$$

This requirement is needed in order to define $v(\tau)$ in a meaningful way, i.e., the limit (6.1) must exist for *every* time sequence t_n that converges to τ (of course, to avoid division by 0, we also require that $t_n \neq \tau$ for all n). Let's consider an example to see that these assumptions are not an issue in typical calculations of velocity.

Galileo Galilei (1564–1642) obtained the equation of motion for freely falling bodies. His equations provide the values of the position x and velocity v as functions of time based on common-sense assumptions and some experimentation (perhaps even dropping various items from the leaning tower of Pisa). In modern notation, these equations are[2]

$$x = \frac{1}{2}gt^2 \quad \text{and} \quad v = gt \tag{6.2}$$

where g is the near-Earth value of gravitational acceleration whose value in the metric system is approximately 9.8.

We use the definition in (6.1) to obtain the formula for v above. Consider Galileo's equation for position with $g = 9.8$

$$x = 4.9t^2$$

Take a time instant $t = \tau$ (or any real number) and *any* sequence t_n of time instants that converges to τ ($t_n \neq \tau$ for all n). Then using the limit theorems in

---

[1] For a more detailed description of this at a more deliberate pace, see my self-study book *Achieving Infinite Resolution*.
[2] We later derive Galileo's equations from Newton's laws of motion, when discussing integration in Chapter 7.

Chapter 3, we have

$$v(\tau) = \lim_{n\to\infty} \frac{4.9t_n^2 - 4.9\tau^2}{t_n - \tau} = \lim_{n\to\infty} \frac{4.9(t_n - \tau)(t_n + \tau)}{t_n - \tau} = 4.9(\tau + \tau) = 9.8\tau$$

This is the velocity equation in (6.2). Notice that except for the condition $t_n \neq \tau$ for all n, we imposed no other restrictions on the sequence t_n; therefore, our calculation gives the same result for all possible sequences that converge to τ.

Slope of the tangent line

We now turn to the problem of finding the slope of the line that is tangent to the graph of a given function $y = f(x)$. This was one of the problems that Leibniz was studying; and while velocity that we discussed above provides a physical context for the concept of derivative, the slope problem provides a visual interpretation. Both concepts are of course different manifestations of the derivative.

Recall from elementary algebra that the *slope* of any straight line with two points (x_1, y_1) and (x_2, y_2) specified on it is

$$\frac{\text{rise}}{\text{run}} = \frac{y_2 - y_1}{x_2 - x_1}$$

Consider the problem of finding the slope of the line that is tangent to the graph of a function $y = f(x)$ at a given point $x = a$. Pick a nearby value $x = x_1$ and notice that we can calculate the slope m_1 of the line that goes through the two points $(x_1, f(x_1))$ and $(a, f(a))$ (called a *secant line*.) using the elementary slope formula above:

$$m_1 = \frac{f(x_1) - f(a)}{x_1 - a}$$

We see that this fraction resembles average velocity; as in the case of position and time, by choosing an arbitrary sequence of points x_n that converge to a, we obtain a sequence of secant lines that ultimately align with the tangent line and thus, their slopes m_n converge to the slope m of the tangent line; see Figure 6.1.

As in the case of velocity, the slope of the tangent line is given by the limit of a fraction or quotient

$$m = \lim_{n\to\infty} \frac{f(x_n) - f(a)}{x_n - a} \tag{6.3}$$

Both the velocity in (6.1) and the slope of the tangent line in (6.3) are given as limits of *difference quotients*. These limits are special cases of the *derivative*.

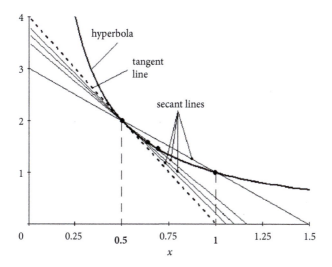

Fig. 6.1 Secant lines approaching a tangent line (dashed)

6.1.2 The derivative

Motivated by the above discussion, we formally define the derivative as follows.

The derivative

Let a be any fixed real number in the domain of the function $y = f(x)$. The *derivative* of $f(x)$ at $x = a$ is a number $f'(a)$ that is defined as the following limit:

$$f'(a) = \lim_{n \to \infty} \frac{f(x_n) - f(a)}{x_n - a} = \lim_{n \to \infty} \frac{\Delta y_n}{\Delta x_n} \qquad (6.4)$$

provided that the limit exists and has the same value $f'(a)$ for *every sequence* x_n in the domain of $f(x)$ that converges to a, and $x_n \neq a$ for all n.

Thus, the actual velocity at any time τ for a position function $x(t)$ is just the derivative $x'(\tau)$; likewise, the slope of the tangent line to the graph of a function $f(x)$ at a point $x = a$ is the derivative $f'(a)$. Additional interpretations of the derivative are discussed later.

Derivative notation and terminology

The notation used to indicate a derivative tends to vary according to the context. There are two common notations, one being the *primes notation* above; the other is the *differential notation* introduced by Leibniz: if $y = f(x)$ then

$$f'(x) = \frac{dy}{dx}$$

This fractional notation isn't a ratio of real numbers or quantities; it is primarily just a symbol. However, in a number of cases such as when dealing with the chain rule or some integration formulas, it is helpful to work with this notation as if it were a fraction (albeit not of real magnitudes). One more notation that is largely confined to physics applications is Newton's "dot notation" that is typically used for time derivatives; so if $x(t)$ is the position of an object at time t, then the velocity is denoted

$$\dot{x} = x'(t) = \frac{dx}{dt}$$

We calculated the derivative of Galileo's position function earlier to derive the corresponding velocity. As another example, we find the slope of the tangent line at any point $x = a$ where $a \neq 0$ by calculating the derivative of a specific function:

$$f(x) = \frac{1}{x} \quad x \neq 0$$

The graph of this function is a hyperbola, and a portion of it for $x > 0$ is shown in Figure 6.1, together with a number of secant lines that pass through the point $(1/2, 2)$ on the graph (where we take $a = 1/2$).

By the definition of derivative,

$$f'(a) = \lim_{n \to \infty} \frac{f(x_n) - f(a)}{x_n - a} = \lim_{n \to \infty} \frac{1/x_n - 1/a}{x_n - a} \tag{6.5}$$

where x_n is an arbitrary sequence of real numbers that converges to a, and $x_n \neq a$ for all n (and also $x_n \neq 0$ since x_n must be in the domain of $1/x$). We simplify the fraction on the right above to

$$\frac{(a - x_n)/ax_n}{x_n - a} = \frac{-(x_n - a)}{(x_n - a)ax_n} = -\frac{1}{ax_n} \tag{6.6}$$

The problematic quantity $x_n - a$ in (6.5) is cancelled in (6.6), so division by zero is avoided. Since $\lim_{n \to \infty} x_n = a$, using the limit theorems of Chapter 3 we get

$$f'(a) = -\frac{\lim_{n \to \infty} 1}{\lim_{n \to \infty} ax_n} = -\frac{1}{a \lim_{n \to \infty} x_n} = -\frac{1}{a^2} \tag{6.7}$$

In particular, when $a = 1/2$, we find $f'(1/2) = -4$, which is the slope of the (dashed) tangent line in Figure 6.1.

Notice once again that in this calculation, as in the velocity calculation earlier, the only restriction on the sequence x_n is that x_n converges to a, and $x_n \neq a$ for all n; otherwise, x_n is completely arbitrary (within the domain of the function).

We now introduce some standard terminology that is associated with the concept of derivative.

Differentiable function If a function $f(x)$ has a derivative at $x = a$, then we say that $f(x)$ is *differentiable at* $x = a$. If $f(x)$ is differentiable at every point of the domain of $f(x)$, then we just say that $f(x)$ is differentiable. If a function is differentiable on an interval I, then its derivative defines another function $f'(x)$ on I. We call $f'(x)$ the *derivative function* of $f(x)$.

In our discussions above, the Galileo position function $x(t) = gt^2/2$ is differentiable at all real values of t, i.e., on the interval $(-\infty, \infty)$, and its derivative function on this interval is the velocity $v(t) = gt$. This follows from the fact that the parameter τ could be any real number without affecting the calculations. Similarly, the function $1/x$ is differentiable on the union $(-\infty, 0) \cup (0, \infty)$, i.e., its domain $x \neq 0$, and its derivative function on this domain is $f'(x) = -1/x^2$. The exact choice of the number a doesn't affect the argument as long as $a \neq 0$.

In the next section, we discuss some formulas that make it easier to calculate the derivatives of some familiar functions. At this point you can prove the following useful formulas using (6.4) as a quick exercise:

$$\frac{d(x)}{dx} = 1 \quad \text{and} \tag{6.8}$$

$$\frac{d(c)}{dx} = 0 \quad \text{for each fixed real number } c \tag{6.9}$$

We close this section with a remark about differentiability; I discuss similar issues later in this chapter. If a function $f(x)$ is not defined at $x = a$ (like $1/x$ above that is undefined at $x = 0$), then its derivative is also undefined at a since the difference quotient cannot be formed, let alone have a limit. But *even if a function is defined at* a, it isn't necessarily differentiable there. Consider

$$f(x) = \sqrt{x}$$

This function is defined for all $x \geq 0$, but it isn't differentiable at $x = 0$. To prove this statement, we only need to show that there is *at least one* sequence x_n in the interval $(0, \infty)$ that converges to 0, but the difference quotient below doesn't converge to a limit.[3] Consider

$$\frac{f(x_n) - f(0)}{x_n - 0} = \frac{\sqrt{x_n}}{x_n} = \frac{1}{\sqrt{x_n}}$$

Let $x_n = 1/n^2$ so that $x_n \to 0$ as $n \to \infty$; however,

$$\frac{1}{\sqrt{x_n}} = \frac{1}{\sqrt{1/n^2}} = n \to \infty$$

It follows that the difference quotient of $\sqrt{x}$ has no limit for $x_n = 1/n^2$ and therefore, $\sqrt{x}$ isn't differentiable at $x = 0$.

6.2 Differentiation rules and higher derivatives

Calculating derivatives using the limit of a difference quotient is fine as long as the function in question is simple enough, but it gets cumbersome and impractical even for functions that are not very complex. To help with calculations, in this section, we derive a number of basic differentiation rules.

6.2.1 Derivatives of sums, products, and quotients

The next result is basic and often used in calculations without explicit mention.

Theorem 188 *Let $f(x)$ and $g(x)$ be functions defined on an interval I that are differentiable at $a \in I$ with derivatives $f'(a)$ and $g'(a)$. Then the following are true:*

(a) *The sum $f + g$ is differentiable at $x = a$ with derivative $f'(a) + g'(a)$.*
(b) *The product fg is differentiable at $x = a$ with derivative $f'(a)g(a)+f(a)g'(a)$.*
(c) *If $g(a) \neq 0$, then the quotient f/g is differentiable at $x = a$ with derivative:*

$$\frac{f'(a)g(a) - f(a)g'(a)}{[g(a)]^a}$$

Proof (a) To apply (6.4) to the sum function $f + g$, we take an arbitrary sequence x_n in I such that $\lim_{n \to \infty} x_n = a$ but $x_n \neq a$ for all n. Then

---

[3] In this case, we can prove that $1/\sqrt{x_n}$ diverges to infinity for *all* sequences of positive numbers that converge to 0. But that would be an unnecessary overkill for proving the *negation* of "differentiable."

$$\lim_{n\to\infty} \frac{(f+g)(x_n) - (f+g)(a)}{x_n - a} = \lim_{n\to\infty} \frac{[f(x_n) - f(a)] + [g(x_n) - g(a)]}{x_n - a}$$

$$= \lim_{n\to\infty} \frac{f(x_n) - f(a)}{x_n - a} + \lim_{n\to\infty} \frac{g(x_n) - g(a)}{x_n - a}$$

The last two limits exist by hypothesis, so the limit on the left-hand side exists, making $f + g$ differentiable. Further, the derivative of $f + g$ is the value of the sum on the right-hand side, i.e.,

$$(f+g)'(a) = \lim_{n\to\infty} \frac{(f+g)(x_n) - (f+g)(a)}{x_n - a} = f'(a) + g'(a)$$

(b) We start as in Part (a) and obtain

$$\lim_{n\to\infty} \frac{(fg)(x_n) - (fg)(a)}{x_n - a} = \lim_{n\to\infty} \frac{f(x_n)g(x_n) - f(a)g(a)}{x_n - a}$$

The last difference quotient is not as readily decomposed into separate difference quotients for f and g; something seems to be missing, and it turns out to be a combined term $f(a)g(x_n)$; to see how, we add and subtract this term in the numerator to obtain

$$\lim_{n\to\infty} \frac{(fg)(x_n) - (fg)(a)}{x_n - a} = \lim_{n\to\infty} \frac{f(x_n)g(x_n) - f(a)g(x_n) + f(a)g(x_n) - f(a)g(a)}{x_n - a}$$

$$= \lim_{n\to\infty} \frac{[f(x_n) - f(a)]g(x_n) + f(a)[g(x_n) - g(a)]}{x_n - a}$$

$$= \lim_{n\to\infty} \frac{f(x_n) - f(a)}{x_n - a} \lim_{n\to\infty} g(x_n) + f(a) \lim_{n\to\infty} \frac{g(x_n) - g(a)}{x_n - a}$$

We now recognize the difference ratios that correspond to $f'(a)$ and $g'(a)$. Further, $\lim_{n\to\infty} g(x_n) = g(a)$ since g is differentiable at $x = a$, requiring $g(x_n) \to g(a)$ so that the limit of the difference quotient can exist to produce $g'(a)$. Since all the limits on the right-hand side exists, it follows that the limit on the left also exists, making the product fg differentiable at $x = a$. Further, the derivative is given in terms of $f'(a)$ and $g'(a)$ as required

$$(fg)'(a) = \lim_{n\to\infty} \frac{(fg)(x_n) - (fg)(a)}{x_n - a} = f'(a)g(a) + f(a)g'(a)$$

(c) Since $f(x)/g(x) = f(x)[1/g(x)]$, we first show that $1/g(x)$ is differentiable and then apply Part (b). To this end, note that

$$\lim_{n\to\infty} \frac{(1/g)(x_n) - (1/g)(a)}{x_n - a} = \lim_{n\to\infty} \frac{1}{x_n - a} \frac{g(a) - g(x_n)}{g(x_n)g(a)}$$

$$= \lim_{n\to\infty} \frac{-[g(x_n) - g(a)]}{x_n - a} \lim_{n\to\infty} \frac{1}{g(x_n)g(a)}$$

As we argued in the proof of (b) above, $\lim_{n\to\infty} g(x_n) = g(a)$ since g is differentiable at $x = a$. The other limit on the right-hand side is $-g'(a)$, so the limit on the left-hand side exists and implies that $1/g(x)$ is differentiable at $x = a$ with derivative

$$\left(\frac{1}{g}\right)'(a) = \lim_{n\to\infty} \frac{(1/g)(x_n) - (1/g)(a)}{x_n - a} = -\frac{g'(a)}{[g(a)]^2}$$

Finally, applying (b) we get

$$\left(\frac{f}{g}\right)'(a) = f'(a)\frac{1}{g(a)} + f(a)\left[-\frac{g'(a)}{[g(a)]^2}\right]$$

which is equivalent to the expression given in (c).   ∎

The above differentiation rules may be rewritten with the aid of the differential notation that makes them easier to remember and more convenient to use, as in the following corollary.

Corollary 189 *Assume that the functions $u = f(x)$ and $v = g(x)$ are differentiable on an interval I.*

(a) The sum rule: $u + v$ is differentiable on I with derivative

$$\frac{d(u + v)}{dx} = \frac{du}{dx} + \frac{dv}{dx}$$

(b) The product rule: uv is differentiable on I with derivative

$$\frac{d(uv)}{dx} = \frac{du}{dx}v + u\frac{dv}{dx} = u'v + uv'$$

(c) The quotient rule: if $v \neq 0$ on I, then u/v is differentiable on I with derivative

$$\frac{d(u/v)}{dx} = \frac{(du/dx)v - u(dv/dx)}{v^2} = \frac{u'v - uv'}{v^2}$$

A simple consequence of the product rule is that if $g(x) = c$ with c being an arbitrary fixed real number, then $cf = gf$ is differentiable, and its derivative is

$$(cf)'(a) = 0f(a) + cf'(a) = cf'(a)$$

The 0 above comes from the fact mentioned earlier that the derivative of a constant is always 0.

This *constant multiples rule* can also be stated in differential notation:

$$\frac{d(cf)}{dx} = c\frac{df}{dx} \tag{6.10}$$

The simple formulas (6.8) and (6.9) together with the product rule and the quotient rule imply the *power rule*[4]

$$\frac{d(x^k)}{dx} = kx^{k-1}, \quad k \in \mathbb{Z} \tag{6.11}$$

For instance, if $k = 2$, then

$$\frac{d(x^2)}{dx} = \frac{d(xx)}{dx} = (1)x + x(1) = 2x$$

and for $k = 3$

$$\frac{d(x^3)}{dx} = \frac{d(xx^2)}{dx} = (1)x^2 + x(2x) = 3x^2$$

For general $k \in \mathbb{N}$, we can use mathematical induction to give a formal proof (see Exercise 217). The extension to negative powers is straightforward with

$$x^k = \frac{1}{x^{-k}}, \quad k < 0$$

where quotient rule applies along with what we just proved for positive k. Keep in mind that if $k < 0$, then x^k is not defined at $x = 0$.

If k is rational, then we can use another differentiation rule that we discuss in the next section to extend (6.11).

---

[4] The usual proof in basic calculus uses the binomial theorem since we can write the difference quotient of x^k alternatively as

$$\frac{(a + \Delta x_n)^k - a^k}{\Delta x_n}, \quad \Delta x_n = x_n - a$$

6.2.2 The chain rule

Recall that another basic way of combining functions is by composing them. In symbols:

$$(g \circ f)(x) = g(f(x))$$

provided that the function $g(x)$ is defined over the range of $f(x)$. This operation, like addition and multiplication, generates a new function from the two given ones.

Now, suppose that both $f(x)$ and $g(x)$ are differentiable functions at a point $x = a$; does it follow that their composition $(g \circ f)(x)$ is also differentiable at a? If it is, then what is its derivative?

The difference quotient of $g \circ f$ at $x = a$ is

$$\frac{g(f(x_n)) - g(f(a))}{x_n - a} \tag{6.12}$$

where as usual, x_n is an arbitrary sequence in the domain of $g \circ f$ such that $x_n \neq a$ for all n. In the above form, the difference quotient in (6.12) doesn't explicitly contain the difference quotients for the derivatives f' or g', but notice that if we multiply and divide by $f(x_n) - f(a)$ to get

$$\frac{g(f(x_n)) - g(f(a))}{x_n - a} = \frac{g(f(x_n)) - g(f(a))}{f(x_n) - f(a)} \frac{f(x_n) - f(a)}{x_n - a}$$

then not only do we recognize the difference quotient for $f'(a)$, but by letting $y = f(x)$, the above produces

$$\frac{g(f(x_n)) - g(f(a))}{x_n - a} = \frac{g(y_n) - g(f(a))}{y_n - f(a)} \frac{f(x_n) - f(a)}{x_n - a}$$

and now the first fraction on the right-hand side is the difference quotient for $g'(f(a))$. Taking limits as $n \to \infty$ then gives

$$(g \circ f)'(a) = g'(f(a))f'(a)$$

in agreement with the chain rule theorem below. However, this argument is not a valid proof of that theorem because in general, there is nothing to prevent the expression $f(x_n) - f(a)$ above to equal 0 for infinitely many indices n even if $x_n \neq a$ for all n. Therefore, we cannot divide by this expression in general.

It is necessary to fix the above argument so that division by $f(x_n) - f(a)$ is not required. We start with a preliminary lemma that provides the correct resolution of the difficulty. The short proof is left as Exercise 219.

Lemma 190 *Assume that a function $f(x)$ is differentiable at a in some interval I and define*

$$\eta_f(x) = \frac{f(x) - f(a)}{x - a} - f'(a) \quad \text{for all } x \in I, x \neq a \tag{6.13}$$

Also define $\eta_f(a) = 0$. Then $\lim_{n \to \infty} \eta_f(x_n) = 0$ for every sequence $x_n \in I$ that converges to a, but $x_n \neq a$ for all n and for all $x \in I$

$$f(x) - f(a) = [f'(a) + \eta_f(x)](x - a)$$

Theorem 191 *(The chain rule) Suppose that $f(x)$ is defined on an interval I, and it is differentiable at $x = a \in I$; and further, $g(x)$ is defined on an interval J that contains the range of f, and it is differentiable at $x = f(a) \in J$. Then the composition $g \circ f$ is defined on I, and it is differentiable at $x = a$ with derivative*

$$(g \circ f)'(a) = g'(f(a))f'(a) \tag{6.14}$$

In differential notation, with $u = f(x)$ and $y = g(u)$

$$\frac{dy}{dx} = \frac{dy}{du}\frac{du}{dx}$$

Proof Given the differentiability hypotheses on f and g, Lemma 190 implies

$$f(x) - f(a) = [f'(a) + \eta_f(x)](x - a), \qquad x \in I \tag{6.15}$$

$$g(y) - g(f(a)) = [g'(f(a)) + \eta_g(y)](y - f(a)), \quad y \in J \tag{6.16}$$

Let x_n be an arbitrary sequence in I that converges to a, but $x_n \neq a$ for all n. Then $\lim_{n \to \infty} \eta_f(x_n) = 0$; and since f is differentiable at a, $f(x_n)$ must converge to $f(a)$ for the limit of the difference quotient to be defined. Therefore, $\lim_{n \to \infty} y_n = f(a)$, and it follows that $\lim_{n \to \infty} \eta_g(y_n) = 0$.

Next,

$$\frac{g(f(x_n)) - g(f(a))}{x_n - a} = \frac{g(y_n) - g(f(a))}{x_n - a}$$

$$= \frac{[g'(f(a)) + \eta_g(y_n)](y_n - f(a))}{x_n - a}$$

where we used (6.16) for the last step. Further, $y_n = f(x_n)$, so using (6.15) we obtain

$$\frac{g(f(x_n)) - g(f(a))}{x_n - a} = \frac{[g'(f(a)) + \eta_g(y_n)][f'(a) + \eta_f(x_n)](x_n - a)}{x_n - a}$$

$$= [g'(f(a)) + \eta_g(y_n)][f'(a) + \eta_f(x_n)]$$

Now we take the limit and use the limit theorems of Chapter 3 to obtain the following:

$$\lim_{n\to\infty} \frac{g(f(x_n)) - g(f(a))}{x_n - a} = [g'(f(a)) + \lim_{n\to\infty} \eta_g(y_n)][f'(a) + \lim_{n\to\infty} \eta_f(x_n)]$$

Each of the limits on the right-hand side is 0, so the limit on the left-hand side exists. It follows that $g \circ f$ is differentiable at $x = a$, and further, its derivative is given by (6.14), as required. ∎

Power rule with rational powers

The chain rule provides a quick way of extending the power rule (6.11) to all rational powers $p = m/n$:

$$\frac{d(x^p)}{dx} = px^{p-1}, \quad p \in \mathbb{Q} \tag{6.17}$$

More restrictions on the values of x exist for (6.17) than there were for (6.11). Generally, if $p < 1$, then (6.17) is not valid at $x = 0$ except for the trivial case $p = 0$, even though x^p is defined at 0 for all $p \geq 0$. Further, if $p = m/n$ in lowest terms, and n is an even integer (and thus m is odd), then it is necessary to restrict the use of (6.17) to positive integers $x > 0$.

To prove (6.17) we define

$$u = x^p = f(x), \quad y = u^n = g(u)$$

so that

$$g(f(x)) = y = (x^p)^n = x^m$$

Now by the chain rule and (6.11)

$$mx^{m-1} = \frac{dy}{dx} = \frac{dy}{du}\frac{du}{dx} = nu^{n-1}\frac{d(x^p)}{dx}$$

from which we infer

$$\frac{d(x^p)}{dx} = \frac{mx^{m-1}}{nu^{n-1}} = \frac{mx^{m-1}}{nx^{pn-p}} = \frac{m}{n}x^{m-1-(m-p)} = px^{p-1}$$

to complete the proof of (6.17).

As an example of the versatility of the chain rule, note that we can use it together with (6.17) to take the derivative of a function like

$$y = \sqrt{x^2 + 1}, \quad x \in \mathbb{R}$$

We set $u = x^2 + 1$ and $y = u^{1/2}$ and readily obtain

$$\frac{dy}{dx} = \frac{1}{2}u^{-1/2}(2x) = \frac{x}{\sqrt{x^2 + 1}}$$

This example is a special case of a more general version of the power rule. The proof of the next result is left as Exercise 218.

Corollary 192 *Let $p \in \mathbb{Q}$ and $f(x)$ be a function such that $[f(x)]^p$ is defined for all x in an interval I. Then $[f(x)]^p$ is differentiable for all $x \in I$ (unless $0 < p < 1$, in which case we require that $f(x) \neq 0$ for all $x \in I$), and its derivative is*

$$\frac{d}{dx}[f(x)]^p = p[f(x)]^{p-1}$$

6.2.3 Derivatives of trigonometric functions

There are a variety of equations that relate the familiar trigonometric functions $\sin x$, $\cos x$, $\tan x$, etc. Once we know the derivative of, say, $\sin x$, we can use these equations together with some of the derivative rules that we discussed above to find the derivatives of all the other trigonometric functions.

We prove shortly that $\sin x$ is differentiable for all real x, and its derivative is

$$\frac{d(\sin x)}{dx} = \cos x$$

Assuming this to be given, using the trigonometric identity $\cos x = \sin(x + \pi/2)$ and the chain rule we conclude that $\cos x$ is differentiable for all x, and its derivative is

$$\frac{d(\cos x)}{dx} = \frac{d}{dx} \sin\left(x + \frac{\pi}{2}\right) = \cos\left(x + \frac{\pi}{2}\right) (1)$$

and since $\cos(x + \pi/2) = -\sin x$, we have

$$\frac{d(\cos x)}{dx} = -\sin x$$

Next, consider $\tan x = (\sin x)/(\cos x)$. The domain of $\tan x$ is restricted to those values of x with $\cos x \neq 0$, i.e., $x \neq \pi k + \pi/2$ for all integers k. For such x, the quotient rule gives

$$\frac{d(\tan x)}{dx} = \frac{(\cos x)\cos x - (\sin x)(-\sin x)}{(\cos x)^2} = \frac{\cos^2 x + \sin^2 x}{\cos^2 x}$$

and since $\cos^2 x + \sin^2 x = 1$, we have

$$\frac{d}{dx}\tan x = \frac{1}{\cos^2 x} = \sec^2 x$$

The derivatives of other trigonometric functions or combinations of trigonometric functions can be found using the sum, product, quotient, and chain rules. For instance, the derivative of

$$\sin(2x) = 2\sin x \cos x$$

can be found using either the chain rule (applied to the left-hand side) or the product rule (applied to the right-hand side) with the former being the quicker.

The derivative of $\sin x$

We now prove that the function $f(x) = \sin x$ is differentiable for all real numbers x, and its derivative is $\cos x$. In the process, we also derive an important limit formula.

Let a be an arbitrary real number. By definition,

$$f'(a) = \lim_{n \to \infty} \frac{f(x_n) - f(a)}{x_n - a} = \lim_{n \to \infty} \frac{\sin x_n - \sin a}{x_n - a}$$

provided that the limit exists for every sequence x_n that approaches a (but $x_n \neq a$ for all n). A modification of the fraction in this limit makes it easier to calculate using trigonometric identities. Define $u_n = x_n - a$ for every n so that $u_n \to 0$ as

$n \to \infty$, but $u_n \neq 0$ for all n. With this substitution, the derivative of $\sin x$ is given by the limit

$$f'(a) = \lim_{n \to \infty} \frac{\sin(a + u_n) - \sin a}{u_n}$$

Using the trigonometric identity $\sin(u + v) = \sin u \cos v + \cos u \sin v$, a couple of limit theorems from Chapter 3, and doing a little algebra, we get

$$f'(a) = \lim_{n \to \infty} \frac{\sin a \cos u_n + \cos a \sin u_n - \sin a}{u_n}$$

$$= \left(\lim_{n \to \infty} \frac{\cos u_n - 1}{u_n} \right) \sin a + \left(\lim_{n \to \infty} \frac{\sin u_n}{u_n} \right) \cos a \qquad (6.18)$$

Next, we use the definitions of trigonometric functions to find the limits in (6.18). Consider Figure 6.2, which contains the geometric representation of the tangent function for an acute angle θ in addition to those of sine and cosine.

In Figure 6.2, we see that in triangle OPQ, the sine of the central angle θ is the length $|PQ|$ of the front side over hypotenuse (the radius of the circle), so $\sin \theta = |PQ|$.

The radian measure of θ is the length of the arc PQ' of the circle of unit radius. Since the length of this arc is greater than the length of PQ, it follows that

$$\sin \theta < \theta \qquad (6.19)$$

Next, in triangle $OP'Q'$, the tangent of the central angle θ is the length of the front side $P'Q'$ over the adjacent side (the radius of the circle): $\tan \theta = |P'Q'|$.

The length $|P'Q'|$ is greater than the length of the arc PQ', but this is not visually obvious. To see why, consider the line segment PS, which is tangent to the circle at

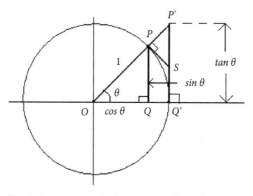

Fig. 6.2 A geometrical representation of sine and tangent functions

the point P and thus, it is also perpendicular to the radius OP. The small triangle SPP' is a right triangle, so the length $\left|SP'\right|$ of its hypotenuse is greater than the length $|SP|$ of the side. It follows that the sum of the lengths of the two line segments $Q'S$ and SP is smaller than the length of the side $Q'P'$, that is, $|Q'S| + |SP| < |Q'P'|$. The number on the left-hand side of this inequality is greater than the length of the arc PQ', the radian measure of θ, so we conclude that

$$\theta < |Q'S| + |SP| < \tan \theta$$

When we put this inequality and (6.19) together, we have

$$\sin \theta < \theta < \tan \theta$$

These inequalities and the fact that $\tan \theta = \sin \theta / \cos \theta$ yield

$$\cos \theta < \frac{\sin \theta}{\theta} < 1$$

By repeating these arguments for the reflections of the triangles in Figure 6.2, we find that these inequalities are also true when $\theta < 0$; alternatively, we may use the fact that $\sin \theta$ is an odd function, while $\cos \theta$ is even.

Now, with our sequence u_n above inserted in these inequalities, we get

$$\cos u_n < \frac{\sin u_n}{u_n} < 1 \tag{6.20}$$

The number 1 on the right-hand side of (6.20) is fixed for all choices of u_n; so if we show that the sequence $\cos u_n$ converges to 1 as $u_n \to 0$, then by the squeeze theorem

$$\lim_{n \to \infty} \frac{\sin u_n}{u_n} = 1 \tag{6.21}$$

This must hold regardless of which sequence u_n is selected that converges to 0, and this is guaranteed if the function $\cos x$ is *continuous* at $x = 0$.[5] This follows immediately from the trigonometric identity $\cos x = \sin(x + \pi/2)$ and Theorem 195 once we show that $\sin x$ is continuous at $x = 0$, i.e., $\lim_{n \to \infty} \sin \theta_n = 0$ for every sequence θ_n that converges to 0.

---

[5] We formally define continuous functions later in this chapter.

To this end, note that by (6.19) and the fact that $\sin x$ is an odd function, we have $\sin(-x) = -\sin x > -x$. It follows that

$$| \sin x | < |x|$$

for all real numbers x (including $x = 0$). Now for all n, we have $0 \le | \sin \theta_n | \le |\theta_n|$, so again the squeeze theorem implies that $\lim_{n \to \infty} \sin \theta_n = 0$ for every sequence θ_n that converges to 0.

To finish the derivation of the formula for the derivative of $\sin x$, we now calculate the other limit in (6.18) as follows: we multiply and divide the fraction by $\cos u_n + 1$ and use the trigonometric Pythagorean identity

$$\frac{\cos u_n - 1}{u_n} = \frac{\cos^2 u_n - 1}{u_n(\cos u_n + 1)} = \frac{-\sin^2 u_n}{u_n(\cos u_n + 1)} = -\frac{\sin u_n}{\cos u_n + 1} \frac{\sin u_n}{u_n}$$

We now take the limit:

$$\lim_{n \to \infty} \frac{\cos u_n - 1}{u_n} = -\lim_{n \to \infty} \left(\frac{\sin u_n}{\cos u_n + 1} \right) \lim_{n \to \infty} \left(\frac{\sin u_n}{u_n} \right)$$

$$= -\frac{\lim_{n \to \infty} \sin u_n}{\lim_{n \to \infty} \cos u_n + 1} (1)$$

$$= 0$$

Inserting this number and using (6.21) in (6.18) gives the formula for the derivative of $\sin x$ at $x = a$

$$f'(a) = \cos a$$

6.2.4 Higher-order derivatives

We saw earlier in this chapter that the rate of change of position $x(t)$ of a moving object is its (instantaneous) velocity $v(t) = x'(t)$. The derivative $v'(t)$ of the velocity function gives the rate of change of velocity and is called the *acceleration* of the object, which we denote by $a(t)$. Since the velocity itself is a derivative, we can then write, using the differential notation

$$a(t) = v'(t) = \frac{dv}{dt} = \frac{d}{dt}\left(\frac{dx}{dt} \right)$$

The last expression on the right-hand side shows that we may alternatively calculate the acceleration by taking the derivative of the position function *twice in a row*. This suggests the alternative primes notation

$$a(t) = x''(t)$$

Now we formally define derivatives of order 2 and higher.

Higher-order derivatives

Suppose that $y = f(x)$ is differentiable and has a derivative function $f'(x)$. If $f'(x)$ is also differentiable, then its derivative is the *second derivative* of $f(x)$ and denoted, in two notations,

$$f''(x) = \frac{d^2 y}{dx^2}$$

The second derivative of $f(x)$ is also called the *derivative of order 2*. We also refer to $f'(x)$ as the derivative of order 1 or the first derivative of $f(x)$. More generally, for each positive integer k, *the derivative of order k* or the *kth derivative* of $f(x)$ is obtained by taking the derivative k times (assuming that we can), and it is denoted

$$f^{(k)}(x) = \frac{d^k y}{dx^k}$$

Higher-order derivatives occur prominently in expansions of functions in the form of Taylor series that we discuss in some detail in Chapter 8.

Finding higher-order derivatives is usually just a matter of repeated differentiation. However, repeated differentiation may not be possible at all points where a function and its derivative are defined. For example, $f(x) = \cos^{4/3} x$ has no undefined points, so its domain is $(-\infty, \infty)$. We find its first derivative using the chain rule

$$f'(x) = -\frac{4}{3} \cos^{1/3} x \sin x \tag{6.22}$$

This function also has domain $(-\infty, \infty)$. However, the second derivative that can be calculated as follows using the chain and product rules

$$f''(x) = \frac{4}{9} \cos^{-2/3} x \sin^2 x - \frac{4}{3} \cos^{4/3} x$$

is not defined where $\cos x = 0$, i.e., on the infinite set of points $x = \pm\pi/2, \pm3\pi/2$, etc. At these points, the factor $\cos^{1/3} x$ in (6.22) is not differentiable. We discuss three different ways where the derivative fails to exist in the next section.

6.2.5 When derivatives fail to exist

If a is a point in the domain of $f(x)$, then $f(a)$ is a well-defined number. But for $f'(a)$ to also exist, *the limit of the difference quotient in (6.4) must exist and have the same value for every sequence x_n that converges to a*. Assuming that $x_n \neq a$ for all n, the fraction

$$\frac{f(x_n) - f(a)}{x_n - a} \qquad (6.23)$$

is a well-defined number for every n. But the *limit* of this fraction as $n \to \infty$ is another matter; here is what can go wrong:

1. *The limit may not exist for some numbers a regardless of the choice of x_n;*
2. *The limit may exist for some sequences x_n that converge to a but not for all such sequences;*
3. *The limit may exist for all sequences that converge to some a but may have different values for different sequences.*

I illustrate each of the above possibilities using an example.

Since graphs help visualize each case, let's also remember that derivatives are slopes of tangent lines. So each of the above three possibilities is a case where it is not possible to draw a unique tangent line with a well-defined slope.

We illustrate Item 1 with the cube-root function:

$$f_1(x) = x^{1/3} = \sqrt[3]{x}$$

The graph is shown in Figure 6.3.

Although the y-axis appears to be a unique tangent line at the origin $x = 0$, it does not have a well-defined slope (or has infinite slope) there. If we draw some tangent lines to the graph of the function both to the left of 0 and to the right,

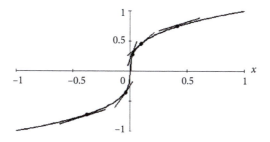

Fig. 6.3 No derivative at the origin: vertical tangent line

the lines appear to be getting vertical and approaching the y-axis as we choose the points of tangency closer to 0. Let's take a look at (6.23) in this case:

$$\frac{\sqrt[3]{x_n} - \sqrt[3]{0}}{x_n - 0} = \frac{x_n^{1/3}}{x_n} = \frac{1}{x_n^{2/3}} = \left(\frac{1}{x_n}\right)^{2/3}$$

Since $x_n \to 0$, it follows that $1/x_n^{2/3} \to \infty$ (the sequence diverges to infinity); so the limit of the difference ratio in (6.23) does not exist, regardless of what sequence x_n we choose that approaches 0.

To illustrate Item 2 above, consider the two-piece function

$$f_2(x) = \begin{cases} \sqrt{x}, & x > 0 \\ x^2, & x \le 0 \end{cases}$$

A graph of this function is shown in Figure 6.4.

Notice that $f_2(0) = 0$. If we draw tangent lines to the graph at various points to the *right* of 0, then they appear to be getting vertical, and their slopes become infinitely large; but if we draw the tangent lines to the *left* of 0, the tangent lines get more horizontal as I approach 0, so the slopes approach 0.

We can verify this dual behavior analytically by examining (6.23). Choose a sequence x_n that converges to 0 and $x_n > 0$ for all n; we can be specific here, so let $x_n = 1/n^2$. Then

$$\frac{f_2(x_n) - f_2(0)}{x_n - 0} = \frac{\sqrt{x_n}}{x_n} = \frac{1}{\sqrt{x_n}} = n$$

The difference quotient diverges ∞ as $n \to \infty$, so the above difference quotient gets infinitely large.

But if we choose any sequence x_n that converges to 0 and $x_n < 0$ for all n, then the limit exists because

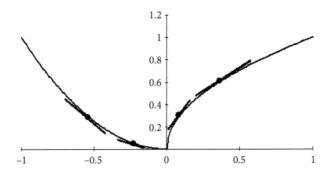

Fig. 6.4 No derivative at the origin: infinite slope from one side

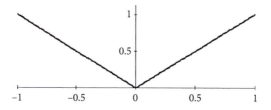

Fig. 6.5 No derivative at the origin: different slopes from opposite sides

$$\lim_{n\to\infty} \frac{f_2(x_n) - f_2(0)}{x_n - 0} = \lim_{n\to\infty} \frac{x_n^2}{x_n} = \lim_{n\to\infty} x_n = 0$$

consistent with the tangent lines from the left of 0 getting horizontal when approaching 0.

We conclude that the derivative of f_2 does not exist at $x = 0$ because for some of the sequences that converge to 0, the difference quotient fails to converge. Geometrically, both the x-axis and the y-axis are tangent to the graph of f_2 at $x = 0$.

Finally, to illustrate Item 3 above, consider the absolute value function $f_3(x) = |x|$, or equivalently,

$$f_3(x) = \begin{cases} x, & x \geq 0 \\ -x, & x \leq 0 \end{cases}$$

The v-shaped graph of $f_3(x)$ appears in Figure 6.5.

In this case, the graph is just a pair of lines, each being its own tangent line; the line on the right has slope 1, while the one on the left has slope -1. As in the discussions for the previous two items, this duality indicates that $f_3(x)$ does not have a derivative at 0. Let's examine (6.23) in this case: for every sequence x_n

$$\frac{f_3(x_n) - f_3(0)}{x_n - 0} = \begin{cases} \frac{x_n}{x_n} = 1, & x_n > 0 \\ \frac{-x_n}{x_n} = -1, & x_n < 0 \end{cases}$$

So if the sequence x_n converges to 0 from the right, then the difference quotient approaches 1; but if x_n converges to zero from the left, then the difference quotient approaches -1. In order for the derivative to exist, the limit must not only exist but always have the *same* value for all sequences that converge to 0.

6.3 Continuous functions

The functions discussed above and illustrated in Figures 6.3, 6.4, and 6.5 may not be differentiable at 0, but they are all *continuous* there. There are no jumps or gaps

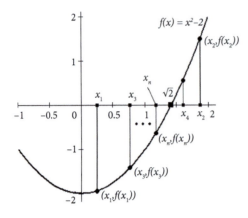

Fig. 6.6 Continuity and sequence limit

in their graphs at $x = 0$ (or anywhere else for that matter). We might imagine sketching their graphs without lifting the tip of the pen from the sheet (or screen). In this section, we formalize this intuitive idea of continuous curve to obtain a precise and useful concept in analysis.

Earlier in our study of the set $\mathbb{R}$ of real numbers we saw that the number line is "continuous" (technically, complete) because the irrational numbers filled in all the gaps among the rational numbers. In this section, we study the continuity of functions whose domains and ranges are contained in $\mathbb{R}$. Similarly to how we defined the continuity of $\mathbb{R}$ using convergent sequences, we define continuity of functions using convergent sequences. In the process, we see where and how the infinity comes into play!

6.3.1 Continuity and limits

Consider the function $f(x) = x^2 - 2$. From the sequence limit theorems in Chapter 3, it follows that

$$\lim_{n\to\infty} f(x_n) = \lim_{n\to\infty} (x_n^2 - 2) = \lim_{n\to\infty} (x_n^2) - \lim_{n\to\infty} 2 = (\lim_{n\to\infty} x_n)^2 - 2$$

If x_n is an arbitrary sequence on the x-axis that converges to $\sqrt{2}$, so that $\lim_{n\to\infty} x_n = \sqrt{2}$ then

$$\lim_{n\to\infty} f(x_n) = (\sqrt{2})^2 - 2 = 0$$

Therefore, the numbers $f(x_n)$ on the y-axis converge to 0. Now remember that the points $(x_n, f(x_n))$ are on the graph of $f(x)$, which means that the graph itself approaches the point $(\sqrt{2}, 0)$ on the x-axis; see Figure 6.6.

The point $(\sqrt{2}, 0)$ is also on the graph of $f(x)$ because the y-coordinate at $x = \sqrt{2}$ is

$$f(\sqrt{2}) = (\sqrt{2})^2 - 2 = 0$$

Therefore, the graph of $f(x)$ crosses the x-axis at the point $(\sqrt{2}, 0)$. If this point weren't on the graph of $f(x)$, then technically, the graph would not cross the x-axis; it would simply jump over it.

This example suggests what is needed in a general definition of continuous functions.

Continuous function

A function $f(x)$ is *continuous at $x = a$* if for *every* sequence x_n of real numbers that approaches a as $n \to \infty$, the image sequence $f(x_n)$ approaches the image $f(a)$. More precisely, $f(x)$ is *continuous at a point a* if $f(a)$ exists; and for *every* sequence x_n of real numbers

$$\lim_{n\to\infty} x_n = a \quad \text{implies} \quad \lim_{n\to\infty} f(x_n) = f(a). \qquad (6.24)$$

If $f(x)$ is continuous at every point of its domain, then we say that $f(x)$ is a *continuous function*. On the other hand, $f(x)$ is *discontinuous at $x = a$* if the first equality in (6.24) holds but *not the second*, for *at least one sequence x_n* that converges to a.

To illustrate the consistency of this definition with our common sense understanding, of continuity, let's look at an example in which it fails, and we see that the resulting "discontinuity" is what we would expect. Consider the following simple, two-piece function

$$f(x) = \begin{cases} 1, & x \geq 0 \\ -1, & x < 0 \end{cases} \qquad (6.25)$$

The graph of this simple function is shown in Figure 6.7; notice that $f(0) = 1$ in this definition as highlighted in the graph by placing a filled circle on the y-axis at 1; the hollow circle at -1 indicates that this point on the y-axis is *not* on the graph of $f(x)$.

Now, consider the definition of continuity above and pick a sequence that converges to 0 from the left, say, $x_n = -1/n$. Then, as we see in Figure 6.7,

$$\lim_{n\to\infty} x_n = 0 \quad \text{and} \quad \lim_{n\to\infty} f(x_n) = \lim_{n\to\infty} (-1) = -1$$

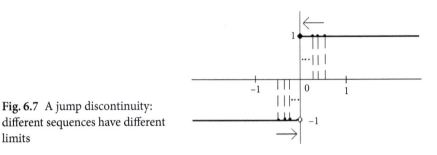

Fig. 6.7 A jump discontinuity: different sequences have different limits

But $f(0) = 1$, which is not equal to the limit -1 above! So our definition of continuity does imply that this function is discontinuous at $x = 0$ as we would expect it to do.

This example also shows that to prove a function is continuous at a point a, the second equality in (6.24) must hold for **every** sequence that converges to a. Indeed, for the function in (6.25), the condition in (6.24) does hold for *some* sequences that converge to 0; if we pick $x_n = 1/n$, then

$$\lim_{n\to\infty} x_n = 0 \quad \text{and} \quad \lim_{n\to\infty} f(x_n) = \lim_{n\to\infty}(1) = 1 = f(0)$$

So if some sequences satisfy the limit condition in (6.24), but *there is even one sequence that doesn't*, then the function is discontinuous.

It is worth emphasizing at this stage if there are no points at which (6.24) fails for a function $f(x)$, then $f(x)$ is continuous. To illustrate this subtle aspect of the definition, consider the reciprocal function $f(x) = 1/x$. We commonly state that the domain of this function is the set of all nonzero real numbers, which we may write as $D=(-\infty, 0)\cup(0, \infty)$. Now, since $1/x$ is continuous at every point of this set D, we would conclude that $1/x$ is a continuous function. This seems at odds with what we see in a typical graph of the reciprocal function, which jumps to infinity as the value of x approaches 0. But there are really no conflicts; the discontinuity occurs at 0, which is not in D.

Of course, if we include 0 and ask if $1/x$ is continuous on $(-\infty, \infty)$, then the answer is no, but mainly because we don't have a value for $1/x$ at 0. If we extend $f(x) = 1/x$ and define it at $x = 0$ by assigning an arbitrary value to $f(0)$, then we quickly discover that for sequences x_n that converge to 0, $f(x_n) = 1/x_n$ has no limit.

Continuity and differentiability

We have discussed examples of functions that are continuous but not differentiable at some point. Therefore, *continuity doesn't imply differentiability*. The next result proves that the converse is true.

Theorem 193 *If $f(x)$ is differentiable at $x = a$, then $f(x)$ is continuous at $x = a$.*

Proof Let x_n be an arbitrary sequence such that $\lim_{n\to\infty} x_n = a$. We must show that $\lim_{n\to\infty} f(x_n) = f(a)$ if $f(x)$ is differentiable at a. We may ignore all terms $x_k = a$ because for all these terms, $f(x_k) = f(a)$ trivially. So suppose that $x_n \neq a$. Then

$$\lim_{n\to\infty} [f(x_n) - f(a)] = \lim_{n\to\infty} \left[\frac{f(x_n) - f(a)}{x_n - a}(x_n - a) \right]$$

$$= f'(a) \lim_{n\to\infty} (x_n - a)$$

$$= 0$$

It follows that $\lim_{n\to\infty} f(x_n) = f(a)$; and since x_n was chosen arbitrarily, $f(x)$ is continuous at $x = a$. ∎

This simple theorem has practical value. For example, to show that $(2x - 1)/\sqrt{x^2 + 1}$ is continuous on $(-\infty, \infty)$, it is not necessary to check its limit for every real number. We note that it is differentiable at all real numbers by the chain rule and the quotient rule, so by the above theorem, it is continuous on $(-\infty, \infty)$.

Examples of discontinuity

We observed earlier that the function $1/x$ is continuous on its domain (nonzero real numbers), but it cannot be extended to a continuous function on all real numbers. But for other functions, the situation can be less severe. For example, the function

$$f(x) = \frac{x^2 - 1}{x - 1} \tag{6.26}$$

is not defined, hence not continuous at $x = 1$. But if $x \neq 1$, then $f(x)$ reduces as follows:

$$f(x) = \frac{(x - 1)(x + 1)}{x - 1} = x + 1$$

so $f(x)$ is continuous at all other real numbers. It just happens to have a hole at $x = 1$.

If x_n is an arbitrary sequence that converges to 1, but $x_n \neq 1$ for all n, then

$$\lim_{n\to\infty} f(x_n) = \lim_{n\to\infty} (x_n + 1) = 2$$

So if we fill the aforementioned hole with the number 2, i.e., define $f(1)=2$, then the extended function is continuous at $x = 1$.

The discontinuity of $f(x)$ in (6.26) is a *removable discontinuity* since by filling in the missing value using the limit (i.e., removing the discontinuity), we extend $f(x)$ to a continuous function. If $f(x)$ is discontinuous at $x = a$, but there are sequences x_n that converge to a without $f(x_n)$ having a limit, then the discontinuity of $f(x)$ at a

is *non-removable* because there are no real numbers that can serve as $f(a)$ and result in a continuous extension of $f(x)$. For example, $1/x$ has no continuous extension to $x = 0$, so it has a non-removable discontinuity at 0.

In spite of their simplicity, removable discontinuities are significant and a more common occurrence than it may seem because differentiation or taking a derivative is essentially removing a discontinuity. It is easy to see how: consider the *difference quotient* of $f(x)$ at $x = a$:

$$Q_f(x) = \frac{f(x) - f(a)}{x - a} \tag{6.27}$$

This function obviously has a singularity at $x = a$; but according to the definition of derivative, this discontinuity is removable whenever $f(x)$ has a derivative at $x = a$.

A function $f(x)$ has a derivative $f'(a)$ if and only if its difference quotient $Q_f(x)$ has a removable discontinuity at $x = a$. By defining the value of $Q_f(x)$ at $x = a$ as

$$Q_f(a) = f'(a)$$

we ensure that the difference ratio $Q_f(x)$ is a continuous function.

This is essentially the definition of derivative reworded without the limit notation; the limit is there implicitly through the removable discontinuity.

The functions that we encounter in standard calculus textbooks, or in typical engineering or scientific models, may be discontinuous at many points, but they are continuous in-between those points. It is difficult to visualize a function that is not continuous at *every* point of a given interval, but it is surprisingly easy to define such a function. Consider the following function, often called *Dirichlet's function* after the German mathematician J. P. G. L. Dirichlet (1805–1859);

$$D(x) = \begin{cases} 1, & \text{if } x \text{ is rational} \\ 0, & \text{if } x \text{ is irrational} \end{cases}$$

Recall that each of the sets of rational numbers and irrational numbers is dense in the set of real numbers. So for every real number a, there is a sequence q_n of rational numbers that converges to a, and we have

$$\lim_{n \to \infty} q_n = a \quad \text{and} \quad \lim_{n \to \infty} D(q_n) = \lim_{n \to \infty} 1 = 1$$

There is also a sequence x_n of irrational numbers that converges to a, and for this sequence, we have:

$$\lim_{n \to \infty} x_n = a \quad \text{and} \quad \lim_{n \to \infty} D(x_n) = \lim_{n \to \infty} 0 = 0$$

Therefore, (6.24) cannot be satisfied in this case because as we have just seen, it cannot hold for *every* sequence that converges to a. It follows that D is discontinuous at a; and since this argument is valid for every real number a, we conclude that Dirichlet's function $D(x)$ is discontinuous everywhere.

It is impossible to draw an actual graph of $D(x)$, but you can imagine it as consisting of the x-axis with all the rational numbers moved up one unit to form a grainy straight line, a fine mist of tiny dots, which hangs above and parallel to the x-axis. This image is not strictly correct, but it may be intuitively helpful.

A more interesting example is the following function: if $0 < x < 1$, then

$$f(x) = \begin{cases} \dfrac{1}{n} & \text{if } x = \dfrac{m}{n} \text{ in lowest terms} \\ 0 & \text{if } x \text{ is irrational} \end{cases}$$

Note that $f(x) \neq 0$ if and only if x is a rational number in the interval $(0, 1)$. We prove that $f(x)$ is continuous at all irrational numbers in this interval and discontinuous at all rational numbers.

A key feature of $f(x)$ is that $m < n$ for all m, n since $m/n = x < 1$. Thus, for every fixed integer $n \geq 2$, there are only a finite number of nonzero (rational) values of $f(x)$. The following table lists these nonzero values for $2 \leq n \leq 6$;

$f(m/n) = 1/n$	$1/2$	$1/3$	$1/4$	$1/5$	$1/6$
m/n (reduced)	$1/2$	$1/3, 2/3$	$1/4, 3/4$	$1/5, 2/5, 3/5, 4/5$	$1/6, 5/6$

For every prime n there are $n - 1$ fractions m/n where $f(m/n) = 1/n$; while if n is not prime, then there are fewer such fractions (in reduced form). If we plot these points in the usual coordinate system, the plot of points vaguely resembles a Christmas tree, so sometimes this function is called the "Christmas tree function."

Now suppose that $q = m/n$ is a rational number in $(0, 1)$. If x_n is a sequence of irrational numbers that converges to q, then $f(x_n) = 0$ for all n. It follows that

$$\lim_{n \to \infty} f(x_n) = 0 \neq \frac{1}{n} = f(q)$$

Thus, f is not continuous at q. Since q was an arbitrary rational number, we conclude that f is discontinuous at all rational numbers in $(0, 1)$.

To see that $f(x)$ is continuous at every irrational number x in $(0, 1)$, notice that if x_n is a sequence of irrational numbers that converges to x, then $f(x_n) = 0 = f(x)$, so we need only check sequences with infinitely many rational terms that converge to x.

Let x_n be a sequence in $(0, 1)$ that converges to the irrational number x with infinitely many rational terms $x_{n_j} = m_j/k_j$. Notice that $k_j \to \infty$ as $j \to \infty$, for if not, then there is a positive integer N such that $k_j \leq N$ for all j, and there would only be a finite number (at most $N-1$) of distinct terms of type m_j/k_j with $m_j < N$. It follows that

$$\lim_{j\to\infty} f(x_{n_j}) = \lim_{j\to\infty} f\left(\frac{m_j}{k_j}\right) = \lim_{j\to\infty} \frac{1}{k_j} = 0$$

Since the irrational terms x_n with $n \neq n_j$ already have $f(x_n) = 0$ by definition, it follows that $\lim_{n\to\infty} f(x_n) = 0$. Therefore, f is continuous at all irrational numbers in $(0, 1)$.

6.3.2 Continuity and algebraic operations

Continuity is essentially the existence of limits according to (6.24); thus, the properties of limits that allow us to distribute limits over sums, products, and quotients translate into similar properties for continuous functions. We summarize these properties as follows (proof is Exercise 225).

Theorem 194 *If the functions $f(x)$ and $g(x)$ are continuous at $x = a$, then so are their sum $f + g$, their product fg, and if $g(a) \neq 0$, also their quotient f/g.*

You may recall that another way to combine functions is to compose them: $g \circ f(x) = g(f(x))$. The following result is often useful in calculus.

Theorem 195 *If $f(x)$ is continuous at $x = a$, and $g(x)$ is continuous at $f(a)$, then the composition $g \circ f$ is continuous at a. Thus, the composition of continuous functions is continuous.*

Proof We use the definition of continuity in (6.24). Let x_n be an arbitrary sequence that converges to a. Since $f(x)$ is continuous at $x = a$, it follows that $\lim_{n\to\infty} f(x_n) = f(a)$. Set $f(x_n) = y_n$ and $f(a) = b$. Since $g(x)$ is continuous at $x = b$, and we just showed that $\lim_{n\to\infty} y_n = b$, it follows that $\lim_{n\to\infty} g(y_n) = g(b)$, or equivalently,

$$\lim_{n\to\infty} g(f(x_n)) = g(f(a))$$

In other words,

$$\lim_{n\to\infty} x_n = a \quad \text{implies} \quad \lim_{n\to\infty} (g \circ f)(x_n) = (g \circ f)(a)$$

which means that $g \circ f(x)$ is continuous at $x = a$. ∎

The condition (6.24) in the definition of continuous functions can be expressed as an operational property, i.e., a rule that enables us to take a limit in or out of the function $f(x)$.

Suppose that x_n is a sequence in an interval I that converges to a limit in I. If $f(x)$ is a continuous function on I, then

$$\lim_{n\to\infty} f(x_n) = f(\lim_{n\to\infty} x_n) \tag{6.28}$$

For example, $f(x) = \sqrt{x}$ is continuous for all $x > 0$ (it is actually differentiable as a power function $x^{1/2}$). So if $\lim_{n\to\infty} x_n = 2$ for some sequence of positive numbers, then we can use this fact and (6.28) to calculate

$$\lim_{n\to\infty} \sqrt{x_n} = \sqrt{\lim_{n\to\infty} x_n} = \sqrt{2}.$$

Here "we moved the limit inside the square root." We can't do this with discontinuous functions; for instance, if $f(x) = (x^2 - 1)/(x - 1)$, and $\lim_{n\to\infty} x_n = 1$, then

$$\lim_{n\to\infty} f(x_n) = \lim_{n\to\infty} \frac{x_n^2 - 1}{x_n - 1} = \lim_{n\to\infty} \frac{(x_n - 1)(x_n + 1)}{x_n - 1} = \lim_{n\to\infty} (x_n + 1) = 2$$

Therefore, if $f(1)$ is defined to be anything other than 2, then the equality in (6.28) *never* holds for *any* sequence x_n that converges to 1.

6.3.3 The intermediate value theorem

At the beginning of this section, we considered the function $f(x) = x^2 - 2$ and how the continuity of this function guarantees that its graph crosses the x-axis at $x = \sqrt{2}$. But the x-axis is not really special; consider this function on the interval $[0, 2]$ and note that $f(0) = -2$ and $f(2) = 2$. Looking back at Figure 6.6, if we draw a horizontal line $y = d$ where d is any number between -2 and 2 (not only $d = 0$), then we expect that this line crosses the graph of $x^2 - 2$ at some point because the graph is continuous. At that point, $f(x) = d$, that is, $f(x)$ takes the value d. This observation holds for all continuous functions defined on closed intervals.

Figure 6.8 displays the graphs of two *continuous* functions. Each function takes the values $f(a)$ and $f(b)$ at each end of the interval; and if $f(a) \ne f(b)$, then $f(x)$ also takes *every value* $y = d$ between $f(a)$ and $f(b)$ for at least one value of the variable x.

You may be wondering why in Figure 6.8 the arrows go from the numbers on the y-axis to numbers on the x-axis rather than the other way around. To see why, consider the graphs in Figure 6.9; notice that in both graphs, every value of x on

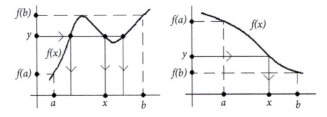

Fig. 6.8 Illustrating the intermediate value property

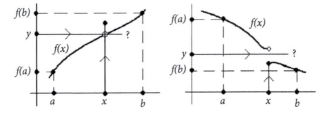

Fig. 6.9 The intermediate value property does not hold with a discontinuity

the x-axis between a and b corresponds to at least one number between $f(a)$ and $f(b)$ on the y-axis. The functions shown are discontinuous, and there are numbers on the y-axis that don't come from any numbers on the x-axis through the graph of $f(x)$.

The property of continuous functions that Figure 6.8 illustrates is stated in the next fundamental result.[6]

Theorem 196 *(Intermediate value property) Assume that $f(x)$ is a continuous function on the interval $[a, b]$, and $f(a) \neq f(b)$. For every given number y between $f(a)$ and $f(b)$, there is a number c (at least one) in the open interval (a, b) such that $f(c) = y$.*

Proof We may assume that $f(a) < y < f(b)$ since if $f(a) > y > f(b)$, then $-f(a) < -y < -f(b)$, and $-f$ is continuous.

The proof is based on an application of the nested intervals property (Theorem 111 in Chapter 4). Our strategy is to progressively reduce the length of the interval by half until a candidate point c is found, and then use proof by contradiction to establish $f(c) = y$.

Let $c_1 = (a + b)/2$. If $f(c_1) = y$, then there is nothing more to do but set $c = c_1$. Otherwise, there are two possible cases:

(i) $f(c_1) < y < f(b)$: in this case, we define $a_1 = c_1$, $b_1 = b$;

---

[6] Topologically, this result is a corollary of the fact that *continuous functions map connected sets into connected sets*. In $\mathbb{R}$, connected sets are precisely the intervals.

(ii) $f(c_1) > y > f(a)$: in this case, we define $a_1 = a$, $b_1 = c_1$;

In either case, the length of the interval $[a_1, b_1]$ is half that of $[a, b]$, i.e., $(b - a)/2$ and

$$a \leq a_1 < b_1 \leq b$$

$$f(a_1) < y < f(b_1)$$

Next, define

$$c_2 = \frac{a_1 + b_1}{2}$$

and repeat the above procedure. After n steps, we have an interval $[a_n, b_n]$ of length $(b - a)/2^n$, and the following relations hold:

$$a \leq a_1 \leq \cdots \leq a_n < b_n \leq \cdots \leq b_1 \leq b$$

$$f(a_n) < y < f(b_n)$$

The conditions of Theorem 111 are satisfied for the sequence of nested intervals

$$[a, b] \supset [a_1, b_1] \supset [a_2, b_2] \supset \cdots$$

so there is a unique number common to all of the above intervals; we claim that this is the number c in the statement of the theorem, i.e., $f(c) = y$.

Suppose this not so, and $f(c) \neq y$. First, assume that $f(c) < y$, and let $\varepsilon = [y - f(c)]/2$. Note that $b_n \to c$ as $n \to \infty$, so given that f is continuous on $[a, b]$ (thus also at $x = c$), it follows that $f(b_n) \to f(c)$. In particular, there is a positive integer N such that

$$|f(b_N) - f(c)| < \varepsilon = \frac{y - f(c)}{2}$$

But this implies that

$$f(b_N) < f(c) + \frac{y - f(c)}{2} = y - \frac{y - f(c)}{2} < y$$

which contradicts the earlier conclusion that $y < f(b_n)$ for all $n \geq 1$. What if $f(c) > y$? We use a similar argument with a_n instead of b_n and $\varepsilon = [f(c) - y]/2$ to arrive at a contradiction. Therefore, we conclude that $f(c) = y$, as claimed. ∎

As the use of the nested intervals property in the proof shows, the validity of the intermediate value property is based on completeness. For instance, $f(x) = x^2 - 2$ is continuous; but if we were using $\mathbb{Q}$ instead of $\mathbb{R}$ as the x-axis, then the graph of

$x^2 - 2$ would slip right through the hole where $\sqrt{2}$ would be without hitting any point of $\mathbb{Q}$!

In applications, the intermediate value property is often used to prove the existence of solutions to nonlinear equations. We discuss this matter later in this chapter in conjunction with the Newton–Raphson method for approximating those solutions.

6.3.4 Boundedness and the extreme value theorem

Consider a bounded function $f(x)$ on a set S, and let m be the absolute minimum of $f(x)$ so that the graph of $f(x)$ comes down as low as m as x ranges over S but no lower; see Figure 6.10.

Similarly, let M be the absolute maximum of $f(x)$ so that the graph reaches up to M as x ranges over S but goes no higher. The numbers m and M are the *extreme values* of $f(x)$. It is reasonable to conclude that the lowest value m must be associated with some number α in S and the highest M to some β in S so that $f(\alpha) \leq f(x) \leq f(\beta)$ for all x in S; see Figure 6.10.

This scenario is intuitively plausible but not always correct. To see what may be missing, consider the functions shown in Figure 6.11.

The left panel in Figure 6.11 shows a simple, bounded function over an interval. This function has both a maximum value M and a minimum value m. But *there*

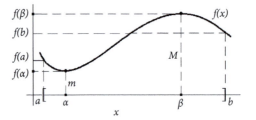

Fig. 6.10 Illustrating the extreme value property

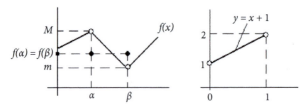

Fig. 6.11 Cases where the extreme value theorem does not hold

are no points in the interval that correspond to its lowest and highest values! The values of the function shown at the numbers labeled α and β in Figure 6.11 are different from both M and m. Notice that the function is discontinuous at the two points shown.

It seems from Figure 6.11 that the highest value M is never really attained by $f(x)$ since there is a hole at the peak. But this high value is reached in the limit. To see how, imagine a sequence x_n on the x-axis that approaches α from either the left or the right, and track the function values $f(x_n)$ on the graph. We see that these numbers do in fact approach M in the sense that

$$\lim_{n\to\infty} f(x_n) = M$$

A similar observation holds for m.

These observations show that continuity is essential for achieving extreme values at specific points of an interval. But continuity alone is not enough; consider the line $y = x + 1$ for $0 < x < 1$ that is shown in the right panel of Figure 6.11 over the open interval $(0, 1)$. This function is continuous for all x, its lowest value $m = 1$ occurs (in the limit) at $x = 0$, and its highest value $M = 2$ at $x = 1$. But neither 0 nor 1 is in $(0, 1)$; here the problem is with the missing endpoints. These observations explain the assumptions in Theorem 198, the *extreme value property*. Before stating and proving the theorem, we present a lemma that is interesting by itself.

Lemma 197 *If a function $f(x)$ is continuous on the closed interval $[a, b]$, then $f(x)$ is bounded, i.e., there are real numbers m and M such that for all $x \in [a, b]$*

$$m \le f(x) \le M \tag{6.29}$$

Proof We use a contradiction argument. If $f(x)$ is not bounded on $[a, b]$, then there is a sequence x_n in $[a, b]$ such that $|f(x_n)| = y_n$ diverges to infinity. The sequence x_n being inside $[a, b]$ is bounded so by the Bolzano–Weierstrass theorem, it has a subsequence x_{n_k} that converges to a point $r \in [a, b]$ (also recall Theorem 59, Chapter 3). Since f is continuous on $[a, b]$, we have

$$|f(r)| = \left|\lim_{k\to\infty} f(x_{n_k})\right| = \lim_{k\to\infty} |f(x_{n_k})| = \lim_{k\to\infty} y_{n_k} = \infty$$

(the limit moves outside absolute value because $|x|$ is continuous on $\mathbb{R}$; see Exercise 223). However, $|f(r)| < \infty$; so to avoid a contradiction, we conclude that $f(x)$ is bounded on $[a, b]$, i.e., (6.29) is true.   ∎

With regard to the hypotheses of Lemma 197, notice that an unbounded function such as $f(x) = 1/x$, which is continuous on the interval $(0, 1]$, doesn't extend to 0 in a continuous fashion because then the resulting continuous function on the closed interval $[0, 1]$ would have to be bounded. Of course, if some $f(x)$ is continuous on a closed $[a, b]$, then it is bounded not only on $[a, b]$ but also on every subset of $[a, b]$, closed or not.

Theorem 198 *(Extreme value property) Let f be a continuous function on the closed interval $[a, b]$. There are real numbers $\alpha, \beta \in [a, b]$ such that*

$$f(\alpha) \le f(x) \le f(\beta) \quad \text{for all } x \in [a, b] \tag{6.30}$$

i.e., f achieves its minimum and maximum values at the points $\alpha, \beta \in [a, b]$.

Proof We prove the existence of β; the existence of α then follows by applying the same argument to $-f$.

By Lemma 197, the image set $f([a, b])$ is bounded above by some number M; so by the least upper bound property (Theorem 112, Chapter 4), it has a supremum

$$s = \sup f([a, b]) = \sup\{f(x) : a \le x \le b\} \le M$$

We must prove that $s = f(\beta)$ for some number $\beta \in [a, b]$.

By Lemma 62 in Chapter 3, there is a sequence y_n in $f([a, b])$ that converges to s. For each index n, let x_n be a number in $[a, b]$ such that $f(x_n) = y_n$. Then the set of all such x_n is a sequence in $[a, b]$ that is bounded, so it has a convergent subsequence by the Bolzano–Weierestrass theorem. Let x_{n_k} be that subsequence, and define

$$\beta = \lim_{k \to \infty} x_{n_k}$$

Note that $\beta \in [a, b]$ (Theorem 59, Chapter 3). We claim that $f(\beta) = s$. Since f is continuous, it follows that

$$\lim_{k \to \infty} f(x_{n_k}) = f\left(\lim_{k \to \infty} x_{n_k} \right) = f(\beta)$$

Further, since y_n converges to s, so does the subsequence y_{n_k} (Theorem 61, Chapter 3). Hence,

$$\lim_{k \to \infty} f(x_{n_k}) = \lim_{k \to \infty} y_{n_k} = s$$

It follows that $s = f(\beta)$, as claimed. ∎

The extreme value property is a fundamental property of continuous functions on closed intervals. It is used to establish the existence of maxima and minima of functions, and we also use it in our study of the integral. Although Theorem 198 guarantees the existence of points α and β, at which minimum and maximum values occur, it doesn't provide a prescription for finding α, β. To help us find these numbers, we need additional results that we discuss in the section on the mean value theorem after a discussion of function limits in the next section.

Theorem 198 can also be proved using either the nested intervals property (Theorem 4.6) or the least upper bound property (Theorem 112), both in Chapter 4. Note that both of these results require the completeness property.

Theorem 198 is also a special case of a general topological fact about *compact sets*, i.e., continuity preserves compactness. Closed and bounded intervals are compact sets. The next corollary is a special case that we can prove here using the extreme value property and the intermediate value property.

Corollary 199 *Let f be a continuous function on the closed interval* $[a, b]$. *Then the image set* $f([a, b]) = \{f(x) : a \le x \le b\}$ *is a closed and bounded interval.*

Proof By Theorem 198, there are numbers $\alpha, \beta \in [a, b]$ such that (6.30) holds. It follows that $f([a, b]) \subset [f(\alpha), f(\beta)]$. To show the reverse inclusion, we observe that f is continuous on the interval of numbers between α and β because this interval is contained in $[a, b]$. Now, let $f(\alpha) \le y \le f(\beta)$, and note that by the intermediate value property (Theorem 196), there is x between α and β such that $y = f(x)$. It follows that $x \in [a, b]$ and thus $y \in f([a, b])$. Since y was arbitrarily chosen in $[f(\alpha), f(\beta)]$, it follows that $[f(\alpha), f(\beta)] \subset f([a, b])$. We conclude that $f([a, b]) = [f(\alpha), f(\beta)]$, which shows that $f([a, b])$ is a closed and bounded interval. ∎

6.3.5 Uniform continuity

In this section, we discuss a concept of continuity that is important in the study of integration and in Chapter 7, in particular, to prove that continuous functions are integrable (Theorem 248). We also discover a hidden occurrence of infinity in the definition of continuous functions that is brought out by our study here and will be seen again in Chapter 8 in our study of uniform convergence. Continuous functions for which this infinity does not occur are the uniformly continuous ones.

Uniformly continuous function

A function $f(x)$ is *uniformly continuous* on a nonempty set S of real numbers if for every pair of sequences x_n and y_n in S

$$\lim_{n \to \infty} (x_n - y_n) = 0 \quad \text{implies} \quad \lim_{n \to \infty} [f(x_n) - f(y_n)] = 0. \qquad (6.31)$$

A more conventional definition of uniform continuity is given in Section 6.4. Notice that the concept of uniform continuity as defined above involves the set S as a whole rather than any particular point in it. It is a bit stronger than continuity because if we choose y_n to be the constant sequence $y_n = a$ for any point $a \in S$, then the above definition reduces to the earlier definition of continuous function. Therefore, every function f that is uniformly continuous on a set S is also continuous at every point of S.

The converse of the last statement is false; for example, the square function x^2 is continuous on $(-\infty, \infty)$, but it is not uniformly continuous. To see why not, let $x_n = n$ and $y_n = n - 1/n$ so that

$$\lim_{n \to \infty} (x_n - y_n) = \lim_{n \to \infty} \frac{1}{n} = 0$$

but

$$\lim_{n \to \infty} [f(x_n) - f(y_n)] = \lim_{n \to \infty} \left(2 - \frac{1}{n^2}\right) = 2$$

An important point that this simple example makes is that *a differentiable function may not be uniformly continuous*. In contrast, recall that a differentiable function is always continuous.

For an example of a bounded function on a bounded set, consider $f(x) = \sin(\pi/x)$ on the interval $(0, 1]$. Here we may use the facts that for all integers n

$$\sin\left(\frac{\pi}{2} + 2n\pi\right) = \sin \frac{\pi}{2} = 1, \qquad \sin(n\pi) = 0$$

to define two sequences

$$x_n = \frac{2}{4n + 1}, \qquad y_n = \frac{1}{n}$$

With these sequences, we obtain

$$\lim_{n \to \infty} (x_n - y_n) = \lim_{n \to \infty} \frac{2}{4n + 1} - \lim_{n \to \infty} \frac{1}{n} = 0$$

whereas

$$\lim_{n \to \infty} [f(x_n) - f(y_n)] = \lim_{n \to \infty} \left[\sin\left(\frac{4n\pi + \pi}{2}\right) - \sin n\pi\right]$$

$$= \lim_{n \to \infty} \sin\left(2n\pi + \frac{\pi}{2}\right)$$

$$= 1$$

If we take a closer look at the last example, we see that it is *not* possible to define $\sin(\pi/x)$ at $x = 0$ in such a way that the extended function is continuous on the closed interval $[0,1]$. The next result shows why continuity on a *closed and bounded* interval is important; it also gives an important criterion that will be useful later in the development of the integral.

Theorem 200 *(Uniform continuity theorem) If f is continuous on a closed and bounded interval $[a, b]$, then f is uniformly continuous.*

Proof We prove this theorem by contradiction. Assume that there is a pair of sequences x_n, y_n in $[a, b]$ such that $\lim_{n\to\infty}(x_n - y_n) = 0$, but $\lim_{n\to\infty}[f(x_n) - f(y_n)] \neq 0$. It follows that there must be an $\varepsilon_0 > 0$ such that

$$|f(x_n) - f(y_n)| \geq \varepsilon_0 \quad \text{for all } n \tag{6.32}$$

Strictly speaking, the inequality in (6.32) holds for some subsequence of x_n and y_n, but we may discard the terms that do not satisfy (6.32) and use what is left as our x_n, y_n.

The sequence x_n being inside $[a, b]$ is bounded so by the Bolzano–Weierstrass theorem; it has a subsequence x_{n_k} that converges to a point $r \in [a, b]$, i.e., $\lim_{k\to\infty} x_{n_k} = r$. Since $\lim_{n\to\infty}(x_n - y_n) = 0$, it follows that $\lim_{k\to\infty} y_{n_k}$ also exists and equals r. Now given that f is continuous on $[a, b]$ and thus also at $x = r$, it follows that

$$\lim_{k\to\infty} f(x_{n_k}) = f(r), \qquad \lim_{k\to\infty} f(y_{n_k}) = f(r)$$

so that $\lim_{k\to\infty}[f(x_{n_k}) - f(y_{n_k})] = 0$. Thus, there is N large enough that for all $n \geq N$

$$\left|f(x_{n_k}) - f(y_{n_k})\right| < \varepsilon_0$$

However, this inequality contradicts (6.32), which means that f must in fact be uniformly continuous. ∎

Theorem 200 in particular implies that x^2 is uniformly continuous on all bounded intervals of type $[-a, a]$ no matter how large a is, but as we showed above, not on $(-\infty, \infty)$. Further, the problem of verifying uniform continuity is reduced to proving continuity, which is usually simpler.

It is necessary now to explain why we use the word "uniform" in the above definition. Recall that a function $f(x)$ is continuous on a nonempty set S if it is continuous at every point of S. Further, $f(x)$ is continuous at a given point $x = a$ if "for every sequence x_n in S that converges to a, the sequence $f(x_n)$ converges to $f(a)$." Let's take a closer look at the statement in quotation marks.

To have $f(x)$ continuous at $x = a$, given an arbitrary number $\varepsilon > 0$, there must exist an index N and a real number $\delta > 0$ such that $|f(x_n) - f(a)| < \varepsilon$ for all $n \geq N$ provided that $|x_n - a| < \delta$ for all $n \geq N$ (so that x_n is near a). We emphasize now that *the integer N and the number δ depend not only on the given ε but also on the given point a.* Therefore, $N = N(a)$ and $\delta = \delta(a)$ are functions of a.

Next, consider the function $f(x) = 1/x$ on the interval $(0, 1)$. On this interval f is a continuous function. We prove this using the definition of continuity in order to point out what makes this function *not uniformly* continuous.

Let $0 < a < 1$, and consider a sequence x_n in $(0,1)$ that converges to a. If $\varepsilon > 0$ is given and fixed, then $|f(x_n) - f(a)| < \varepsilon$ if

$$\left| \frac{1}{x_n} - \frac{1}{a} \right| < \varepsilon$$

$$|x_n - a| < \varepsilon a x_n$$

Since $x_n \approx a$ for large index values of n, we might choose $N(a)$ large enough that $x_n \leq a(1 + \varepsilon)$, so for all $n \geq N$, we get

$$|x_n - a| < a^2 \varepsilon (1 + \varepsilon)$$

Next, we may define $\delta(a) = a^2 \varepsilon (1 + \varepsilon)$ for fixed ε to conclude that if $n \geq N(a)$, then $|x_n - a| < \delta(a)$, and $|f(x_n) - f(a)| < \varepsilon$.

This argument for proving the continuity of $1/x$ on $(0,1)$ highlights the role of the position of a in $(0,1)$. For concreteness, suppose that $x_n = a + 1/n$ to get a specific form for $N(a)$. Then $x_n \leq a(1 + \varepsilon)$ if

$$a + \frac{1}{n} < a(1 + \varepsilon)$$

$$n > \frac{1}{a\varepsilon}$$

For this choice of x_n, we may choose $N(a) = \lceil 1/(a\varepsilon) \rceil$, i.e., the least integer that is greater than or equal to $1/(a\varepsilon)$. We conclude that as we choose a closer to 0, the value of $N(a)$ goes to infinity. This is not the case for every choice of x_n, but the important point is that *there are some sequences for which* $N(a) \to \infty$ as we choose a closer to 0.

This discussion illustrates why $f(x) = 1/x$ is not *uniformly* continuous on $(0,1)$; we need to pick larger N (and smaller δ) for some values of a than for others; further, there are sequences x_n for which $N(a)$ can get infinitely large. This cannot happen if f is uniformly continuous since the definition of uniform continuity given above doesn't depend on any specific point a.

The hidden occurrence of infinity mentioned earlier refers to $N(a) \to \infty$. This $N(a)$ is analogous to the ε-index of a sequence that we discussed in Chapter 3. It occurs again more prominently in Chapter 8 when we talk about *convergence of sequences of functions being uniform, or not.*

6.4 Limits via neighborhoods: The ε and δ

In a standard calculus textbook, there is usually a chapter or a section on limits. But these limits seem different from what we discussed above. There is no mention of sequences usually until the chapter on infinite series, where of course, converging sequences of partial sums need to be defined as we have already seen.

In this section, we discuss the type of limit that is used in a standard coverage of calculus for defining continuity and derivatives. We will also see that this concept is entirely equivalent to the one involving sequences that we discussed earlier: each one implies the other.

We will also use both definitions in the rest of the book, depending on which is more convenient or more insightful in expressing or proving various concepts.

The traditional definition renders itself more naturally to the study of topological concepts, such as *compactness*. Topological spaces often have neighborhood systems that are too big for sequential convergence to handle, so instead we use *neighborhoods*, which, as sets, are more flexible. Some spaces of functions in analysis are of this sort, so it is necessary and useful to have some understanding of the neighborhood-based approach.

On the other hand, sequences are sufficient for metric spaces of any dimension (including infinite), and they are more convenient to use in situations where neighborhoods are awkward. For instance, using sequences, we can define limits and derivatives without having to worry about "one-sided limit" scenarios; and showing that a limit does not exist is a matter of finding just one sequence that fails to do the job at hand.

Consider the function

$$f(x) = \frac{x^2 - 1}{x - 1} = \frac{(x - 1)(x + 1)}{x - 1} = x + 1 \tag{6.33}$$

As long as $x \neq 1$, the above function is given by $x + 1$ whose graph is that of the straight line $y = x + 1$ with a hole or gap where the point $(1, 2)$ would be. Although this function is not continuous at $x = 1$, it does have a limit at this point: if x_n is any sequence that approaches 1 as $n \to \infty$ (but $x_n \neq 1$ for all n), then

$$\lim_{n \to \infty} \frac{x_n^2 - 1}{x_n - 1} = \lim_{n \to \infty} (x_n + 1) = 2$$

Note that this limit exists even though the fraction itself is not defined at $x = 1$. So we may define the limit of a function, whether it is continuous or not, as follows:

A function $f(x)$ has a *limit* L at $x = a$ if for *every* sequence x_n such that $x_n \neq a$ for all n,

$$\lim_{n \to \infty} x_n = a \quad \text{implies} \quad \lim_{n \to \infty} f(x_n) = L. \tag{6.34}$$

In the above definition, the value of the function $f(x)$ at $x = a$ is irrelevant, and like the function in (6.33), it need not even be defined at a. This is routinely the case for the difference quotient that defines a derivative. And as before, the term "every" in the definition is important when using sequences to present a fact about functions.

The ε paired with a δ

Let us take a closer look at the two limits in (6.34). They say that for every real number $\varepsilon > 0$ there is a positive integer N such that

$$|f(x_n) - L| < \varepsilon \quad \text{for all } n > N$$

provided that x_n is "close enough" to the point a; or, more precisely, provided that there is some (small enough) real number $\delta > 0$ such that[7]

$$|x_n - a| < \delta \quad \text{for all } n > N$$

We can streamline the above description *by dropping any mention of sequences* and just define:

The function limit

The function $f(x)$ approaches the number L as x approaches the number a in a nonempty set S if for every real number $\varepsilon > 0$, there is a real number $\delta > 0$ such that

$$|f(x) - L| < \varepsilon \quad \text{whenever} \quad x \in S \text{ and } 0 < |x - a| < \delta \tag{6.35}$$

The customary notation that summarizes (6.35) is

$$\lim_{x \to a} f(x) = L \tag{6.36}$$

---

[7] The symbols ε and δ have become standard; beyond that, there is nothing special about them.

The inequality $0 < |x - a|$ in (6.35) simply says that $x - a \neq 0$ or $x \neq a$, which says that we do not need to require $f(x)$ be defined at $x = a$. As we saw earlier, this is necessary for defining derivatives using the difference quotient (as in the standard calculus books):

$$f'(a) = \lim_{x \to a} \frac{f(x) - f(a)}{x - a}$$

In this definition, it is essential that $x \neq a$ so that the quotient on the right-hand side is defined. But when defining the continuity of a function $f(x)$ at $x = a$, where $f(a)$ is actually defined, we must consider the value $x = a$. We say that $f(x)$ is continuous at $x = a$ if

$$\lim_{x \to a} f(x) = f(a)$$

Or more precisely, for every real number $\varepsilon > 0$, there is a real number $\delta > 0$ such that

$$|f(x) - f(a)| < \varepsilon \quad \text{whenever} \quad |x - a| < \delta \tag{6.37}$$

You may want to carefully compare this statement with (6.35) above!

Note also that there are no explicit references to infinity in this definition of continuity. This does not mean that infinity is not involved, but its presence is suppressed for brevity. When we write $x \to a$, there is an implicit reference to infinity here that is best described by $x_n \to a$ as $n \to \infty$. In the latter description, we must be careful to add "for every sequence x_n" to not give the impression that any one sequence is enough. So as you can see, the simple notation $x \to a$ contains a lot of information!

There is also the hidden occurrence of infinity that we alluded to in the definition of uniformly continuous functions above. To make the connection, we define uniform continuity in terms of ε and δ concept:

f is uniformly continuous on a nonempty set S if for each given $\varepsilon > 0$ there is $\delta > 0$ such that

$$|x - y| < \delta \quad \text{implies} \quad |f(x) - f(y)| < \varepsilon \tag{6.38}$$

We see that in this definition, there is no mention of specific points in S, so the δ here depends solely on ε. This is not the case for the δ in (6.37).

To illustrate, consider the function $f(x) = 1/x$ on the interval $(0,1)$ that we studied earlier using the sequential definition of uniform continuity. Now (6.37) reads as follows:

$$\left|\frac{1}{x} - \frac{1}{a}\right| < \varepsilon \quad \text{whenever} \quad |x - a| < \delta \tag{6.39}$$

For $a, x \in (0, 1)$, the first inequality on the left may be written as

$$|x - a| < \varepsilon a x$$

The right-hand side is close to a^2 if δ is chosen so small that $x \approx a$. If we choose

$$\delta = \delta(a) = \varepsilon a^2 (1 + \varepsilon)$$

then (6.39) holds with this choice of δ. Now by choosing a closer to 0, the value of $\delta(a)$ also goes to 0, so the choice of a affects the value of δ. We see then that in this sense, $1/x$ is not *uniformly* continuous on $(0,1)$.

There is a sense of proximity or "neighborhood" involved in the function limit idea that gives it a meaning beyond the collection of all sequences. The information contained in (6.35) can also be conveyed via intervals and diagrams. First, let's use the definition of absolute value to translate the inequalities as

$$-\varepsilon < f(x) - L < \varepsilon \quad \text{whenever} \quad -\delta < x - a < \delta \quad \text{and } x - a \neq 0, \quad \text{or}$$

$$L - \varepsilon < f(x) < L + \varepsilon \quad \text{whenever} \quad a - \delta < x < a + \delta \quad \text{and } x \neq a \tag{6.40}$$

In terms of intervals, (6.40) can be stated as

$$f(x) \in (L - \varepsilon, L + \varepsilon) \quad \text{whenever} \quad x \in (a - \delta, a + \delta) \quad \text{and } x \neq a \tag{6.41}$$

The interval $(a - \delta, a + \delta)$ is a *basic neighborhood* of x;[8] likewise, $(L - \varepsilon, L + \varepsilon)$ is a basic neighborhood of $f(x)$.

Inequalities (6.40) and (6.41) are illustrated in Figure 6.12.

The arrows going from the y-axis to the x-axis in Figure 6.12 highlight the important fact that *ε is given and δ is determined by it*; further, if you look closely, then you see that the symmetric interval $(L - \varepsilon, L + \varepsilon)$ on the y-axis is reflected onto a non-symmetric interval on the x-axis; this is typically the case. We can choose any δ small enough that the also symmetric interval $(a - \delta, a + \delta)$ fits inside the non-symmetric interval, as shown in Figure 6.12.

We may now translate (6.40) or (6.41) in words as

f(x) must be in the (small) interval $(L - \varepsilon, L + \varepsilon)$, or neighborhood of the number L, if x is in the (small) interval $(a - \delta, a + \delta)$, or neighborhood of the number a. In

---

[8] In topology, each open set is a neighborhood of every point that it contains. Further, each open set is a union of basic (or basis) open sets that are simpler but make up all open sets. In the set of real numbers with its usual topology, the open intervals are basic sets or neighborhoods.

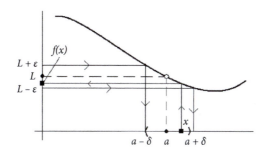

Fig. 6.12 Illustrating limit via neighborhoods

other words, we can bring f(x) as close to the number L as we may want simply by bringing the variable x close enough to a.

The equivalence of function limits and sequence limits

Now that we have defined the neighborhood-based concept of limit, we can go on to prove limit theorems involving algebraic operations and other results that we have already proved for sequence-based limits. For instance, assume that f and g are functions whose limits $\lim_{x \to a} f(x)$ and $\lim_{x \to a} g(x)$ exist for some real number a. To prove the statement

$$\lim_{x \to a}[f(x) + g(x)] = \lim_{x \to a} f(x) + \lim_{x \to a} g(x) \qquad (6.42)$$

let $\varepsilon > 0$ be an arbitrary number, and define $L = \lim_{x \to a} f(x)$ and $M = \lim_{x \to a} g(x)$. Then there are numbers $\delta, \delta' > 0$ such that

$$|f(x) - L| < \frac{\varepsilon}{2} \quad \text{whenever } 0 < |x - a| < \delta$$

$$|f(x) - M| < \frac{\varepsilon}{2} \quad \text{whenever } 0 < |x - a| < \delta'$$

Let δ be the smaller of δ, and δ' and note that whenever $0 < |x - a| < \delta$, by the triangle inequality

$$|f(x) + g(x) - (L + M)| \leq |f(x) - L| + |g(x) - M|$$

$$< \frac{\varepsilon}{2} + \frac{\varepsilon}{2} = \varepsilon$$

Since the above δ exists for every $\varepsilon > 0$, we have proved that (6.42) is true.

Alternatively, instead of repeating earlier results in Chapter 3, we can show that the two concepts of limit are equivalent: each one implies the other. We prove this equivalence next.

Theorem 201 *(Equivalence of limits) A function f(x) has a limit L in the sense of (6.35) if and only if L is its limit in the sense of (6.34); i.e.,*

$$\lim_{x \to a} f(x) = L \quad \text{if and only if} \quad \lim_{n \to \infty} f(x_n) = L$$

for every sequence x_n in the domain of f(x) such that $\lim_{n \to \infty} x_n = a$, but $x_n \neq a$ for all n.

Proof First, assume that $\lim_{x \to a} f(x) = L$ so that for every $\varepsilon > 0$, there is a $\delta > 0$ such that

$$|f(x) - L| < \varepsilon \quad \text{if } 0 < |x - a| < \delta \tag{6.43}$$

Now, let x_n be an arbitrary sequence that converges to a, but $x_n \neq a$. Then there is a positive integer N such that $0 < |x_n - a| < \delta$ for all $n \geq N$, and by (6.43), $|f(x_n) - L| < \varepsilon$ for all $n \geq N$. It follows that $\lim_{n \to \infty} f(x_n) = L$.

Conversely, assume that $\lim_{n \to \infty} f(x_n) = L$ for every sequence x_n that converges to a, but $x_n \neq a$. We use a contradiction argument: suppose that $\lim_{x \to a} f(x) \neq L$, i.e., L isn't the limit in the sense of (6.35). We show that this assumption leads to a contradiction.

By our assumption, there is some $\varepsilon_0 > 0$ such that for every number $\delta > 0$ (no matter how small), there is some number x with $0 < |x - a| < \delta$, but $|f(x) - L| \geq \varepsilon_0$.[9]

In particular, if $\delta = 1$, then there is a number u_1 in the interval $(a - 1, a + 1)$ such that

$$|f(u_1) - L| \geq \varepsilon_0 \quad \text{where } 0 < |u_1 - a| < 1$$

Next, pick a strictly smaller δ, say, $\delta = 1/2$, and find u_2 in the interval $(a - 1/2, a + 1/2)$ such that

$$|f(u_2) - L| \geq \varepsilon_0 \quad \text{where } 0 < |u_2 - a| < \frac{1}{2}$$

Note that u_2 may be equal to u_1; the important thing is that u_2 is in a smaller neighborhood of a, namely, the interval $(a - 1/2, a + 1/2)$, and thus closer to it.

---

[9] This may be a good time to take a look back at the discussion of negations of statements in the logic review section in Chapter 2.

We continue this way; and for every index n, obtain a number u_n such that

$$|f(u_n) - L| \geq \varepsilon_0 \quad \text{where } 0 < |u_n - a| < \frac{1}{n}$$

The last inequality implies that $u_n \to a$ as $n \to \infty$ since $1/n \to 0$. So $\lim_{n \to \infty} u_n = a$. But there is no positive integer N such that $|f(u_n) - L| < \varepsilon_0$ for $n \geq N$. Therefore, $\lim_{n \to \infty} f(u_n)$ is not L. But this contradicts our converse assumption that $\lim_{n \to \infty} f(x_n) = L$ for every sequence x_n that converges to a, but $x_n \neq a$. To avoid this contradiction, we conclude that $\lim_{x \to a} f(x) = L$. ∎

To illustrate how we can use the above theorem, we prove (6.42) using Theorem 201.

Let u_n be an arbitrary sequence that converges to the number a, but $u_n \neq a$ for all n. For the existing limits of f and g, Theorem 201 implies

$$\lim_{x \to a} f(x) = \lim_{n \to \infty} f(u_n) \quad \text{and} \quad \lim_{x \to a} g(x) = \lim_{n \to \infty} g(u_n)$$

Next, by the sum of limits theorem in Chapter 3, if $s_n = f(u_n)$, and $s_n' = g(u_n)$, then

$$\lim_{n \to \infty} f(u_n) + \lim_{n \to \infty} g(u_n) = \lim_{n \to \infty} s_n + \lim_{n \to \infty} s_n' = \lim_{n \to \infty} (s_n + s_n') = \lim_{n \to \infty} [f(x_n) + g(x_n)]$$

Since the above equality holds for any arbitrarily chosen sequence u_n, Theorem 201 once more implies that (6.42) is true. Note the *two applications* of Theorem 201 in this argument: once to translate the neighborhood limit to a sequential one, and then again to translate the conclusion back into the original neighborhood limit.

Other types of neighborhood limits

A function $f(x)$ that is defined for all x in an interval (a, b) may also have a limit at a or at b, even though $f(x)$ may not be defined at a or b. Consider what happens to

$$f(x) = \frac{x^2 - 2x}{\sqrt{x}(x + 1)} \tag{6.44}$$

as $x \to 0$. This function is defined only for positive values of x, so we can only ask what happens to $f(x)$ as $x \to 0$ *from the right of 0*, i.e., $x > 0$. Cases like this lead us to define one-sided limits.

The left and right limits Assume that a function $f(x)$ is defined on an interval (a, b). A number L is the *right limit of $f(x)$ as x approaches a from the right*, i.e., within the interval (a, b) if for every real number $\varepsilon > 0$, there is a real number $\delta > 0$ such that

$$|f(x) - L| < \varepsilon \quad \text{whenever} \quad a < x < a + \delta < b \qquad (6.45)$$

We summarize (6.35) as

$$\lim_{x \to a^+} f(x) = L \qquad (6.46)$$

Likewise, L is the *left limit of $f(x)$ as x approaches b from the left* if for every real number $\varepsilon > 0$, there is a real number $\delta > 0$ such that

$$|f(x) - L| < \varepsilon \quad \text{whenever} \quad a < b - \delta < x < b \qquad (6.47)$$

In symbols,

$$\lim_{x \to b^-} f(x) = L \qquad (6.48)$$

The inequalities in (6.43) can be written as

$$f(x) \in (L - \varepsilon, L + \varepsilon) \quad \text{if} \quad x \in (a - \delta, a) \text{ and } x \in (a, a + \delta)$$

or equivalently,

$$|f(x) - L| < \varepsilon \quad \text{if} \quad a - \delta < x < a \text{ and } a < x < a + \delta$$

From this observation, we may conclude the following useful fact.

Theorem 202 *Assume that $f(x)$ is defined on the set $S = (a, c) \cup (c, b)$. Then $\lim_{x \to c} f(x)$ exists and has a value L if and only if both $\lim_{x \to c^+} f(x)$ and $\lim_{x \to c^-} f(x)$ exist and*

$$\lim_{x \to c^+} f(x) = \lim_{x \to c^-} f(x) = L$$

The calculation of right and left limits involving sums, products, etc., is done the same way as the main (two-sided) limit, and the proof is straightforward using Theorem 201.

We can now state and find the limit of $f(x)$ in (6.44) properly as follows:

$$\lim_{x \to 0^+} \frac{x^2 - 2x}{\sqrt{x}(x + 1)} = \lim_{x \to 0^+} \frac{x(x - 2)}{\sqrt{x}(x + 1)} = \lim_{x \to 0^+} \frac{\sqrt{x}(x - 2)}{x + 1} = 0$$

The often used concept of *limits at infinity* is defined in the usual way.

Limits at ∞ and $-\infty$ Let D be the domain of a function f, and assume that $D \cap (u, \infty)$ is nonempty for all positive real numbers u. The function $f(x)$ approaches the number L as x approaches ∞ if for every real number $\varepsilon > 0$, there is a real number $M > 0$ such that

$$|f(x) - L| < \varepsilon \quad \text{whenever} \quad x \in D \text{ and } x > M \tag{6.49}$$

We denote this as

$$\lim_{x \to \infty} f(x) = L \tag{6.50}$$

Similarly, if $D \cap (-\infty, v)$ is nonempty for all negative real numbers v, then

$$\lim_{x \to -\infty} f(x) = L \tag{6.51}$$

means for every real number $\varepsilon > 0$, there is a real number $M > 0$ such that

$$|f(x) - L| < \varepsilon \quad \text{whenever} \quad x \in D \text{ and } x < -M \tag{6.52}$$

For instance,

$$\lim_{x \to \infty} \frac{1}{x} = 0 \quad \text{and} \quad \lim_{x \to -\infty} \frac{1}{x} = 0$$

Infinite limits are similarly defined as follows.

Infinite Limits. Assume that a function $f(x)$ is defined on an interval (a, b) where $a = -\infty$ is allowed, as is $b = \infty$. We write

$$\lim_{x \to a^+} f(x) = \infty \tag{6.53}$$

if for every real number $M > 0$, there is a real number $\delta > 0$ such that

$$f(x) > M \quad \text{whenever} \quad a < x < a + \delta < b \tag{6.54}$$

Likewise,

$$\lim_{x \to a^+} f(x) = -\infty \tag{6.55}$$

if for every real number $M > 0$, there is a real number $\delta > 0$ such that

$$f(x) < -M \quad \text{whenever} \quad a < x < a + \delta < b \tag{6.56}$$

The left limit version as $x \to b^-$ are defined analogously. If $a < c < b$, and

$$\lim_{x \to c^+} f(x) = \pm\infty = \lim_{x \to c^+} f(x)$$

then we write

$$\lim_{x \to c} f(x) = \pm\infty$$

For instance,

$$\lim_{x \to 0^+} \frac{1}{x} = \infty \quad \text{and} \quad \lim_{x \to 0^-} \frac{1}{x} = -\infty$$

Note that $\lim_{x \to 0}(1/x)$ doesn't exist. However, it is true that

$$\lim_{x \to 0} \frac{1}{x^2} = \infty$$

6.5 The mean value theorem

We may compute the velocity of a freely falling body taking measurements of its position at various times. This procedure generates discrete data points, and from these, we can only compute the average velocity. But we believe that the falling body is moving *continuously* in space and time, so in-between the recorded data points, there are infinitely many unrecorded ones. Thus, if the *average* velocity from, say, $t = 1$ to $t = 2$ is 14.7 m/s, then its *actual* velocity is smaller at $t = 1$ than at $t = 2$. Based on these criteria, we conclude that *at some point in time between 1 second and 2 seconds, the object is moving at exactly 14.7 m/s*. For example, if in Galileo's equation we set the velocity equal to 14.7, then $14.7 = 9.8t$. Solving for t gives $t = 1.5$ seconds, the instant of time at which the object's velocity is *exactly* 14.7. The same reasoning applies to any time interval as long as the motion of the falling body is not altered.

But how do we prove the existence of such an instant of time without Galileo's velocity equation? Given a position function $x(t)$, can we be certain that at some instance t between two instances t_1 and t_2 of time the actual velocity $x'(t)$ equals the average velocity over the interval $[t_1, t_2]$?

The mean value theorem

Let $f(x)$ be differentiable at every point of some interval $a < x < b$. Then, as we saw earlier (Theorem 193), $f(x)$ is continuous over the same interval since differentiability is a stronger requirement than continuity. Also assume that $f(x)$ is continuous at the two points a and b so that its value does not jump at $x = a$ or $x = b$. Figure 6.13 illustrates such a configuration.

The variation, or the change in $f(x)$ as x goes from a to b, is $f(b) - f(a)$; and of course, the change in x is $b - a$. If we divide these two changes, then we get the ratio

$$\frac{f(b) - f(a)}{b - a}$$

This difference ratio represents the slope of the secant line that passes through the points $(a, f(a))$ and $(b, f(b))$ on the graph of $f(x)$ as shown in Figure 6.13, and of course, the average velocity if x is time and f is position.

If the function $f(x)$ is continuous, then tracing the arc of the curve in Figure 6.13, we see points at which the tangent line has the same slope as the secant line (in Figure 6.13, these points are labeled c and c'). Having the same slopes at some point means that the tangent line is parallel to the secant line at that point. If the x-coordinate of the point is $x = c$, then

$$\frac{f(b) - f(a)}{b - a} = f'(c) \tag{6.57}$$

You can see why we need differentiability here: there must be a tangent line with a well-defined slope for the above discussion to make sense!

As self-evident as (6.57) may seem in Figure 6.13, its proof for all possible functions requires some build up. We have identified two essential features of the function $f(x)$ that satisfies (6.57): continuity and differentiability. Since the latter

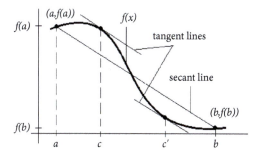

Fig. 6.13 Illustrating the mean value theorem

implies the former, we may be tempted to declare the essential requirement to be differentiability; however, the existence of derivatives at the points a and b serves no useful purpose since we need $f'(c)$ only at points like c in-between a and b. On the other hand, the continuity at the endpoints is necessary as we discover when we go through the process of proving the validity of the equality in (6.57).

We begin with the observation that if $f(a) = f(b)$ in (6.57), then $f'(c) = 0$, i.e., the tangent line is horizontal. Recall from calculus that where the derivative of a function is zero, that function has a "critical point." Such a point is often associated with the existence of an *extremum*, i.e., a local maximum or minimum, but it may be neither; e.g., consider $fx) = x^3$ at $x = 0$.

With this in mind, consider the *converse* question: suppose that $f'(x)$ exists at a number c between a and b, and $f(c)$ is a local maximum or minimum. Must $f'(c) = 0$? This question has an affirmative answer. To prove this claim, we first need a precise definition of different types of extrema (maxima and minima).

Local and absolute extrema

Let f be a function that is defined on an interval I. A function $f(x)$ has a *local maximum* at a point $p \in I$ if there is a number $\delta > 0$ such that $(p - \delta, p + \delta)$ is contained in I, and $f(x) \leq f(p)$ for all $x \in (p - \delta, p + \delta)$. The function f has an *absolute (or global) maximum* at p if $f(x) \leq f(p)$ for all $x \in I$.

Similarly, $f(x)$ has a *local minimum* at a point p in D if there is a number $\delta > 0$ such that $(p - \delta, p + \delta)$ is contained in I, and $f(x) \geq f(p)$ for all $x \in (p - \delta, p + \delta)$. Also, f has an *absolute (or global) minimum* at p if $f(x) \geq f(p)$ for all $x \in I$.

An absolute maximum or minimum is also a local one since any positive δ will work, but a local maximum or minimum is generally not absolute. For example, consider $f(x) = x^3 - 3x$, whose graph is shown in Figure 6.14 on the interval $I = [-2, 3]$.

As we see in Figure 6.14, this function has a local maximum at $x = -1$ where $f(-1) = 2$ for all points near -1. But the absolute maximum of f occurs at $x = 3$ where $f(3) = 18$. On the other hand, f has a local minimum at $x = 1$ with $f(1) = -2$, and this is also an absolute minimum of f because -2 is the smallest possible value of $f(x)$ on I.

Now we prove our claim above as a lemma to upcoming results.

Lemma 203 *Suppose that f is defined on the interval $[a, b]$, and c is a point of the open interval (a, b). If f is differentiable at c and has either a local minimum or a local maximum at c, then $f'(c) = 0$.*

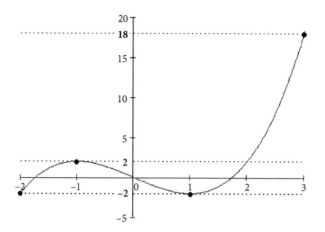

Fig. 6.14 Local and absolute extreme points

Proof We prove this lemma with c a local maximum. The minimum case is proved similarly (Exercise 227).

By definition,

$$f'(c) = \lim_{n \to \infty} \frac{f(x_n) - f(c)}{x_n - c}$$

for every sequence x_n that converges to c, but $x_n \neq c$ for all n. Further, since f has a local maximum at c, let δ be a number with the property that

$$f(x) \leq f(c) \quad \text{for } c - \delta < x < c + \delta \tag{6.58}$$

Since $x_n \to c$ as $n \to \infty$, there is a positive integer N such that for all $n \geq N$

$$-\delta < x_n - c < \delta$$

If $f'(c) \neq 0$, then it is either positive or negative. If $f'(c) > 0$, then there is a positive integer N' such that for $n \geq N'$

$$\frac{f(x_n) - f(c)}{x_n - c} > 0 \tag{6.59}$$

Letting N'' be the larger of N and N' (6.58) implies that (6.59) holds only if $x_n - c < 0$ for all $n \geq N''$. But since c is not an endpoint a or b, we can choose our sequence x_n such that $x_n > c$ for all $n \geq N''$ (e.g., $x_n = c + 1/(n + M)$ for a large enough number M so that $x_n < c + \delta$) and for such a sequence

$$\frac{f(x_n) - f(c)}{x_n - c} \leq 0$$

To avoid this contradiction with (6.59), we conclude that $f'(c) \leq 0$. Now suppose that $f'(c) < 0$. Then, as in the preceding argument,

$$\frac{f(x_n) - f(c)}{x_n - c} < 0$$

for all large indices n so that we must have $x_n - c > 0$ for all such n. But again, we may choose a sequence x_n so that $x_n < c$ and arrive at a contradiction. Therefore, the remaining case $f'(c) = 0$ is the only possible case, and the proof is complete.   ∎

With the aid of Lemma 203, we prove the next result, attributed to Michel Rolle (1652–1719).

Theorem 204 *(Rolle's theorem) Assume that a function f is continuous on the interval $[a, b]$ and differentiable on the open interval (a, b). If $f(a) = f(b)$, then there is a number $c \in (a, b)$ such that $f'(c) = 0$.*

Proof The assumptions allow for a trivial case where f is constant, i.e., $f(x) = f(a)$ for all $x \in [a, b]$. In this case, $f'(x) = 0$ for all x, and there is nothing further to do. So assume that $f(x)$ is not constant. Then there is a number $r \in (a, b)$ such that either $f(r) > f(a)$ or $f(r) < f(a)$.

If $f(r) > f(a) = f(b)$ then f has a maximum at some $c \in (a, b)$, so by Lemma 203, $f'(c) = 0$. Similarly, if $f(r) < f(a) = f(b)$, then f has a minimum at some $c \in (a, b)$; so again by Lemma 203, $f'(c) = 0$.   ∎

In Theorem 204, it is necessary that f be continuous at the endpoints a and b. For example, if $f(x) = x$ for $0 \leq x < 1$, and $f(1) = 0$, then f is continuous on $[0, 1)$, differentiable on $(0, 1)$ and further, $f(0) = f(1) = 0$. But $f'(x) = 1$ for all $x \in (0, 1)$.

On the other hand, it is not necessary for f to be differentiable at the endpoints so that Rolle's theorem may be applied to functions like the following:

$$f(x) = \sqrt{1 - x^2} \quad x \in [-1, 1]$$

In this case, f is continuous on $[-1, 1]$, with $f(-1) = f(1) = 0$; and also, f is differentiable on $(-1, 1)$ with derivative

$$f'(x) = \frac{-x}{\sqrt{1 - x^2}}$$

calculated using the chain rule. Notice that $f'(0) = 0$ as predicted by Rolle's theorem.

The next result, a fundamental theorem in calculus, is a corollary of Rolle's theorem and a generalization of it in that $f(a)$ need not equal $f(b)$. We will come across this theorem a number of times in this and later chapters when we need information about a function that may be obtained from its derivative.

Theorem 205 (*Mean value theorem*) *If a function f is continuous on the interval* $[a, b]$ *and differentiable on the open interval* (a, b)*, then there is a number* $c \in (a, b)$ *such that*

$$f(b) - f(a) = f'(c)(b - a) \qquad (6.60)$$

Proof We apply Rolle's theorem to the function

$$g(x) = f(x) - f(a) - \frac{f(b) - f(a)}{b - a}(x - a)$$

that is designed to satisfy all the hypotheses of Theorem 204: g is continuous on $[a, b]$ with $g(a) = g(b) = 0$; and further, g is differentiable on (a, b) with derivative

$$g'(x) = f'(x) - \frac{f(b) - f(a)}{b - a}$$

By Rolle's theorem, there is $c \in (a, b)$ such that $g'(c) = 0$; and for this c, we have

$$f'(c) = \frac{f(b) - f(a)}{b - a}$$

which is equivalent to (6.60).   ∎

The name "mean value" or average value is exactly average (or mean) velocity if f is a position function. Finding the number c is possible in simple examples; let's find such a point (or points) in the interval $-2 < x < 2$ where the function $f(x) = x^3 - 2x - 1$ satisfies the hypotheses of Theorem 205:

$$\frac{f(2) - f(-2)}{2 - (-2)} = f'(c), \quad -2 < c < 2$$

$$\frac{3 - (-5)}{4} = 3c^2 - 2$$

The last equality above reduces to the equation $3c^2 = 4$, which is easy to solve:

$$c = \pm \frac{2}{\sqrt{3}} \simeq \pm 1.155$$

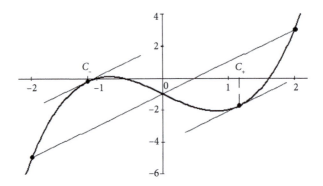

Fig. 6.15 Verifying the mean value property

See Figure 6.15, which shows a sketch of the graph of $f(x)$, the secant line through the points $(-2, -5)$ and $(2, 3)$, and the two tangent lines that are parallel to it. Both $c_+ = 2/\sqrt{3}$ and $c_- = -2/\sqrt{3}$ are between $x = -2$ and $x = 2$, so both of them are valid choices for c in (6.57).

If the derivative of $f(x)$ doesn't exist *even at a single point* between $x = a$ and $x = b$, then the mean value theorem is false.

Consider the function $f(x) = |x|$ for $-1 < x < 2$. Recall that this function has a derivative at every real number x as long as $x \neq 0$; in fact,

$$f'(x) = \begin{cases} 1 & \text{if } x > 0 \\ -1 & \text{if } x < 0 \end{cases}$$

But since $f'(0)$ doesn't exist, $f(x)$ isn't differentiable on the interval $-1 < x < 2$. Set $a = -1$ and $b = 2$ in (6.57) to get

$$\frac{f(2) - f(-1)}{2 - (-1)} = \frac{|2| - |-1|}{3} = \frac{1}{3}$$

This equals neither 1 nor -1, so there is no number c between -1 and 2 that satisfies (6.57).

| Consequences of the mean value theorem |

In calculus, we discover that the derivative provides a good deal of information about the behaviors of functions and the shapes of their graphs. What we usually don't realize at the time is that this information often comes courtesy of the mean value theorem. Now we can explain what goes on behind the scenes, so to speak.

Recall that a function $f(x)$ is increasing (respectively, nondecreasing) from $x = a$ to $x = b$ if for every pair of numbers u, v between a and b

$$u < v \quad \text{implies} \quad f(u) < f(v) \quad (\text{respectively, } f(u) \leq f(v)) \qquad (6.61)$$

Similarly, $f(x)$ is decreasing (respectively, nonincreasing) from $x = a$ to $x = b$ if for every pair of numbers u, v between a and b

$$u < v \quad \text{implies} \quad f(u) > f(v) \quad (\text{respectively } f(u) \geq f(v)) \qquad (6.62)$$

Further, if a function is always nonincreasing or always nondecreasing, then it is a monotone function.

These definitions, though precise, are not easy to use in practice. The mean value theorem provides a simpler and quicker way of understanding the behavior of a function by checking the *sign* of its derivative, as we see in the next corollary.

Corollary 206 *Assume that the function f is differentiable on an interval I.*
  (i) If $f'(x) > 0$ (or $f'(x) \geq 0$) for all x in I, then the function f is increasing (or nondecreasing) on I;
  (ii) If $f'(x) < 0$ (or $f'(x) \leq 0$) for all x in I, then the function f is decreasing (or nonincreasing) on I;
  (iii) If $f'(x) = 0$ for all x in I, then f is a constant function on I.

Proof I prove (i) and (ii) as follows by applying (i) to $-f$. Part (iii) is Exercise 232. Let $u, v \in I$ and $u < v$. Note that f is differentiable on $[u, v]$, so it satisfies the hypotheses of Theorem 205. Therefore, there is a number $c \in (u, v)$ such that

$$f(v) - f(u) = f'(c)(v - u)$$

Since $f'(c) > 0$, by assumption the right-hand side is positive. It follows that $f(v) - f(u) > 0$, so $f(u) < f(v)$; and by (6.61), f is increasing. The last inequality changes to $\leq$ if $f'(x) \geq 0$; so that again by (6.61), f is nondecreasing. ∎

The converses of Parts (i) and (ii) in the above corollary are false. For example, $f(x) = x^3$ is increasing on $(-\infty, \infty)$, but $f'(0) = 0$.

Corollary 206 is the basis of the familiar *first-derivative test* for maxima and minima in calculus, which we list as another corollary; its proof is Exercise 233.

Corollary 207 *(The first-derivative test) Let c be a critical point in the interval (a, b), i.e., a point where $f'(c) = 0$, and let $\delta > 0$ be small enough so that $(c - \delta, c + \delta) \subset (a, b)$.*
  (a) If $f'(x) > 0$ for $c - \delta < x < c$, and $f'(x) < 0$ for $c < x < c + \delta$, then f has a local maximum at $x = c$;
  (b) If $f'(x) < 0$ for $c - \delta < x < c$, and $f'(x) > 0$ for $c < x < c + \delta$, then f has a local minimum at $x = c$.

We close this section with the following generalization of the mean value theorem (attributed to Cauchy) that is used to prove important results like l'Hospital's rule later in this chapter and Lagrange's formula for the remainder of the Taylor series in Chapter 8. It is also a mean value theorem for parametric curves, as we discover after the theorem.

Theorem 208 *(Generalized mean value theorem) Let f and g be continuous functions on the interval* $[a, b]$ *that are differentiable in* (a, b)*. There is a number* $c \in (a, b)$ *such that*

$$[f(b) - f(a)]g'(c) = [g(b) - g(a)]f'(c) \tag{6.63}$$

Proof The argument here is similar to that in the proof of Theorem 205. We define a function

$$h(x) = [f(b) - f(a)]g(x) - [g(b) - g(a)]f(x)$$

This function is continuous on $[a, b]$ and differentiable on (a, b). Further,

$$h(a) = f(b)g(a) - f(a)g(b) = h(b)$$

Now, Rolle's theorem implies that there is $c \in (a, b)$ such that $h'(c) = 0$, which is equivalent to (6.63). ∎

If $g'(x) \neq 0$ for all $x \in (a, b)$, then by Corollary 206, $g(a) \neq g(b)$, so we may write (6.63) in the form

$$\frac{f(b) - f(a)}{g(b) - g(a)} = \frac{f'(c)}{g'(c)} \tag{6.64}$$

This happens in particular when $g(x) = x$, which yields Theorem 205. But more generally, Theorem 208 is a mean value theorem for parametric curves. Consider a curve C whose parametric representation is

$$x = g(t), \quad y = f(t), \quad a \leq t \leq b$$

The slope of the line that joins a point $A = (g(a), f(a))$ to $B = (g(b), f(b))$ is given by the left-hand side of (6.64). The right-hand side gives the slope of C itself at the point $f(t)/g(t)$ in the direction from A to B as t goes from a to b. It follows that there is a point on the graph of C at which the slope of C is equal to the slope of the line joining A and B, rather like the mean value theorem.

6.6 Indeterminate forms and l'Hospital's rule

Consider the two fractions

$$Q_1(x) = \frac{x^2 - 1}{x - 1} \quad \text{and} \quad Q_2(x) = \frac{x^3 - 1}{x - 1}$$

Suppose that we are asked to calculate the limits $\lim_{x \to 1} Q_1(x)$ and $\lim_{x \to 1} Q_2(x)$. Because the denominator is zero when $x = 1$, we cannot calculate the limits by simple substitution; but from algebra, we may recall that

$$x^2 - 1 = (x - 1)(x + 1)$$
$$x^3 - 1 = (x - 1)(x^2 + x + 1)$$

Using these factorizations, we calculate

$$\lim_{x \to 1} Q_1(x) = \lim_{x \to 1} \frac{(x - 1)(x + 1)}{x - 1} = \lim_{x \to 1} (x + 1) = 2$$

$$\lim_{x \to 1} Q_2(x) = \lim_{x \to 1} \frac{(x - 1)(x^2 + x + 1)}{x - 1} = \lim_{x \to 1} (x^2 + x + 1) = 3$$

In these calculations, we managed to cancel out from the numerator and the denominator of each fraction the term $x - 1$ that has a zero value at $x = 1$.

Each of the fractions has the form $0/0$ when $x = 1$; although $0/0$ isn't a well-defined number, *it can be assigned different values using limits*. It is called an *indeterminate form* because it cannot have a single numerical value.

Indeterminate forms of $0/0$ type occur routinely in calculus; difference quotients are by definition $0/0$ forms whose limits define derivatives. In fact, the fractions $Q_1(x)$ and $Q_2(x)$ above are the difference quotients of the functions x^2 and x^3, respectively, at $x = 1$, so their limits are precisely the derivatives of these functions at $x = 1$. Of course, not every $0/0$ form is a difference quotient; e.g., the fraction $(x^2 - 1)/(x^3 - 1)$.

The existence of a limit for a $0/0$ indeterminate form is by no means guaranteed. For example,

$$\frac{x^2 - 1}{x^2 - 2x + 1} = \frac{(x - 1)(x + 1)}{(x - 1)^2} = \frac{x + 1}{x - 1} \quad \text{if } x \neq 1$$

But the last fraction has no limit as $x \to 1$. This is easily shown by considering the two sequences $u_n = 1 + 1/n$ and $v_n = 1 - 1/n$, both of which converge to 1 as

$n \to \infty$; but

$$\lim_{n\to\infty} \frac{u_n + 1}{u_n - 1} = \lim_{n\to\infty} \frac{2 + 1/n}{1/n} = \lim_{n\to\infty} (2n + 1) = \infty$$

$$\lim_{n\to\infty} \frac{v_n + 1}{v_n - 1} = \lim_{n\to\infty} \frac{2 - 1/n}{-1/n} = \lim_{n\to\infty} (-2n + 1) = -\infty$$

The question that we want to answer in this section is *how do we find the limits of indeterminate forms like 0/0?* This question is not easy to answer if we cannot simplify the fraction. For instance, consider the following problem:

$$\frac{x^2 - 2x}{2 \sin x} \to? \quad \text{as } x \to 0 \tag{6.65}$$

The fraction is a 0/0 form at $x = 0$, but how do we calculate its limit? Does the limit even exist?

6.6.1 L'Hospital's rule: The 0/0 form

An answer to our question above on how to find the limit of an indeterminate form comes in the form of *l'Hospital's rule*. This is an important result that facilitates both practical and theoretical calculations. Its name, however, turns out to be a historical anomaly: the result was actually the work of Johann Bernoulli (1667–1748) who provided it as part of mathematical lectures paid for by the Marquis Guillaume de l'Hospital (1661–1704), who published it in 1696 in the first calculus text ever written.[10]

The basic idea is as follows. Consider the fraction $f(x)/g(x)$ and a real number a such that both $\lim_{x\to a} f(x) = 0$ and $\lim_{x\to a} g(x) = 0$. Then, defining $f(a) = g(a) = 0$

$$\lim_{x\to a} \frac{f(x)}{g(x)} = \lim_{x\to a} \frac{f(x) - f(a)}{g(x) - g(a)} = \lim_{x\to a} \frac{[f(x) - f(a)]/(x - a)}{[g(x) - g(a)]/(x - a)} = \frac{f'(a)}{g'(a)}$$

Several assumptions are required to make this argument work, such as differentiability and avoiding division by 0. Ignoring these assumptions for the moment, the equality

$$\lim_{x\to a} \frac{f(x)}{g(x)} = \frac{f'(a)}{g'(a)} \tag{6.66}$$

---

[10] Another common spelling of l'Hospital's name is l'Hôpital. We use the spelling that was used on the cover of his calculus book.

can be used to resolve the limit issues with (6.65):

$$\lim_{x\to 0} \frac{x^2 - 2x}{2\sin x} = \frac{2(0) - 2}{2\cos 0} = -1$$

However, the equality in (6.66) cannot be applied to finding the limit of the following fraction as x approaches 0

$$\frac{x^2 - 2x}{2\sqrt{x}}$$

because here, $\sqrt{x}$ is not defined unless we use a one-sided limit $x \to 0$ *from the right* $(x > 0)$ or $x \to 0^+$. Further, if in (6.66) $f'(a) = g'(a) = 0$, then the quotient on the right-hand side must be properly defined if the limit on the left side exists.

Because of the many technical but important issues like these that arise frequently in the applications of l'Hospital's rule, we state and prove a general version of the rule using Cauchy's generalization of the mean value theorem.

We begin with the following result. Recall that $x \to a^+$ means $x > a$ and $x - a \to 0$.

Theorem 209 *(L'Hospital's rule, the 0/0 case) Assume that functions f and g are differentiable on (a, b) and that $g'(x) \neq 0$ for all $x \in (a, b)$.*
(a) If

$$\lim_{x\to a^+} f(x) = 0 = \lim_{x\to a^+} g(x) \tag{6.67}$$

and further,

$$\lim_{x\to a^+} \frac{f'(x)}{g'(x)} = L \tag{6.68}$$

where L is a real number or possibly ∞ or $-\infty$, then

$$\lim_{x\to a^+} \frac{f(x)}{g(x)} = L \tag{6.69}$$

(b) If a^+ is replaced by b^- in (6.67) and (6.68), then (6.69) is true with a^+ replaced by b^-.
(c) If a^+ is replaced by $-\infty$ or b^- by ∞ in (6.67) and (6.68), then (6.69) is true with the same substitutions.

Proof (a) Since we are interested in limits that don't depend on function values at specific points, we define

$$f(a) = g(a) = 0 \tag{6.70}$$

based on (6.67) to make f and g continuous on the interval $[a, b)$. Now let x_n be an arbitrary sequence in (a, b) such that $\lim_{n\to\infty} x_n = a$, but $x_n \neq a$ for all n. For each n, both f and g are continuous on $[a, x_n]$ and differentiable on (a, x_n); so by Theorem 208, there is a number c_n between a and x_n such that

$$[f(x_n) - f(a)]g'(c_n) = [g(x_n) - g(a)]f'(c_n)$$

or, given (6.70),

$$\frac{f(x_n)}{g(x_n)} = \frac{f'(c_n)}{g'(c_n)} \tag{6.71}$$

The quotient on the left is well defined because $g(x) \neq g(a) = 0$ for all $x \in (\alpha, \beta)$, since by hypothesis, $g'(x) \neq 0$.

Note that $c_n \to a$ as $n \to \infty$ because $0 < |c_n - a| < |x_n - a|$, and by hypothesis, $\lim_{n\to\infty} x_n = a$, so $\lim_{n\to\infty} |x_n - a| = 0$. Therefore, by the squeeze theorem, $\lim_{n\to\infty} |c_n - a| = 0$.

Now, taking the limit in (6.71) as $n \to \infty$, and using (6.68) and Theorem 201, gives

$$\lim_{n\to\infty} \frac{f(x_n)}{g(x_n)} = \lim_{n\to\infty} \frac{f'(c_n)}{g'(c_n)} = \lim_{x\to a} \frac{f'(x)}{g'(x)} = L$$

Since x_n was an arbitrarily chosen sequence converging to a, a second application of Theorem 201 establishes (6.69).

(b) The argument for this part is analogous to the above for (a); see Exercise 234.

(c) Assume that

$$\lim_{x\to-\infty} f(x) = 0 = \lim_{x\to-\infty} g(x) \tag{6.72}$$

and also

$$\lim_{x\to-\infty} \frac{f'(x)}{g'(x)} = L \tag{6.73}$$

We prove that

$$\lim_{x\to-\infty} \frac{f(x)}{g(x)} = L \tag{6.74}$$

We prove this by transforming this case to a problem of the type in Part (a) of the theorem. Consider the substitution

$$x = -\frac{1}{t}$$

and notice that $x \to -\infty$ if and only if $t \to 0^+$. Further, $x \in (-\infty, b)$ implies that

$$t \in \left(0, -\frac{1}{b}\right) \quad \text{if } b < 0$$

$$t \in (0, \infty) \quad \text{if } b = 0$$

$$t \in \left(-\infty, -\frac{1}{b}\right) \cup (0, \infty) \quad \text{if } b > 0$$

Define the new functions

$$\phi(t) = f\left(-\frac{1}{t}\right) \quad \text{and} \quad \gamma(t) = g\left(-\frac{1}{t}\right)$$

and observe that for all values of b, these functions are defined on an interval of type $(0, c)$. By (6.72)

$$\lim_{t \to 0^+} \phi(t) = \lim_{x \to -\infty} f(x) = 0, \qquad \lim_{t \to 0^+} \gamma(t) = \lim_{x \to -\infty} g(x) = 0$$

Further, by the chain rule, $\phi(t)$ and $\gamma(t)$ are differentiable on $(0, c)$, and

$$\phi'(t) = f'(x)\left(-\frac{1}{t^2}\right), \qquad \gamma'(t) = g'(x)\left(-\frac{1}{t^2}\right)$$

Therefore, by (6.73)

$$\lim_{t \to 0^+} \frac{\phi'(t)}{\gamma'(t)} = \lim_{x \to -\infty} \frac{f'(x)}{g'(x)} = L$$

It follows that $\phi(t)$ and $\gamma(t)$ satisfy the hypotheses of Part (a), so by (6.69)

$$\lim_{x \to -\infty} \frac{f(x)}{g(x)} = \lim_{t \to 0^+} \frac{\phi(t)}{\gamma(t)} = \lim_{t \to 0^+} \frac{\phi'(t)}{\gamma'(t)} = L$$

and (6.74) is now proved. The remaining case where b^- is changed to ∞ is similarly proved using Part (b). ∎

We may use Theorem 209 to find the limit

$$\lim_{x\to 0^+} \frac{\sin(\sqrt{2x})}{2\sqrt{x}} = \lim_{x\to 0^+} \frac{(2x)^{-1/2}\cos(\sqrt{2x})}{x^{-1/2}} = \lim_{x\to 0^+} 2^{-1/2}\cos(\sqrt{2x}) = \frac{1}{\sqrt{2}}$$

Similarly,

$$\lim_{x\to\infty} x\sin\frac{1}{x} = \lim_{x\to\infty} \frac{\sin(1/x)}{1/x} = \lim_{x\to\infty} \frac{\cos(1/x)(-1/x^2)}{-1/x^2} = \lim_{x\to\infty} \cos\frac{1}{x} = 1 \quad (6.75)$$

Noteworthy about this example is the fact that as $x \to \infty$, the expression $x\sin(1/x)$ takes the form $\infty \times 0$, which is another type of indeterminate form. As we did in the example, this type of indeterminate form may be transformed into a 0/0 form to which Theorem 209 can be applied.

It is necessary to keep in mind that *Theorem 209 is not applicable if the fraction is not a 0/0 form.*

For instance,

$$\lim_{x\to 1} \frac{x^2 - 1}{x + 1} = 0$$

by direct substitution since the denominator does not vanish at $x = 1$, and the fraction is continuous. But an improper use of Theorem 209 gives the false result

$$\lim_{x\to 1} \frac{x^2 - 1}{x + 1} = \lim_{x\to 1} \frac{2x}{1} = 2$$

Similarly,

$$\lim_{x\to\infty} x\cos\frac{1}{x} = \infty$$

since $\lim_{x\to\infty} \cos(1/x) = \cos 0 = 1$, and we cannot apply Theorem 209; indeed, the limit of $x\cos(1/x)$ is not an indeterminate form. A calculation like that in (6.75) would give a false result:

$$\lim_{x\to\infty} x\cos\frac{1}{x} = \lim_{x\to\infty} \frac{\cos(1/x)}{1/x} = \lim_{x\to\infty} \frac{-\sin(1/x)(-1/x^2)}{-1/x^2} = -\lim_{x\to\infty} \sin\frac{1}{x} = 0$$

6.6.2 L'Hospital's rule: The ∞/∞ form

A kind of mirror image version of Theorem 209 involves indeterminate forms that are of type ∞/∞. This case is important, as it simplifies many calculations.

Suppose that

$$\lim_{x \to a^+} |f(x)| = \lim_{x \to a^+} |g(x)| = \infty$$

If we define the reciprocal functions

$$F(x) = \frac{1}{f(x)} \quad \text{and} \quad G(x) = \frac{1}{g(x)}$$

then

$$\frac{f(x)}{g(x)} = \frac{G(x)}{F(x)}$$

and

$$\lim_{x \to a^+} F(x) = \lim_{x \to a^+} G(x) = 0$$

An application of Theorem 209 to $G(x)/F(x)$ now gives

$$\lim_{x \to a^+} \frac{G(x)}{F(x)} = \lim_{x \to a^+} \frac{G'(x)}{f'(x)}$$

and using the chain rule, it follows that

$$\lim_{x \to a^+} \frac{f(x)}{g(x)} = \lim_{x \to a^+} \left(\frac{-g'(x)}{[g(x)]^2} \right) \left(\frac{[f(x)]^2}{-f'(x)} \right) = \lim_{x \to a^+} \frac{g'(x)}{f'(x)} \left[\lim_{x \to a^+} \frac{f(x)}{g(x)} \right]^2$$

Now, we algebraically simplify to get

$$\lim_{x \to a^+} \frac{f(x)}{g(x)} = \lim_{x \to a^+} \frac{f'(x)}{g'(x)}$$

This argument shows that Theorem 209 extends to the ∞/∞ case but under more restrictions than those imposed in Theorem 209. The most severe of these restrictions is the necessity for $f'(x) \neq 0$ in an interval (a, c) where $c \leq b$. This condition implies that $f'(x) \neq 0$ in (a, c) in addition to $g'(x) \neq 0$ that is already assumed in Theorem 209. However, requiring $f(x) \neq 0$ does not seem necessary, and there is no obvious reason for it.

It is indeed possible to avoid such extra conditions by using a more technical argument that we give as the proof for the ∞/∞ case.

Theorem 210 (*L'Hospital's rule, the ∞/∞ case*) *Assume that functions f and g are differentiable on (a, b) and that $g'(x) \neq 0$ for all $x \in (a, b)$. We allow for $a = -\infty$ or $b = \infty$ here.*

(a) If

$$\lim_{x \to a^+} |f(x)| = \lim_{x \to a^+} |g(x)| = \infty \tag{6.76}$$

and further,

$$\lim_{x \to a^+} \frac{f'(x)}{g'(x)} = L \tag{6.77}$$

where L is a real number or possibly ∞ or −∞, then

$$\lim_{x \to a^+} \frac{f(x)}{g(x)} = L \tag{6.78}$$

(b) If a^+ is replaced by b^- in (6.76) and (6.77), then (6.78) is true with a^+ replaced by b^-.

(c) If a^+ is replaced by $-\infty$ or b^- by ∞ in (6.76) and (6.77), then (6.78) is true with the same substitutions.

Proof (a) To simplify the writing, assume that $\lim_{x \to a^+} g(x) = \infty$; the case where $g(x) \to -\infty$ uses a similar argument.

We begin by assuming $L < \infty$ (though we allow for $L = -\infty$), and let r, r' be a pair of numbers such that $L < r < r'$. Define

$$Q_1(x) = \frac{f'(x)}{g'(x)} \quad \text{for } x \in (a, b)$$

Since

$$\lim_{x \to a^+} Q_1(x) = L < r$$

there is a real number $c_1 \in (a, b)$ such that

$$Q_1(x) < r \quad \text{for } x \in (a, c_1)$$

See Exercise 235. Let x_0 be a fixed number in (a, c_1), and note that since $g(x) \to \infty$ as $x \to a^+$, and $g'(x) \neq 0$, it follows that $g'(x) > 0$; and there is $c_2 \in (a, x_0)$ such that $g(x) > g(x_0)$, as well as $g(x) > 0$ for $x \in (a, x_0)$. Now by

Theorem 208, there is $\zeta \in (x, x_0)$ such that

$$\frac{f(x) - f(x_0)}{g(x) - g(x_0)} = \frac{f'(\zeta)}{g'(\zeta)} = Q_1(\zeta) < r$$

From this inequality, we obtain

$$f(x) - f(x_0) < r[g(x) - g(x_0)]$$

Dividing by $g(x)$ and rearranging terms gives

$$\frac{f(x)}{g(x)} < \frac{f(x_0)}{g(x)} + r\left(1 - \frac{g(x_0)}{g(x)}\right) \quad \text{for } x \in (a, c_2) \tag{6.79}$$

With numerators fixed and $g(x) \to \infty$ as $x \to a^+$, it follows that

$$\lim_{x \to a^+} \frac{f(x_0)}{g(x)} = \lim_{x \to a^+} \frac{g(x_0)}{g(x)} = 0$$

so that if we define

$$Q_2(x) = \frac{f(x_0)}{g(x)} + r\left(1 - \frac{g(x_0)}{g(x)}\right)$$

then

$$\lim_{x \to a^+} Q_2(x) = r < r'$$

Therefore (Exercise 235), there is $c_3 \in (a, c_2)$ such that

$$Q_2(x) < r'$$

for all $x \in (a, c_3)$. Now by (6.79)

$$\frac{f(x)}{g(x)} < r' \quad \text{for } x \in (a, c_3) \tag{6.80}$$

We have two cases to consider for L; first, if $L = -\infty$, then (6.80) holds for every real number $r' > L$, and it follows that

$$\lim_{x \to a^+} \frac{f(x)}{g(x)} = -\infty = L$$

which proves (6.78) for $L = -\infty$. If L is finite ($L > -\infty$), and $\varepsilon > 0$ is arbitrary, then by setting $r' = L + \varepsilon$ in (6.80), we obtain

$$\frac{f(x)}{g(x)} < L + \varepsilon \quad \text{for } x \in (a, c_3) \tag{6.81}$$

We now look for a lower bound $L - \varepsilon$ for $f(x)/g(x)$ with $L > -\infty$. We choose $r'' < L$ and use an argument similar to the preceding for $L < r'$ to obtain a $c_3' \in (a, b)$ such that

$$\frac{f(x)}{g(x)} > r'' \quad \text{for } x \in (a, c_3') \tag{6.82}$$

Setting $r'' = L - \varepsilon$ in this inequality and recalling (6.81) gives

$$L - \varepsilon < \frac{f(x)}{g(x)} < L + \varepsilon \quad \text{for } x \in (a, c)$$

where c is the smaller of c_3 and c_3'. This proves (6.78) for finite L. If $L = \infty$ then (6.82) is valid for all real numbers r'', so we conclude that

$$\lim_{x \to a^+} \frac{f(x)}{g(x)} = \infty = L$$

and prove (6.78) for $L = \infty$.

(b) The argument for this part is analogous to the above for (a).

(c) Notice that the arguments in the proofs of (a) and (b) are valid if $a = -\infty$ or $b = \infty$. Therefore, no additional arguments are needed, and the proof is complete. ∎

Putting Parts (a) and (b) in Theorems 209 and 210 together, we obtain the following two-sided limits version of l'Hospital's rule (see Theorem 202).

Corollary 211 *Assume that f and g are differentiable on the set $S = (a, c) \cup (c, b)$ for given real numbers a, b, c where $-\infty \le a \le c \le b \le \infty$, and $a < b$. Further, assume $g'(x) \ne 0$ for all $x \in S$ and*

$$\lim_{x \to c} \frac{f'(x)}{g'(x)} = L \quad \text{where } -\infty \le L \le \infty$$

If either

$$\lim_{x \to c} f(x) = \lim_{x \to c} g(x) = 0$$

or

$$\lim_{x \to c} |f(x)| = \lim_{x \to c} |g(x)| = \infty$$

then

$$\lim_{x \to c} \frac{f(x)}{g(x)} = L$$

With regard to the notation in the above corollary, it is possible that $c = \infty$, in which case $b = \infty$ also, and we have $-\infty \le a < c = \infty$. Likewise, if $c = -\infty$, then $a = c$, and we get $-\infty < b \le \infty$.

To illustrate Corollary 211, we calculate

$$\lim_{x \to \infty} \frac{2x + \sqrt{x}}{3 - x} = \lim_{x \to \infty} \frac{2 - (1/2)x^{-1/2}}{-1} = \lim_{x \to \infty} \left(-2 + \frac{1}{2\sqrt{x}} \right) = -2$$

and

$$\lim_{x \to 0} \frac{\sin(2x)}{x^2 - x} = \lim_{x \to 0} \frac{2\cos(2x)}{2x - 1} = -2$$

Multiple applications of l'Hospital's rule

We conclude with the observation that for some pairs of functions f and g, it is possible that their derivative ratios are also $0/0$ or ∞/∞ forms, e.g.,

$$\lim_{x \to c} f'(x) = \lim_{x \to c} g'(x) = 0 \tag{6.83}$$

or

$$\lim_{x \to c} |f'(x)| = \lim_{x \to c} |g'(x)| = \infty$$

In this case, if f and g are twice differentiable with $g''(x) \ne 0$ and

$$\lim_{x \to c} \frac{f''(x)}{g''(x)} = L \tag{6.84}$$

then the conclusion of Corollary 211 is still valid, as are the conclusions of Theorems 209 and 210 in the one-sided limits cases. We can apply these results multiple times as long as their hypotheses remain valid. Consider for example,

$$\lim_{x \to 0} \frac{x^2 + \cos x}{\sin^2 x}$$

Here, with $g(x) = \sin^2 x$ and $g'(x) = 2 \sin x \cos x$, we have $g(0) = g'(0) = 0$. When we examine the fraction on the right-hand side above, we see that $g'(x) \ne 0$ for all

$x \in (-\pi/2, 0) \cup (0, \pi/2)$; so we may apply Corollary 211 to get

$$\lim_{x \to 0} \frac{x^2 + \cos x}{\sin^2 x} = \lim_{x \to 0} \frac{2x - \sin x}{2 \sin x \cos x} = \lim_{x \to 0} \frac{2x - \sin x}{\sin(2x)}$$

One application of l'Hospital's rule produced the fraction on the right-hand side that is still a 0/0 indeterminate form, corresponding to (6.83). We examine $g''(x)$ to see if Corollary 211 can be applied again:

$$g''(x) = 2 \cos(2x) \neq 0 \quad \text{if } x \in \left(-\frac{\pi}{4}, 0\right) \cup \left(0, \frac{\pi}{4}\right)$$

We may apply l'Hospital's rule once more to get

$$\lim_{x \to 0} \frac{2x - \sin x}{\sin(2x)} = \lim_{x \to 0} \frac{2 - \cos x}{2 \cos(2x)} = \frac{2 - 1}{2(1)} = \frac{1}{2}$$

6.7 The Newton–Raphson approximation method

The method that we discuss in this section is an essential topic in numerical analysis and scientific computing. We discuss it here for two reasons: one is that it provides a good illustration of the use of derivative and the intermediate value theorem. Further, being a recursive method, it is based on sequences and therefore, of interest in this book.

Also relevant to our purpose in this book, the Newton–Raphson method brings out the hidden infinity in spectacular fashion as we discover later in this section. Although there are approximation methods without infinity issues, they tend to be either more complicated to use or slower to converge to a solution. Working with infinity lurking in the background may have its risks, but it comes with significant rewards in this case.

When studying mathematical models in science and engineering, we often find it necessary to solve nonlinear equations. An equation of this type may be written as

$$f(x) = 0 \qquad (6.85)$$

where $f(x)$ is some nonlinear function, i.e., its graph is *not* a straight line. We will be looking for a value of x that makes the value of the function on the left-hand side zero, so we may also call the solution a *zero* (or a *root*) of $f(x)$. Here is a concrete example:

$$\cos x = x \quad \text{or equivalently,} \quad x - \cos x = 0 \qquad (6.86)$$

The zeros of this equation are the x-coordinates of points where $f(x) = x - \cos x = 0$ (equivalently, where $y = \cos x$ intersects the line $y = x$).

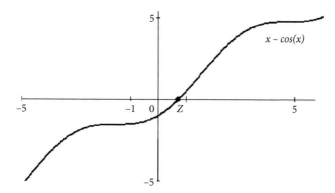

Fig. 6.16 The zero or x-intercept of a smooth curve

Note that $f(x) = x - \cos x$ is continuous (everywhere) since it is a difference of two continuous functions. Further,

$$f(0) = 0 - \cos 0 = -1 < 0, \qquad f\left(\frac{\pi}{2}\right) = \frac{\pi}{2} - \cos\left(\frac{\pi}{2}\right) = \frac{\pi}{2} > 0$$

so the intermediate value theorem implies that there is a zero z between 0 and $\pi/2 \simeq 1.57$; see Figure 6.16.

Can we find the *exact value* of the solution z?

The answer is *no*, and this is true typically of nonlinear equations, not just for the one in (6.86). But *we can approximate the solution as closely as we want* by finding as many digits as needed or desired. One of the simplest, yet most powerful, recursive methods or algorithms known (due the fast rate at which it improves the approximation at each step) was introduced by Newton in 1669, but he didn't publish it. John Wallis did it for him in 1685. The improved version that we discuss here is due to Joseph Raphson (1648–1715), who published it in 1690.[11]

6.7.1 The Newton–Raphson recursion

Let's go back to (6.85). We will be using tangent lines, so we assume that $f(x)$ is differentiable, at least near the solution z. Pick a point x_1 near z, and find the equation

---

[11] Although this is an old result, special cases of it were known much earlier, like the divide and average rule for estimating $\sqrt{2}$ mentioned earlier. The Persian mathematicians Nasir al-Din Tusi (1201–1274) and Jamshid al-Kashi (1380–1429) also studied and implemented special cases for approximating the roots of integers prior to Newton. The Newton–Raphson method itself is a special type of *fixed point* method, a large class of recursions that lie in the overlap between analysis and dynamical systems theory.

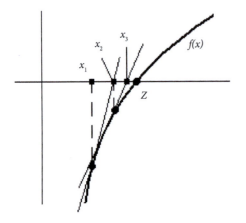

Fig. 6.17 Illustrating the Newton–Raphson method

of the tangent line at $x = x_1$

$$y = f(x_1) + f'(x_1)(x - x_1)$$

This line crosses the x-axis when $y = 0$; we label this crossing point x_2 so that

$$f(x_1) + f'(x_1)(x_2 - x_1) = 0$$

See Figure 6.17.

When we solve this equation for x_2, we get

$$x_2 = x_1 - \frac{f(x_1)}{f'(x_1)}$$

provided that $f'(x_1) \neq 0$. Now that we have found x_2, which according to Figure 6.17 is closer to z than our initial guess x_1 was, we discard x_1 and repeat the above process with x_2 instead. Find the tangent line $x = x_2$ and calculate the point x_3 where this line crosses the x-axis, which comes out to be

$$x_3 = x_2 - \frac{f(x_2)}{f'(x_2)}$$

if $f'(x_2) \neq 0$. Repeating n times, we get

$$x_{n+1} = x_n - \frac{f(x_n)}{f'(x_n)}, \quad n = 1, 2, 3, \ldots \tag{6.87}$$

This is the *Newton–Raphson recursion*. If we continue the iteration indefinitely, then a sequence x_n of successive approximations is generated. This sequence is considered a solution of the recursion, or the difference equation in (6.87).

| Convergence of the Newton–Raphson recursion |

A solution x_n of the Newton–Raphson recursion must converge to a solution of the equation in (6.85) in order to be considered a sequence of approximations. The next theorem specifies the conditions that imply the existence of such a sequence as well as its fast convergence property.

Theorem 212 *Let f be a twice differentiable function on an interval* $[a, b]$, *i.e.,* $f''(x)$ *exists for all* $x \in [a, b]$. *Further, assume that* $f(a)f(b) < 0$, *and there are constants* m, M *such that*

$$|f''(x)| \leq M \quad and \quad |f'(x)| \geq m > 0 \quad for\ all\ x \in [a, b] \qquad (6.88)$$

Then there is an interval $I \subset [a, b]$ *containing a zero* z *of* $f(x)$ *such that for every* $x_1 \in I$, *the sequence* x_n *generated by (6.87) is in I, and* $\lim_{n \to \infty} x_n = z$. *Furthermore, for all* n

$$|x_{n+1} - z| \leq \frac{M}{2m}|x_n - z|^2 \qquad (6.89)$$

Proof Given that $f(a)f(b) < 0$ by the intermediate value theorem, the graph of $f(x)$ crosses the x-axis at some point $z \in (a, b)$. By the second inequality in (6.88), $f'(x) \neq 0$ for all $x \in [a, b]$; so by Corollary 206, $f(x)$ is either increasing or decreasing, and it follows that z is the unique zero of f in (a, b).

In discussing Taylor series in Chapter 8, we establish that for each $c \in [a, b]$, there is a number ζ between c and z such that

$$f(z) = f(c) + f'(c)(z - c) + \frac{1}{2}f''(\zeta)(z - c)^2$$

Since $f(z) = 0$, this equality can be written as

$$-f(c) = f'(c)(z - c) + \frac{1}{2}f''(\zeta)(z - c)^2 \qquad (6.90)$$

If we define

$$r = c - \frac{f(c)}{f'(c)}$$

then by (6.90),

$$r = c + (z - c) + \frac{f''(\zeta)}{2f'(c)}(z - c)^2$$

Rearranging the above equality, taking absolute values, and using the inequalities in (6.88), we get

$$|r - c| = \frac{|f''(\zeta)|}{2|f'(c)|}|z - c|^2 \leq \frac{M}{2m}|z - c|^2 \qquad (6.91)$$

Now choose a number δ such that $0 < \delta < 2m/M$, and δ is small enough that

$$I = [z - \delta, z + \delta] \subset [a, b]$$

Let $x_1 \in I$ generate a sequence x_n using the recursion (6.87). If $x_k \in I$ for some index k, then

$$|x_{k+1} - z| \leq \frac{M}{2m}|z - x_k|^2 \tag{6.92}$$

Since $x_k \in I$, it follows that $|x_k - z| \leq \delta$, and therefore,

$$|x_{k+1} - z| \leq \frac{M}{2m}\delta^2 < \delta$$

Thus, $x_{k+1} \in I$ too, and we may conclude (by mathematical induction) that $x_n \in I$ for all n if we choose $x_1 \in I$. Further, by its construction, this sequence x_n satisfies (6.89). It remains to prove that $\lim_{n\to\infty} x_n = z$.

Since x_n satisfies (6.92), we have

$$|x_{n+1} - z| \leq \left(\frac{M}{2m}|x_n - z|\right)|x_n - z| < \left(\frac{M}{2m}\delta\right)|x_n - z|$$

Using this again,

$$|x_{n+1} - z| < \left(\frac{M}{2m}\delta\right)\left(\frac{M}{2m}\delta|x_{n-1} - z|\right) = \left(\frac{M}{2m}\delta\right)^2 |x_{n-1} - z|$$

By induction,

$$|x_{n+1} - z| < \left(\frac{M}{2m}\delta\right)^n |x_1 - z|$$

The δ we defined δ above implies that $M\delta/(2m) < 1$, so the difference between x_n and z goes to 0 as $n \to \infty$, i.e., $\lim_{n\to\infty} x_n = z$, as claimed. ∎

Given the function f, we can determine M and m that satisfy (6.88) and thus obtain a value for δ. This number gives us the size of I and therefore, "how close" x_1 has to be for the sequence x_n to converge to z. Note also that the interval $[a, b]$ in the statement of Theorem (212) was already small enough to assure that z is the only solution of the equation $f(x) = 0$ in $[a, b]$.

If we think of the difference $|x_n - z|$ as the *approximation error* E_n *in step* n *of* the process, then in the next step by (6.89)

$$E_{n+1} \leq \frac{M}{2m}E_n^2$$

The appearance of the square means that accuracy doubles at each step in the sense that if $E_n \leq 10^{-k}\alpha$ for some number α (independent of n), then $E_{n+1} \leq$

$10^{-2k}\beta$ for some β. Thus, a k-digit accuracy doubles to $2k$ digits in just one step. For this reason, the convergence of x_n in the Newton–Raphson method is called *second order* or *quadratic*.

We now apply the method to estimate the solution of

$$f(x) = x - \cos x = 0$$

accurate to six decimal places. Using the derivative formulas discussed above, we get

$$f'(x) = 1 + \sin x$$

Let $x_1 = 0$, for no particular reason other than the fact that it is a nice round number that is not far from z in Figure 6.16. Then

$$x_2 = 0 - \frac{f(0)}{f'(0)} = -\frac{0 - \cos 0}{1 + \sin 0} = 1$$

This in turn gives

$$x_3 = 1 - \frac{1 - \cos 1}{1 + \sin 1} \simeq 0.75036387$$

We keep one or more decimal places beyond the desired level of accuracy (six decimal places). The chart below shows the values that generated by carrying out some additional steps of this process:

n	1	2	3	4	5	6
x_n	0	1	0.75036387	0.73911289	0.73908513	0.73908513

Since the first six decimal places do not change going from step 5 to step 6, we conclude that to six decimal places,[12]

$$z \simeq 0.739085$$

is the solution of the equation in (6.86), which was obtained in just five steps starting from $x_1 = 0$.

6.7.2 Estimating roots of real numbers

A practical and simple application of the Newton–Raphson method is to efficiently estimate roots of real numbers to a high degree of accuracy. Let r be any positive

[12] Strictly speaking, this is not always a valid criterion for ending the iteration process because it is possible that after several steps, the numbers will shift again and approach some other value. But given the estimated location of z in this example, we can be confident that we have reached the desired accuracy in this case.

real number and m an integer greater than 1. The mth root of r, that is, $\sqrt[m]{r} = r^{1/m}$, is then a root of the polynomial $f(x) = x^m - r$. The derivative is $f'(x) = mx^{m-1}$, so the Newton–Raphson recursion for this function is:

$$x_{n+1} = x_n - \frac{x_n^m - r}{mx_n^{m-1}} = \frac{(m-1)x_n^m + r}{mx_n^{m-1}} \tag{6.93}$$

Note that in the special case $m = r = 2$, (6.93) can be written as

$$x_{n+1} = \frac{x_n^2 + 2}{2x_n} = \frac{1}{2}\left(x_n + \frac{2}{x_n}\right)$$

The right-hand side of the above recursion is the divide and average rule for estimating $\sqrt{2}$ that we discussed earlier.

We now use (6.93) to estimate $\sqrt[3]{2}$ correct to 6 decimal places. Here $m = 3$, and $r = 2$. Pick $x_1 = 1$ (a convenient number not far from the target $\sqrt[3]{2}$, but any positive value works). Then

$$x_2 = \frac{2x_1^3 + 2}{3x_1^2} = \frac{4}{3} \simeq 1.3333333$$

I rounded the answer to seven decimal places. Next,

$$x_3 \simeq \frac{2(1.3333333)^3 + 2}{3(1.3333333)} \simeq 1.2638889$$

Repeating, we find that

$$x_4 \simeq 1.2599335, \qquad x_5 \simeq 1.2599211, \qquad x_6 \simeq 1.2599211$$

As a check, we see that multiplying the last number above three times yields the following result:

$$1.2599211^3 = 2.00000024$$

which differs from 2 by less than $0.000001 = 10^{-6}$. It is worth mentioning that you can do the above calculations without a calculator, or at most, using a basic calculator that can only add, multiply, and divide numbers. Further, *all of the approximations of an irrational are rational numbers.*

6.7.3 Choosing the initial value

If we sketch the graph of a function $f(x)$ using a computing device, then we can often pick a point that is close to a zero of $f(x)$ as the starting point of iteration.

If $f(x)$ is continuous, then a zero exists by the intermediate value theorem provided that there are numbers a and b such that $f(a)f(b) < 0$, i.e., $f(a)$ and $f(b)$ have opposite signs. For instance, in the case of (6.86), as we saw previously,

$$f(0) = -1 < 0, \qquad f\left(\frac{\pi}{2}\right) = \frac{\pi}{2} > 0$$

so there is a zero of $f(x)$ between $x = 0$ and $x = \pi/2$.

The bisection algorithm

The intermediate value theorem above is not sufficient to ensure the existence of a unique solution unless a and b are so close that there are no other zeros of $f(x)$ between them.[13] It is often possible to narrow things down without graphing $f(x)$. For example, getting back to (6.86), we calculate f at the midpoint of the interval $[0, \pi/2]$ to get

$$f\left(\frac{\pi}{4}\right) = \frac{\pi}{4} - \cos\frac{\pi}{4} = \frac{\pi}{4} - \frac{1}{\sqrt{2}} \simeq 0.08 > 0$$

which means that the zero z is between 0 and $\pi/4 \simeq 0.79$. We can repeat this halving method by calculating the value of f at the midpoint of $[0, \pi/4]$, i.e., $f(\pi/8)$, and keep going until we eventually get close enough to the point z. This idea is sufficiently useful to be given a name: *the bisection method or bisection algorithm* because of checking the midpoint at each step.

Noteworthy is the fact that the bisection algorithm does not require $f(x)$ to be differentiable, and it always works for any continuous $f(x)$. Why then bother with the Newton–Raphson method?

There are two reasons: first, the bisection method is much slower in estimating the zero z to a desired accuracy than Newton's method is. Second, during the bisection process, we may inadvertently pass by a very close estimate of the zero; for instance, in the above discussion, $\pi/4$ is rather close to z; but if we continue the bisection process by going to the next midpoint $\pi/8 \simeq 0.45$, we see that it has moved away from z. There is no general way of knowing when to stop.

In practice, the bisection method is used only to get a number close enough to z to serve as x_1 in the Newton–Raphson recursion.

Beware of the hidden infinity!

Suppose that we know that a function has a unique zero in an interval $[a, b]$. Why do we need to get close to z to use Newton–Raphson method? Inequalities (6.88) in Theorem 212 provide an answer, but they are not necessary conditions. We now consider what happens if we *don't* start sufficiently close to the zero.

---

[13] An extreme case is the function $\sin(1/x)$ that is at least twice differentiable for $x > 0$. To find a zero of the equation $\sin(1/x) = 2/3$, we must specify which of the infinitely many zeros we are interested in.

Looking again at the graph of $f(x) = x - \cos x$ in Figure 6.16, nothing jumps out as odd; all is nice and smooth with not a trace of abnormalities. But looking at the Newton–Raphson recursion itself, we spot a division by the derivative $f'(x_n)$. If this quantity gets close to 0 for some n while $f(x_n)$ is not near 0, then the next value x_{n+1} spikes and goes off track.

There are a number of places on the graph of $f(x) = x - \cos x$ in (6.86) where this occurs; in Figure 6.16, the places where the curve is flat (with horizontal tangent line) are where the derivative is near 0 or equal to it.

We can be more precise in this case: since $f'(x) = 1 + \sin x$, we calculate that $f'(x) = 0$ where $\sin x = -1$, that is,

$$x = -\frac{\pi}{2} + 2k\pi = (4k - 1)\frac{\pi}{2}, \quad \text{all } k \text{ in } \mathbb{Z} \tag{6.94}$$

We see two of these points in Figure 6.16. To understand what these zero points of the derivative do to the recursion algorithm, we define the function

$$F_{NR}(x) = x - \frac{f(x)}{f'(x)}$$

We call $F_{NR}(x)$ the *Newton–Raphson recursion function (or recursion map)*. Figure 6.18 illustrates the graph of F_{NR} near the two zero points of the derivative in Figure 6.16.

The recursion function in this case is

$$F_{NR}(x) = x - \frac{x - \cos x}{1 + \sin x} = \frac{x \sin x + \cos x}{1 + \sin x} \tag{6.95}$$

In Figure 6.16, we see a part of the graph of this function that consists of three separate pieces and contains z and two singularities of $F_{NR}(x)$ where $\sin x = -1$.

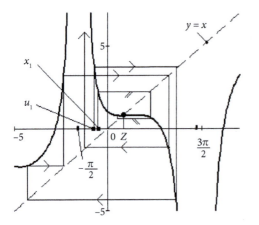

Fig. 6.18 Exposing hidden infinities in the Newton–Raphson method

The sequence generated by the iteration of $F_{NR}(x)$ may jump from one piece to another, if it does not actually land on one of the singularities.

If we start from x_1 shown in Figure 6.18, then $x_2 = F_{NR}(x_1)$ is obtained by moving vertically from x_1 to the curve, then horizontally until we reach the identity line $y = x$. This is where x_2 is on the x-axis. Starting from this point, we get to $x_3 = F(x_2)$ again by moving vertically to the curve, then horizontally to $y = x$, and so on. Repeating this process gives the successive values

$$x_2 = F_{NR}(x_1)$$
$$x_3 = F_{NR}(x_2) = F_{NR}(F_{NR}(x_1))$$
$$x_4 = F_{NR}(x_3) = F_{NR}(F_{NR}(F_{NR}(x_1)))$$

$$\vdots$$

Figure 6.18 shows that we get close enough to z in just three iterations that it is hard to show further approximations x_n distinctly from z.

By contrast, the iterations starting from the initial point u_1 that is near x_1 result in a series of numbers that do not approach z; if we follow the cobweb of arrows $\longrightarrow$ from u_1 on the x-axis to the curve, then to $y = x$ to get u_2, then back to the curve again and repeat, then we see that the path can move far from z, and there is no reason to think it will ever approach it.

The following table lists the results obtained using the Newton–Raphson method with a few starting points from -0.75 to -0.8, about *halfway* between 0 and the nearest infinite singularity at $-\pi/2 \simeq -1.57$, so we aren't starting close to a singularity. As you can see, there is significant unpredictability about the outcome even though the graph of $x - \cos x$ is perfectly smooth and showing no hint of infinities of any kind in this range! *This curve is not even flat between -0.75 and -0.8!*

x_1	Outcome
-0.75	reaching 0.739 in about 215 iterations
-0.76	reaching 0.739 in about 7 iterations
-0.77	iteration values exceed 10^{12} in 76 iterations
-0.78	iteration values exceed 10^{12} in 31 iterations
-0.79	reaching 0.739 in about 9 iterations
-0.80	iteration values exceed 10^{12} in 110 iterations

As we see, singularities can throw a sequence generated by the Newton–Raphson method if we don't choose the initial value close enough to the solution. But things can go off track even when there are no singularities.

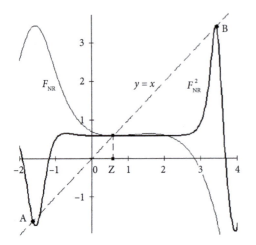

Fig. 6.19 Newton–Raphson method goes off track without a singularity

Take the function $f(x) = x - a \cos x$ where $0 < a < 1$. Then the derivative $f'(x) = 1 + a \sin x > 0$ for all real values of x, so

$$F_{NR}(x) = x - \frac{x - a \cos x}{1 + a \sin x} = \frac{ax \sin x + a \cos x}{1 + a \sin x}$$

We may also write

$$F_{NR}(x) = \frac{x \sin x + \cos x}{1/a + \sin x}$$

which makes it easier to compare the function F_{NR} with that in (6.95). Here, $1/a > 1$, and division by 0 is avoided.

Let $a = 0.6875$. Then $f(x)$ has a unique zero at $z \approx 0.576415$, as you can quickly verify using the Newton–Raphson method or using a calculator. A graph of $F_{NR}(x)$ for this case is shown in Figure 6.19 together with the graph of $F_{NR}^2(x) = F_{NR}(F_{NR}(x))$ as the thick curve in the same coordinate system.

The fixed point of $F_{NR}(x)$ is also the zero z of $f(x)$ identified as z in Figure 6.19. The points $A \approx -1.64554$, and $B = F_{NR}(A)$ are fixed points[14] of $F_{NR}^2(x)$ but not of $F_{NR}(x)$. Therefore, the pair of points $\{A, B\}$ is a cycle of $F_{NR}(x)$; if we take $x_1 = A$, then $x_2 = B$, and the sequence x_n cycles back and forth between A and B without converging to the zero z of $f(x)$.[15]

---

[14] A fixed point of any function $f(x)$ is a point p where $f(p) = p$; i.e., the graph of $f(x)$ crosses the identity line $y = x$ at $x = p$.

[15] If x_1 differs slightly from A or B, then this cycle may not be visited, and the sequence x_n may eventually converge to z. However, the number of terms of x_n before it gets sufficiently close to z may be quite large, depending on the exact value of x_1.

6.8 Exercises

Exercise 213 (a) Use (6.4) to show that the identity function $f(x) = x$ is differentiable for all real x, and show that $f'(x) = 1$.

   (b) Repeat (a) for a constant function $f(x) = c$ where c is any fixed real number, and show that $f'(x) = 0$. Your calculation should make it clear that the specific value of c doesn't matter.

Exercise 214 Explain why the function $\sqrt[3]{x-1}$ is defined for all real numbers x, but it is not differentiable at $x = 1$.

Exercise 215 Suppose that a is an isolated point of the domain of a function $f(x)$, i.e., there is a small interval $(a - \varepsilon, a + \varepsilon)$ such that $f(x)$ is undefined for all $x \in (a - \varepsilon, a + \varepsilon)$, except for $x = a$. Can $f(x)$ be differentiable at a? Explain.

Exercise 216 Prove the constant multiples rule (6.10) by using the definition of derivative, i.e., by taking the limit in (6.4).

Exercise 217 Prove the power rule (6.11) using the product and quotient rules and mathematical induction.

Exercise 218 (a) Prove Corollary 192.

   (b) For what values of x is the function $\sqrt[3]{x-1}$ in Exercise 214 differentiable, and what is its derivative? Explain.

Exercise 219 Prove Lemma 190.

Exercise 220 (a) Find the derivative of the following function for $x > 0$:

$$f(x) = x \sin \frac{1}{x}$$

   (b) If x_n is an arbitrary sequence of positive real numbers that converges to 0, then prove that

$$\lim_{n \to \infty} f(x_n) = 0$$

   (c) Motivated by (b), define $f(0) = 0$. With this extension, prove that $f(x)$ is continuous at $x = 0$ but not differentiable there (choose a suitable sequence x_n of positive numbers that converges to 0, but the difference quotient doesn't work out).

Exercise 221 Find the second derivative of the function $f(x) = \sec x - \sin(x^2)$.

Exercise 222 (a) Find all distinct higher-order derivatives of $x^7 - 2x^5 - x^4 + x^2 + 7$, and calculate their values at $x = 1$;

(b) Find all distinct higher-order derivatives of $1/x$, and calculate their values at $x = 1$ (a formula for the nth derivative is needed);

(c) Find all distinct higher-order derivatives of $\sin x$ (only a few of these derivatives are distinct, or non-repeating);

(d) Find all distinct higher-order derivatives of $\sin(2x)$, and calculate their values at $x = \pi/4$.

Exercise 223 *(a) Prove that the absolute value function $|x|$ is continuous on $\mathbb{R}$ (recall that $||a| - |b|| \le |a - b|$).*

(b) Prove that the absolute value function is not differentiable at $x = 0$.

Exercise 224 *Consider the following Dirichlet-type function:*

$$D_0(x) = \begin{cases} x, & \text{if } x \text{ is rational} \\ 0, & \text{if } x \text{ is irrational} \end{cases}$$

Prove that $D_0(x)$ is continuous at $x = 0$ but at no other real number. Is $D_0(x)$ differentiable at $x = 0$? Explain using the difference quotient of D_0.

Exercise 225 *Prove Theorem 194.*

Exercise 226 *Let a be a fixed real number and f and g functions whose limits $\lim_{x \to a} f(x)$ and $\lim_{x \to a} g(x)$ exist. Use Theorem 201 and the limit theorems from Chapter 3 to prove the following statement:*

$$\lim_{x \to a}[f(x)g(x)] = \lim_{x \to a} f(x) \lim_{x \to a} g(x)$$

Exercise 227 *Complete the proof of Lemma 203.*

Exercise 228 *(a) For the function $f(x) = x^3 - 2x - 1$, find the number(s) c that satisfy the conclusion of Theorem 205 on the interval $[0, 3]$.*

(b) For what values of α is $f(x) = x^3 + \alpha x - 1$ increasing on its domain $(-\infty, \infty)$? Prove your claim.

Exercise 229 *(a) If f and g are uniformly continuous on a set S, then prove that $f + g$ and cf are also uniformly continuous for every real constant c.*

(b) Using a (simple) example, show that the product of two uniformly continuous functions (possibly identical) may not be uniformly continuous.

(c) If f and g are uniformly continuous and bounded on a set S, then prove that the product fg is uniformly continuous on S. Note that we may write

$$f(x)g(x) - f(y)g(y) = f(x)[g(x) - g(y)] + g(y)[f(x) - f(y)]$$

Exercise 230 *A function f(x) is said to be Lipschitz on a nonempty set S if there is a constant M > 0 such that*

$$|f(x) - f(y)| \le M|x - y| \quad \text{for all } x, y \in S$$

(a) Prove that the functions $|x|$ and $\sin x$ are Lipschitz on $(-\infty, \infty)$.
(b) Explain why $f(x) = 1/x$ is Lipschitz on $[1, \infty)$ but not on $(0, 1)$.

Exercise 231 *(a) Prove that if f(x) is Lipschitz on S, then f(x) is uniformly continuous on S.*
(b) Prove that if f(x) is Lipschitz and differentiable on an open interval (a, b), then its derivative $f'(x)$ is bounded on (a, b).
(c) Use (b) to prove that $f(x) = \sqrt{x}$ is not Lipschitz on (0,1).
(d) Explain why $\sqrt{x}$ is uniformly continuous but not Lipschitz on [0,1].

Exercise 232 *(a) Suppose that a function f is differentiable on an interval I. Prove that f is constant on I if and only if $f'(x) = 0$ for all $x \in I$.*
(b) Explain why the function

$$f(x) = \begin{cases} 2 & \text{if } x > 0 \\ 1 & \text{if } x < 0 \end{cases}$$

is differentiable on the set $D = (-\infty, 0) \cup (0, \infty)$ and $f'(x) = 0$ for all $x \in D$. Notice that f isn't constant on D; does this invalidate (a)?

Exercise 233 *(a) Prove Corollary 207.*
(b) Consider the function

$$f(x) = x^4 \left(2 + \sin \frac{1}{x}\right)$$

if $x \neq 0$, and define $f(0) = 0$. Prove that f(x) is differentiable on $(-\infty, \infty)$, and $f'(0) = 0$. Explain why f has a local minimum at $x = 0$, but we can't use Corollary 207 to prove it.

Exercise 234 *Prove Part (b) of Theorem 209.*

Exercise 235 *Let $q(x)$ be a function defined on an interval (a, b) where $-\infty \leq a < b \leq \infty$. If $\lim_{x \to a^+} q(x) < r$ for some real number r, then prove that there is $c \in (a, b)$ such that $q(x) < r$ for all $x \in (a, c)$.*

Exercise 236 *Prove the following using l'Hospital's rule:*

$$\text{(a)} \quad \lim_{x \to \infty} \frac{6x^3 - 2x - 1}{5 + x^2 - 2x^3} = -3 \qquad \text{(b)} \quad \lim_{x \to 0^+} \frac{1 - \cos \sqrt{x}}{2 \sin x} = \frac{1}{4}$$

Exercise 237 *Find the following limits, using l'Hospital's rule where appropriate:*

$$\text{(a)} \quad \lim_{x \to \infty} \sqrt{x} \sin \frac{1}{x} \qquad \text{(b)} \quad \lim_{x \to \infty} x^2 \sin \frac{1}{x} \qquad \text{(c)} \quad \lim_{x \to 0^+} \sqrt{x} \sin \frac{1}{x}$$

Exercise 238 *(a) Find the following limit without using l'Hospital's rule:*

$$\lim_{x \to \infty} \frac{\sqrt{x^2 + 1}}{x}$$

(b) Verify that all the hypotheses of Corollary 211 are satisfied in (a); what happens when you use that corollary to calculate the limit in (a)?
(c) Without using l'Hospital's rule, show that

$$\lim_{x \to \infty} \frac{x + \sin x}{x} = 1$$

Explain why Corollary 211 cannot be used to calculate this limit.

Exercise 239 *(a) Prove that*

$$\lim_{x \to 0^+} \frac{x}{1 - \cos x} = \infty$$

(b) Does the following limit exist? Explain.

$$\lim_{x \to 0} \frac{x}{1 - \cos x}$$

Exercise 240 *(a) Write the nonlinear equation $x^3 = 2x + 1$ as $f(x) = 0$, and sketch a graph of $f(x)$ between $x = -2$ and $x = 2$ to see that it has three zeros.*
(b) Calculate the derivative of $f(x)$ in (a) and figure out the Newton–Raphson recursion;
(c) Use the recursion in (b) to calculate the positive zero of $f(x)$ accurate to 6 decimal places with $x_1 = 1$.

(d) Verify by direct substitution that $f(-1) = 0$ so that -1 is a zero of $f(x)$ too. Divide $f(x)$ by $x + 1$ to get the quotient $x^2 - x - 1$ and conclude that $f(x) = (x + 1)(x^2 - x - 1)$. Obtain the roots of the quotient as the other two zeros of $f(x)$:

$$\frac{1 \pm \sqrt{5}}{2}$$

The positive one of these is the "golden ratio" φ, and the negative one is the conjugate of φ that we denote by $\overline{\varphi}$. Which of these did you estimate in (c)? How would you change x_1 in order to approximate the other?

Exercise 241 *Use the recursion in (6.93) with the initial value $x_1 = 1$ to estimate each of the following roots correct to at least six decimal places:*

$$\sqrt{5}, \qquad \sqrt[5]{2}$$

Exercise 242 *This exercise continues the study begun in Exercise 240.*

(a) Solve the equation $f'(x) = 0$ to calculate the exact values of the singularities of the Newton–Raphson function $F_{NR}(x)$.

(b) Determine $F_{NR}(x)$ and sketch its graph carefully to visually identify its two singularities as well as its crossings with the identity line $y = x$ that correspond to the zeros -1, $\overline{\varphi}$, and φ of $f(x)$.

(c) Use your figure in (b) to pick a good initial value to estimate the negative zero $\overline{\varphi}$ accurately to six decimal places. Why can we not use -1 as an initial value? Consider using the same reasoning to find a range of possible initial values that will always lead to φ.

(d) It would seem reasonable that an initial value between -1 and $\overline{\varphi}$ should generate a sequence that converges to either -1 or to $\overline{\varphi}$. But the initial value $x_1 = -0.81$ quickly takes us to the positive zero φ of $f(x)$! Check this out by iterating $F_{NR}(x)$ a few times; how would you explain this unexpected outcome?

7

Integration

The concept of integral was developed by Newton and Leibniz in the seventeenth century and further used and enhanced throughout the eighteenth century by mathematicians of the era. The integral was used in calculations mainly as what we now call the antiderivative, although Euler was known to have used what is nowadays called a Riemann sum to approximate the value of the antiderivative when an elementary closed form was not known.

The modern theory of integration began with the work of Cauchy in the early nineteenth century who systematically developed the concept of integral as the limit of a Riemann sum. Riemann's name is associated with the sum in part because he finished the investigations that Cauchy had started, and he was the first to develop integration theory in sufficient detail to be able to obtain a criterion that was both necessary and sufficient for a function to be integrable. This in turn introduced a new, important class of functions, the *integrable functions*, that properly included the bounded, continuous functions.

The work of Cauchy and Riemann was only the starting point for the development of powerful theories of integral by many others, including Lebesgue, whose work led to the creation of measure theory and mathematical probability theory.

We discuss the Riemann integral and related concepts in this section using the approach developed by Jean Gaston Darboux (1842–1917) in 1875. This approach is technically easier to follow, and it also lends itself more easily to generalizations beyond the Riemann integral.

7.1 From velocity to position, from curve to area

The purpose of this section is to motivate the various concepts and methods that we use later in this section to define the Riemann integral. You may move ahead to the next section if these ideas are familiar or not of interest.

Acceleration is defined as the rate of change of the velocity of a moving body with respect to variation in time. *Average acceleration* is defined the same way that average velocity is:

$$\frac{\text{change in velocity}}{\text{change in time}} = \frac{\Delta v}{\Delta t}$$

Real Analysis and Infinity. Hassan Sedaghat, Oxford University Press.
© H. Sedaghat (2022). DOI: 10.1093/oso/9780192895622.003.0007

In general, acceleration is a variable quantity like velocity itself, so it is different from average acceleration over any fixed time interval, no matter how short. But in an interesting case, it is constant, and we use that to illustrate the core of the integration process.

Deriving Galileo's equations from constant acceleration

The acceleration of a freely falling body near the surface of the Earth has a nearly constant value of $g = 9.8$ in the metric system. Thus,

$$\frac{\Delta v}{\Delta t} = g$$

is fixed over any time interval Δt. Let $v(0) = 0$, i.e., we just drop the object starting from rest at time $t = 0$. Then t seconds later,

$$g = \frac{v(t) - v(0)}{t - 0} = \frac{v(t)}{t}$$

From this we get

$$v(t) = gt \tag{7.1}$$

Recall that this is Galileo's equation for the velocity of a falling object. We needed only elementary algebra to derive it.

The above derivation of velocity from acceleration is especially simple since the acceleration g is constant; it doesn't work if acceleration is variable. The derivation of position $x(t)$ from velocity cannot be done the same way because the velocity in (7.1) is not constant. But since that velocity is *continuous*, we are justified in thinking that

> *If we measure the velocity over a small enough time interval, then it is approximately constant because its value does not change significantly from the beginning of the interval to its end.*

This observation takes us back to the previous case with constant acceleration. Now the velocity is *nearly constant* over a small time interval, so we can repeat the above calculation to obtain close approximations to the position; of course, we must do it over many adjacent time intervals of short duration, one interval at a time, till we reach the final time instant.

To obtain an estimate of the position $x(t)$ at time t, we divide up (or partition) the time interval from 0 to t into N subintervals of equal length (each t/N seconds long) to get

$$t_0 = 0, \quad t_0 + \frac{t}{N} = t_1, \quad t_1 + \frac{t}{N} = t_2, ..., t_{N-1} + \frac{t}{N} = t_N = t \tag{7.2}$$

For each $n = 1, 2, 3, ..., N$, let $v(t_n)$ be the velocity of the object n time-subintervals after the object is released. Now, over the time-subinterval $[t_{n-1}, t_n]$, the object moves a little bit, from $x(t_{n-1})$ to $x(t_n)$. If the duration $\Delta t = t_n - t_{n-1}$ is small (that is, *if N is large*), then the velocity changes very little from $v(t_{n-1})$ at the beginning of the short subinterval $[t_{n-1}, t_n]$. This reduces the approximation error in finding the velocity when using the average velocity formula:

$$\frac{x(t_n) - x(t_{n-1})}{t_n - t_{n-1}} \approx v(t_{n-1}) \tag{7.3}$$

We may write this as

$$x(t_n) - x(t_{n-1}) \approx v(t_{n-1})(t_n - t_{n-1}) \tag{7.4}$$

For $n = 1, 2, 3, ..., N$, this gives us a set of differences:

$$x(t_1) - x(t_0) \approx v(t_0)(t_1 - t_0)$$
$$x(t_2) - x(t_1) \approx v(t_1)(t_2 - t_1)$$
$$x(t_3) - x(t_2) \approx v(t_2)(t_3 - t_2)$$
$$\vdots$$
$$x(t_N) - x(t_{N-1}) \approx v(t_{N-1})(t_N - t_{N-1})$$

Adding all of the above, and canceling all equal terms with opposite signs on the left-hand side gives

$$x(t_N) - x(t_0) \approx v(t_0)(t_1 - t_0) + v(t_1)(t_2 - t_1) + \cdots + v(t_{N-1})(t_N - t_{N-1})$$

Since $t_N = t$ and $x(t_0) = x(0)$, we can rewrite the above as

$$x(t) \approx x(0) + \sum_{n=1}^{N} v(t_{n-1})\Delta t \tag{7.5}$$

Further, we have $\Delta t = t/N$, and by (7.1),

$$v(t_{n-1}) = gt_{n-1} = g\frac{(n-1)t}{N} = gt\frac{n-1}{N}$$

If we insert these quantities into (7.5), then

$$x(t) \approx x(0) + \sum_{n=1}^{N} gt \frac{n-1}{N} \frac{t}{N}$$

$$= x(0) + \sum_{n=1}^{N} gt^2 \frac{n-1}{N^2}$$

$$= x(0) + gt^2(0) + gt^2 \left(\frac{1}{N^2}\right) + gt^2 \left(\frac{2}{N^2}\right) + \cdots + gt^2 \left(\frac{N-1}{N^2}\right)$$

$$= x(0) + gt^2 \frac{1 + 2 + 3 + \cdots + N - 1}{N^2} \tag{7.6}$$

The sum in the numerator is an arithmetic progression; let S be its total, and notice that

$$
\begin{array}{ccccccccccccc}
2S & = & 1 & + & 2 & + & 3 & + & \cdots & + & N-2 & + & N-1 \\
& + & N-1 & + & N-2 & + & N-3 & + & \cdots & + & 2 & + & 1 \\
2S & = & N & + & N & + & N & + & \cdots & + & N & + & N
\end{array}
$$

Therefore, $2S$ is just N added to itself $N - 1$ times, or

$$2S = N(N - 1)$$

$$1 + 2 + 3 + \cdots + N - 1 = \frac{N(N-1)}{2}$$

Inserting this in (7.6) and setting $x(0) = 0$ (so that we measure the distance from the point where the object is dropped), we get

$$x(t) \approx gt^2 \frac{N-1}{2N} \tag{7.7}$$

We can improve the above approximation by choosing N larger, because larger N makes the duration $\Delta t = t/N$ between t_{n-1} and t_n smaller. In turn, this reduces the change in velocity from $v(t_{n-1})$ to $v(t_n)$. Taking the limit as $N \to \infty$ reduces Δt to zero and gives us a precise value for the position in (7.5) as

$$x(t) = \lim_{N \to \infty} gt^2 \frac{N-1}{2N} = gt^2 \lim_{N \to \infty} \left(\frac{1}{2} - \frac{1}{2N}\right) = gt^2 \left(\frac{1}{2} - \lim_{N \to \infty} \frac{1}{2N}\right) = \frac{1}{2}gt^2$$

Recall that this is Galileo's equation of motion.[1]

---

[1] You may have noticed that Galileo's equations do not contain the masses of objects, so (in the absence of air resistance) a feather falls just as fast as a brick. This is indeed true: such objects have been videotaped falling at the same speed *in a vacuum*, in one instance by an astronaut on the surface of the moon! See my self-study book *Achieving Infinite Resolution* for more details on these interesting, modern-day experiments.

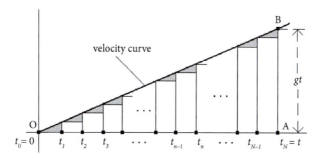

Fig. 7.1 A partition of the region under the velocity curve

The above discussion is a simple overview of the process of integration:

(1) Partitioning the time interval into many intervals of very short duration,
(2) calculating the change in position for each short time period and adding the results, and
(3) taking a limit to reduce the duration of each small interval to 0 to eliminate calculation errors.

We have much more to say about this interesting three-step process below.

Position as area

We now consider a geometric way of deriving the equation of position from velocity. In Figure 7.1, we see a graph of the velocity function $v(t) = gt$ as a straight line.

On the horizontal (time) axis in Figure 7.1, we see a partition of the interval from 0 to t into N subintervals of equal length. The nth vertical strip of width Δt has area

$$v(t_{n-1})\Delta t = gt_{n-1}\Delta t = g\frac{(n-1)t}{N}\frac{t}{N} = gt^2\frac{n-1}{N^2}$$

This is recognizable as one term of the sum in (7.6). We see in Figure 7.1 that *this sum is the total area of all the strips, so it approximates (underestimates in this case) the area of triangle OAB*:

$$\sum_{n=1}^{N} v(t_{n-1})\Delta t \approx \text{area of triangle OAB} = \frac{(gt)(t)}{2} = \frac{1}{2}gt^2$$

The error of this approximation is the sum of the areas of all the small shaded regions on top of the strips.

In Figure 7.2, the number of strips N is doubled to reduce the approximation error. We see in this figure that the sum of the little shaded areas on top of each

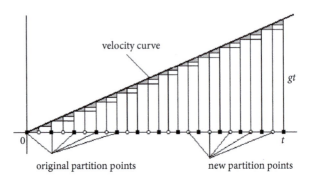

Fig. 7.2 Refining a partition reduces the approximation error

strip is clearly less than the similarly shaded area in Figure 7.1. As we increase N, we indeed reduce the error.

Letting $N \to \infty$ the width Δt of each strip approaches zero, and the strips thin out to line segments. The error goes to zero, so we end up with the area of triangle OAB in the limit. Going back to (7.5) with $x(0) = 0$, we may conclude that

$$x(t) = \lim_{N \to \infty} \sum_{n=1}^{N} v(t_{n-1})\Delta t = \frac{1}{2}gt^2$$

which is what we derived earlier using arithmetic rather than geometry.[2]

The geometric considerations show that:

The position function $x(t)$ is the area under the graph of the velocity function $v(t)$, which we defined as the derivative $x'(t)$. In fact, the same is true about velocity and acceleration: velocity $v(t)$ is the area under the graph of (constant) acceleration g, which is the derivative $v'(t)$.

Figure 7.3 illustrates this fact.

In the right panel of Figure 7.3, we see that the *area* of the shaded rectangle is gt, which is none other than the velocity *function* in the left panel; the area in the right panel increases with t as a linear function of t that shows up as the straight line of slope g in the left panel. The area of the shaded triangle in the left panel also increases with t but does so as a quadratic function $gt^2/2$ that we recognize as the equation of a parabola.

We can think of the *velocity as the "area function" of acceleration* and of *position as the "area function" of velocity*.

---

[2] In the arithmetic calculation, we didn't need to *assume* that the approximation error goes to zero. This observation is important because it proves (rigorously) that our intuitive assumption is indeed valid: the error does vanish as $N \to \infty$.

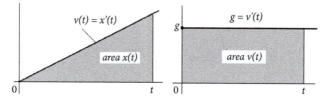

Fig. 7.3 Velocity and position as area functions

Thus, *area* is a geometric representation of "backward" derivatives, better known as *antiderivatives*. Just as the slope of the tangent line gives the rate of change of a function, the function itself gives the rate of change of the area under its graph.

We also need to comment about *signs*. The term "area" makes sense only as a positive quantity; and to ensure this, we may stipulate that the function is non-negative. But this is unnecessarily restrictive; for example, both velocity and acceleration functions meaningfully take on negative values, as in an object that is slowing down (negatively accelerating, or decelerating) or an object that is moving back and forth (velocity changes sign).

We might consider defining "negative area" to allow for a greater range of functions; but rather than doing that, we use a new word: the *integral*. The position function $x(t)$ is an integral of the velocity $v(t)$, and the velocity is the integral of acceleration.

7.2 Riemann integration and integrable functions

In this section, we define the concept of integral. The trio of steps mentioned in the last section are the main pillars on which the idea of integral rests, so we discuss each in a separate subsection.

7.2.1 Step 1: Interval partitions

Consider an interval $[a, b]$ of real numbers and a finite set of numbers $P = \{x_0, x_1, x_2, \ldots, x_N\}$ in this interval (not necessarily equally spaced) such that

$$a = x_0 < x_1 < x_2 < \cdots < x_{N-1} < x_N = b$$

Partition The set P of $N + 1$ points above is a *partition* of $[a, b]$. If the points in P are equally spaced, then P is a *regular partition*.

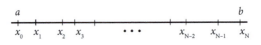

Fig. 7.4 A partition of the interval $[a, b]$

Figure 7.4 illustrates a partition visually.

There are practical as well as theoretical reasons for allowing the partition points to be arbitrarily placed between a and b. I will discuss the theory below, but for a practical reason, consider that if we were using rectangular strips to estimate the area under the graph of a function that changes slowly over a part of $[a, b]$ and rapidly over another, as in Figure 7.5; then we might save some time and effort by taking fewer subintervals under the flat portion and more under the steep portion so as to get roughly the same errors (shaded) per strip in approximating the area without doing unnecessary calculations.

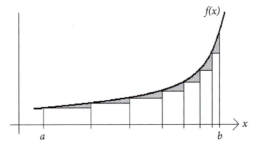

Fig. 7.5 A non-regular partition may be more efficient

Subinterval The set of all real numbers in between each pair of numbers in a partition is a *subinterval* of $[a, b]$.

The $N + 1$ points of a partition P generate N subintervals:

$$[x_0, x_1], [x_1, x_2], ..., [x_{N-1}, x_N]$$

These intervals do not overlap except at their endpoints; we allow their lengths to be different in general, so the ith subinterval $[x_{i-1}, x_i]$ has length

$$x_i - x_{i-1} = \Delta x_i \quad i = 1, 2, ..., N$$

The partition mesh The length of the *largest* of subinterval is often called the the *mesh (or norm) of the partition*, which is commonly denoted by $\|P\|$. If the partition is regular, then

$$\|P\| = \frac{b - a}{N} \quad \text{(mesh of a regular partition)}$$

The above fraction is evidently the minimum possible value for the mesh of any partition with N subinterval given that the total must be the length $b - a$ of the whole interval:

$$\sum_{i=1}^{N} \Delta x_i = (x_1 - x_0) + (x_2 - x_1) + \cdots + (x_{N-1} - x_{N-2}) + (x_N - x_{N-1})$$

$$= x_N - x_0$$

Refinements By adding one or more partition points to an existing partition P of $[a, b]$, we *refine* the partition.

For example, for the interval $[0, 1]$, a refinement of

$$P = \left\{0, \frac{1}{2}, 1\right\} \tag{7.8}$$

is the partition

$$P' = \left\{0, \frac{1}{2}, \frac{2}{3}, \frac{3}{4}, 1\right\}$$

Note that both of these partitions have mesh $1/2$.

Union of partitions If P and P' are two partitions of $[a, b]$, then their *union* $P \cup P'$ is just the ordinary union of the two finite sets P and P'.

If P is a refinement of P' then clearly $P \cup P'$ doesn't add new points to P, so $P \cup P' = P$. Otherwise, $P \cup P'$ refines P and P'. For example, if P is the partition in (7.8), and

$$P' = \left\{ 0, \frac{1}{4}, \frac{3}{4}, 1 \right\}$$

then their union is

$$P \cup P' = \left\{ 0, \frac{1}{4}, \frac{1}{2}, \frac{3}{4}, 1 \right\}$$

This partition refines both P and P'; also $P \cup P'$ is a regular partition in this case with mesh $1/4$, whereas each of P and P' has mesh $1/2$.

7.2.2 Step 2: Riemann sums

Suppose that $f(x)$ is any *bounded* function that is defined over an interval $[a, b]$ and let

$$P = \{x_0, x_1, x_2, \cdots , x_{N-1}, x_N\}$$

be a partition of $[a, b]$.

A *Riemann sum* of $f(x)$ relative to a partition P is the quantity

$$\sum_{i=1}^{N} f(x_i^*)\Delta x_i = f(x_1^*)(x_1 - x_0) + f(x_2^*)(x_2 - x_1) + \cdots + f(x_N^*)(x_N - x_{N-1}) \quad (7.9)$$

where the numbers $x_1^*, x_2^*, ..., x_N^*$ are arbitrarily chosen in the corresponding subintervals of P; more precisely,

$$x_{i-1} \leq x_i^* \leq x_i \quad \text{for } i = 1, 2, ..., N$$

You may recognize that the sums of thin strips under the velocity curve that we discussed earlier were examples of Riemann sums with $x_i^* = x_{i-1}$ for $i = 1, 2, ..., N$ (the left-hand endpoints of the subintervals $[x_{i-1}, x_i]$). We could alternatively choose the right-hand endpoints $x_i^* = x_i$. There are practical and theoretical reasons for being flexible here; from a practical of view, we may find it more useful in some cases to pick points other than the left- or right-hand endpoints. For example, we might pick the *midpoints*

$$x_i^* = \frac{x_{i-1} + x_i}{2}$$

because these often result in smaller errors per partition. For example, consider $f(x) = 2x$ for $0 \leq x \leq 2$, and take the regular partition $P = \{0, 1, 2\}$. To approximate

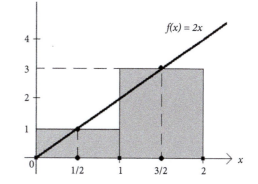

Fig. 7.6 Riemann sum using midpoints

the area under the graph of this $f(x)$, consider the three Riemann sums below for the two subintervals $[0, 1]$ and $[1, 2]$ of P as follows:

Left-hand endpoints: $2(0)(1 - 0) + 2(1)(2 - 1) = 0 + 2 = 2$

Right-hand endpoints: $2(1)(1 - 0) + 2(2)(2 - 1) = 2 + 4 = 6$

Midpoints: $2\left(\dfrac{1}{2}\right)(1 - 0) + 2\left(\dfrac{3}{2}\right)(2 - 1) = 1 + 3 = 4$

The area in question is that of a right triangle with a base of length 2 and a height of length 4, which gives a total area

$$\frac{1}{2}(2)(4) = 4$$

See Figure 7.6. In this case, where the function is a straight line, the midpoints give the exact answer because the overestimation errors (dark triangles above the line) cancelled the underestimation errors (clear triangles below the line).

Upper and lower Riemann sums

Allowing x_i^* to be flexible is a useful option to have when we try to answer a question like the following about approximation of area:

Given any function $f(x)$ and any partition P, can we choose the x_i^* so that the Riemann sum underestimates (or overestimates) the area?

If $f(x)$ is *continuous*, then there is a satisfying answer to the above question, thanks to Theorem 198 (the extreme value theorem):

For a continuous function, in every partition of the interval, there is always a set of x_i^* that give an underestimating Riemann sum and another set of x_i^* that give an overestimating Riemann sum.

It is not hard to see why. If $f(x)$ is continuous on the interval $[a, b]$, then it is continuous at every point of this interval. In particular, $f(x)$ is continuous on every subinterval $[x_{i-1}, x_i]$, regardless of how we partition $[a, b]$.

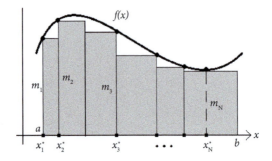

Fig. 7.7 A lower Riemann sum

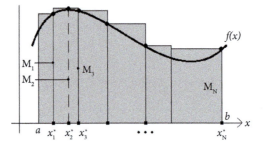

Fig. 7.8 An upper Riemann sum

Apply Theorem 198 to pick a point x_i^* in $[x_{i-1}, x_i]$ for each index i to be where $f(x)$ achieves its *minimum* value m_i. Then the *Lower Riemann sum*

$$R_L(P) = \sum_{i=1}^{N} f(x_i^*)\Delta x_i = \sum_{i=1}^{N} m_i \Delta x_i$$

underestimates the area; see Figure 7.7 where $R_L(P)$ is shaded.

Similarly, by Theorem 198, there is x_i^* where $f(x)$ has its *maximum* value M_i in $[x_{i-1}, x_i]$ so that the *Upper Riemann sum*

$$R_U(P) = \sum_{i=1}^{N} f(x_i^*)\Delta x_i = \sum_{i=1}^{N} M_i \Delta x_i$$

overestimates the area; see Figure 7.8 where $R_U(P)$ is shaded.

In practical terms, the upper and lower Riemann sums $R_L(P)$ and $R_U(P)$ are rarely used for approximating areas. Their significance lies in the properties that make them useful for defining the integral.

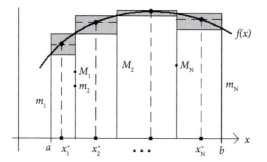

Fig. 7.9 A generic Riemann sum together with the upper and lower sums

Next, note that since $m_i \leq f(x) \leq M_i$ for $x \in [x_{i-1}, x_i]$ and $i = 1, 2, ..., N$, *every Riemann sum relative to the same partition P is between $R_L(P)$ and $R_U(P)$*:

$$R_L(P) \leq \sum_{i=1}^{N} f(x_i^*)\Delta x_i \leq R_U(P) \tag{7.10}$$

regardless of where in $[x_{i-1}, x_i]$ we choose x_i^*, for every $i = 1, 2, ..., N$. Figure 7.9 illustrates these inequalities: the areas of strips with dashed tops are the terms of the Riemann sum; their areas, individually and collectively, are larger than the areas of the strips with heights m_1, m_2, etc. (terms of the lower sum) but smaller than the areas of the strips with heights M_1, M_2, etc. (terms of the upper sum).

These lower and upper sums also have nice properties that make them suitable for defining limits, which we turn to next to complete our definition of the integral.

7.2.3 Step 3: Taking limits (invoking infinity)

You may recall that the extreme sums $R_L(P)$ and $R_U(P)$ were defined for continuous functions where we could use the extreme value theorem. But as long as the function $f(x)$ is bounded, we can define an upper and a lower sum regardless of whether $f(x)$ is continuous or not.

Upper and lower Darboux sums

Let $f(x)$ be a bounded function on the interval $[a, b]$, and P be a partition of $[a, b]$. The *lower sum* $L(P)$ and the *upper sum* $U(P)$ of $f(x)$ relative to P are defined as

$$L(P) = \sum_{i=1}^{N} m_i\Delta x_i \qquad U(P) = \sum_{i=1}^{N} M_i\Delta x_i$$

continued

continued

---

where m_i and M_i are the minimum and maximum values of $f(x)$ on the subinterval $[x_{i-1}, x_i]$ for $i = 1, 2, ..., N$, i.e.,

$$m_i \leq f(x) \leq M_i \quad \text{for all } x \text{ in } [x_{i-1}, x_i]$$

It is a subtle fact that $L(P)$ and $U(P)$ are not necessarily Riemann sums because we cannot generally write them in the form (7.9) unless $f(x)$ is continuous. But they still satisfy the analog of (7.10); that is, regardless of the choice of x_i^*

$$L(P) \leq \sum_{i=1}^{N} f(x_i^*)\Delta x_i \leq U(P) \tag{7.11}$$

because $m_i \leq f(x) \leq M_i$ for every $i = 1, 2, ..., N$; see Figure 7.9. Therefore, for each partition P, every Riemann sum is between $L(P)$ and $U(P)$; and of course, thanks to the extreme value theorem

---

If $f(x)$ **is continuous, then** $L(P) = R_L(P)$ **and** $U(P) = R_U(P)$.

The lower and upper sums have the following important property.

Theorem 243 *(Monotone property of Darboux sums) Suppose that P and P' are partitions of an interval $[a, b]$. If $P \supset P'$ (i.e., P refines P'), then*

$$L(P') \leq L(P) \leq U(P) \leq U(P') \tag{7.12}$$

Proof Every partition point in P' is also in P, so each subinterval $[x_{j-1}, x_j]$ of P' is a union of one or more subintervals of P:

$$[x_{j-1}, x_j] = [x_{i-1}, x_i] \cup [x_i, x_{i+1}] \cup \cdots \cup [x_{i+k-1}, x_{i+k}]$$

Further, the subintervals do not overlap except at their endpoints, and we have

$$\Delta x_i + \Delta x_{i+1} + \cdots + \Delta x_{i+k} = x_{i+k} - x_{i-1} = x_j - x_{j-1} = \Delta x_j$$

Looking at the lower sum relative to P'

$$L(P') = \sum_{j=1}^{N'} m_j'\Delta x_j$$

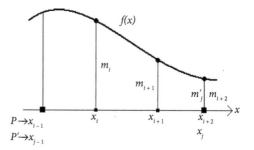

Fig. 7.10 Illustrating the
monotone property: lower sums

we see that for each term j

$$m_j'\Delta x_j = m_j'\Delta x_i + m_j'\Delta x_{i+1} + \cdots + m_j'\Delta x_{i+k}$$

$$\leq m_i\Delta x_i + m_{i+1}\Delta x_{i+1} + \cdots + m_{i+k}\Delta x_{i+k} \tag{7.13}$$

This is true because the lowest value of $f(x)$ on the set $[x_{j-1}, x_j]$ is the minimum of the lowest values of $f(x)$ on each of the subsets $[x_{i-1}, x_i], \ldots [x_{i+k-1}, x_{i+k}]$. Since each term of $L(P')$ consists of a sum of terms of $L(P)$ as in (7.13), it follows that $L(P') \leq L(P)$, as claimed in (7.12).

For the upper sums, we likewise have

$$U(P') = \sum_{j=1}^{N'} M_j'\Delta x_j$$

where for each term j

$$M_j'\Delta x_j \geq M_i\Delta x_i + M_{i+1}\Delta x_{i+1} + \cdots + M_{i+k}\Delta x_{i+k}$$

because the largest value of $f(x)$ on the set $[x_{j-1}, x_j]$ is the maximum of the largest values of $f(x)$ on each of the subsets $[x_{i-1}, x_i], \ldots, [x_{i+k-1}, x_{i+k}]$. It follows that $U(P') \geq U(P)$. This completes the proof of (7.12) because $L(P) \leq U(P)$ for every partition P by (7.11). ∎

The inequalities in (7.11) and (7.12) ensure that *by refining a partition, we narrow the range of variation of the Riemann sums.* Figure 7.10 illustrates the situation for the lower sums ($k = 2$ in the figure).

Figure 7.11 illustrates the case for the upper Darboux sums.

It is worth mentioning that there is a maximum possible value for all the upper sums and a minimum possible value for all the lower sums. Since $f(x)$ is bounded,

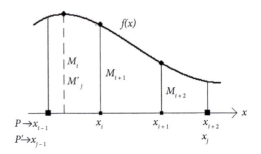

Fig. 7.11 Illustrating the
monotone property: upper sums

there are numbers m and M such that

$$m \leq f(x) \leq M \quad \text{for all } x \text{ in } [a, b]$$

Clearly, $m \leq m_i$ and $M \geq M_i$ for all i, so

$$U(P) = \sum_{i=1}^{N} M_i \Delta x_i \leq \sum_{i=1}^{N} M \Delta x_i = M \sum_{i=1}^{N} \Delta x_i = M(b - a)$$

$$L(P) = \sum_{i=1}^{N} m_i \Delta x_i \geq \sum_{i=1}^{N} m \Delta x_i = m \sum_{i=1}^{N} \Delta x_i = m(b - a)$$

To summarize,

$$m(b - a) \leq L(P) \leq U(P) \leq M(b - a) \tag{7.14}$$

We may identify $m(b - a)$ and $M(b - a)$ as the lower and upper sums of the *coarsest* possible partition of $[a, b]$, namely, the two-point partition $\{a, b\}$:

$$L(\{a, b\}) = m(b - a) \quad \text{and} \quad U(\{a, b\}) = M(b - a)$$

The following corollary of Theorem 243 is another important step toward the definition of the integral.

Corollary 244 *Let $f(x)$ be a bounded function on the interval $[a, b]$. For every pair of partitions P and P' of $[a, b]$ (not necessarily refinements of each other), it is true that*

$$L(P) \leq U(P') \tag{7.15}$$

Proof Define the partition $P'' = P \cup P'$ so that P'' is a refinement of both P and P'. Then by Theorem 243

$$L(P) \leq L(P'') \leq U(P'') \leq U(P')$$

which shows that (7.15) is true.   ∎

Next, let $f(x)$ be a (bounded) function on an interval $[a, b]$ and consider all possible partitions of $[a, b]$. Suppose that we collect all the possible lower sums and upper sums for *every partition P* and define the following:

$$\mathcal{L} = \{L(P) : P \text{ is a partition of } [a, b]\}$$
$$\mathcal{U} = \{U(P) : P \text{ is a partition of } [a, b]\}$$

Note that $\mathcal{U}$ and $\mathcal{L}$ are sets of real numbers.

By (7.15), every number in $\mathcal{L}$ is bounded above by some (any) number in $\mathcal{U}$, so $\mathcal{L}$ has a least upper bound, or supremum, which we denote by a real number I_L. Similarly, every number in $\mathcal{U}$ is bounded below by some (any) number in $\mathcal{L}$, so $\mathcal{U}$ has a greatest lower bound, or infimum, which we denote by the real number I_U. The two numbers I_L and I_U are given special names, and the reason for these names becomes clear soon.

Lower and upper integrals

The number $I_L(f)$ is *the lower integral* of $f(x)$ and $I_U(f)$ *the upper integral.* For reference

$$I_L(f) = \sup(\mathcal{L}) = \sup\{L(P) : P \text{ is a partition of } [a, b]\}$$
$$I_U(f) = \inf(\mathcal{L}) = \inf\{U(P) : P \text{ is a partition of } [a, b]\}$$

We may suppress a mention of f and simply write I_U or I_L where confusion is unlikely.

The next result verifies an expected inequality.

Lemma 245 For every bounded function f on the interval $[a, b]$, both $I_L(f)$ and $I_U(f)$ exist and

$$I_L(f) \leq I_U(f) \qquad\qquad (7.16)$$

Proof The existence of I_L and I_U is guaranteed by completeness. To prove the inequality, note that there are partitions P_n (e.g., refining a given partition) such that the *increasing* sequence of numbers $L(P_n)$ in $\mathcal{L}$ approaches I_L; and likewise, there are partitions P_n' such that the *decreasing* sequence of numbers $U(P_n')$ in $\mathcal{U}$ approaches I_U.

Now, for every fixed positive integer m, (7.15) implies that

$$U(P'_m) \geq L(P_n) \quad \text{for all } n$$

Therefore, $U(P'_m)$ is an upper bound for all $L(P_n)$, and as such, it can be no less that the *least* upper bound I_L; that is,

$$U(P'_m) \geq I_L \quad \text{for all } m$$

It follows that I_L is a lower bound for $U(P'_m)$, and as such, it can be no greater that the *greatest* lower bound I_U. Therefore, $I_U \geq I_L$, and (7.16) is true.   ■

We are finally ready to define the integral. Since the existence of the numbers I_L and I_U requires the completeness of the set of real numbers, this definition makes rather explicit the role of completeness (and thus of infinity) in the concept of the integral.

The integral

A bounded function $f(x)$ is *(Riemann) integrable on the interval* $[a, b]$ if its lower and upper integrals are equal: $I_L(f) = I_U(f)$. This common value is the *(Riemann) integral of $f(x)$* and is denoted

$$\int_a^b f(x)\,dx$$

If $I_L(f) < I_U(f)$, then $f(x)$ is not (Riemann) integrable, i.e., its integral does not exist. If $\int_a^b f(x)\,dx$ exists, then we define

$$\int_b^a f(x)\,dx = -\int_a^b f(x)\,dx$$

Figure 7.12 illustrates the above definition.

The definition of integral above will be consistent with (7.14) if

$$\int_a^a f(x)\,dx = 0$$

which is what we get if we set $b = a$ or let b approach a in (7.14). Alternatively, we may formally extend the concept of partition slightly so that the only partition of

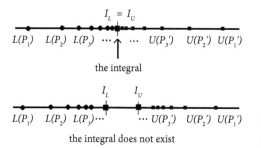

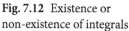

Fig. 7.12 Existence or
non-existence of integrals

the degenerate, point interval $[a, a] = \{a\}$, is $P_0 = \{a\}$, which has a single "subinterval" of length $\Delta x = 0$. Then $L(P_0) = 0 = R(P_0)$; and since P_0 is the only possible
partition, $I_L = I_U = 0$.

Darboux's approach to the development of the Riemann integral is equivalent
to Riemann's original definition. The basic idea is simple: the length of each subin-
terval $[x_{i-1}, x_i]$ of P is no greater than the mesh $\|P\|$, so $\|P\| \to 0$ simply means
that we are adding more in-between points and refining P over and over. Then
Theorem 243 implies that the upper sums move down and approach I_U, while the
lower sums move up and approach I_L.

Naturally, the upper and lower sums squeeze all the Riemann sums that are
sandwiched between them. So if $f(x)$ is integrable as we defined above, then the
confined Riemann sums have nowhere to go but approach the common value
$\int_a^b f(x)dx$ of I_L and I_U. In this precise sense, we can say that the Riemann sums
approach the integral as $\|P\| \to 0$, as in typical calculus textbooks.

The definition of integral is rarely used in practice to calculate integrals since
it is difficult to work with even with simple functions. We will soon show that
are many functions that are integrable in the sense defined above (including all
continuous functions), but as a quick illustration of the definition of integral, we
find the integral of a simple *point function*:

$$f(x) = 0 \quad \text{if } 0 \leq x \leq 2, x \neq 1$$
$$f(1) = d \quad d > 0$$

If P is any partition of $[0, 2]$, then there is a subinterval $[x_{j-1}, x_j]$ that contains 1.
But every subinterval also contains points other than 1, so the smallest values m_i
of $f(x)$ are all zeros for every i, including $i = j$. See Figure 7.13.

On the other hand, although the largest values M_i of $f(x)$ are zeros for $i \neq j$, it is
true that $M_j = d > 0$. Therefore, for every partition P, there is j such that

$$L(P) = 0, \quad U(P) = d(x_j - x_{j-1})$$

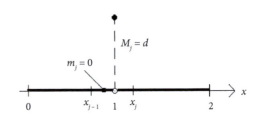

Fig. 7.13 A point function

Since P is an arbitrary partition, we conclude that $I_L = 0$, and

$$0 \le I_U \le U(P) \le d\,\|P\|$$

So by choosing a sequence of partitions P_k whose norms go to zero (i.e., letting $\|P\| \to 0$), we conclude that $U(P_k)$ goes to 0 and squeezes I_U to 0 too. Therefore, the point function f is integrable, and $\int_0^2 f(x)dx = 0$.

Intuitively, if we think of the point function as a strip of zero width, then we expect that such a "strip" will have zero area too. But of course, there is no such thing as a "zero-width strip"; the technical machinery developed above is designed to deal with the thorny issues about infinity that intuition conveniently glosses over.

We now define an important class of bounded functions.

The class of Riemann integrable functions The set of all Riemann integrable functions on an interval $[a, b]$ is denoted by $R[a, b]$.

We have seen that the point function is Riemann integrable on any interval. Later, we also prove that all continuous functions on closed intervals as well as many types of discontinuous functions are Riemann integrable. To help us do that, we now state and prove a necessary and sufficient condition for integrability, whose original version was proved by Riemann.

Theorem 246 *(Riemann integrability criterion) A bounded function $f: [a, b] \to \mathbb{R}$ is (Riemann) integrable if and only if for every $\varepsilon > 0$, there is a partition P of $[a, b]$ such that*

$$U(P) - L(P) < \varepsilon \qquad (7.17)$$

Further, if (7.17) holds for some partition, then it holds for all refinements of that partition.

Proof First, assume that for every $\varepsilon > 0$, there is a partition P that satisfies (7.17). Since $L(P) \leq I_L \leq I_U \leq U(P)$, it follows that

$$0 \leq I_U - I_L \leq U(P) - L(P) < \varepsilon$$

Since ε may be arbitrarily close to 0, the fixed real number $I_U - I_L$ must be 0. Therefore, f is integrable.

Conversely, suppose that f is integrable, and let $\varepsilon > 0$ be arbitrary. Then we can find partitions P_1 and P_2 of $[a, b]$ such that

$$\int_a^b f(x)dx - L(P_1) < \frac{\varepsilon}{2}$$

$$U(P_2) - \int_a^b f(x)dx < \frac{\varepsilon}{2}$$

Let $P = P_1 \cup P_2$, and notice that

$$U(P) \leq U(P_2) < \int_a^b f(x)dx + \frac{\varepsilon}{2} < L(P_1) + \varepsilon \leq L(P) + \varepsilon$$

Therefore, $U(P) - L(P) < \varepsilon$, and this partition P satisfies (7.17).

Finally, if P satisfies (7.17) for some $\varepsilon > 0$, and P' is a refinement of P, then

$$U(P') - L(P') \leq U(P) - L(P) < \varepsilon$$

and (7.17) holds for P' too.   ∎

The above theorem can be stated in sequential way that may make it easier to understand. In particular, it yields an expression of the integral as a limit of a sequence of real numbers. These features are closer to what we encounter in calculus, where we are often introduced to the integral as the "limit of a Riemann sum corresponding to a sequence of (usually regular) partitions whose mesh approach 0."

Theorem 247 *(Riemann integrability criterion, sequential) A bounded function $f : [a, b] \rightarrow \mathbb{R}$ is (Riemann) integrable if and only if there is a sequence P_n of* *partitions of $[a, b]$ such that*

$$\lim_{n \to \infty} [U(P_n) - L(P_n)] = 0 \tag{7.18}$$

Further, if P_n is any sequence of partitions that satisfies (7.18), then

$$\lim_{n \to \infty} L(P_n) = \int_a^b f(x)dx \quad and \quad \lim_{n \to \infty} U(P_n) = \int_a^b f(x)dx \tag{7.19}$$

Proof First, suppose that f is integrable on $[a, b]$, and let $\varepsilon = 1/n$. For each $n = 1, 2, 3, \ldots$ Theorem 246 guarantees the existence of a partition P_n that satisfies (7.17) for each n. Define $s_n = U(P_n) - L(P_n)$, and note that by (7.17)

$$0 \leq s_n < \frac{1}{n}$$

Since the above inequalities hold for every n, the squeeze theorem implies $\lim_{n \to \infty} s_n = 0$, i.e., (7.18) holds.

Conversely, if there is a sequence of partitions P_n that satisfies (7.18), then for arbitrary $\varepsilon > 0$ we can find n large enough that $1/n < \varepsilon$ (recall the Archimedean property of real numbers). Then $P = P_n$ is the partition that satisfies (7.17) for the given ε; so by Theorem 246, f is integrable on $[a, b]$.

It remains to prove that the equalities in (7.19) hold for an integrable f. Note that

$$L(P_n) \leq I_L = \int_a^b f(x)dx = I_U \leq U(P_n) \tag{7.20}$$

from which we conclude that

$$\int_a^b f(x)dx - L(P_n) \leq U(P_n) - L(P_n)$$

Taking the limit as $n \to \infty$ and using (7.18) proves the first equality in (7.19). The second follows similarly from (7.18) and (7.20). ∎

The limits in (7.19) are analogous to limits of sums of rectangular strips that either underestimate the area under the graph of f or overestimate it. For instance, suppose that f is a nondecreasing function. If P_n is a regular partition, then

$$L(P_n) = \sum_{i=1}^{N} f(x_{i-1})\Delta x = \frac{b-a}{N} \sum_{i=1}^{N} f(x_{i-1}) = \|P_n\| \sum_{i=1}^{N} f(x_{i-1})$$

$$U(P_n) = \sum_{i=1}^{N} f(x_i)\Delta x = \frac{b-a}{N} \sum_{i=1}^{N} f(x_i) = \|P_n\| \sum_{i=1}^{N} f(x_i)$$

These expressions are closer to what we come across in calculus texts, and Theorem 247 shows that the usually imprecise statements that we find in those texts are justified by our results here.

To illustrate, suppose that we want to calculate the integral of

$$f(x) = 2x - 3$$

on the interval $[0, 1]$. We can partition the interval $[0, 1]$ into N subintervals of equal size, $\Delta x = 1/N$, to get the partition

$$P_N = \left\{0, \frac{1}{N}, \frac{2}{N}, \ldots, \frac{N-1}{N}, 1\right\}$$

and set up the upper sum (which is a Riemann sum in this case)

$$U(P_N) = \sum_{i=1}^{N} f(x_i)\Delta x = \frac{1}{N} \sum_{i=1}^{N} (2x_i - 3)$$

Since

$$2x_i - 3 = \frac{2i}{N} - 3$$

we obtain

$$U(P_N) = \frac{1}{N}\left(\frac{2}{N} \sum_{i=1}^{N} i - 3N\right) = \frac{2}{N^2} \sum_{i=1}^{N} i - 3$$

While discussing Galileo's equations earlier, we found that

$$\sum_{i=1}^{N} i = \frac{N(N+1)}{2}$$

Therefore,

$$U(P_n) = \frac{N+1}{N} - 3$$

Taking the limit as $N \to \infty$, we obtain from (7.19)

$$\int_0^1 (2x - 3)dx = \lim_{N \to \infty} U(P_N) = 1 - 3 = -2$$

This method of calculating the integral grows in difficulty rapidly if we apply it to more complex functions. The fundamental theorem of calculus that we discuss below helps us calculate integrals more efficiently using antiderivatives.

7.2.4 Integrability of continuous functions and monotone functions

What types of functions are Riemann integrable, i.e., satisfy Theorem 246? In this section, we show that the class $R[a, b]$ of integrable functions includes familiar functions, including continuous functions and monotone functions (continuous or not).

Continuous functions are integrable

Our first result shows that all continuous functions are integrable. Although this result doesn't tell us how to integrate continuous functions, it does show that there is a large supply of integrable functions.

Theorem 248 *If $f(x)$ is continuous on the closed interval $[a, b]$, then it is Riemann integrable on $[a, b]$.*

Proof Note that f is bounded (Lemma 197 in Chapter 6), and for every partition $P = \{x_0, \ldots, x_N\}$ of $[a, b]$

$$U(P) - L(P) = \sum_{i=1}^{N}(M_i - m_i)\Delta x_i$$

where M_i is the maximum value of $f(x)$ on the ith subinterval $[x_{i-1}, x_i]$, and m_i is its minimum value.

Let $\varepsilon > 0$. Note that $f(x)$ is uniformly continuous on the closed and bounded interval $[a, b]$ (Theorem 200, Chapter 6). Therefore, there is $\delta > 0$ that satisfies the ε and δ version (6.38):

$$|f(x) - f(y)| < \frac{\varepsilon}{b - a} \quad \text{for all } x, y \in [a, b] \text{ such that } |x - y| < \delta \qquad (7.21)$$

Now since f is continuous, by the extreme value property, there are $\alpha_i, \beta_i \in [x_{i-1}, x_i]$ such that

$$m_i = f(\alpha_i), \qquad M_i = f(\beta_i)$$

so by (7.21), there is $\delta > 0$ such that

$$M_i - m_i < \frac{\varepsilon}{b - a} \quad \text{for all } i = 1, \ldots, N$$

if the partition P is chosen such that $\|P\| < \delta$. For such a P

$$U(P) - L(P) < \sum_{i=1}^{N} \frac{\varepsilon}{b - a} \Delta x_i = \frac{\varepsilon}{b - a}(b - a) = \varepsilon$$

Therefore, by Theorem 246, f is integrable on $[a, b]$. ∎

Theorem 248 implies that $R[a, b]$ is large enough to contain all continuous functions on $[a, b]$, which is significant. But notice that Theorem 248 doesn't tell us *how to find the integral* of a continuous function. We discuss that issue in the section on the fundamental theorem of calculus.

Monotone functions are integrable

We now consider monotone functions, which are functions that are either nondecreasing or nonincreasing. We should emphasize that a monotone function need not be continuous and can, in fact, have infinitely many points of discontinuity; we discuss an example after the next theorem. Nevertheless, all monotone functions on a closed interval are integrable, as we now prove.

Theorem 249 *If $f(x)$ is monotone on the closed interval $[a, b]$, then it is Riemann integrable on $[a, b]$.*

Proof We assume that $f(x)$ is nondecreasing (the nonincreasing case is proved similarly; see Exercise 278). Since $f(a) \leq f(x) \leq f(b)$ for all $x \in [a, b]$, it follows that f is bounded. Let $\varepsilon > 0$.

For each partition $P = \{x_0, x_1, \cdots, x_N\}$ of $[a, b]$, the minimum m_i and maximum M_i of $f(x)$ on each subinterval $[x_{i-1}, x_i]$ occur at an endpoint:

$$m_i = f(x_{i-1}), \qquad M_i = f(x_i)$$

Hence,

$$U(P) - L(P) = \sum_{i=1}^{N} [f(x_i) - f(x_{i-1})] \Delta x_i$$

Recall that $\Delta x_i = x_i - x_{i-1} \leq \|P\|$ for every index i, so we have

$$U(P) - L(P) \leq \sum_{i=1}^{N} [f(x_i) - f(x_{i-1})] \|P\| = \|P\| \sum_{i=1}^{N} [f(x_i) - f(x_{i-1})]$$

The last sum telescopes down to a single difference:

$$\sum_{i=1}^{N}[f(x_i) - f(x_{i-1})] = f(x_N) - f(x_0) = f(b) - f(a)$$

Therefore,

$$U(P) - L(P) \le \|P\| \, [f(b) - f(a)].$$

We may choose a partition P so that $\|P\| \, [f(b) - f(a)] < \varepsilon$, then Theorem 246 implies that f is Riemann integrable on $[a, b]$.   ∎

To illustrate Theorem 249, consider the function below that is defined on the interval $[0, 1]$:

$$f(x) = \frac{1}{2^{n-1}} \quad \text{for} \quad \frac{1}{2^n} < x \le \frac{1}{2^{n-1}}, \quad f(0) = 0 \tag{7.22}$$

This is a nondecreasing function with infinitely many points of discontinuity; a rough sketch is shown in Figure 7.14. By Theorem 249, this function is Riemann integrable.

The integral is the total area of all the strips in Figure 7.14. The area of the largest rectangular strip is "base times height" or $(1/2)(1)$. Similarly, the strip next to it has area

$$\left(\frac{1}{2} - \frac{1}{2^2}\right)\frac{1}{2} = \left(\frac{1}{2^2}\right)\frac{1}{2} = \frac{1}{2^3}$$

and so on; adding the areas of all the strips gives the value of the integral as an infinite series:

$$\int_0^1 f(x)dx = \frac{1}{2} + \frac{1}{2^3} + \frac{1}{2^5} + \cdots$$

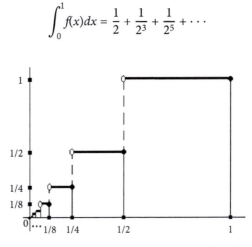

Fig. 7.14 An increasing function with infinitely many discontinuities

The infinite series on the right-hand side is geometric with first term $1/2$ and common ratio $1/2^2 = 1/4$, so we calculate its sum as

$$\int_0^1 f(x)dx = \frac{1/2}{1-1/4} = \frac{2}{3}$$

7.2.5 A non-Riemann integrable function

Given the preceding discussion, a bounded function that is not Riemann integrable can be neither continuous nor monotone; in short, it is pathological in nature. Dirichlet's function $D(x)$ that we defined in Chapter 6 as an example of a nowhere continuous function is also the standard example of a non-integrable function.[3] We first restrict $D(x)$ to a closed interval $[a, b]$:

$$D_{[a,b]}(x) = \begin{cases} 1, & \text{if } x \in [a,b] \text{ and } x \text{ is rational} \\ 0, & \text{if } x \in [a,b] \text{ and } x \text{ is irrational} \end{cases}$$

Let P be any partition of $[a, b]$. Since rational numbers are dense, every subinterval of P contains a rational number q where $D_{[a,b]}(q) = 1$, that is, $M_i = 1$ for all i. Similarly, the irrational numbers are dense too, so every subinterval contains an irrational number r where $D_{[a,b]}(r) = 0$; therefore, $m_i = 0$ for all i. It follows that

$$L(P) = \sum_{i=1}^{N} m_i \Delta x_i = 0$$

$$U(P) = \sum_{i=1}^{N} M_i \Delta x_i = \sum_{i=1}^{N} \Delta x_i = x_N - x_0 = b - a$$

Since the above two values are the same for every partition of $[a, b]$, it follows that the sets $\mathcal{L}$ and $\mathcal{U}$ each contain just one number

$$\mathcal{L} = \{0\}, \quad \mathcal{U} = \{b - a\}$$

Therefore, $I_L = 0$ and $I_U = b - a$; these two numbers not being equal, we conclude that $D_{[a,b]}(x)$ is not Riemann integrable.

---

[3] The Dirichlet function may be integrable via a different integral. For example, it is *Lebesgue integrable*, and its integral over the entire set $\mathbb{R}$ is zero. The reason it is Lebesgue integrable is basically that the set of rational numbers is so small compared to the set of all real numbers that it has *measure zero*; whatever a function does on such a set contributes nothing to the (Lebesgue) integral.

7.3 Properties of the integral

In this section, we state and prove the basic operational properties of the Riemann integral that was defined in the last section. These statements also provide information about integrable functions. Recall that the set of all Riemann integrable functions is denoted $R[a, b]$, so to say that f is integrable is equivalent to stating $f \in R[a, b]$.

Theorem 250 *(Linear property) Let $f, g \in R[a, b]$. Then*

(a) $f + g \in R[a, b]$ *and*

$$\int_a^b (f + g)(x)dx = \int_a^b f(x)dx + \int_a^b g(x)dx$$

(b) *If c is a fixed real number, then $cf \in R[a, b]$, and*

$$\int_a^b cf(x)dx = c \int_a^b f(x)dx$$

Proof (a) Let $P = \{x_0, x_1, \ldots, x_N\}$ be an arbitrary partition of $[a, b]$. Let

$$M_i = \sup\{f(x) : x \in [x_{i-1}, x_i]\}$$
$$M_i' = \sup\{g(x) : x \in [x_{i-1}, x_i]\}$$

For all $x \in [x_{i-1}, x_i]$

$$f(x) + g(x) \le M_i + M_i'$$

so

$$\sup\{f(x) + g(x) : x \in [x_{i-1}, x_i]\} \le M_i + M_i'$$

It follows that

$$U(P, f + g) \le \sum_{i=1}^N (M_i + M_i')\Delta x_i = U(P, f) + U(P, g) \tag{7.23}$$

Since f and g are integrable, Theorem 246 implies that for arbitrary $\varepsilon > 0$, there are partitions P', P'' of $[a, b]$ such that

$$U(P', f) < \int_a^b f(x)dx + \frac{\varepsilon}{2} \quad \text{and} \quad U(P'', g) < \int_a^b g(x)dx + \frac{\varepsilon}{2}$$

If we choose $P = P' \cup P''$, then by (7.23)

$$U(P, f + g) < \int_a^b f(x)dx + \int_a^b g(x)dx + \varepsilon$$

Therefore,

$$I_U(f + g) < \int_a^b f(x)dx + \int_a^b g(x)dx + \varepsilon$$

for every $\varepsilon > 0$; and we conclude, by letting ε go to 0, that

$$I_U(f + g) \leq \int_a^b f(x)dx + \int_a^b g(x)dx \qquad (7.24)$$

A similar argument (left as Exercise 279) gives

$$I_L(f + g) \geq \int_a^b f(x)dx + \int_a^b g(x)dx \qquad (7.25)$$

Now (7.24) and (7.25) together with Lemma 245 imply that

$$I_L(f + g) \leq I_U(f + g) \leq \int_a^b f(x)dx + \int_a^b g(x)dx \leq I_L(f + g)$$

It follows that $I_L(f + g) = I_U(f + g)$; so $f + g$ is integrable, and its integral is

$$\int_a^b (f + g)(x)dx = \int_a^b f(x)dx + \int_a^b g(x)dx$$

(b) First let $c \geq 0$. For an arbitrary partition $P = \{x_0, x_1, \ldots, x_N\}$ of $[a, b]$, if $M_i = \sup\{f(x) : x \in [x_{i-1}, x_i]\}$, then $cf(x) \leq cM_i$ for every $x \in [x_{i-1}, x_i]$, so

$$\sup\{cf(x) : x \in [x_{i-1}, x_i]\} \leq cM_i$$

It follows that

$$U(P, cf) \leq cU(P, f)$$

For every $\varepsilon > 0$, we can select the partition P so refined that

$$U(P, f) \leq I_U(f) + \varepsilon = \int_a^b f(x)dx + \varepsilon$$

The equality holds since f is integrable. Therefore,

$$I_U(cf) \leq U(P, cf) < c \int_a^b f(x)dx + c\varepsilon$$

Letting ε go to zero, it follows that

$$I_U(cf) \leq c \int_a^b f(x)dx$$

By a similar argument,

$$I_L(cf) \geq c \int_a^b f(x)dx$$

Since by Lemma 245 $I_L(cf) \leq I_U(cf)$, we conclude that

$$\int_a^b cf(x)dx = I_L(cf) = I_U(cf) = c \int_a^b f(x)dx$$

Next, if $c < 0$, then $c = (-1)|c|$, and since we have already proved (b) for $|c| > 0$, the proof will be complete once we show that (b) is true for $c = -1$. By (a) with $g(x) = -f(x)$, we have

$$\int_a^b f(x)dx + \int_a^b (-f)(x)dx = \int_a^b [f + (-f)](x)dx = 0$$

from which we obtain

$$\int_a^b (-f)(x)dx = -\int_a^b f(x)dx$$

This proves (b) for $c = -1$ and completes the proof. ■

The above theorem establishes two useful facts: first, the set $R[a, b]$ is a vector space (or linear space) since for every pair of functions $f, g \in R[a, b]$ and arbitrary real constants α, β, the linear combination $\alpha f + \beta g$ is integrable and thus in $R[a, b]$; second, the Riemann integral is a linear operator on $R[a, b]$ because

$$\int_a^b (\alpha f + \beta g)(x)dx = \alpha \int_a^b f(x)dx + \beta \int_a^b g(x)dx$$

Theorem 250 can be extended by induction to any finite number of functions $f_1, f_2, ..., f_k \in R[a, b]$ so that $f_1 + f_2 + \cdots + f_k \in R[a, b]$, and

$$\int_a^b (f_1 + \cdots + f_k)(x)dx = \int_a^b f_1(x)dx + \cdots + \int_a^b f_k(x)dx \qquad (7.26)$$

We now prove a useful fact with help of (7.26), and our earlier result that every point function

$$p_c(x) = \begin{cases} r & \text{if } x = c \\ 0 & \text{if } x \neq c \end{cases}$$

is integrable on every closed interval $[a, b]$ that contains c, and its integral is zero regardless of the value of r:

$$\int_a^b p_c(x)dx = 0$$

Corollary 251 *Let $f, g \in R[a, b]$, and assume that $f(x) = g(x)$ for all $x \in [a, b]$ except at k points $c_i \in [a, b]$, where $f(c_i) \neq g(c_i)$ for $i = 1, 2, ..., k$. Then*

$$\int_a^b f(x)dx = \int_a^b g(x)dx$$

Proof For $i = 1, ..., k$, define the point functions $p_i(x)$ with $p_i(c_i) = f(c_i) - g(c_i)$, and $p_i(x) = 0$ if $x \neq c_i$, and note that

$$f(x) = g(x) + \sum_{i=1}^k p_i(x)$$

for all $x \in [a, b]$, including at $x = c_i$ for $i = 1, ..., k$. Since each $p_i(x)$ is integrable, so is their sum by (7.26), which further implies that

$$\int_a^b f(x)dx = \int_a^b g(x)dx + \sum_{i=1}^k \int_a^b p_i(x)dx = \int_a^b g(x)dx \qquad \blacksquare$$

The vector space $R[a, b]$ has an even richer structure because it also contains the product fg of every pair $f, g \in R[a, b]$, thus making it an algebra. To prove this fact, we first prove another theorem that is of interest in its own right and implies the integrability of the product fg as a simple consequence.

Theorem 252 *(Composition property) Let f be a bounded, Riemann integrable function on $[a, b]$ with range contained in an interval $[c, d]$. If φ is continuous on $[c, d]$, then $\varphi \circ f$ is integrable on $[a, b]$.*

Proof Since φ is continuous on $[c, d]$, it is bounded (Lemma 197 in Chapter 6) and uniformly continuous on $[c, d]$ (Theorem 200, Chapter 6). The latter property implies that for every $\varepsilon' > 0$ there is a number $\delta > 0$, such that

$$|\varphi(x) - \varphi(y)| < \varepsilon' \quad \text{for all } x, y \in [c, d] \text{ such that } |x - y| < \delta$$

Since the above inequality holds for all smaller values of δ, we may assume that $\delta < \varepsilon'$. Now, by Theorem 246, there is a partition $P = \{x_0, x_1, ..., x_N\}$ of $[a, b]$ so refined that

$$U(P, f) - L(P, f) < \delta^2 \tag{7.27}$$

To complete the proof, define the infimum and supremum of $\varphi \circ f$ on each subinterval $[x_{i-1}, x_i]$ as

$$m'_i = \inf\{\varphi(f(x)) : x \in [x_{i-1}, x_i]\}$$
$$M'_i = \sup\{\varphi(f(x)) : x \in [x_{i-1}, x_i]\}$$

for $i = 1, 2, ..., N$. Likewise, let m_i and M_i be the infimum and supremum, respectively, of f on each $[x_{i-1}, x_i]$.

Let $\varepsilon > 0$ be arbitrary. Note that

$$U(P, \varphi \circ f) - L(P, \varphi \circ f) = \sum_{i=1}^{N} (M'_i - m'_i) \Delta x_i$$

Since $|f(x) - f(y)| \leq M_i - m_i$ for all $x, y \in [x_{i-1}, x_i]$, it follows that

$$|\varphi(f(u)) - \varphi(f(v))| < \varepsilon'$$

for all indices i with $M_i - m_i < \delta$; let A be the set of all indices in $\{1, 2, ..., N\}$ with this property and B be the set of indices that don't have this property, i.e.,

$M_i - m_i \geq \delta$ for $i \in B$. Then $M'_i - m'_i \leq \varepsilon'$ for all $i \in A$. If $i \in B$, then

$$M'_i - m'_i \leq 2 \sup\{|\varphi(x)| : x \in [c, d]\}$$

Let s be the quantity on the right-hand side of the above inequality. It follows that

$$U(P, \varphi \circ f) - L(P, \varphi \circ f) = \sum_{i \in A}(M'_i - m'_i)\Delta x_i + \sum_{i \in B}(M'_i - m'_i)\Delta x_i$$

$$\leq \varepsilon' \sum_{i \in A} \Delta x_i + s \sum_{i \in B} \Delta x_i$$

$$\leq \varepsilon'(b - a) + s \sum_{i \in B} \Delta x_i$$

Recall that $M_i - m_i \geq \delta$ for $i \in B$; so by (7.27),

$$\sum_{i \in B} \Delta x_i \leq \frac{1}{\delta} \sum_{i \in B}(M_i - m_i)\Delta x_i \leq \frac{1}{\delta}[U(P, f) - L(P, f)] < \delta < \varepsilon'$$

so that

$$U(P, \varphi \circ f) - L(P, \varphi \circ f) < \varepsilon'(b - a) + s\varepsilon' = \varepsilon'(b - a + s)$$

Now we select $\varepsilon' = \varepsilon/(b - a + s)$ to get

$$U(P, \varphi \circ f) - L(P, \varphi \circ f) < \varepsilon$$

Given the arbitrary nature of ε, Theorem 246 implies that $\varphi \circ f$ is integrable on $[a, b]$. ∎

We can now prove that the product of two Riemann integrable functions is Riemann integrable and more.

Corollary 253   (a) If $f \in R[a, b]$, then $|f|, F2 \in R[a, b]$.
(b) If $f, g \in R[a, b]$, then $fg \in R[a, b]$.

Proof (a) The functions $\varphi(x) = |x|$ and $\psi(x) = x^2$ are continuous on $(-\infty, \infty)$, so for every closed interval $[a, b]$ and function $f \in R[a, b]$, Theorem 252 implies that the compositions

$$\varphi \circ f(x) = |f(x)| \quad \text{and} \quad \psi \circ f(x) = [f(x)]^2$$

are integrable on $[a, b]$.

(b) Note that

$$fg = \frac{1}{2}[(f+g)^2 - (f-g)^2]$$

By Theorem 250, $f+g$ and $f-g$ are both integrable; so by (a), both $(f+g)^2$ and $(f-g)^2$ are also integrable. It follows (again by Theorem 250) that fg is integrable on $[a, b]$. ∎

The next result establishes another useful property of the integral, namely, it preserves inequalities.

Theorem 254 *(Monotone property) If $f, g \in R[a, b]$ and $f(x) \leq g(x)$ for all $x \in [a, b]$, then*

$$\int_a^b f(x)dx \leq \int_a^b g(x)dx \tag{7.28}$$

Proof By Theorem 247, there are sequences P_n' and P_n'' of partitions of $[a, b]$ such that

$$\lim_{n\to\infty} U(P_n', f) = \int_a^b f(x)dx \quad \text{and} \quad \lim_{n\to\infty} U(P_n'', g) = \int_a^b g(x)dx$$

Let $P_n = P_n' \cup P_n''$. Then

$$U(P_n', f) \leq U(P_n, f) \leq \int_a^b f(x)dx$$

so that by the squeeze theorem,

$$\lim_{n\to\infty} U(P_n, f) = \int_a^b f(x)dx$$

Similarly,

$$\lim_{n\to\infty} U(P_n, g) = \int_a^b g(x)dx$$

Now,

$$U(P_n, f) = \sum_{i=1}^N M_i' \Delta x_i \leq \sum_{i=1}^N M_i'' \Delta x_i = U(P_n, g)$$

where, since $f(x) \leq g(x)$ for all $x \in [a, b]$,

$$M_i' = \sup\{f(x) : x \in [x_{i-1}, x_i]\} \leq \sup\{g(x) : x \in [x_{i-1}, x_i]\} = M_i''$$

for all i. Now (7.19) and the monotone property of limits (preserving inequalities) implies that

$$\int_a^b f(x)dx = \lim_{n\to\infty} U(P_n,f) \leq \lim_{n\to\infty} U(P_n,g) = \int_a^b g(x)dx \qquad \blacksquare$$

The following property of the integral involving the absolute value is often useful in analysis; in particular, we use it in the proof of the fundamental theorem of calculus below.

Corollary 255 *If $f \in R[a,b]$, then*

$$\left| \int_a^b f(x)dx \right| \leq \int_a^b |f(x)|dx \qquad (7.29)$$

Proof By the definition of absolute value, for every $x \in [a,b]$

$$-|f(x)| \leq f(x) \leq |f(x)|$$

By Corollary 253, $|f(x)|$ is integrable on $[a,b]$, so by Theorem 254

$$-\int_a^b |f(x)|dx \leq \int_a^b f(x)dx \leq \int_a^b |f(x)|dx$$

These inequalities are equivalent to the inequality in (7.29). $\blacksquare$

A subtle note about (7.29) is this: if $a \leq c < d \leq b$, then $\left| \int_d^c f(x)dx \right| \geq 0$ (integrating from d to c now), but $\int_d^c |f(x)|dx \leq 0$; so the inequality in (7.29) doesn't hold as stated. Of course, we do have

$$\left| \int_d^c f(x)dx \right| = \left| \int_c^d f(x)dx \right| \leq \int_c^d |f(x)|dx = -\int_d^c |f(x)|dx$$

We use this observation in the proof of the second half of the fundamental theorem of calculus below.

We close this section with another property of the integral that is familiar from calculus.

Theorem 256 *(Decomposition property) Let $f : [a,b] \to \mathbb{R}$ be a bounded function, and assume that $a < c < b$. Then f is integrable on $[a,b]$ if and only if f is integrable on both $[a,c]$ and $[c,b]$. In this case,*

$$\int_a^b f(x)dx = \int_a^c f(x)dx + \int_c^b f(x)dx \tag{7.30}$$

Proof First we show that the upper integrals of f over the three intervals satisfy the equality

$$I_U[a, b] = I_U[a, c] + I_U[c, b] \tag{7.31}$$

Note if P' and P'' are partitions of $[a, c]$ and $[c, b]$, respectively, then $P = P' \cup P''$ is a partition of $[a, b]$ that also contains the number c. Further, if P is a partition of $[a, b]$, and $c \in P$, then we may decompose $P = P' \cup P''$ where P' and P'' are partitions of $[a, c]$ and $[c, b]$. For these partitions

$$U(P) = U(P') + U(P'') \geq I_U[a, c] + I_U[c, b]$$

since the upper integrals are the infima of the upper sums.

If $\mathcal{P}$ is an arbitrary partition of $[a, b]$, then $\mathcal{P} \cup \{c\}$ is a refinement of $\mathcal{P}$ that contains c. Therefore,

$$U(\mathcal{P}) \geq U(\mathcal{P} \cup \{c\}) \geq I_U[a, c] + I_U[c, b]$$

Taking the infimum over all partitions $\mathcal{P}$ gives the lower integral over $[a, b]$;

$$I_U[a, b] \geq I_U[a, c] + I_U[c, b] \tag{7.32}$$

Next, we prove the reverse inequality is also true. For every $\varepsilon > 0$, there are P' and P'' that are partitions of $[a, c]$ and $[c, b]$, respectively, such that

$$U(P') < I_U[a, c] + \frac{\varepsilon}{2} \quad \text{and} \quad U(P'') < I_U[c, b] + \frac{\varepsilon}{2}$$

Thus, for a partition $P = P' \cup P''$ of $[a, b]$,

$$I_U[a, b] \leq U(P) = U(P') + U(P'') < I_U[a, c] + I_U[c, b] + \varepsilon$$

This inequality being true for all $\varepsilon > 0$, we conclude that

$$I_U[a, b] \leq I_U[a, c] + I_U[c, b]$$

This inequality is the reverse of that in (7.32), so we have now proved (7.31). A similar argument proves the analogous equality for the lower integrals:

$$I_L[a, b] = I_L[a, c] + I_L[c, b] \tag{7.33}$$

Now, if f is integrable on the subintervals $[a, c]$ and $[c, b]$, then by (7.31) and (7.33),

$$\int_a^c f(x)dx + \int_c^b f(x)dx = I_L[a, b] \leq I_U[a, b] = \int_a^c f(x)dx + \int_c^b f(x)dx$$

from which we conclude that $I_L[a, b] = I_U[a, b]$, i.e., f is integrable on $[a, b]$, and its integral satisfies the equality in (7.30).

On the other hand, if f is integrable on $[a, b]$, then $I_L[a, b] = I_U[a, b]$, so by (7.31) and (7.33)

$$I_L[a, c] + I_L[c, b] = I_U[a, c] + I_U[c, b]$$

which can be written as

$$I_L[a, c] - I_U[a, c] = I_U[c, b] - I_L[c, b]$$

Since the left-hand side of the above is non-positive, and the right-hand side is non-negative, it follows that both sides must equal zero. It follows that f is integrable on both $[a, c]$ and $[c, b]$, and again, the equality in (7.30) holds. ∎

As a simple illustration, we can use the decomposition property to write

$$\int_{-1}^1 |x|dx = \int_{-1}^0 (-x)dx + \int_0^1 xdx$$

The integrals on the right-hand side can be quickly calculated (see Exercise 280 or by using the fundamental theorem of calculus below).

A question that arises looking at the statement of Theorem 256 is that since the number c appears twice on the right-hand side of (7.30) but only once on the left, as part of the interval $[a, b]$, isn't there an imbalance? A quick answer is provided by Corollary 251, which shows that the values of a function at individual points of an interval contribute nothing to the value of its integral.

7.4 The Fundamental Theorem of Calculus

In the previous two sections, we defined a Riemann integrable function, its integral, and the fundamental properties of the integral. Now we focus on the relationship between the three important concepts of integrability, continuity, and differentiability. This relationship is encapsulated in the so-called fundamental theorem of calculus, which besides being an interesting result in its own right, also turns out to be quite useful in calculating the integrals of a substantial class of functions via the concept of antiderivative.

Antiderivatives

Since the derivative of x^3 is $3x^2$, we may think of x^3 as a reverse derivative, or an antiderivative of $3x^2$. Of course, $x^3 + 2$ and $x^3 - \sqrt{5}$ are also antiderivatives because when we take their derivatives, the constants drop out, and we always get $3x^2$. More generally

---

Antiderivative Let $f(x)$ be a function on an interval I of real numbers. A function $F(x)$ is an *antiderivative* of $f(x)$ on I if $F'(x) = f(x)$ for all x in I. Since the derivative of a constant C is zero, we see that if $F(x)$ is an antiderivative of $f(x)$, then so is $F(x) + C$.

If $F_1(x)$ and $F_2(x)$ are both antiderivatives of the *same function* $f(x)$, then for all x

$$\frac{d}{dx}[F_1(x) - F_2(x)] = F_1'(x) - F_2'(x) = f(x) - f(x) = 0$$

It follows that $F_1(x) - F_2(x) = C$ must be a constant. We conclude that *all possible antiderivatives of $f(x)$* are of type $F(x) + C$, where $F(x)$ is any particular antiderivative of $f(x)$. Because every possible real value of C is admissible, C, often called the "(arbitrary) constant of integration."

Each derivative formula that we discussed above yields an antiderivative formula; for example, since the derivative of $\sin x$ is $\cos x$, we conclude that the antiderivatives of $\cos x$ are $\sin x + C$. Likewise, to calculate the antiderivative of x^p where p is a non-zero real number, we first recall that by the Power Rule

$$\frac{d}{dx}x^p = px^{p-1}$$

This shows that an antiderivative of x^{p-1} is x^p/p. By adding 1 to p, we get this useful formula:

All the antiderivatives of x^p are given by

$$\frac{x^{p+1}}{p+1} + C \quad \text{if } p \neq -1$$

For instance, the antiderivatives of x^2 are $x^3/3 + C$. The above formula leaves out one value of p, namely, $p = -1$. The reason is, of course, because we can't divide by

REAL ANALYSIS AND INFINITY   345

0. But *the function whose derivative is* $1/x$ is the natural logarithm that we discuss later in this chapter.

A point worth emphasizing about antiderivatives is this:

If $F(x)$ is an antiderivative of $f(x)$, then (by definition) $F(x)$ is differentiable, i.e., $F'(x)$ exists; in fact, $F'(x) = f(x)$.

Some functions simply do not have antiderivatives; for instance,

$$f(x) = \begin{cases} 1 & \text{if } x > 0 \\ -1 & \text{if } x < 0 \end{cases}$$

cannot have an antiderivative on any open interval containing 0 regardless of how we define $f(0)$.[4]

To see why not, consider that at $x = 0$ we must have $F'(0) = f(0)$. But for $x < 0$, it is given that $F'(x) = f(x) = -1$, so that $F(x) = -x + C$ on $(-\infty, 0)$. Similarly, $F(x) = x + C$ on $(0, \infty)$.

Now if we take a sequence x_n that converges to 0 from the left, i.e., $x_n < 0$, then

$$\frac{F(x_n) - F(0)}{x_n - 0} = \frac{(-x_n + C) - (0 + C)}{x_n} = -1$$

for all n so that

$$\lim_{n \to \infty} \frac{F(x_n) - F(0)}{x_n - 0} = -1$$

This suggests that if we define $f(0) = -1$, then $F'(0) = f(0)$ agrees with the above limit. However, taking a sequence x_n in $(0, \infty)$ that converges to 0 leads to

$$\lim_{n \to \infty} \frac{F(x_n) - F(0)}{x_n - 0} = 1$$

and it follows that $F'(0)$ cannot be defined.

Evidently, the jump discontinuity of $f(x)$ at 0 is the cause of the problem here; we will soon discover that if a function $f(x)$ is continuous on an interval I, then not only does it have an antiderivative $F(x)$, but also $F(x)$ is smooth, i.e., it has no corners or sharp points.

---

[4] The absolute value function $|x|$ is an antiderivative of this function on any interval that doesn't contain $x = 0$.

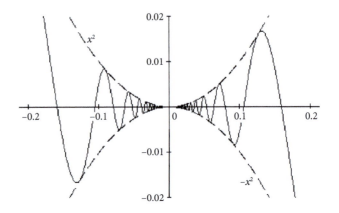

Fig. 7.15 A differentiable function of a discontinuous derivative

On the other hand, continuity is a sufficient condition that is not necessary for the existence of an antiderivative; for example, consider

$$F(x) = x^2 \sin \frac{1}{x} \quad \text{if } x \neq 0, \quad F(0) = 0$$

See Figure 7.15. This figure also shows that $F(x)$ is continuous everywhere, including at 0; see Exercise 282.

Now, if $x \neq 0$, then using the derivative rules and formulas, we calculate

$$F'(x) = 2x \sin \frac{1}{x} - \cos \frac{1}{x} \doteq f(x) \tag{7.34}$$

The derivative function $f(x)$ is discontinuous at $x = 0$; see Figure 7.16 for a graph of $f(x)$.[5]

But we can show that $F'(0)$ exists, albeit discontinuously. For every sequence x_n that converges to 0, and $x_n \neq 0$, then by the squeeze theorem and the fact that $|\sin \theta| \leq 1$

$$\lim_{n \to \infty} \frac{F(x_n) - F(0)}{x_n - 0} = \lim_{n \to \infty} x_n \sin \frac{1}{x_n} = 0$$

So $F'(0)$ exists and equals 0. So if we define $f(0) = 0$, then we see that the function $f(x)$ in (7.34), which is discontinuous at 0, can very well have a *continuous* antiderivative!

---

[5] $F(x)$ here is an example of a function that is differentiable everywhere but not continuously; that is, its derivative exists but is discontinuous at some point ($x = 0$ in this case).

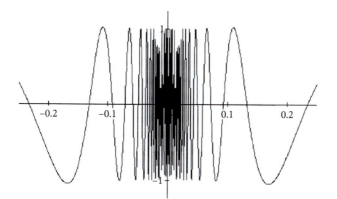

Fig. 7.16 The discontinuous derivative of the function in Figure 7.15

Fundamental theorem of calculus: Part 1. Integrating derivatives

Our discussion of antiderivatives above suggests that the existence of antiderivative is a weaker condition than continuity, whereas the existence of derivatives is a stronger or more restrictive condition, in the following sense:

The derivative of $f(x)$ exists $\Rightarrow$ $f(x)$ is continuous $\Rightarrow$ the antiderivative of $f(x)$ exists

We saw earlier why the first implication is true, and we prove the second later. There are two halves of the fundamental theorem that say different things, so they are usually presented separately in standard calculus textbooks, and I follow the same practice here. We start with the first half, which is based on the mean value theorem (see Chapter 6).

Theorem 257 (*Fundamental theorem of calculus: using the antiderivative*) *If $f \in R[a, b]$, and F is an antiderivative of f on $[a, b]$, then*

$$\int_a^b f(x)dx = F(b) - F(a) \tag{7.35}$$

Proof Let $P = \{x_0, x_1, ..., x_N\}$ be any partition of $[a, b]$. Being an antiderivative of f, the function F is differentiable on $[a, b]$, and thus on every subinterval $[x_{i-1}, x_i]$. Thus, by the mean value theorem, for each $i = 1, 2, ..., N$ there is $t_i \in (x_{i-1}, x_i)$ such that

$$F(x_i) - F(x_{i-1}) = F'(t_i)(x_i - x_{i-1}) = f(t_i)\Delta x_i$$

Adding, we get

$$\sum_{i=1}^{N} f(t_i)\Delta x_i = \sum_{i=1}^{N} [F(x_i) - F(x_{i-1})]$$

$$= F(x_N) - F(x_0)$$

$$= F(b) - F(a)$$

If $U(P)$ and $L(P)$ are the upper and lower Darboux sums, then

$$L(P) \leq \sum_{i=1}^{N} f(t_i)\Delta x_i = F(b) - F(a) \leq U(P)$$

Since these inequalities are true for all partitions, we conclude that

$$I_L \leq F(b) - F(a) \leq I_U$$

But $I_L = I_U$ because f is integrable, so (7.35) is true. ∎

This theorem makes calculating some integrals very easy. For example, an antiderivative of $f(x) = x^2$ is $F(x) = x^3/3$, so

$$\int_{-1}^{1} x^2\,dx = F(1) - F(-1) = \frac{(1)^3}{3} - \frac{(-1)^3}{3} = \frac{1}{3} + \frac{1}{3} = \frac{2}{3}$$

More generally, over any interval $[a, b]$, we have

$$\int_{a}^{b} x^2\,dx = F(b) - F(a) = \frac{b^3}{3} - \frac{a^3}{3} = \frac{b^3 - a^3}{3}$$

Still more generally, for every $p \neq -1$ using the power rule for the antiderivative of x^p above, we obtain

$$\int_{a}^{b} x^p\,dx = \frac{b^{p+1} - a^{p+1}}{p + 1} \tag{7.36}$$

It would be quite difficult to obtain this integration formula directly from the definition of the integral (calculating I_L and I_U).

It is worth mentioning that in (7.36) we know that x^p is integrable (an important hypothesis in Theorem 257) because it is continuous. In general, showing that a function $f(x)$ is integrable is not an easy task and often practically infeasible if $f(x)$ is not continuous or monotone.

We may be tempted to think that if $f(x)$ has an antiderivative, then it should be integrable. But this assumption is not generally valid; consider $f(x) = 1/\sqrt{x} = x^{-1/2}$ for $x > 0$. This function has an antiderivative $F(x) = 2\sqrt{x}$, which is continuous on $[0, \infty)$. Now take the interval $[0, 1]$, define $f(0)$ arbitrarily, and note that

$$F(1) - F(0) = 2\sqrt{1} - 2\sqrt{0} = 2$$

However, $1/\sqrt{x}$ is *not* bounded on $[0, 1]$ and thus *not integrable*, regardless of how we define $f(0)$.

It so happens that with proper hypotheses added, an extension of the Riemann integral to unbounded functions is possible where the value 2 can be assigned to the *improper integral* $\int_0^1 (1/\sqrt{x})dx$; we discuss this issue in a later section.

The familiar *integration by parts formula* of calculus is now a quick result of the machinery we've built so far. Its proof is a good illustration of the combined use of the various results that we proved earlier. After reading the proof, you have the opportunity to review that material by taking a minute to trace back the various results mentioned to even earlier theorems and definitions that were used to prove them.

Corollary 258 *(Integration by parts) Let f and g be differentiable functions on an interval* $[a, b]$*, and assume that their derivatives are integrable, i.e.,* $F', g' \in R[a, b]$*. Then*

$$\int_a^b f(x)g'(x)dx = f(b)g(b) - f(a)g(a) - \int_a^b F'(x)g(x)dx \qquad (7.37)$$

Proof Since differentiable functions are continuous, Theorem 248 implies that f and g are integrable. Further, by hypothesis, $F', g' \in R[a, b]$; so by Corollary 253, the products fg' and Fg are integrable. By the product rule for derivatives (Chapter 6)

$$\frac{d}{dx}[f(x)g(x)] = F'(x)g(x) + f(x)g'(x) \qquad (7.38)$$

Since the right-hand side is integrable by Theorem 250, it follows that $d[f(x)g(x)]/dx$ is integrable too. Notice that the product fg inside the brackets is an antiderivative of the left-hand side of (7.38), so taking integrals and using Theorems 257 and 250, we obtain

$$f(b)g(b) - f(a)g(a) = \int_a^b F'(x)g(x)dx + \int_a^b f(x)g'(x)dx$$

The above equality is equivalent to (7.37), so the proof is complete.   ■

The integration by parts formula is usually written using the u and v notation as follows: if we define

$$u = f(x), \qquad dv = g'(x)dx$$
$$du = F'(x)dx, \qquad v = g(x)$$

then (7.37) may be written as

$$\int_a^b u\,dv = uv\Big|_a^b - \int_a^b v\,du$$

Fundamental theorem of calculus: Part 2. Existence of antiderivative

Theorem 257 requires that $f(x)$ be not only integrable but also have an antiderivative on the interval of interest. These are independent properties because as we have seen, one does not generally imply the other. Theorem 257 doesn't say anything if either property is missing.

There is a second part of the fundamental theorem of calculus that connects these properties. In particular, we discover that on closed and bounded intervals, continuous functions are not only integrable, but they also have antiderivatives.

Theorem 259 (*Fundamental theorem of calculus: existence of antiderivative*) *Let* $f \in R[a, b]$, *and define*[6]

$$F(x) = \int_a^x f(t)dt, \quad x \in [a, b] \tag{7.39}$$

(i) *The function* $F : [a, b] \rightarrow \mathbb{R}$ *is continuous on* $[a, b]$ *(even if f is not)*
(ii) *If f is continuous at some* $c \in [a, b]$, *then F is differentiable at c, and*

$$F'(c) = f(c)$$

Proof (i) Let x_n be an arbitrary sequence in $[a, b]$ that converges to c. To show that $\lim_{n \to \infty} F(x_n) = F(c)$, we first use the decomposition property (Theorem

---

[6] $\int_a^b f(t)dt$ and $\int_a^b f(x)dx$ *say the same thing since the symbol that is used inside the integral is a temporary variable that is "integrated away." The answer is a number or a parameter that does not contain the "dummy variable" inside the integral.*

256) to obtain

$$F(x_n) = \int_c^{x_n} f(t)dt + \int_a^c f(t)dt = \int_c^{x_n} f(t)dt + F(c)$$

and then conclude that

$$|F(x_n) - F(c)| = \left| \int_c^{x_n} f(t)dt \right|$$

Since f is bounded on $[a, b]$, there is a number $M > 0$ such that $|f(t)| \leq M$ for all $t \in [a, b]$. There are two possibilities: if $x_n \geq c$ for some index n, then by Corollary 255

$$|F(x_n) - F(c)| \leq \int_c^{x_n} |f(t)|dt \leq M(x_n - c) = M|x_n - c|$$

On the other hand, if $x_n \leq c$ for some n, then again by Corollary 255

$$|F(x_n) - F(c)| = |F(c) - F(x_n)| \leq \int_{x_n}^c |f(t)|dt \leq M(c - x_n) = M|x_n - c|$$

It follows that for all n

$$|F(x_n) - F(c)| \leq M|x_n - c|$$

Given $\varepsilon > 0$, there is N large enough that $|x_n - c| < \varepsilon/M$ for $n \geq N$; therefore,

$$|F(x_n) - F(c)| < \varepsilon \quad \text{for } n \geq N$$

Since x_n was an arbitrary sequence converging to c, it follows that $\lim_{n\to\infty} F(x_n) = F(c)$, i.e., F is continuous at c.

(ii) Suppose that f is continuous at c. To show that F is differentiable at c with derivative $f(c)$, let x_n be an arbitrary sequence in $[a, b]$ that converges to c, but $x_n \neq c$ for all n. Since $f(c)$ is a constant,

$$\int_c^{x_n} f(c)dt = f(c)(x_n - c)$$

so we have

$$\left| \frac{F(x_n) - F(c)}{x_n - c} - f(c) \right| = \frac{1}{|x_n - c|} \left| F(x_n) - F(c) - f(c)(x_n - c) \right|$$

$$= \frac{1}{|x_n - c|} \left| \int_c^{x_n} f(t)dt - \int_c^{x_n} f(c)dt \right|$$

$$= \frac{1}{|x_n - c|} \left| \int_c^{x_n} [f(t) - f(c)]dt \right|$$

Given $\varepsilon > 0$, since f is continuous at c, there is $\delta > 0$ such that $|f(t) - f(c)| < \varepsilon$ if $|t - c| < \delta$. Also, there is N large enough that $|x_n - c| < \delta$ for $n \geq N$, and since all values of t inside the integral are between c and x_n for each n, it follows that $|t - c| \leq |x_n - c| < \delta$ for all $n \geq N$. Therefore,

$$|f(t) - f(c)| < \varepsilon \quad \text{for } n \geq N$$

Next, as in (i), we need to consider two possibilities; if $x_n > c$ for some index n, then $|x_n - c| = x_n - c$, and by Corollary 255

$$\frac{1}{|x_n - c|} \left| \int_c^{x_n} [f(t) - f(c)]dt \right| \leq \frac{1}{x_n - c} \int_c^{x_n} |f(t) - f(c)|dt$$

$$< \frac{1}{x_n - c} \varepsilon(x_n - c) = \varepsilon$$

If $x_n < c$ for some n, then $|x_n - c| = c - x_n$, and again by Corollary 255

$$\frac{1}{|x_n - c|} \left| \int_c^{x_n} [f(t) - f(c)]dt \right| = \frac{1}{c - x_n} \left| \int_{x_n}^c [f(t) - f(c)]dt \right|$$

$$\leq \frac{1}{c - x_n} \int_{x_n}^c |f(t) - f(c)|dt$$

$$< \frac{1}{c - x_n} \varepsilon(c - x_n) = \varepsilon$$

Thus, for all $n \geq N$,

$$\left| \frac{F(x_n) - F(c)}{x_n - c} - f(c) \right| < \varepsilon$$

This shows that

$$\lim_{n \to \infty} \frac{F(x_n) - F(c)}{x_n - c} = f(c)$$

Since the above limit has the same value $f(c)$ regardless of which sequence x_n we choose that converges to c, with $x_n \neq c$ for all n, we have shown that F is differentiable at c with $F'(c) = f(c)$. ∎

If f is *not* continuous at *any* point $r \in [a, b]$, then the function F in (7.39) *doesn't represent an antiderivative* of f on $[a, b]$. For instance, if

$$f(x) = \begin{cases} 1 & \text{if } x \geq 0 \\ -1 & \text{if } x < 0 \end{cases}$$

then

$$F_0(x) = \int_0^x f(t)dt = \left\{ \begin{array}{ll} x & \text{if } x \geq 0 \\ -x & \text{if } x < 0 \end{array} \right\} = |x|$$

Recall that $F_0(x) = |x|$ is not differentiable at $x = 0$, so F_0 isn't an antiderivative, although it is continuous everywhere. More pointedly perhaps, consider

$$F_1(x) = \int_{-1}^x f(t)dt$$

If $x > 0$, then

$$F_1(x) = \int_{-1}^0 (-1)dt + \int_0^x 1dt = 1 + x$$

The derivative of $F_1(x)$ is 1, not $f(x)$ above.

On the other hand, in Theorem 259(i), it is actually the case that $F(x)$ is *uniformly continuous* on the closed interval $[a, b]$; see Chapter 6.

An immediate consequence of Theorem 259 is the following.

Corollary 260 *If f is continuous on the interval $[a, b]$, then it has an antiderivative given by the function F in (7.39) that is differentiable on $[a, b]$ with the property that $F(a) = 0$.*

To illustrate a use of the above corollary, consider $f(x) = \sin(x^2)$. Being a composition of two continuous functions, $f(x)$ is continuous on $(-\infty, \infty)$; and in addition, it is bounded, since the sine function is bounded. Therefore, f is continuous, hence integrable on every closed interval $[a, b]$, and it has an antiderivative

$$F(x) = \int_0^x \sin(t^2)dt$$

Note that $F(x)$ is defined for all real numbers x. The graph of a part of $F(x)$ is shown in Figure 7.17.

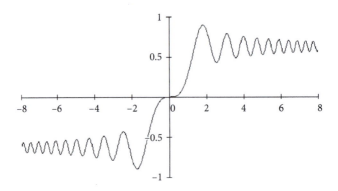

Fig. 7.17 The graph of a function defined as an integral

An elementary or closed-form formula for F is not known, but its value can be approximated to any desired level of accuracy for each fixed x using various methods, including Taylor series, which that we discuss later in Chapter 8.

There is one more thing that we can do at this point: find the exact location of the maxima and minima of $F(x)$ because we have a formula for its derivative, i.e., $f(x)$; see Exercise 285.

The following is a useful extension of Corollary 260.

Corollary 261 *Assume that f is continuous on the interval $[a, b]$, and g is differentiable on an interval I with $g(I) \subset [a, b]$. If*

$$F(x) = \int_a^{g(x)} f(t)dt$$

then F is differentiable on $[a, b]$ with derivative

$$F'(x) = f(g(x))g'(x)$$

Proof Define $\varphi(x) = \int_a^x f(t)dt$, and note that $F(x) = \varphi(g(x))$. By Corollary 260, $\varphi(x)$ is differentiable on $[a, b]$, and its derivative is $\varphi'(x) = f(x)$. Now, by the chain rule, F is differentiable, and its derivative is given as in the statement of this corollary. ∎

For example, the function $g(x) = \sqrt{x}$ is differentiable for all $x > 0$, so

$$\frac{d}{dx} \int_0^{\sqrt{x}} \sin(t^2)dt = \sin(\sqrt{x})^2 \frac{d}{dx}\sqrt{x} = \frac{\sin x}{2\sqrt{x}}, \quad x \in (0, \infty)$$

In this example, we also discover that although we don't have a formula for the function $\int_0^{\sqrt{x}} \sin(t^2)dt$, we *can* find a simple formula for its derivative.

We now derive the *change of variables rule*, also known as the "substitution rule" in calculus. It is a good application of the fundamental theorem of calculus because it uses both halves of that theorem.

Corollary 262 *(Change of variables, substitution rule) Assume that g is a differentiable function on the interval $[a, b]$ and that its derivative g' is integrable, i.e., $g' \in R[a, b]$. If f is a continuous function on $I = g([a, b])$, then*

$$\int_a^b f(g(x))g'(x)dx = \int_{g(a)}^{g(b)} f(u)du \qquad (7.40)$$

Proof By Corollary 199, I is a closed and bounded interval that contains $g(a)$ and $g(b)$, and since f is continuous on I, Theorem 248 implies that it is integrable on I. Thus, the integral on the left-hand side of (7.40) is defined. Further, since $f \circ g$ is continuous (hence, integrable), and $g' \in R[a, b]$, Corollary 253 implies that the product $(f \circ g)g'$ is integrable on $[a, b]$. Thus, the integral on the right-hand side of (7.40) is also defined. It remains to show that the two integrals are equal.

First, the trivial case: if I is a single point, then g is constant on $[a, b]$; therefore, $g(a) = g(b)$, and $g'(x) = 0$, so both of the integrals in (7.40) are zero. Next, assume that I is nontrivial and define the function

$$F(x) = \int_{g(a)}^x f(u)du \qquad (7.41)$$

Since f is continuous on I, the second part of the fundamental theorem of calculus (Theorem 259) implies that F is differentiable, and $F'(x) = f(x)$ for all $x \in I$. So by the chain rule

$$\frac{d}{dx}F(g(x)) = F'(g(x))g'(x) = f(g(x))g'(x)$$

for all $x \in [a, b]$. Now, by the first part of the fundamental theorem of calculus (Theorem 257),

$$\int_a^b f(g(x))g'(x)dx = F(g(b)) - F(g(a))$$

and by (7.41),

$$F(g(b)) - F(g(a)) = \int_{g(a)}^{g(b)} f(u)du$$

This shows that the integrals in (7.40) are equal and completes the proof. ∎

The change of variables rule is an integration method that is quite often used in practice. To use (7.40), we identify a given integral with the left-hand side of (7.40) by guessing the g and substituting $u = g(x)$ and $du/dx = g'(x)$, or $du = g'(x)dx$. If there is a formula for the antiderivative in (7.41), then we can use it to calculate the original integral.

For example, to find the integral

$$\int_0^1 x \sin(x^2 + \pi)dx$$

we identify $g(x) = u = x^2 + \pi$, $f(u) = \sin u$, and calculate $du/dx = 2x$. Substituting u and $(1/2)du = xdx$ in (7.40) gives

$$\int_0^1 x \sin(x^2 + \pi)dx = \int_\pi^{\pi+1} \sin u \frac{1}{2} du = -\frac{1}{2}[\cos(\pi + 1) - \cos \pi]$$

The answer can be simplified using the trigonometric identity $\cos(x + y) = \cos x \cos y - \sin x \sin y$ to get $\cos(\pi + 1) = \cos \pi \cos 1 = -\cos 1$. With this value, we obtain the answer

$$\int_0^1 x \sin(x^2 + \pi)dx = \frac{1 - \cos 1}{2}$$

For another example, consider the integral

$$\int_0^{\pi/2} \frac{\sin(2x)}{\sqrt{1 + \sin^2 x}} dx$$

Let $g(x) = u = 1 + \sin^2 x$ to obtain $du/dx = 2 \sin x \cos x = \sin(2x)$ so we may substitute $du = \sin(2x)dx$. These substitutions give

$$\int_{g(0)}^{g(\pi/2)} \frac{1}{\sqrt{u}} du = \int_1^2 u^{-1/2} du$$

where we see that $f(u) = 1/\sqrt{u}$ in this case. Since an antiderivative of $u^{-1/2}$ is $2u^{1/2} = 2\sqrt{u}$, we obtain

$$\int_0^{\pi/2} \frac{\sin(2x)}{\sqrt{1 + \sin^2 x}} dx = 2\sqrt{2} - 2\sqrt{1} = 2(\sqrt{2} - 1)$$

Not every integral can or should be approached using a change of variables (x to u) method. For instance, the integral

$$\int_0^\pi x \sin x \, dx$$

doesn't yield to a useful change of variables and should be calculated using integration by parts. Another example is provided by the integral

$$\int_{\pi/3}^{\pi/2} \frac{\sin(2x)}{\sqrt{1 - \sin^2 x}} \, dx$$

which can be done using change of variables, but it is easier to do it using trigonometric identities. In this case, $1-\sin^2 x = \cos^2 x$, so that $\sqrt{1 - \sin^2 x} = \cos x$ if $\pi/3 \le x \le \pi/2$. Thus,

$$\int_{\pi/3}^{\pi/2} \frac{\sin(2x)}{\sqrt{1 - \sin^2 x}} \, dx = \int_{\pi/3}^{\pi/2} \frac{2 \sin x \cos x}{\cos x} \, dx$$

$$= 2 \int_{\pi/3}^{\pi/2} \sin x \, dx$$

$$= -2 \left(\cos \frac{\pi}{2} - \cos \frac{\pi}{3} \right) = 1$$

Mean values

We now discuss the average values of functions over intervals. This is an interesting application of integration beyond area calculation that does not require an exposure to science or engineering.

First, we discuss the *mean value theorem for integrals* as a consequence of the fundamental theorem of calculus. It is a good application of that theorem because it uses *both parts*, thus illustrating the significance of the two parts working together.

Corollary 263 *(Mean value theorem for integrals) If f is a continuous function on the interval $[a, b]$, then there exists a number $c \in [a, b]$ such that*

$$\int_a^b f(x)dx = f(c)(x - c) \tag{7.42}$$

Proof By hypothesis, f is continuous, so by Theorem 259, an antiderivative of f is $F(x) = \int_a^x f(t)dt$ for $x \in [a, b]$. Since F is differentiable on $[a, b]$, the mean value theorem for derivatives implies the existence of some number $c \in [a, b]$ such that

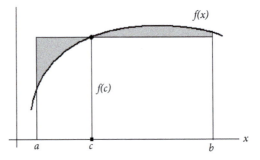

Fig. 7.18 Illustrating the mean value theorem for integrals

$$F(b) - F(a) = F'(c)(b - a)$$

But $F'(c) = f(c)$, and by Theorem 257 (the first part of the fundamental theorem), the left-hand side of the above equality is the integral $\int_a^b f(x)dx$ and (7.42) holds. ∎

In terms of *areas*, Corollary 263 says that (for a positive function), the area under the graph of $f(x)$ is equal to the area of a suitably chosen rectangle of height $f(c)$ and base length $b - a$; see Figure 7.18, the shaded area overestimate by the rectangle is equal in size to the also shaded area underestimate.

Corollary 263 is generally false if $f(x)$ is *not* continuous; looking at Figure 7.18 again, if there were a hole in the graph of $f(x)$ right above where c is located, then $f(c)$ would have a higher or lower value, making it greater or less than required in (7.42).

The number $f(c)$ in (7.42) is the *mean or average value of the function $f(x)$* over the interval $[a, b]$, though this is not readily apparent. To see why the term "mean value" is apt, consider the usual idea of average value. The average of N numbers $y_1, y_2, ..., y_N$ is

$$\frac{y_1 + y_2 + \cdots + y_N}{N}$$

If $y = f(x)$ is a function defined on an interval $[a, b]$, then its average value involves all of its values, that is, *an (uncountable) infinity of numbers*. We can't add up that many numbers, but the average of any finite number of values of the function $f(x)$ at points $x_1, x_2, ..., x_N$ in an interval $[a, b]$ is

$$\frac{f(x_1) + f(x_2) + \cdots + f(x_N)}{N} \tag{7.43}$$

To go from here to the average of the whole function, suppose that the N points $x_1, x_2, ..., x_N$ are equally spaced so that with $x_1 = a$ and $x_N = b$, we have a regular partition of the interval $[a, b]$. Each subinterval of this partition has length

$$\Delta x = \frac{b - a}{N}$$

Other than $b - a$, all the ingredients of a Riemann sum are present in (7.43). So to complete the picture, we multiply and divide the fraction in (7.43) by $b - a$ and write it as

$$\frac{1}{b - a} \frac{b - a}{N} [f(x_1) + f(x_2) + \cdots + f(x_N)] = \frac{1}{b - a} \sum_{i=1}^{N} f(x_i) \Delta x$$

If $f(x)$ is integrable (for instance, if it is continuous), then the Riemann sum on the right-hand side converges to the integral of $f(x)$ as we refine the partition by increasing N indefinitely. This leads to the following definition:

Average Value of a function If $f(x)$ is integrable over an interval $[a, b]$, then its average value over this interval is

$$\frac{1}{b - a} \int_a^b f(x) dx$$

For example, the average value of $f(x) = 1 - 1/x^2$ over the interval $[1, 3]$ is calculated as follows:

$$\frac{1}{3 - 1} \int_1^3 \left(1 - \frac{1}{x^2}\right) dx = \frac{2}{3} \simeq 0.67$$

since an antiderivative of $f(x)$ is the function $x + 1/x$.

We can use the average value to find a number c that works in (7.42). For $f(x) = 1 - 1/x^2$ over $[1, 3]$, we have

$$f(c) = \frac{1}{3 - 1} \int_1^3 \left(1 - \frac{1}{x^2}\right) dx$$

$$1 - \frac{1}{c^2} = \frac{2}{3}$$

We use a little algebra to solve the above equation for c:

$$\frac{1}{c^2} = \frac{1}{3}$$

$$c^2 = 3$$

The last equation has two possible solutions: $c = \pm\sqrt{3}$. We are looking for a c in the interval $[1, 3]$ where only $c = \sqrt{3} \simeq 1.73$ works.

Note that the average value of f over the interval $[1/2, 2]$ is 0. Since $f(x)$ crosses the x-axis at $x = 1$, it follows that $c = 1$. Can you tell why $c = 1$ for intervals of type $[1/n, n]$ for all positive integers n?

7.5 Logarithmic and exponential functions

In elementary coverage, the exponential function is defined analogously to the power function, and then the logarithm is defined as its inverse function. This is a natural idea when we think of integer powers like $2^3 = (2)(2)(2)$ but requires defining roots (zeros of polynomials) when extending to rational powers like $2^{1/2} = \sqrt{2}$. Then the technical difficulty in using this approach in a rigorous way increases further when extending to real number powers, like $2^{\sqrt{2}}$.

An alternative approach is to go in the reverse direction, starting with the logarithm and then define the exponential function as its inverse function. This approach is more natural in real analysis, not only because it resolves the issue of what the antiderivative of $1/x$ is (the exception to the power rule) but also because it is technically easier to implement after the integral has been defined and its properties established. In particular, the entire process starts with real numbers, so that we need not be concerned about different types of numbers (integer, rational, irrational).

We start in this section with the definition of *logarithm as a function defined by an integral* in the sense that we have already discussed in the second half of the fundamental theorem of calculus. We then rigorously derive its well-known properties and use them to define the exponential function as the inverse function and to derive their properties.

7.5.1 The natural logarithm function

Recall that we obtained the power rule for integrals in (7.36) by observing that an antiderivative of the power function x^p is given by

$$\frac{x^{p+1}}{p + 1}, \quad \text{if } p \neq -1$$

This expression is undefined when $p = -1$ since we can't divide by zero. However, the function $f(x) = x^{-1} = 1/x$ is continuous for all $x > 0$, so it is integrable on every closed interval $[a, b]$ where $0 < a < b$. Therefore, by Theorem 259, an

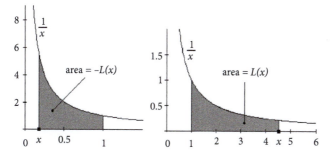

Fig. 7.19 Defining the natural logarithm using an integral

antiderivative of $1/x$ exists; we call it the function $L : (0, \infty) \to \mathbb{R}$ where

$$L(x) = \int_1^x \frac{1}{t}\,dt, \quad x > 0 \tag{7.44}$$

Note that $L(1) = 0$, and by Theorem 259, $L'(x) = 1/x$ so $L(x)$ *is differentiable on its domain* $(0, \infty)$.

Further, since $x > 0$, we conclude that $L(x)$ *is an increasing function*. As we see in the right panel of Figure 7.19, for each fixed $x > 1$, the value of $L(x)$ is the area under the graph of $1/x$; but if $x < 1$, then the area under the graph of $1/x$ is the negative of the value of $L(x)$ since for $0 < x < 1$

$$L(x) = \int_1^x \frac{1}{t}\,dt = -\int_x^1 \frac{1}{t}\,dt < 0$$

We now prove that $L(x) \to \infty$ as $x \to \infty$.

For every real number x there is a greatest integer $n = \lfloor x \rfloor$ that is less than or equal to x. Consider the regular partition $P_n = \{1, 2, 3, \ldots, n\}$ of the interval $[1, n]$. Note that all subintervals of P_n have length $\Delta x = 1$. Let's calculate the lower Riemann sum $R_L(P_n)$ corresponding to P_n as an underestimation of $L(n)$:

$$L(n) \geq R_L(P_n) = \sum_{i=1}^n \frac{1}{x_i}\Delta x = \sum_{i=1}^n \frac{1}{i+1} \tag{7.45}$$

As $n \to \infty$, the right-hand side diverges to ∞ as the harmonic series, so $L(n) \to \infty$ as $n \to \infty$. Because $L(x)$ is an increasing function, $L(x) \geq L(n)$. We conclude that $L(x) \to \infty$ as $x \to \infty$, as was claimed.

The lower sum $R_L(P_n)$ is shaded in Figure 7.20; we see that the total areas of the shaded strips is indeed less than the actual area $L(n)$ under the graph of $1/x$ from $x = 1$ to $x = n$.

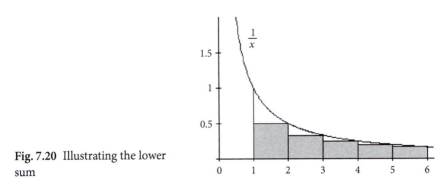

Fig. 7.20 Illustrating the lower sum

A consequence of this observation is that $L(x) > 1$ for x sufficiently large, so the intermediate value theorem and the increasing nature of $L(x)$ imply that there is a *unique* real number e such that

$$e > 1 \quad \text{and} \quad L(e) = 1.$$

We prove in Chapter 200 that this number $e \approx 2.7182815$ is irrational (in fact, it is known to be transcendental). It forms the base of the *natural logarithm*, a topic that we discuss further below.[7]

$L(x)$ is the *natural logarithm function, or the logarithm to the base e*. It is commonly denoted $\ln x$.

The values of $\ln x$ can be accurately approximated using Riemann sums or other methods such as infinite series; we discuss some of them below. A graph of $\ln x$ is shown in Figure 7.21.

We summarize our results above as the following theorem for easy reference.

Theorem 264 *(The natural logarithm) The natural logarithm function is defined by*

$$\ln x = \int_1^x \frac{1}{t}\, dt$$

---

[7] The Scottish mathematician John Napier (1550–1617) is credited with the discovery of logarithm around 1614; he found it to be a powerful computational aid. The term "natural logarithm" is due to Nicholas Mercator (1620–1687), a German mathematician who should not be confused with the more famous cartographer Gerhard (Gerardus) Mercator who lived in the 1500s and created the familiar Mercator projection. The symbol e seems to be Euler's.

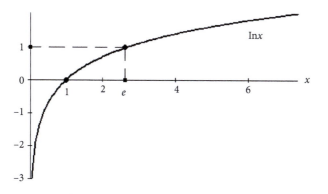

Fig. 7.21 The natural logarithm function

(a) *The domain of* ln *x is* $(0, \infty)$*; and on this set,* ln *x is an increasing and differentiable (hence continuous) function with derivative*

$$\frac{d}{dx}(\ln x) = \frac{1}{x}$$

(b) $\lim_{x \to \infty} \ln x = \infty$*, and there is a unique number* $e \approx 2.7182815$ *such that* $\ln e = 1.$

From the above discussion, we also see that the missing case of the power rule has now been supplied as

$$\int_a^b x^{-1}dx = \ln b - \ln a, \quad 0 < a < b \tag{7.46}$$

Logarithm and the harmonic series

The inequality in (7.45) points to a relationship between the logarithm and the harmonic series. It is worth exploring this further to discover that the harmonic series $\sum_{n=1}^{\infty} 1/n$ diverges slowly at a "logarithmic rate."

Going back to Figure 7.20, observe that

$$\frac{1}{2} + \frac{1}{3} + \frac{1}{4} + \frac{1}{5} + \frac{1}{6} < \ln 6$$

since the area of the dark strips adds up to less than the area under $1/x$. This is clearly not limited to the number 6 and can be extended to any positive integer k, so

$$\sum_{n=2}^{k} \frac{1}{n} < \ln k$$

Therefore, adding 1 to both sides gives the following:

$$\sum_{n=1}^{k} \frac{1}{n} < 1 + \ln k \qquad (7.47)$$

The strips in Figure 7.20 happen to be rectangles with heights $1/x_n$ where x_n is the right-hand endpoint of the interval $[n, n+1]$ for $n = 1, 2, 3, \dots$ The areas of these rectangles underestimate the area under $1/x$. Now, imagine using the strips whose heights are calculated using the left-hand endpoints of the same integer intervals; the areas of these strips overestimate the area under $1/x$ so that

$$\sum_{n=1}^{k} \frac{1}{n} > \ln(k + 1) \qquad (7.48)$$

For instance, if $k = 5$, then the sum of the areas of the five overestimating strips exceeds the area under $1/x$ from 1 to 6. Now, from (7.47) and (7.48), it follows that

$$\ln(k + 1) < \sum_{n=1}^{k} \frac{1}{n} < 1 + \ln k \qquad (7.49)$$

These inequalities show that

> For large values of k, the partial sums of the harmonic series increase logarithmically or at a logarithmic pace.

If we use the average of the bounds in (7.49) as an approximation to the partial sum, then we have the estimate

$$\sum_{n=1}^{k} \frac{1}{n} \approx \frac{1}{2} + \ln k \quad \text{(large } k\text{)}$$

This approximation improves as $k \to \infty$; for instance, if $k = 100,000$, then $\ln(k + 1) \simeq 11.512935$, $\ln k \simeq 11.512925$, and

$$\sum_{n=1}^{k} \frac{1}{n} \simeq 12.090, \quad \frac{1}{2} + \ln k \simeq 12.013$$

Additional properties of the natural logarithm

We can use our definition of the natural logarithm to discover some additional properties of ln x.

Theorem 265 (a) For all real numbers $u, v > 0$

$$\ln(uv) = \ln u + \ln v \tag{7.50}$$

and

$$\ln \frac{u}{v} = \ln u - \ln v \tag{7.51}$$

(b) $\lim_{x \to 0^+} \ln x = -\infty$; *hence, the range of* ln x *is all real numbers:* $(-\infty, \infty)$

Proof (a) Fix v and consider u to be variable to define the function

$$g(u) = \ln(uv)$$

With v held constant, we use the chain rule to obtain

$$g'(u) = \frac{d}{du} \ln(uv) = \frac{1}{uv} \frac{d}{du} (uv) = \frac{1}{uv} v = \frac{1}{u}$$

This means that $g(u)$ is an antiderivative of $1/u$, that is, $g(u) = \ln u + C$ where C is a constant of integration. Notice that

$$g(1) = \ln 1 + C = C$$

and

$$g(1) = \ln(1v) = \ln v$$

so

$$\ln(uv) = g(u) = \ln u + g(1) = \ln u + \ln v$$

and the equality in (7.50) follows. The proof of (7.51) and are left as Exercise 289. ■

A property of the natural logarithm that, among other things, is essential for the development of exponential functions, is the following:

$$\ln u^p = p \ln u \tag{7.52}$$

where p is any fixed *rational* number. This restriction of p to rational numbers is temporary and needed because *we have not yet defined what we mean by raising a real number u to an irrational power*, like $3^{\sqrt{2}}$ or $(\sqrt{2})^{\pi}$. This issue will be resolved when we develop a general concept of exponential function that is designed essentially for this purpose.

To prove (7.52), we start with $p = n$, a positive integer. Then repeated applications of (7.50) n times give

$$\ln u^n = \ln(u \cdots u) = \ln u + \cdots + \ln u = n \ln u \qquad (7.53)$$

Therefore, (7.52) is true when p is a positive integer. It is also true if $p = 0$ because[8]

$$\ln u^0 = \ln 1 = 0 = 0 \ln u$$

If $p = -n$ is a negative integer, then by (7.74) and (7.53)

$$\ln u^{-n} = \ln \left(\frac{1}{u}\right)^n = n \ln \frac{1}{u} = -n \ln u$$

so (7.52) still holds. Next, let $p = 1/n$ where n is a positive integer and notice that since $u^{1/n} = \sqrt[n]{u}$, it follows that

$$n \ln \sqrt[n]{u} = \ln \left(\sqrt[n]{u}\right)^n = \ln u$$

which after dividing by n gives (7.52) again. Finally, if $p = m/n$ is any rational number, then

$$\ln u^{m/n} = \ln \left(\sqrt[n]{u}\right)^m = m \ln \sqrt[n]{u} = \frac{m}{n} \ln u$$

so we see that (7.52) is true for all rational p, as claimed.

7.5.2 The natural exponential function

The fact that $\ln x$ is an increasing function implies that it is one-to-one and therefore has an inverse function. Let's go back to the L notation for a moment: $\ln x = L(x)$. The inverse function $L^{-1} : \mathbb{R} \to (0, \infty)$ satisfies the relations

$$L^{-1}(L(x)) = x \quad \text{and} \quad L(L^{-1}(y)) = y$$

for all x in $(0, \infty)$ and y in $\mathbb{R}$. See Figure 7.22.

---

[8] To see why $u^0 = 1$, note that $u^0 = u^{1-1} = u^1 u^{-1} = u(1/u) = 1$. We need to exclude the value $u = 0$ because we are using the fraction $1/u$. In fact, 0^0 is an "indeterminate form" that can have no specific value.

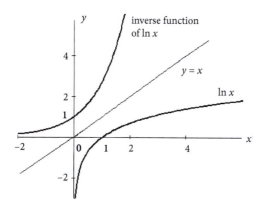

Fig. 7.22 The natural logarithm and its inverse function

Now let's denote $E(y) = L^{-1}(y)$. Then we can rewrite the above two inverse function relations as

$$E(\ln x) = x \quad \text{and} \quad \ln E(y) = y \tag{7.54}$$

These relations allow us to derive the properties of the inverse function $E(y)$ using the properties of $\ln x$. First, since the range of $E(y)$ is $(0, \infty)$ (the same as the domain of $\ln x$), we have

$$E(y) > 0 \quad \text{for all } y \text{ in } \mathbb{R}$$

Next, setting $x = 1$ and $x = e$ in the equality on the left in (7.54) gives

$$E(\ln 1) = 1, \qquad E(\ln e) = e$$

which we can rewrite as

$$E(0) = 1, \qquad E(1) = e$$

Next, we show that $E(y)$ is differentiable and find its derivative using the inverse function relations (7.54).

Theorem 266 *The function E is differentiable on $(-\infty, \infty)$ with derivative $E'(y) = E(y)$.*

Proof Let y be a fixed real number and y_n any sequence of real numbers that converges to y with $y_n \neq y$ for all n. If $x = E(y)$ and $x_n = E(y_n)$ for each n, then by (7.54), $y = \ln x$ and $y_n = \ln x_n$ and

$$\frac{E(y_n) - E(y)}{y_n - y} = \frac{x_n - x}{\ln x_n - \ln x} = \frac{1}{(\ln x_n - \ln x)/(x_n - x)}$$

Recall that $y_n \neq y$, so $\ln x_n \neq \ln x$ for all n; since $\ln x$ is a one-to-one function, we also see that $x_n \neq x$ for all n. Taking the limit as $n \to \infty$ we get

$$\lim_{n \to \infty} \frac{E(y_n) - E(y)}{y_n - y} = \frac{1}{\lim_{n \to \infty}(\ln x_n - \ln x)/(x_n - x)} = \frac{1}{d/dx(\ln x)} = x$$

This equality holds for every sequence y_n that that converges to y, with $y_n \neq y$ for all n, so it is the derivative $E'(y)$. Since $x = E(y)$, we have $E'(y) = E(y)$. ∎

The function $E(y)$ has the distinctive property that it is its own derivative. Another important consequence of the above lemma is that $E(y)$ *is continuous and increasing on its domain* $\mathbb{R}$.

We now have reason to believe that $E(y)$ is the natural exponential function of calculus e^y. To explain this exponential notation, it is first necessary to explain what we mean by raising the number e to a real number power y, which may be irrational.

Recall from (7.52) that if $q = m/k$ is a fixed rational number, then e^q is defined as $e^{m/k} = \sqrt[k]{e^m}$. Further,

$$\ln e^q = \ln E(q)$$

and because $\ln x$ is a one-to-one function, we must have

$$e^q = E(q), \quad q \text{ rational}$$

Now recall that the rational numbers are dense in $\mathbb{R}$, so if $x \in \mathbb{R}$, then there is a sequence of rational numbers q_n that converge to x; that is, $x = \lim_{n \to \infty} q_n$. Since E is continuous

$$E(x) = E(\lim_{n \to \infty} q_n) = \lim_{n \to \infty} E(q_n) = \lim_{n \to \infty} e^{q_n}$$

This shows that if we define

$$e^x = \lim_{n \to \infty} e^{q_n}$$

then the interpretation of $E(x) = e^x$ as the "xth power of e" is extended to all real x (rational or irrational).

The properties of the natural exponential function

Also as promised, with the help of the above extension, we can extend (7.52) to all real p. Let $p = \lim_{n \to \infty} q_n$ and note that

$$\ln u^p = \ln \left(\lim_{n \to \infty} u^{q_n} \right)$$

$$= \lim_{n \to \infty} \ln(u^{q_n}), \quad \text{since } \ln x \text{ is continuous}$$

$$= \lim_{n \to \infty} q_n \ln u, \quad \text{using (7.52) on the rational powers } q_n$$

$$= p \ln u$$

We state this property as a theorem:

Theorem 267 *For all real numbers u, p with $u > 0$*

$$\ln u^p = p \ln u$$

According to Theorem 267, $\ln(u^2) = 2 \ln u$. Notice that $\ln(u^2)$ is defined for all $u < 0$, but $\ln u$ is not. This doesn't contradict the theorem since in Theorem 267, we have $u > 0$.

The apparent inconsistency is removed if we notice that $u^2 = |u|^2$ so, since $|u| > 0$ if $u \neq 0$, Theorem 267 gives the correct identity:

$$\ln(u^2) = 2 \ln |u|$$

which is valid for $u < 0$ also.

Using the properties of $\ln x$, we can now derive the corresponding properties of e^x.

Theorem 268 *For all real numbers u, v, p, the following identities hold:*

$$e^{u+v} = e^u e^v, \qquad e^{-u} = \frac{1}{e^u}, \qquad e^{pu} = (e^u)^p \tag{7.55}$$

Proof Recall that $\ln x \to \infty$ as $x \to \infty$, and $\ln x \to -\infty$ as $x \to 0^+$. Since $\ln x$ is continuous, the intermediate value theorem implies that it is onto $\mathbb{R}$. So for arbitrary real numbers u and v, there are positive real numbers s and t such that $u = \ln s$ and $v = \ln t$; and s, t are uniquely determined by u, v because $\ln x$ is one-to-one. Now, from (7.54) and using (7.50), we derive

$$E(u + v) = E(\ln s + \ln t) = E(\ln st) = st$$

Since $s = E(\ln s) = E(u)$, and likewise, $t = E(v)$, we have shown that

$$E(u + v) = E(u)E(v) \qquad (7.56)$$

This identity translates into the more familiar form

$$e^{u+v} = e^u e^v$$

that establishes the first identity. You may prove the remaining two identities as Exercise 290. ∎

The power function and its derivative revisited

In Chapter 6, we noted that the power function x^p is not well-defined for all real numbers if p is not a positive integer or 0. If p is negative, then x^p is not defined at $x = 0$ (think of $x^{-1} = 1/x$). If p is a rational number that is not an integer, the x^p may be undefined for all $x \leq 0$ (think of $x^{-1/2} = 1/\sqrt{x}$). But at least we were able to define x^p for all rational p using roots and fractions as necessary.

If p is irrational, then it is not obvious that x^p can be defined even for $x > 0$. For instance, how do we define $x^{\sqrt{2}}$?

Theorem 267 provides a way of defining x^p for all real numbers p as long as $x > 0$ (the domain of $\ln x$). Notice that

$$\ln(x^p) = p \ln x$$

is well-defined for all $x > 0$ and can be written in inverse form as an exponential function:

$$x^p = e^{p \ln x} \qquad (7.57)$$

Thus, for example, $x^{\sqrt{2}} = e^{\sqrt{2}\ln x}$ and $x^{-\pi} = e^{-\pi \ln x}$. We have already defined irrational powers of e, so these are well-defined quantities.

With the definition of power functions extended as above, an important question follows: is the derivative of a power function still given by the formula in Chapter 6?

The answer is yes, and we use the chain rule to prove it:

$$\frac{d}{dx}(x^p) = \frac{d}{dx}(e^{p \ln x}) = e^{p \ln x}\frac{d}{dx}(p \ln x) = x^p p \frac{1}{x} = px^{p-1}$$

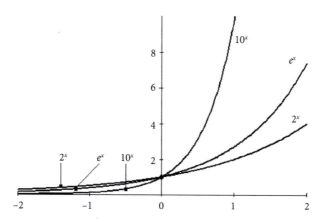

Fig. 7.23 Exponential functions with different bases

7.5.3 The general exponential and logarithmic functions

We now discuss the extensions of the natural exponential and logarithmic functions. It is more convenient to start with the general exponential functions.

The standard way to extend the concept of "raising to a power" to all real numbers is by defining a continuous function a^x that has the properties in (7.55). Then we can use this continuity and the fact that rational numbers are dense among the real numbers to bump up from rational powers and bases to all real powers and (positive) bases, as we did for e^x. Following the lead in (7.57), we define

$$a^x = e^{x \ln a} \tag{7.58}$$

The function $a^x = e^{\ln x}$ is *the (general) exponential function in base a* where $a > 0$ and $a \neq 1$. If $a = e$, then we recover the natural exponential function (in base e) since $\ln e = 1$.

Notice that the domain of a^x is all of $\mathbb{R}$ since the exponent $x \ln a$ in $e^{x \ln a}$ sweeps $\mathbb{R}$ as x does. Further, $a^x > 0$ for all real x because $e^{x \ln a} > 0$. Figure 7.23 shows the graphs of a^x for three different values of a.

Using (7.58), we can derive the properties of a^x rather quickly. For example, we know that a^x is differentiable because $e^{x \ln a}$ is, and the derivative formula is found using the chain rule:

$$\frac{d}{dx}a^x = \frac{d}{dx}e^{x \ln a} = e^{x \ln a}\frac{d}{dx}(x \ln a) = a^x \ln a \tag{7.59}$$

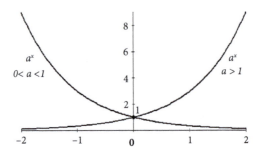

Fig. 7.24 Exponential functions with bases larger and smaller than 1

The algebraic properties of exponents are likewise a breeze; for instance, for all real numbers u, v

$$a^{u+v} = e^{(u+v)\ln a} = e^{u\ln a + v\ln a} = e^{u\ln a}e^{v\ln a} = a^u a^v$$

We can quickly gain significant information from (7.59) about the function a^x. Specifically, the derivative is negative if $a < 1$ since $\ln a < 0$, and the derivative is positive if $a > 1$; see Figure 7.24.

The following is an immediate consequence of our earlier results.

Corollary 269 *The general exponential function a^x is continuous and monotone on $\mathbb{R}$. It is decreasing if $a < 1$ and increasing if $a > 1$.*

The *monotone feature* of a^x implies that it has an inverse function whose domain is the range $(0, \infty)$ of a^x. Like the case $a = e$, the inverse function is a logarithmic function that we call *the logarithm to the base a*

$$y = \log_a x \quad \text{if} \quad x = a^y$$

Of course, $\ln x = \log_e x$ is the logarithm to the base e. This definition gives us the useful inverse function relations:

$$\log_a(a^x) = x, \qquad a^{\log_a x} = x$$

From these we may derive the various properties of the base a logarithm function, which are identical to those of the natural logarithm $\ln x$. Just as we defined $a^x = e^{x\ln a}$ we can define $\log_a x$ in terms of the natural logarithm. From the inverse function relations we have:

$$x = a^{\log_a x} = e^{(\log_a x)(\ln a)}$$

The natural logarithms of each side must then be equal:

$$(\log_a x)(\ln a) = \ln x$$

$$\log_a x = \frac{\ln x}{\ln a} \qquad (7.60)$$

It is worth noticing that if $a < 1$, then $\ln a < 0$; so the graph of $\log_a x$ is a flipped version of the graph of $\ln x$.

We obtain the properties of $\log_a x$ quickly using (7.60). For example, since $1/\ln a$ is a constant and $\ln x$ is differentiable, it follows that $\log_a x$ is differentiable too, and its derivative formula is

$$\frac{d}{dx} \log_a x = \frac{1}{x \ln a}$$

It worth mentioning here that since scientific and engineering calculations usually involve powers of 10, it is more convenient to use the base 10 logarithm in those contexts than the natural logarithm; we often suppress the base to avoid notational clutter. Using (7.60) we have

$$\log_{10} x = \log x = \frac{\ln x}{\ln 10} \simeq 0.43 \ln x \qquad (7.61)$$

Hyperbolic functions

Based on the analogy with the complex exponential e^{ix} that defines the trigonometric functions $\cos x$ and $\sin x$ as the real and imaginary parts of e^{ix}, respectively, we may define the *real* analogs as follows:

$$\cosh x = \frac{e^x + e^{-x}}{2}, \qquad \sinh x = \frac{e^x - e^{-x}}{2} \qquad (7.62)$$

The name of the first function is commonly pronounced as "cosh x" and of the second as "sinch x." Also in analogy with trigonometric function $\tan x$, we define

$$\tanh x = \frac{\sinh x}{\cosh x}$$

The graphs of these functions are illustrated in Figure 7.25. It is an interesting coincidence that the graph of $\tanh x$ (thick curve) resembles the graph of the inverse tangent function $\tan^{-1} x$ in a qualitative sense.

These functions have algebraic and analytical properties that are quite similar (though not identical) to those of the trigonometric functions. For example, both

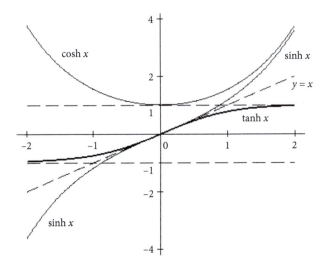

Fig. 7.25 The three hyperbolic functions

have all real numbers $(-\infty, \infty)$ as their domain, and they satisfy the analog of the Pythagorean identity:

$$\cosh^2 x - \sinh^2 x = 1 \quad \text{for all } x \in (-\infty, \infty) \qquad (7.63)$$

as can be verified by substituting the expressions defined in (7.62) in the right-hand side of (7.63) and simplifying the result (see Exercise 293). Similarly, we can show that

$$\cosh(2x) = 2\cosh^2 x - 1, \qquad \sinh(2x) = 2\sinh x \cosh x$$

which are analogous to the trigonometric double-angle identities. The derivative formulas are also directly obtained from (7.62):

$$\frac{d}{dx}\cosh x = \sinh x, \qquad \frac{d}{dx}\sinh x = \cosh x \qquad (7.64)$$

from which we readily obtain the antiderivative formulas using the second half of the fundamental theorem of calculus (Theorem 259); see Exercise 293.

The functions $\cosh x$, $\sinh x$, and $\tanh x$ are called *hyperbolic* again by analogy with the trigonometric (or circular) functions. If we set $u = \cosh x$ and $v = \sinh x$, then by (7.63) the point (u, v) is on a branch[9] of the hyperbola $u^2 - v^2 = 1$—the same way the point $(\cos x, \sin x)$ defines a point on the unit circle $u^2 + v^2 = 1$.

---

[9] This is the positive or right-hand branch, to be precise, since $\cosh x \geq 1$ for all x.

7.6 The improper Riemann integral

Earlier we discussed the function $f(x) = 1/\sqrt{x}$ that is continuous for $x > 0$, and has an antiderivative $F(x) = 2\sqrt{x}$ that is defined and continuous for all $x \geq 0$. But $f(x)$ is unbounded (has an infinite discontinuity) at 0, so it is not Riemann integrable on any interval that contains 0, like $[0,1]$, regardless of what value we assign to $f(0)$.

On the other hand, unlike $f(x)$, its antiderivative $F(x)$ is defined at 0; in fact, $F(1) - F(0) = 2$, so one side of (7.35) in Theorem 257 is well-defined. Our aim in this section is to fix this disparity using *limits*.

| **Using limits to deal with infinity: Examples** |

Since the difficulties arise at 0 where $1/\sqrt{x}$ is undefined, suppose that we move a little away from it: $1/\sqrt{x}$ is continuous (hence integrable) on the interval $[a, 1]$ for $0 < a < 1$, and the first part of Theorem 257 gives

$$\int_a^1 \frac{1}{\sqrt{x}}dx = 2\sqrt{1} - 2\sqrt{a} = 2 - 2\sqrt{a}$$

If we take a sequence of values for a that converges to 0, say $1/n$, then for each n

$$\int_{1/n}^1 \frac{1}{\sqrt{x}}dx = 2 - 2\sqrt{\frac{1}{n}}$$

As $n \to \infty$, we see that in the limit

$$\lim_{n\to\infty} \int_{1/n}^1 \frac{1}{\sqrt{x}}dx = \lim_{n\to\infty}\left(2 - 2\sqrt{\frac{1}{n}}\right) = 2 - 2\sqrt{\lim_{n\to\infty}\frac{1}{n}} = 2$$

This argument is valid for *any* sequence x_n of positive real numbers that converges to 0 because $\sqrt{x}$ is a continuous function for $x \geq 0$. Therefore, for every sequence x_n in $(0, 1)$ that converges to 0,

$$\lim_{n\to\infty} \int_{x_n}^1 \frac{1}{\sqrt{x}}dx = 2 = F(1) - \lim_{n\to\infty} F(x_n)$$

This discussion suggests that the fundamental theorem of calculus may hold *in the limit* for an unbounded function when a continuous antiderivative exists.

The continuity of the antiderivative is important. To illustrate, consider the function $f(x) = 1/x$ that also has an antiderivative: $F(x) = \ln x$, but

$$\lim_{n \to \infty} \int_{1/n}^{1} \frac{1}{x} dx = \lim_{n \to \infty} \left(\ln 1 - \ln \frac{1}{n} \right) = \lim_{n \to \infty} \left(\ln n \right) = \infty$$

The difference between $f(x) = 1/x$ and $f(x) = 1/\sqrt{x}$ is this: the antiderivative of $1/\sqrt{x}$ is $2\sqrt{x}$, which is defined at $x = 0$; but the antiderivative of $1/x$, namely, $\ln x$, is *not* defined at 0. Therefore, *where the antiderivative $F(x)$ is continuous, we should be able to define the integral as a finite number via that limit* even if the function $f(x)$ is not Riemann integrable. This is the main idea behind the improper Riemann integral; even the terminology of limits (convergence or divergence) carries over to the standard definition.

7.6.1 Unbounded functions

We now give the formal definition to clarify the preceding discussion.

The improper integral of unbounded functions

Assume that a function $f(x)$ defined on the interval $(a, b]$ is Riemann integrable on $[r, b]$ for all $r \in (a, b)$. We define the *improper Riemann integral of $f(x)$* on $[a, b]$ as

$$\lim_{n \to \infty} \int_{x_n}^{b} f(x)dx$$

if the limit exists and has the *same value* for *every* sequence x_n in $(a, b]$ that converges to a. In this case, we say that the improper integral *converges* (to the limit). Otherwise, the improper integral *diverges*. The above definition may be stated alternatively using the following notation:

$$\int_{a}^{b} f(x)dx = \lim_{r \to a^+} \int_{r}^{b} f(x)dx$$

which is consistent with the definition of a one-sided limit that we discussed in Chapter 6.

Note that if $f(x)$ is actually continuous on $(a, b]$, then according to the second half of the fundamental theorem of calculus (Theorem 259)

$$\int_x^b f(x)dx = F(b) - F(x), \quad x \in (a, b]$$

where F is an antiderivative of f. Now, if $\lim_{x \to a^+} F(x)$ exists, i.e., if F is continuous on $[a, b]$, then the above definition can be stated as

$$\int_a^b f(x)dx = F(b) - \lim_{x \to a^+} F(x)$$

just as we discussed in our earlier example where $F(x) = \sqrt{x}$.

Despite the slight abuse of notation, the left-hand side in the above definition does not say that f is Riemann integrable on $[a, b]$. We are simply saying that the integral on the left converges to the answer that we get on the right.

A similar definition exists for the case where $f(x)$ is defined on a half-open interval $[a, b)$ but not bounded at b. We write

$$\int_a^b f(x)dx = \lim_{n \to \infty} \int_a^{x_n} f(x)dx$$

whenever the limit on the right exists and has the same value for every sequence x_n in $[a, b)$ that converges to b. Equivalently, we may also use the left-limit notation to write

$$\int_a^b f(x)dx = \lim_{r \to b^-} \int_a^c f(x)dx$$

Combinations of the above two cases are used to deal with other possible cases where $f(x)$ is unbounded at some point *inside* $[a, b]$.

If $f(x)$ is unbounded at $x = c$ where $a < c < b$, then we consider the improper integrals on the two sides of c in the way defined above and define

$$\int_a^b f(x)dx = \lim_{n \to \infty} \int_a^{x_n} f(x)dx + \lim_{n \to \infty} \int_{y_n}^b f(x)dx$$

If both of the two limits above exist for all sequences x_n in $[a, c)$ and all sequences y_n in $(c, b]$ that converge to c, then the integral $\int_a^b f(x)dx$ converges to whatever number the sum of the two limits is. If either one of the two limits fails to exist, then $\int_a^b f(x)dx$ diverges. In the one-sided limit notation, we have

$$\int_a^b f(x)dx = \lim_{r \to b^-} \int_a^r f(x)dx + \lim_{r \to a^+} \int_r^b f(x)dx$$

For example, consider

$$\int_{-1}^1 \frac{1}{x^2} dx$$

The function $1/x^2$ is unbounded at $x = 0$, so we define the above integral as

$$\int_{-1}^1 \frac{1}{x^2} dx = \lim_{r \to 0^-} \int_{-1}^r \frac{1}{x^2} dx + \lim_{r \to 0^+} \int_r^1 \frac{1}{x^2} dx$$

$$= \lim_{r \to 0^-} \left(1 - \frac{1}{r}\right) - \lim_{r \to 0^+} \left(1 + \frac{1}{r}\right)$$

Both $\lim_{r \to 0^-}(1/r)$ and $\lim_{r \to 0^+}(1/r)$ diverge, so we conclude that the improper integral diverges too.

It is worth mentioning that if we missed the discontinuity at 0 and used the antiderivative $-1/x$ of $1/x^2$ (incorrectly) over the interval $[-1, 1]$, then we would get

$$\int_{-1}^1 \frac{1}{x^2} dx = -\frac{1}{1} + \frac{1}{-1} = -2$$

This answer is wrong for another reason too: $1/x^2 > 0$ for all x (except $x = 0$ where it is not defined), so the monotone property of integrals requires a non-negative answer for the integral, not -2.

7.6.2 Unbounded intervals

When we visually examine a plot of $1/x$ alongside that of $1/x^2$, it is their qualitative similarity that strikes us more than their differences: they are both decreasing, become infinitely large as $x \to 0$, and go to 0 monotonically as $x \to \infty$. When we superimpose one plot over the other, we also notice that $1/x$ approaches 0 more slowly than $1/x^2$ without either one of them actually reaching 0. The regions under the graphs of these functions are unbounded in both cases.

But there is a subtle quantitative difference between the two functions that manifests itself in a limit. To see what that is, we calculate the integrals below, which we can also think of as areas under the graphs of each function on the same interval $[1, r]$ where $r > 1$:

$$\int_1^r \frac{1}{x}dx = \ln r, \qquad \int_1^r \frac{1}{x^2}dx = 1 - \frac{1}{r}$$

Notice that as $r \to \infty$

$$\lim_{r \to \infty} \int_1^r \frac{1}{x}dx = \infty, \qquad \lim_{r \to \infty} \int_1^r \frac{1}{x^2}dx = 1$$

The limit on the left is not surprising if we think of the integral inside the limit as the area of a region that becomes unbounded as $r \to \infty$. It is the limit on the right that seems paradoxical since it states that the area of an unbounded region is 1 square unit.

This situation is not inconsistent with the theory that we developed above; in fact, it is of exactly the same nature as the improper integral $\int_0^1 (1/\sqrt{x})dx = 2$ being finite. In both cases, we have manifestations of infinity that are intuitively paradoxical but logically consistent and therefore possible outcomes. Indeed, we cannot justify ruling out one case without having to rule out the other. Because of this technical aspect, we need to develop the theory of improper Riemann integration relying on logic and theory rather than intuition.[10]

To that end, here are the formal definitions.

The improper Riemann integral, unbounded intervals

Let a be a fixed real number and assume that $f(x)$ is Riemann integrable on the intervals $[a, r]$ for all real numbers $r > a$. The *improper Riemann integral* of f on the interval $[a, \infty)$ is defined as

$$\int_a^\infty f(x)dx = \lim_{n \to \infty} \int_a^{x_n} f(x)dx$$

provided that the limit exists and has the *same value* for *every* sequence x_n in $[a, \infty)$ that diverges to ∞. In this case, we say that the integral on the left-hand side *converges*. Otherwise, the integral *diverges*. We may equivalently write

$$\int_a^\infty f(x)dx = \lim_{r \to \infty} \int_a^r f(x)dx$$

---

[10] A somewhat lighter discussion of these issues using graphical and numerical examples can be found in my self-study book *Achieving Infinite Resolution*.

To illustrate, recall that the derivative of $\tan^{-1} x$ is $1/(x^2 + 1)$, so

$$\int_0^\infty \frac{2}{x^2 + 1} \, dx = 2 \lim_{r \to \infty} \int_0^r \frac{1}{x^2 + 1} \, dx = 2 \lim_{r \to \infty} \tan^{-1} r = \pi$$

We conclude that the improper integral converges (to the number π). On the other hand, since $2x/(x^2 + 1)$ is the derivative of $\ln(x^2 + 1)$, we have

$$\int_0^\infty \frac{2x}{x^2 + 1} \, dx = \lim_{r \to \infty} \int_0^r \frac{2x}{x^2 + 1} \, dx = \lim_{r \to \infty} \ln(r^2 + 1) \qquad (7.65)$$

For the sequence $x_n = \sqrt{n - 1}$, we find that $\lim_{n \to \infty} x_n = \infty$ and

$$\lim_{n \to \infty} \ln(x_n^2 + 1) = \lim_{n \to \infty} \ln n = \infty$$

Therefore, the improper integral in (7.65) diverges.

We can often use known results, together with the properties of the integral, to prove convergence or divergence of improper integrals. For example, notice that

$$0 \le \frac{4}{x^3 + x^2 - 7} \le \frac{4}{x^2 + 1}$$

for all $x \in [2, \infty)$ since $x^3 - 7 \ge 1$ in this interval. The monotone property of the Riemann integral now implies that

$$0 \le \int_2^r \frac{4}{x^3 + x^2 - 7} \, dx \le 4 \int_2^r \frac{1}{x^2 + 1} \, dx = 4 \tan^{-1} r - 4 \tan^{-1} 2 < 4 \tan^{-1} r$$

for all $r > 2$. Next, we take the limit as $r \to \infty$ and use the monotone property of limits to conclude that

$$\int_2^\infty \frac{4}{x^3 + x^2 - 7} \, dx = 4 \lim_{r \to \infty} \int_2^r \frac{1}{x^3 + x^2 - 7} \, dx < 4 \lim_{r \to \infty} \tan^{-1} r = 2\pi$$

We proved that an integral converges without having a formula for the antiderivative of the function inside. Because we don't have the exact antiderivative, we don't get the exact value of the improper integral, but we can see that the value is less than 2π. In Exercise 305, you use similar arguments to prove convergence or divergence of some integrals.

The improper integral over an unbounded interval like $(-\infty, b]$ is defined similarly for a function f that is integrable over intervals of type $[r, b]$ for all real numbers $r < b$. Specifically,

$$\int_{-\infty}^{b} f(x)dx = \lim_{n \to \infty} \int_{x_n}^{b} f(x)dx \qquad (7.66)$$

if the limit exists and has the same value for every sequence x_n in $(-\infty, b]$ that diverges to $-\infty$. In this case, the integral on the left-hand side of (7.66) *converges*. Otherwise, it *diverges*.

We may equivalently write (7.66) as

$$\int_{-\infty}^{b} f(x)dx = \lim_{r \to -\infty} \int_{r}^{b} f(x)dx \qquad (7.67)$$

To illustrate this case, we calculate

$$\int_{-\infty}^{0} e^x dx = \lim_{r \to -\infty} \int_{r}^{0} e^x dx = \lim_{r \to -\infty} (1 - e^r) = 1$$

to find that the above improper integral converges to 1.

The definitions are completed by including the double-infinite interval. If f is defined on $(-\infty, \infty)$, and there is a real number c such that both $\int_{-\infty}^{c} f(x)dx$ and $\int_{c}^{\infty} f(x)dx$ converge, then the following integral *converges*:

$$\int_{-\infty}^{\infty} f(x)dx = \int_{-\infty}^{c} f(x)dx + \int_{c}^{\infty} f(x)dx \qquad (7.68)$$

Otherwise, the integral on the left-hand side above *diverges*.

For illustration, we use our earlier observation involving $\tan^{-1} x$ to calculate

$$\int_{-\infty}^{\infty} \frac{1}{x^2 + 1} dx = \int_{-\infty}^{0} \frac{1}{x^2 + 1} dx + \int_{0}^{\infty} \frac{1}{x^2 + 1} dx = -\frac{\pi}{2} + \frac{\pi}{2} = 0$$

It is important to realize that this is *not* the same as the limit

$$\lim_{r \to \infty} \int_{-r}^{r} f(x)dx \qquad (7.69)$$

The existence (or not) of the limit in (7.69) has no bearing on the existence (or not) of the integral on the left side of (7.68); see Exercise 303.

Properties of integral not preserved by improper integrals

At this point, it is necessary to mention that some properties of the Riemann integral don't hold for the improper integral.

Earlier in Corollary 253, we saw that if f is integrable on an interval $[a, b]$, then so are $|f|$ and f^2. But as we have already seen, $f(x) = 1/\sqrt{x}$ has a convergent improper integral on $[0, 1]$ with $\int_0^1 f(x)dx = 2$, whereas $f^2(x) = 1/x$ has a divergent improper integral, i.e., its integral fails to have a finite value. Also recall that while $1/x$ has a divergent improper integral on $[1, \infty)$, its square $1/x^2$ has a convergent integral $\int_1^\infty (1/x^2)dx = 1$.

To show that f having a convergent improper integral doesn't imply that the integral of $|f|$ also converges, consider

$$\int_\pi^\infty \frac{\sin x}{x}dx$$

In Exercise 302, you can prove that this integral converges. Here with $f(x) = \sin x/x$, we prove that the integral of $|f|$ diverges. We find a sequence $r_n \in (\pi, \infty)$ such that $r_n \to \infty$ and

$$\lim_{n\to\infty} \int_\pi^{r_n} \frac{|\sin x|}{x}dx = \infty$$

Based on the periodic nature of $\sin x$, let $r_n = (n + 1)\pi$, and note that

$$\int_\pi^{r_n} \frac{|\sin x|}{x}dx = \sum_{j=1}^n \int_{j\pi}^{(j+1)\pi} \frac{|\sin x|}{x}dx$$

For each $j = 1, 2, \ldots, n$, since $j\pi \le x \le (j + 1)\pi$, we conclude that

$$\frac{1}{x} \ge \frac{1}{(j + 1)\pi}$$

so that

$$\int_\pi^{r_n} \frac{|\sin x|}{x}dx \ge \sum_{j=1}^n \frac{1}{(j + 1)\pi} \int_{j\pi}^{(j+1)\pi} |\sin x|dx \qquad (7.70)$$

Further, since $|\sin x|$ is always non-negative, we may use the decomposition property of the Riemann integral to conclude that

$$\int_{j\pi}^{(j+1)\pi} |\sin x|dx \ge \int_{j\pi+\pi/6}^{(j+1)\pi-\pi/6} |\sin x|dx = \int_{(j+1/6)\pi}^{(j+5/6)\pi} |\sin x|dx$$

since the interval $[j\pi, (j+1)\pi]$ can be decomposed as a union of non-overlapping intervals:

$$\left[j\pi, \left(j+\frac{1}{6}\right)\pi\right] \cup \left[\left(j+\frac{1}{6}\right)\pi, \left(j+\frac{5}{6}\right)\pi\right] \cup \left[\left(j+\frac{5}{6}\right)\pi, (j+1)\pi\right]$$

Next, note that

$$|\sin x| \geq \frac{1}{2} \quad \text{for } x \in \left[\left(j+\frac{1}{6}\right)\pi, \left(j+\frac{5}{6}\right)\pi\right]$$

so

$$\int_{j\pi}^{(j+1)\pi} |\sin x| dx \geq \int_{(j+1/6)\pi}^{(j+5/6)\pi} \frac{1}{2}dx = \frac{1}{2}\left[\left(j+\frac{5}{6}\right)\pi - \left(j+\frac{1}{6}\right)\pi\right] = \frac{\pi}{3}$$

Going back to (7.70), we obtain

$$\int_{\pi}^{r_n} \frac{|\sin x|}{x} dx \geq \sum_{j=1}^{n} \frac{1}{3(j+1)}$$

Now, taking the limit gives

$$\lim_{n\to\infty} \int_{\pi}^{r_n} \frac{|\sin x|}{x} dx = 3 \lim_{n\to\infty} \sum_{j=1}^{n} \frac{1}{j+1} = \infty$$

because as we found in Chapter 5, the harmonic series diverges to ∞.

An infinite boundary, finite area paradox

In earlier discussion in Chapter 5 involving infinite series, we came across the Tower of Boxes where we found constructs having finite volume but infinite surface area. The improper Riemann integral is another topic in analysis that is ripe for the occurrence of infinity paradoxes. This may not be so striking given that improper integrals involve infinity explicitly, but there is really no explaining away the paradoxical nature of the results. Of course, we won't have a problem as long as we don't try to associate physical objects to the mathematical constructs.

Specifically, consider the integral $\int_{1}^{\infty}(1/x^2)dx$ whose value we calculated above to be 1. Since $1/x^2$ is non-negative, we may interpret the integral's value as the area under the graph of $1/x^2$ from $x = 1$ to infinity. This is a region in the first quadrant of the coordinate system that is bounded by the curve $1/x^2$ on top, the x-axis at the bottom, and a segment of the line $y = 1$ on the right. Apart from this segment, the rest of the boundary or perimeter of the region of area 1 is infinitely long, as the region itself is unbounded.

In our discussion of the Tower of Boxes, we discovered that the object didn't need to be unbounded or have infinite extent in order to have a paradoxical-looking situation (finite volume but infinite surface area). In Chapter 9, we encounter a bounded (non-fractal) region of finite area but with perimeter that can be arbitrarily large. Infinity is full of surprises!

7.7 The integral test for infinite series

If k is a positive integer, then we discovered earlier (Exercise 172, Chapter 5) that the series

$$\sum_{n=1}^{\infty} \frac{1}{n^k}$$

converges for $k = 2, 3, 4, \ldots$ and diverges for $k = 1$. In this section, we prove that this fact can be extended to all real numbers $p > 0$. The extension is significant as it includes fractional values of p that correspond to roots of n. The tool that we use to prove this fact uses the improper Riemann integral on the infinite interval $[a, \infty)$ for a fixed real number a.

Theorem 270 *(The integral test for series) Consider the infinite series $\sum_{k=1}^{\infty} a_k$ whose terms a_k form a decreasing sequence of non-negative real numbers after some index m. If f(x) is a nondecreasing function such that $f(k) = a_k$ for $k \geq m$, then*

$$\sum_{k=1}^{\infty} a_k < \infty \quad \text{if and only if} \quad \int_{m}^{\infty} f(x)dx < \infty$$

Proof First, notice that $f(x) \geq 0$ for all $x \in [m, \infty)$; if not, then $f(r) < 0$ for some $r \geq m$. But if we pick an integer $k > r$ (using the Archimedean property), then $f(k) = a_k \geq 0$, and it follows that $f(k) > f(r)$ with $k > r$, which contradicts the fact that f is nondecreasing.

Next, f is integrable on every interval $[m, b]$ with $b > m$ by Theorem 249. Let $P_n = \{m, m + 1, \ldots, n\}$ for $n \geq m + 1$, and note that since f is nondecreasing, for every k between $m + 1$ and n

$$\sup\{f(x) : x \in [k - 1, k]\} = f(k - 1) = a_{k-1}$$
$$\inf\{f(x) : x \in [k - 1, k]\} = f(k) = a_k$$

See Figure 7.26, which illustrates the situation with $m = 1$. It follows that

$$\sum_{k=m+1}^{n} a_k = L(P_n) \leq \int_{m}^{n} f(x)dx \leq U(P_n) = \sum_{k=m}^{n-1} a_k$$

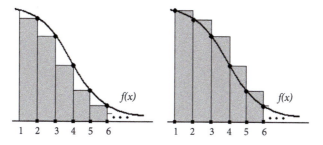

Fig. 7.26 The bounds furnished by the integral test

Taking the limit as $n \to \infty$, we conclude that if the integral converges, then

$$\sum_{k=m+1}^{\infty} a_k \le \int_m^{\infty} f(x)\,dx < \infty$$

in which case the original series converges too because

$$\sum_{k=1}^{\infty} a_k = \sum_{k=1}^{m} a_k + \sum_{k=m+1}^{\infty} a_k < \infty$$

Otherwise, the integral diverges, and since

$$\sum_{k=m}^{\infty} a_k \ge \int_m^{\infty} f(x)\,dx$$

the series diverges too. ∎

To illustrate Theorem 270, consider the infinite series

$$\sum_{k=1}^{\infty} \frac{k}{e^k} = \frac{1}{e^1} + \frac{2}{e^2} + \frac{3}{e^3} + \cdots \qquad (7.71)$$

Here $f(k) = k/e^k = ke^{-k}$ so we consider the function $f(x) = xe^{-x}$. If $x > 0$, then $xe^{-x} > 0$; so $f(x)$ is a positive function. Further, as a product of two differentiable functions, $f(x)$ is differentiable with derivative

$$F'(x) = (1)e^{-x} + x(-e^{-x}) = (1 - x)e^{-x}$$

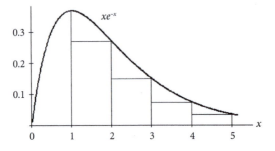

Fig. 7.27 Using the integral test to prove convergence

Since $F'(x) < 0$ when $x > 1$, it follows that $f(x)$ is decreasing on the interval $[1, \infty)$; see Figure 7.27.

Next, we check the improper integral:

$$\int_1^\infty xe^{-x}dx = \lim_{r \to \infty} \int_1^r xe^{-x}dx = \lim_{r \to \infty}(2e^{-1} - re^{-r} - e^{-r})$$

where we used integration by parts. Taking the limit as $r \to \infty$, we obtain

$$\int_1^\infty xe^{-x}dx = 2e^{-1} - \lim_{r \to \infty} re^{-r} - \lim_{r \to \infty} e^{-r} = \frac{2}{e}$$

The limits are both zero; the first is proved to be 0 using l'Hospital's rule ∞/∞ case that we discussed in Chapter 6:

$$\lim_{r \to \infty} re^{-r} = \lim_{r \to \infty} \frac{r}{e^r} = \lim_{r \to \infty} \frac{1}{e^r} = 0$$

Since the integral converges, it follows that the series in (7.71) converges too.

A useful corollary of Theorem 270 is the following, whose proof is left as an exercise.

Corollary 271 *(The p-series test) Let p be a fixed real number. Then*

$$\sum_{n=1}^\infty \frac{1}{n^p} < \infty \quad \text{if and only if } p > 1$$

For example, with $p = 1/2$, the following series diverges

$$\sum_{n=1}^\infty \frac{1}{\sqrt{n}} = \infty$$

whereas with $p = e$, the series below converges

$$\sum_{n=1}^{\infty} \frac{1}{n^e} < \infty$$

Corollary 271 is often used through the comparison test. For example, consider the series

$$\sum_{k=1}^{\infty} \frac{k}{k^3 + 1} \tag{7.72}$$

that is not easy to apply the integral test to directly. But since

$$\frac{k}{k^3 + 1} < \frac{k}{k^3} = \frac{1}{k^2}$$

and $\sum_{n=1}^{\infty} 1/n^2 < \infty$ by Corollary 271, we conclude that the series in (324) also converges.

Two points are worth emphasizing about Theorem 270.

First, with regard to the number m, it is often the case that the terms of a sequence don't conveniently decrease starting from 1 but may be decreasing *eventually*, i.e., after a certain index that we call m. For instance, the sequence

$$\sum_{k=1}^{\infty} \frac{1}{(2k-7)^2} \tag{7.73}$$

is increasing for the first three terms and decreases thereafter; further, the function $f(x) = 1/(2x - 7)^2$ that naturally corresponds to the above series is increasing for $x < 7/2 = 3.5$ and not even integrable on any interval that contains $x = 7/2$. But it is decreasing and integrable over all bounded intervals contained in $[4, \infty)$. So with

$$\int_4^{\infty} \frac{1}{(2x-7)^2} dx = \frac{1}{2}\left(1 - \lim_{r \to \infty} \frac{1}{2r - 7}\right) = \frac{1}{2}$$

we conclude the series in (7.73) converges too.

Another point worth mentioning about Theorem 270 is that the integral provides an upper bound for the value of the infinite series. For example, the series in (7.71) converges to a number that is less than $2/e$; similarly, the series in (7.73) converges to a limit that is bounded as follows:

$$\sum_{k=1}^{\infty} \frac{1}{(2k-7)^2} = \frac{1}{(2-7)^2} + \frac{1}{(4-7)^2} + \frac{1}{(6-7)^2} + \sum_{k=1}^{\infty} \frac{1}{(2k-7)^2}$$

$$\leq \frac{1}{25} + \frac{1}{9} + 1 + \frac{1}{2} = \frac{743}{450} \approx 1.651$$

7.8 Exercises

Exercise 272 *Consider the interval* $[-1, 2]$.

(a) *Plot the partition P defined by the following set on the x-axis:*

$$P = \{-1, 0, 1, 2\}$$

What is the mesh of this partition?

(b) *Plot the partition P′ defined by the following set on the x-axis:*

$$P' = \left\{-1, -\frac{1}{2}, \frac{2}{3}, 1, 1.25, 1.3, 2\right\}$$

What is the mesh of this partition?

(c) *Is P′ a refinement of P, or P of P′?*

(d) *Write down the union* $P \cup P'$ *explicitly as a partition. What is its mesh?*

Exercise 273 *Let* $f(x) = x^2$, *and consider the partition* $P = \{0, 1, 2\}$ *of the interval* $[0, 2]$.

(a) *Calculate the three Riemann sums using the left, right, and midpoints of subintervals. Which sum gives a better approximation of the area under the graph of this function? Consider making a careful sketch of the graph of* x^2 *and marking the rectangular strips on it to see which of the three numbers looks best (the exact area is* $8/3$*).*

(b) *Repeat (a) for the exponential function* $f(x) = 2^x$ *(the exact value is* $3/\ln 2$*).*

Exercise 274 *Consider* $f(x) = x^2 + 1$ *and the partition* $P = \{-2, -1, 0, 1, 2\}$ *of the interval* $[-2, 2]$. *Sketch graphs like those in Figures 7.7 and 7.8 for this case, and calculate both* $R_L(P)$ *and* $R_U(P)$. *For comparison, also find the Riemann sum with* x_i^* *being the midpoint of each subinterval.*

Exercise 275 *Suppose that* $f(x)$ *is a **nondecreasing** function on an interval* $[a, b]$. *Explain why the following equalities with the indicated values of* x_i^* *are valid:*

$$L(P) = R_L(P) = \sum_{i=1}^{N} f(x_{i-1}) \Delta x_i$$

$$U(P) = R_U(P) = \sum_{i=1}^{N} f(x_i) \Delta x_i$$

Sketching a graph may be helpful here. How do you think $R_L(P)$ *and* $R_U(P)$ *change for a **nonincreasing** function* $f(x)$*?*

Exercise 276 *Prove that if f(x) is continuous, then $L(P) = R_L(P)$ and $U(P) = R_U(P)$.*

Exercise 277 *Suppose that f(x) = C for all x in an interval $[a, b]$. Show that for every partition P of the interval, all Darboux sums are equal:*

$$L(P) = U(P) = C(b - a)$$

Thus the constant function f(x) = C is integrable, and its integral is

$$\int_a^b C dx = C(b - a)$$

Exercise 278 *Prove Theorem 249 in the case f is a nonincreasing function.*

Exercise 279 *Complete the proof of Theorem 250.*

Exercise 280 *Let α, β be real numbers with $\alpha \neq 0$. Use a calculation similar to that in the example after Theorem 247 to prove that*

$$\int_0^1 (\alpha x + \beta) dx = \frac{\alpha}{2} + \beta$$

Exercise 281 *Complete the proof of Theorem 256 by proving (7.33).*

Exercise 282 *Prove that the function below is differentiable everywhere, including at x = 0 with derivative F'(0) = 0:*

$$F(x) = x^2 \sin \frac{1}{x} \quad \text{if } x \neq 0, \quad F(0) = 0$$

Note that this is an example of a function that is differentiable everywhere, but its derivative F'(x) is discontinuous at x = 0. Can F'(x) be made continuous by redefining F'(0) to be some suitable number?

Exercise 283

(a) *By taking the derivative, verify that an antiderivative of f(x) = |x| is*

$$F(x) = \frac{1}{2}x|x|$$

(b) *Use Theorem 259 and (a) to find*

$$\int_{-1}^2 |x| dx$$

Exercise 284 *Let $f(x) = \lfloor x \rfloor$ be the floor function, i.e., the greatest integer that is less than or equal to the real number x; for instance, $\lfloor \sqrt{2} \rfloor = 1$ and $\lfloor -0.43 \rfloor = -1$.*

(a) *Sketch the graph of f for $-2 \leq x \leq 3$ and explain why f is integrable on $[-2, 3]$.*

(b) *Define $F(x) = \int_0^x \lfloor t \rfloor dt$. Explain why F is continuous and further, non-increasing if $-2 \leq x \leq 0$ and nondecreasing if $0 \leq x \leq 3$ (note that $F(x) = 3 + \int_{-2}^x \lfloor t \rfloor dt$).*

(c) *Is $F(x)$ an antiderivative of $\lfloor x \rfloor$ on $[-2, 3]$? Explain.*

(d) *Prove that a formula for F on $[-2, 3]$ is the following, and then sketch its graph:*

$$F(x) = \begin{cases} -2x - 1 & \text{if } -2 \leq x < -1 \\ -x & \text{if } -1 \leq x < 0 \\ 0 & \text{if } 0 \leq x < 1 \\ x - 1 & \text{if } 1 \leq x < 2 \\ 2x - 3 & \text{if } 2 \leq x \leq 3 \end{cases}$$

Note that F is not differentiable at $x = n$ where n is an integer in $[-2, 3]$.

Exercise 285

(a) *Find the values of x at which $F(x) = \int_0^x \sin(t^2) dt$ has either a maximum or a minimum; see Figure 7.17.*

(b) *Which of these numbers correspond to the maxima?*

(c) *Why does F have neither a maximum nor a minimum at $x = 0$ even though $F'(0) = 0$?*

Exercise 286 *Find the formula for the following derivative*

$$\frac{d}{dx} \int_{-2x}^{\sqrt{x}} \cos(t^2) dt$$

Note that the integral may be decomposed into two integrals of the type in Corollary 261.

Exercise 287 *Find the average value of $f(x) = x^2$ first over the interval $[-1, 1]$ and then over $[0, 2]$, which has the same length. In each case, find the point(s) c and the height(s) $f(c)$ of the rectangle that satisfies the equality in (7.42).*

Exercise 288

(a) Find the average value of $\sin x$ for one of its humps, say, $0 \leq x \leq \pi$. What is the average value for two adjacent humps over $0 \leq x \leq 2\pi$? Can you tell what it is without explicitly calculating it?

(b) Repeat (a) for $x \sin x$.

Exercise 289

(a) Use the fact that $u(1/u) = 1$ and the equality in (7.50) to prove that for all $u > 0$

$$\ln\left(\frac{1}{u}\right) = -\ln u \qquad (7.74)$$

Now use (7.74) to prove that

$$\ln\frac{u}{v} = \ln u - \ln v$$

(b) Prove that $\lim_{x \to 0^+} \ln x = -\infty$ (consider what happens to $u = 1/x$ as $x \to 0^+$ and use (a).)

Exercise 290 Use an argument similar to the one used to obtain (7.56) to prove each of the following identities:

$$E(-u) = \frac{1}{E(u)} \qquad E(pu) = [E(u)]^p$$

where p is a fixed real number. In the exponential notation, these identities translate into

$$e^{-u} = \frac{1}{e^u} \qquad e^{pu} = (e^u)^p$$

Exercise 291 Prove the following properties of the general exponential function:

$$a^{-u} = \frac{1}{a^u} \qquad a^{pu} = (a^u)^p \qquad (ab)^u = a^u b^u$$

Assume that $p \in \mathbb{R}$ and $b > 0$ are fixed real numbers.

Exercise 292 *Use the identity in (7.60) to show that if $a > 0$, $a \neq 1$, then*

$$\log_a 1 = 0, \qquad \log_a a = 1$$

Also verify the following logarithm identities:

$$\log_a(uv) = \log_a u + \log_a v, \qquad \log_a \frac{1}{u} = -\log_a u, \qquad \log_a(u^p) = p \log_a u$$

Assume that u, v are positive real variables, and p is any fixed real number.

Exercise 293

(a) *Prove each of the following hyperbolic identities:*

$$\cosh^2 x - \sinh^2 x = 1$$
$$\cosh(2x) = 2\cosh^2 x - 1 = 2\sinh^2 x + 1$$
$$\sinh(2x) = 2\sinh x \cosh x$$

(b) *Explain why both $\cosh x$ and $\sinh x$ are differentiable on $(-\infty, \infty)$ and show that their derivatives are given by*

$$\frac{d}{dx}\cosh x = \sinh x, \qquad \frac{d}{dx}\sinh x = \cosh x$$

(c) *Explain why the hyperbolic tangent $\tanh x$ is differentiable on $(-\infty, \infty)$ and find a formula for its derivative. Use your derivative formula to prove that $\tanh x$ is increasing on $(-\infty, \infty)$.*

Exercise 294 *We defined $\ln x$ as an integral. But since $\ln x$ is a continuous function for $x > 0$ it is also integrable on every closed interval contained in $(0, \infty)$; show that*

$$\int_1^e \ln x \, dx = 1$$

using integration by parts. Also use the above equality to prove (without integrating) that

$$\int_1^e \log x \, dx = \log e$$

Exercise 295 *Explain why in each case below, the function involved is differentiable, and use appropriate formulas to prove each equality:*

$$\frac{d}{dx}\ln\sqrt{x^2+1} = \frac{x}{x^2+1}, \qquad \frac{d}{dx}(\pi^x x^\pi) = \pi^x x^\pi \ln\pi + \pi^{x+1}x^{\pi-1},$$

$$\frac{d}{dx}(x\log x - x + e) = \log x + \log e - 1 \approx \log x - 0.5657$$

To speed up calculations for the first formula, use the fact that $\ln u^{1/2} = (1/2)\ln u$.

Exercise 296 *Calculate each of the integrals below:*

(a) $\displaystyle\int_{-1}^{0} x^2 e^x dx$
(b) $\displaystyle\int_{-1}^{0} xe^{x^2} dx$
(c) $\displaystyle\int_{1}^{2} [2^x - \log(x^2)]dx$

(d) $\displaystyle\int_{1}^{e} \frac{\ln x}{x}dx$
(e) $\displaystyle\int_{2}^{e} \frac{1}{x\ln x}dx$
(f) $\displaystyle\int_{0}^{1} \frac{x-1}{x+1}dx$

Exercise 297

(a) *Calculate the following integral where $p \in \mathbb{R}$ and $r > 0$:*

$$\int_{1}^{r} x^p \cosh(\ln x)dx$$

(b) *Show also that:*

$$\int_{1}^{e} \frac{\cosh(\ln x)}{x}dx = \sinh 1, \qquad \int_{1}^{e} \frac{\cos(\ln x)}{x}dx = \sin 1$$

(c) *Use integration by parts to find a formula for the following integral if $p \neq -1$*

$$\int_{1}^{r} x^p \cos(\ln x)dx$$

How does your answer here compare with that in (a) for $p = 0$ and $r = e$?

Exercise 298 *Calculate an antiderivative for the function $1/(e^x + 1)$; you can use the following steps:*

(a) *Show that*

$$\int_{0}^{x} \frac{1}{e^t+1}dt = \int_{0}^{x} \frac{e^t}{e^t+1}dt - \int_{0}^{x} \tanh\frac{t}{2}dt$$

(b) *Using the definition of* $\tanh x$ *to calculate the second antiderivative, show that*

$$\int_0^x \frac{1}{e^t + 1} dt = \ln\left(\frac{e^x + 1}{2}\right) - 2\ln\left(\cosh\frac{x}{2}\right)$$

Exercise 299

(a) *Explain why the following integrals are improper, and determine if they converge or diverge; find the value of the integral when convergent:*

$$(a) \int_{-1}^{2} \frac{1}{\sqrt[3]{x}} dx \qquad (b) \int_0^1 \ln x\, dx \qquad (c) \int_0^{1/2} \frac{e^x}{x^e} dx$$

In (c), notice that $e^x > 1$ *if* $x > 0$, *and recall the monotone property of the integral and of the limit.*

(b) *Explain why the following integral is not improper:*

$$\int_0^{2\pi} \frac{\sin x}{x} dx$$

However, show that the integral below is improper, and prove that it diverges:

$$\int_0^{2\pi} \frac{x}{\sin x} dx$$

Exercise 300 *Prove that the following integral converges if* $p \geq 0$ *and diverges if* $p < 0$; *further, if* $p > 0$, *show that the improper integral converges to* 0:

$$\int_0^1 x^p \ln x\, dx$$

Exercise 301 *Prove that the following improper integral converges if* $p > 1$ *and diverges if* $p \leq 1$:

$$\int_2^\infty \frac{1}{x(\ln x)^p} dx$$

Exercise 302 (a) *Prove that the following integral converges:*

$$\int_\pi^\infty \frac{|\cos x|}{x^2} dx$$

(b) *Explain why the result in (a) implies that the following integral converges:*

$$\int_{\pi}^{\infty} \frac{\cos x}{x^2} \, dx$$

(c) *Use the result in (b) and integration by parts to prove that the following integral converges too:*

$$\int_{\pi}^{\infty} \frac{\sin x}{x} \, dx$$

Exercise 303 *Show that*

$$\lim_{r \to \infty} \int_{-r}^{r} x \, dx = 0, \qquad \lim_{r \to \infty} \int_{0}^{r} x \, dx = \infty$$

The second limit leads to the (correct) conclusion that $\int_{-\infty}^{\infty} x \, dx$ diverges.

Exercise 304 *Prove that the integral below converges, and find its limit or value:*

$$\int_{-\infty}^{\infty} \frac{2e^x}{e^{2x} + 1} \, dx$$

Exercise 305 *Determine whether the following integrals converge or diverge:*

(a) $\displaystyle \int_{0}^{\infty} \frac{x + \sin x}{x^2 + \cos x} \, dx$ (b) $\displaystyle \int_{1}^{\infty} \frac{\ln x + \sin x}{x^2 + \cos x} \, dx$ (c) $\displaystyle \int_{0}^{\infty} \frac{\ln x + \sin x}{x^2 + \cos x} \, dx$

The integral in (c) is defined as

$$\int_{0}^{\infty} \frac{\ln x + \sin x}{x^2 + \cos x} \, dx = \lim_{r \to 0^+} \int_{r}^{1} \frac{\ln x + \sin x}{x^2 + \cos x} \, dx + \lim_{s \to \infty} \int_{1}^{s} \frac{\ln x + \sin x}{x^2 + \cos x} \, dx$$

The second integral is the same as that in (b). Decide whether the first integral on the right-hand side converges or diverges.

Exercise 306 *Prove the p-series test, Corollary 271.*

Exercise 307

(a) *Determine all values of $p, q \geq 0$ such that the following series converges:*

$$\sum_{k=2}^{\infty} \frac{1}{k^q (\ln k)^p}$$

Note that in the special case $q = 1$, the series is sometimes called the "logarithmic p-series." If $q \neq 1$, then consider comparing the above series with an appropriate ordinary p-series; if $0 \leq q < 1$, then you can use the limit comparison test from Chapter 5 if comparing to the series $\sum_{k=2}^{\infty} 1/k^{q+\varepsilon}$ where $q + \varepsilon \leq 1$.

(b) *In the case $q = 1$ and for all p such that the series converges, specify the upper bound of the series as derived from the integral test.*

Exercise 308 *Prove that the following series diverges for all $p \geq 0$:*

$$\sum_{k=3}^{\infty} \frac{1}{k[\ln(\ln(k))]^p}$$

Note. You can prove and then use the fact that $\ln(\ln(k)) < \ln k$ for $k \geq 3$ and use Exercise 307. Alternatively, you can first use the integral test to prove that the series

$$\sum_{k=3}^{\infty} \frac{1}{k(\ln k)[\ln(\ln(k))]^p}$$

diverges, and then compare this series with the original series.

Exercise 309 *Prove that each of the following series converge, and give an upper bound for the limit in each case:*

(a) $\displaystyle\sum_{k=1}^{\infty} \frac{k^2}{2^k}$ (b) $\displaystyle\sum_{k=4}^{\infty} \frac{\ln(2k-7)}{k(2k-7)}$

8

Infinite Sequences and Series of Functions

We introduced infinite series of numbers in Chapter 5 as limits of sequences of finite sums, applying the concept of sequence that we discussed in Chapter 3. Recall that the "infinite" in infinite series comes through the use of *limit*, not by adding infinitely many numbers. The primary interest in sequences and series from the standpoint of analysis is whether they *converge* or not. Since then, we have developed powerful methods for dealing with *functions* from the set of real numbers into itself. In this chapter, we put that theory to use in our study of sequences and series of functions.

As in earlier discussions in Chapters 3 and 5, we must show how infinity needs to be handled consistently; in fact, when dealing with functions, we discover that we must be extra careful so as to avoid errors and contradictions that were historically made by many prominent mathematicians, including the likes of Newton and Cauchy. The latter is known as one of the founding fathers of modern analysis (as an extension of calculus) because he identified many issues and problems that earlier mathematicians (including Euler) had missed; Cauchy is credited with resolving many of these problems. However, he also made some mistakes, especially when dealing with infinite sequences and series of functions, typically by assuming continuity or uniformity without realizing it since he did not explicitly state them as hypotheses.

The mathematician most prominently credited with putting on firm ground the theory that we discuss in this chapter is Karl Weierstrass (1815–1897). He is considered to have been the first to make a distinction between the two types of convergence that we discuss in this chapter: pointwise vs. uniform. For his contributions, he is known as the other founding father of modern analysis (along with Cauchy). One of his doctoral students was Georg Cantor, who is already familiar to us from earlier chapters.

One of the interesting by-products of studying infinite series of constants was their enabling us to deal with irrational numbers in a rigorous way. We learned how to represent and to approximate these numbers using sequences and series of rational numbers. Recall that the rational numbers were built up from integers using algebraic operations only (no limits involved, hence no infinity).

In this chapter, we do something similar for functions. We show that transcendental functions like $\ln x$, $\sin x$, etc., can be approximated using infinite sequences

Real Analysis and Infinity. Hassan Sedaghat, Oxford University Press.
© H. Sedaghat, (2022). DOI: 10.1093/oso/9780192895622.003.0008

of *polynomials*. The latter are simple algebraic combinations of the power functions x, x^2, x^3, ... which we may loosely liken to integers. Polynomial sequences play a role in approximating transcendental functions that is similar to sequences of rational numbers approximating irrational numbers, in the sense that both rational numbers and polynomials are defined algebraically without invoking infinity; but using them, plus the concept of limit suitably defined, we reach the otherwise enigmatic (yet indispensable) irrational numbers and transcendental functions.

The standard process of representing functions by converging sequences of polynomials involves infinite series of power functions that are appropriately called *power series*; they include the Taylor series of calculus. We discover that a large variety of transcendental functions, including functions of interest in applied mathematics, can be expanded in the form of a power series.

We discuss issues relating to the three basic properties of continuity, differentiability, and integrability of power series as limits of sequences of polynomials. Our discussion reveals the importance of uniform convergence in preserving and extending the aforementioned three properties of polynomials to power series.

After a brief discussion of Fourier series, mainly to compare trigonometric series with power series and to obtain a solution of the "Basel problem," we close the chapter with a proof of the Weierstrass approximation theorem, an important result in analysis that shows that even non-differentiable functions can be uniformly approximated by polynomials. The proof that we discuss in this chapter also furnishes a set of such approximating polynomials that were introduced by Bernstein.

8.1 The geometric series as a function series

Recall the geometric series

$$1 + x + x^2 + x^3 + \cdots = \frac{1}{1-x} \quad \text{if } -1 < x < 1 \tag{8.1}$$

Here the fraction on the right-hand side is a simple rational function that equals the sum of *all* power functions as long as x is a real number between -1 and 1. In our first encounter with infinite series, x was a constant, namely, the *common ratio*; but we are now thinking of x as a *variable* that ranges in the open interval $(-1, 1)$. To distinguish this extended version from the series of constants, we refer to the series in (8.1) as the *geometric power series*. The geometric power series has many of the same basic features as the general power series that we discuss later in this chapter.

The left-hand side of (8.1) contains an occurrence of infinity that needs to be defined independently of the right-hand side. *The equality in (8.1) does not define the left-hand side;* rather, it is saying that on the left-hand side is a certain function of x that happens to be equal to the rational function on the right as long as $-1 < x < 1$.

This distinction is important because the right-hand side of (8.1) is well defined as long as $x \neq 1$ but its left-hand side is defined only for $-1 < x < 1$ because the geometric power series diverges if $x \geq 1$ or $x \leq -1$ (that is, $|x| \geq 1$).

To define the left-hand side of (8.1) we use the sequence of partial sums. The series is defined where its sequence of partial sums converges. For the series in (8.1)

$$s_0(x) = 1$$
$$s_1(x) = 1 + x$$
$$s_2(x) = 1 + x + x^2$$
$$s_3(x) = 1 + x + x^2 + x^3$$

$$\vdots$$

Notice that each partial sum $s_n(x)$ is a polynomial whose degree n is the index of the partial sum. Polynomials are defined on all of $\mathbb{R}$, so the partial sums $s_0(x), s_1(x), s_2(x), \ldots$ are all defined for all real numbers x. Figure 8.1 shows the partial sum polynomials $s_1(x)$ to $s_4(x)$, together with the graph of the rational function (dashed) over the interval $[-1, 1]$.

In Figure 8.1, we see that between -1 and 1 the partial sums line up with their rational function limit more closely, especially in the middle, as we add more terms (power functions of higher degrees). The partial sums drift away from the rational

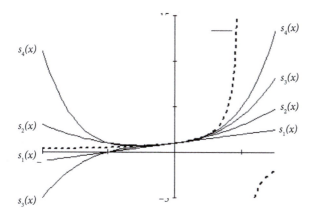

Fig. 8.1 Several partial sums of the geometric power series

function near 1 or -1; and the departure is more noticeable near $x = 1$, where the rational function has a non-removable, infinite discontinuity.

We also see in Figure 8.1 that different things happen near the two endpoints; in particular, the infinite discontinuity of the limit function $1/(1 - x)$ at $x = 1$ clearly affects how the partial sums converge to the limit function, but the divergence near $x = -1$ does not seem to involve infinity. But in fact it does; as we discover below, *infinity plays essentially the same role at both of the endpoints ± 1*, and the infinite discontinuity at the right endpoint is not the root cause. To explain this, we use the concept of ε-index that we discussed earlier.

Convergence near the right endpoint

Recall that the common domain of all the partial sums is the set of all real numbers, including $x = 1$. But the limit function $1/(1 - x)$ diverges to ∞ or $-\infty$ at $x = 1$, which raises the question as to how partial sums manage to converge to the limit function near (and to the left) of $x = 1$. Figure 8.2 shows that convergence is not as rapid near this point as at other points.

If we look closely at Figure 8.2, we notice that as the value of x approaches 1, the convergence seems to slow down (but not fully stop as long as x remains to the left of 1). To be more precise, let r be any number in $(-1, 1)$; and suppose that when $x = r$, the approximation $s_{10}(r)$, which is indicated by the dot on the partial sum curve $s_{10}(x)$ at $x = r$, just reaches within ε of the limit curve, that is,

$$\frac{1}{1 - r} - s_{10}(r) = \varepsilon$$

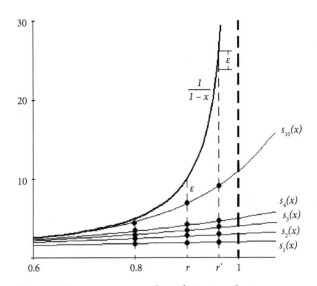

Fig. 8.2 Convergence near the infinite singularity

The exact value of ε is not important; think of it as the usual small positive quantity that measures how closely $s_{10}(r)$ approximates $1/(1-r)$. The vertical sequence of dots on the various partial sum curves right above the value r approach a point on the curve $1/(1-r)$; in particular, the dot on the curve $s_{10}(x)$ at $x = r$ is within ε of the dot on the limit curve, i.e., the dot is ε-*close* to the number $1/(1-r)$. But if we check the similar sequence of dots above $x = r'$, we see that these points fail to come within ε by a wide margin. We need to find $s_n(r')$ for a larger value of n in order to reach within ε, or be ε-close of the limit. On the other hand, at $x = 0.8$, the value $s_4(0.8)$ already seems within ε of the limit curve, and we do not actually need to go all the way to $n = 10$ if $x = 0.8$.

Let's summarize this observation:

Given a fixed (small) $\varepsilon > 0$, the *index* n of $s_n(x)$ needed to approximate the limit function $1/(1-x)$ within a distance ε (i.e., be ε-close) *depends on the value of x* in $(-1, 1)$.

Next, if we insert $x = r$ into the equality in (8.1), we get an infinite series of constants shown as vertically rising dots right above r on partial sum curves in Figure 8.2, together with their limit:

$$1 + r + r^2 + r^3 + \cdots = \frac{1}{1-r} \qquad (8.2)$$

From our discussion of converging sequences of numbers, recall that for each (arbitrarily small) value of $\varepsilon > 0$, we must find an ε-index N_ε, a positive integer that is just big enough that for all indices $n \geq N_\varepsilon$

$$\left| \frac{1}{1-r} - \left(1 + r + r^2 + \cdots + r^n\right) \right| < \varepsilon \qquad (8.3)$$

so that the partial sum is ε-close to the limit.

We can actually remove the absolute value sign since the sum of the infinite series is greater than the sum of any finite number of its terms. The calculation is made simpler by a formula that we earlier derived for the finite geometric sum or progression:

$$1 + r + r^2 + \cdots + r^n = \frac{1 - r^{n+1}}{1 - r}$$

So now (8.3) reduces to

$$\frac{1}{1-r} - \frac{1 - r^{n+1}}{1 - r} < \varepsilon$$

Simplify the left-hand side to get

$$\frac{r^{n+1}}{1-r} < \varepsilon$$

Since we are looking for the index value N_ε, we solve the above inequality for n as follows:

$$r^{n+1} < (1-r)\varepsilon$$

Taking the natural logarithm releases the exponent:

$$(n+1)\ln r < \ln((1-r)\varepsilon)$$

The direction of the inequality did not change because the natural logarithm is an increasing function. Now we divide by $\ln r$ and solve for n to get

$$n > \frac{\ln((1-r)\varepsilon)}{\ln r} \tag{8.4}$$

Here the inequality changed from < to > because $\ln r < 0$ if $0 < r < 1$. The *smallest integer that is greater than or equal to the number on the right-hand side of the above inequality is the ε-index* for the numerical sequence $s_n(r)$:

$$N_\varepsilon(r) = \left\lceil \frac{\ln((1-r)\varepsilon)}{\ln r} \right\rceil \tag{8.5}$$

Thus, $s_n(r)$ is within ε of $1/(1-r)$ or is ε-close to it for all $n \geq N_\varepsilon(r)$. We use the notation $N_\varepsilon(r)$ because the number N_ε *depends on the choice of r;* thus, $N_\varepsilon(r)$ is no longer just a number but a function that changes as r does (we keep ε fixed).

There is no analog of this extra parameter for infinite sequences and series of constants, but we did encounter it in our discussion of uniform continuity in Chapter 6. The setting was similar in the sense that in general, the sequence index for continuity at a given point $x = a$ depends on the number a; and this dependence could cause the sequence index diverge to infinity.[1]

To illustrate the effect of this new parameter for a converging sequence of functions, let $\varepsilon = 0.01$. The following table lists the values that we get for N_ε in (8.5) for each specified number r in the interval $(0, 1)$:

r	0.5	0.6	0.7	0.8	0.9	0.95	0.99
$N_\varepsilon(r)$	8	11	17	28	41	149	917

---

[1] Equivalently, the δ parameter goes to 0 in the ε, δ formalism of continuity. This may lead to the effective value of δ becoming infinitesimally small as the value of a changes (ε fixed).

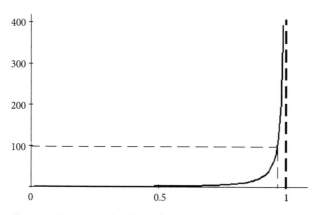

Fig. 8.3 Exposing a hidden infinity in pointwise convergence

The table shows that in order for the partial sum $s_N(r)$ (and therefore, $s_n(r)$ for $n \geq N_\varepsilon$) to be ε-close to the limit $1/(1-r)$ with $\varepsilon = 0.01$, *the value of N_ε increases substantially as r gets closer to 1.* For example, if $r = 0.5$, then the approximation to within 0.01 is achieved by just eight terms of the sum; whereas for $r = 0.95$, we require at least 149 terms to achieve the same accuracy. This result is consistent with, and sharpens, what is shown in Figure 8.1, where the graphed partial sums $s_1(x)$ to $s_4(x)$ and even $s_{10}(x)$ are nowhere near the limit curve when the value of x is very close to 1.

The table also suggests that $N_\varepsilon(r) \to \infty$ as $r \to 1$. This is true regardless of the value of ε because

$$\lim_{r \to 1^-} \frac{\ln((1-r)\varepsilon)}{\ln r} = \infty \qquad (8.6)$$

(see Figure 8.3). Note that if $\varepsilon < 1$, then as $r \to 1$ from the left, both $(1-r)\varepsilon$ and r are in the interval $(0, 1)$ so that the fraction inside the limit is always positive.

This occurrence of infinity in the limiting value of the ε-index N_ε was hinted at by the infinite discontinuity of the limit function $1/(1-x)$ at $x = 1$. Is this infinite discontinuity responsible for the infinite limit in (8.6)? The answer is no, at least *not primarily*. To explain the subtle but important reason, we take a closer look at what happens near $x = -1$ where the limit function $1/(1-x)$ is quite smooth and bounded, but the ε-index still diverges to ∞.

Convergence near the left endpoint

Note that where $x = -1$, the limit function $1/(1-x)$ is well-defined and equals $1/2$, so there is no occurrence of infinity whatsoever in the function; see Figure 8.1. The divergence of partial sums that we see in Figure 8.1 as we move to the left past -1 is now due to the behavior of the polynomials $s_n(x)$; they split from each other

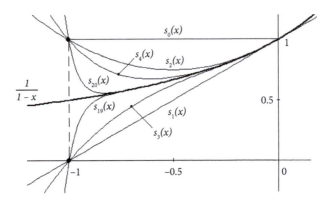

Fig. 8.4 Several partial sums near the left-hand endpoint

depending on whether n is even or odd. We might conclude that in this case, the divergence is not tied to any quantity that diverges to infinity, but we would be wrong!

Let $-1 < r < 0$ and fix the value of $x = r$. By (8.1), the partial sums $s_n(x)$ at this specified value do converge to the number $1/(1 - r)$. This is illustrated in Figure 8.4 where the partial sum polynomials $s_1(x)$ to $s_4(x)$ as well as $s_{19}(x)$ and $s_{20}(x)$ are plotted together with a portion of the limit function (the thick curve). Worth noticing in this diagram is the fact that all even-indexed partial sums pass through the point $(-1, 1)$, while all the odd-indexed ones pass through $(-1, 0)$; I leave the quick proof of this to you.

In Figure 8.4, we see that the numbers $s_n(-1)$ oscillate by jumping from -1 to 1 and back; so they cannot converge to any number, including $1/2$ (the value of the limit function at $x = -1$). Going back to (8.3), since r is negative, we retain the absolute value and continue with our calculation as before. We are led to the inequality

$$\frac{|r|^{n+1}}{1 - r} < \varepsilon$$

which can be solved for n just as before to yield the ε-index that, again, depends on the choice of the number r:

$$N_\varepsilon(r) = \left\lceil \frac{\ln((1 - r)\varepsilon)}{\ln |r|} \right\rceil \tag{8.7}$$

Thus, the number $s_n(r)$ is ε-close to $1/(1 - r)$ if $n \geq N_\varepsilon(r)$.

The sign of r plays a role in the numerator because if $r < 0$, then $1 - r > 1$, so $\ln(1 - r) > 0$. But then $\ln |r| < 0$ because $|r| < 1$, which seems to give a (nonsensical) negative value for N_ε. But if $\varepsilon < 1/2$, then since $r > -1$ we get

$$(1 - r)\varepsilon < 2\varepsilon < 1$$

so $\ln((1-r)\varepsilon) > 0$; and as before, we get a positive value for N_ε. Taking $\varepsilon < 1/2$ is consistent with our purpose since we generally assume that ε is an *arbitrarily small positive number* when calculating limits.

Now, as in the previous case with $x = 1$, if we set $\varepsilon = 0.01$, then using (8.7) we may create a table for N_ε corresponding to different values of r as follows:

r	-0.5	-0.6	-0.7	-0.8	-0.9	-0.95	-0.99
$N_\varepsilon(r)$	7	9	12	19	38	77	390

As we see from the above table, the value of N_ε increases substantially as r is chosen closer to -1. As in (8.6), we can prove that the value of N_ε blows up to infinity:

$$\lim_{r \to -1^+} \frac{\ln((1-r)\varepsilon)}{\ln |r|} = \infty$$

as long as $\varepsilon < 1/2$.[2] It is remarkable that this happens even though the limit function $1/(1-x)$ and all of the partial sums $s_n(x)$ are continuous at $x = -1$ and at all nearby points. Infinity occurs in the ε-index because to get $s_n(x)$ ε-close to $1/(1-x)$ when x is near -1, it is necessary to pick an arbitrarily large index n.

Convergence over sets. *The divergence to infinity of the value of N_ε in order to bring the sequence of functions $s_n(x)$ ε-close to the limit function $1/(1-x)$ shows that for a fixed $\varepsilon > 0$, there is no one N_ε large enough that works for **all** numbers in the interval $(-1, 1)$, as we may choose a value r for x that is arbitrarily close to 1 or to -1. Nevertheless, as long as the values $x = \pm 1$ are excluded, convergence does occur for each individual value of x. Therefore, convergence at a single point must be distinguished from convergence on an entire set, such as the interval $(-1, 1)$.*

This issue has profound consequences for infinite sequences and series of functions; we discuss some of these consequences later in this chapter by answering questions like When does an infinite series of continuous functions (recall that all polynomials are continuous) converge to a limit function that is also continuous? Can an infinite series of functions be integrated or differentiated term by term, one term at a time, without producing contradictory or nonsensical results?

8.2 Convergence of an infinite sequence of functions

In the last section, we saw that the partial sums of the geometric power series were in fact polynomial functions. In this section, we define what we mean by

---

[2] Computationally speaking, N_ε here is not as supercharged as in the case $x = 1$. Unlike that case, near $x = -1$, the numerator $\ln(1-r)$ approaches $\ln 2$ as $r \to -1^+$ and only the factor $1/\ln |r|$ blows up. The effect of infinity at $x = 1$ is simply to kick N_ε up about twice as hard. We also see this effect when we compare the tables of values in the two cases.

a *convergent sequence of functions*, whether it is a partial sum sequence, a polynomial sequence, or whatever. Aside from being the foundations for studying infinite series of functions, this topic is quite interesting in its own right; and its applications to the study of spaces of functions go far beyond the topic of infinite series.

Recall the ε and N definition of limits for sequences of constants in Chapter 3; when we apply the same idea to sequences of functions, we discover that there are different ways in which sequences of functions may converge to limits, each way being important in its own context. We discuss two general types of convergence in this section that are of fundamental importance in analysis.

8.2.1 Pointwise convergence

Consider a sequence of functions $f_n(x)$ where the variable x is in some nonempty (typically infinite) set S of real numbers. An example is the sequence of partial sums where $f_n(x)$ is a partial sum like $s_n(x) = 1 + x + \cdots + x^n$.

First, we consider each value of x separately; by setting $x = c$ where c is a fixed real number in the domain of every function $f_n(x)$, we get a sequence of *constants* $c_n = f_n(c)$. Now this is an ordinary sequence of numbers not a sequence of functions, so we can use the ideas in Chapter 3. If c_n converges according to our earlier definition, then its limit is a real number; let's call it $f(c)$ since it was calculated for the fixed value c. The equation

$$\lim_{n \to \infty} f_n(c) = f(c) \tag{8.8}$$

states that our sequence of functions $f_n(x)$ converges to *the number $f(c)$ at the chosen point $x = c$.*

From Chapter 3 recall that the equality in (8.8) means this: for each $\varepsilon > 0$, there is an index N such that

$$|f_n(c) - f(c)| < \varepsilon \quad \text{for all } n \geq N$$

As before, the ε-index N_ε denotes the least such value of N. If the equality in (8.8) holds for some value c of x, then it may or may not hold for *other* values of x.

Consider the sequence $f_n(x) = nx$, each of which is a straight line that passes through the origin with slope n, and let $S = (-\infty, \infty)$ be the set of all real numbers. At the value $x = 0$, we see that $f_n(0) = 0$ for every n, so $\lim_{n \to \infty} f_n(0) = 0$. But if $c \neq 0$, then at $x = c$, the sequence $f_n(c)$ diverges to infinity because

$$\lim_{n \to \infty} f_n(c) = \lim_{n \to \infty} nc = \begin{cases} \infty & \text{if } c > 0 \\ -\infty & \text{if } c < 0 \end{cases}$$

This sequence of linear functions converges *only* at the single value $x = 0$.

If (8.8) holds for *every* fixed value of x (or point) in S, then for each x the limit $f(x)$ is a real number that depends on the choice of x. Putting all of these real numbers together defines a function $f : S \to \mathbb{R}$ and gives us the following definition.

Pointwise convergence and limit

A sequence of functions $f_n(x)$ *converges pointwise* to a function $f(x)$ on a set S if for each x in S the following is true:

$$\lim_{n \to \infty} f_n(x) = f(x)$$

This means that for every $\varepsilon > 0$, there is a positive integer N that generally depends on ε and x such that

$$|f_n(x) - f(x)| < \varepsilon \quad \text{for all } n \geq N$$

For each fixed r in S, the least integer N for which the above inequality holds is the ε-index $N_\varepsilon(r)$ relative to S. The limit function $f(x)$ is called the *pointwise limit* of the sequence of functions $f_n(x)$. We also write

$$f_n \xrightarrow{p} f \quad \text{on } S$$

In the previous section, we discussed the pointwise convergence of the sequence of partial sums of the geometric power series. Specifically, we showed that the sequence of partial sums $s_n(x)$ converges pointwise to the rational function $R(x) = 1/(1 - x)$ for each x in the set $S = (-1, 1)$. In symbols

$$\lim_{n \to \infty} s_n(x) = \frac{1}{1 - x} \quad \text{or} \quad s_n \xrightarrow{p} R \quad \text{on } (-1, 1)$$

We also calculated the ε-index $N_\varepsilon(r)$ and showed that it diverges to infinity as $x \to 1^-$; this is a common occurrence for pointwise converging sequences of functions, but there are also many cases where it doesn't occur.

For example, consider the sequence of functions

$$f_n(x) = \frac{n + x}{nx}, \quad x \in (0, \infty), \ n = 1, 2, 3, \ldots \tag{8.9}$$

Note that

$$\lim_{n\to\infty} f_n(x) = \lim_{n\to\infty} \left(\frac{1}{x} + \frac{1}{n}\right) = \frac{1}{x}$$

Therefore, the pointwise limit of this sequence is $1/x$ for every $x > 0$. Next, consider a fixed number $r \in (0, \infty)$ and note that

$$\left|f_n(r) - \frac{1}{r}\right| = \frac{1}{n}$$

is independent of x for every n. Therefore, in this case, the ε-index

$$N_e = \left\lceil \frac{1}{\varepsilon} \right\rceil$$

does not depend on the specific number r, i.e., it is constant for each given ε.

Limits in which N_e is independent of the value of x (and therefore, doesn't diverge to ∞) are special and important; we discuss them a little later.

Shortcomings of pointwise convergence

We now consider a sequence of differentiable functions that contains no discontinuities or explicit occurrences of infinity to make an important new point: *the pointwise limit function of a sequence of continuous functions may be discontinuous.*

Consider the sequence of inverse tangent functions:

$$f_n(x) = \tan^{-1}(nx) \quad n = 1, 2, 3, \ldots \tag{8.10}$$

Figure 8.5 shows parts of the graphs of some of these functions. Note that f_n is differentiable for every n, and its derivative $f'_n(x) = n/(1 + n^2x^2)$ is defined on $(-\infty, \infty)$.

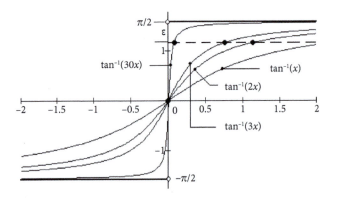

Fig. 8.5 Continuous functions converging pointwise to a discontinuous function

With the increasing value of n, we see in Figure 8.5 that the S-shaped curves become aligned with the three-piece, discontinuous function:[3]

$$f(x) = \begin{cases} \pi/2 & \text{if } x > 0 \\ 0 & \text{if } x = 0 \\ -\pi/2 & \text{if } x < 0 \end{cases} \qquad (8.11)$$

This function is the pointwise limit of the sequence in (8.10). To prove it, fix the value of the variable $x = c$. If $c > 0$, then $nc \to \infty$ as $n \to \infty$, so

$$\lim_{n\to\infty} f_n(c) = \lim_{n\to\infty} \tan^{-1}(nc) = \frac{\pi}{2}$$

Similarly, if $c < 0$, then $nc \to -\infty$ as $n \to \infty$; so $\lim_{n\to\infty} f_n(c) = -\pi/2$. If $c = 0$, then $nc = 0$ for all n, so $\tan^{-1}(nc) = \tan^{-1}(0) = 0$. Therefore, $f(x)$ in (8.11) is indeed the pointwise limit, and we may write

$$f_n(x) = \tan^{-1}(nx) \xrightarrow{p} f(x) = \frac{\pi}{2} \quad \text{on } (-\infty, \infty)$$

Notice that there are no explicit occurrences of infinity near the origin. Now consider the ε-index N_ε for a (small) positive ε. Figure 8.5 shows one such ε; we see that if we choose c greater than 1.5, then $f(c) - f_n(c) < \varepsilon$ for $n \geq 2$ (so $N_\varepsilon = 2$ for this ε); while if c is between 1 and 1.5, then $f(c) - f_n(c) < \varepsilon$ for $n \geq 3$ (so $N_\varepsilon = 3$).

As c is chosen closer and closer to 0, a greater value of n is required to bring $f_n(c)$ within ε of $f(c)$; in Figure 8.5, we see that if c is 0.5 or 0.25, then $n \geq 30$ guarantees that $f(c) - f_n(c) < \varepsilon$ (though a smaller n might work too).

To calculate the formula for N_ε, it is enough to consider either the positive halves or the negative halves since $f(x)$ and all $f_n(x)$ are symmetric with respect to the origin (they are odd functions). Looking at the positive halves, since $f(c) = \pi/2$ is larger than $f_n(c)$ for each $c > 0$, we can drop the absolute value and write

$$f(c) - f_n(c) < \varepsilon$$

$$\frac{\pi}{2} - \tan^{-1}(nc) < \varepsilon$$

$$\frac{\pi}{2} - \varepsilon < \tan^{-1}(nc)$$

Next, take the tangent of the last inequality to cancel $\tan^{-1}$ and get[4]

---

[3] Remember that $\tan^{-1}(x)$ approaches $\pi/2$ as $x \to \infty$ and to $-\pi/2$ as $x \to -\infty$. Also worth a mention is the fact that the tangent line to the graph of $\tan^{-1}(nx)$ at the origin is just the line nx that we discussed earlier.

[4] Since $\tan x$ is an increasing function when x is between $-\pi/2$ and $\pi/2$, the inequality direction remains unchanged after taking the tangent.

$$\tan\left(\frac{\pi}{2} - \varepsilon\right) < nc$$

Since $c > 0$, we can divide by c to obtain

$$\frac{1}{c}\tan\left(\frac{\pi}{2} - \varepsilon\right) < n$$

As usual, N_ε is the smallest integer that is greater than the expression on the left-hand side so

$$N_\varepsilon(c) = \left\lceil \frac{1}{c}\tan\left(\frac{\pi}{2} - \varepsilon\right) \right\rceil \tag{8.12}$$

Notice that the factor $\tan(\pi/2 - \varepsilon)$ is a positive constant per fixed value of ε; for instance, if $\varepsilon = 0.01$, then $\tan(\pi/2 - \varepsilon) \simeq 92.62$. As c gets closer to zero, $1/c$ diverges to infinity, taking $N_\varepsilon(c)$ along with it. Once again we discover the occurrence of infinity in the ε-index because to be ε-close to the limit function f, the terms of the sequence $f_n(x) = \tan^{-1}(nx)$ may require arbitrarily large index n.

The occurrence of infinity in the ε-index of some pointwise converging sequences is analogous to that we discussed in Chapter 6 for continuous functions that are not *uniformly* continuous; see the discussion after Theorem 200. This occurrence of infinity is responsible for significant shortcomings such as loss of continuity in the limit function. For instance, although every function $f_n(x) = \tan^{-1}(nx)$ above is differentiable everywhere, the pointwise limit function $f(x)$ is not even continuous, let alone differentiable at $x = 0$. We conclude that

The properties of being continuous, differentiable, or integrable are not preserved by pointwise convergence.

Consider integrability: suppose that $f_n \xrightarrow{p} f$ on an interval $[a, b]$, and every f_n is integrable. Then the limit function f need not be integrable; for instance, recall from the geometric power series example that every partial sum $s_n(x)$ is continuous, hence integrable on $[0, 1]$; but the limit $f(x) = 1/(1 - x)$ is not bounded, hence not integrable (with a divergent improper integral $\int_0^1 1/(1 - x)dx = \infty$).

Even when f is integrable, pointwise convergence is not enough to allow the following operation, namely, taking the limit in or out of the integral:

$$\lim_{n\to\infty} \int_a^b f_n(x)dx = \int_a^b f(x)dx = \int_a^b \lim_{n\to\infty} f_n(x)dx \tag{8.13}$$

As we discover later, being able to do this is crucial to proving that term by term integration of the infinite series expansion of a function f gives the integral of f.

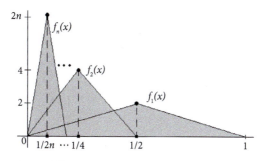

Fig. 8.6 The integral and pointwise limit are not interchangeable

A simple example where (8.13) fails is illustrated in Figure 8.6. We discuss this example because it illustrates the problem that occurs if the ε-index depends on the value of x, without significant technical detail.

In Figure 8.6, we see a sequence of functions $f_1(x), f_2(x), \ldots, f_n(x), \ldots$ consisting of slanted and flat segments: the first function $f_1(x)$ is just the sides of the shaded triangle having a height of 2; the interval $[0,1]$ on the x-axis is its base. The second function $f_2(x)$ consists of the sides of the shaded triangle of height 4 with the interval $[0,1/2]$ as its base, together with the flat line segment from $1/2$ to 1; so we keep in mind that $f_2(x) = 0$ on the interval $[1/2,1]$. Generally, the nth function $f_n(x)$ consists of the sides of the shaded triangle of height $2n$ and the interval $[0, 1/n]$ as its base, plus a flat segment $f_n(x) = 0$ on $[1/n, 1]$. Notice that the lengths of the flat segments increase as n does, to eventually cover the interval $(0,1]$ as $n \to \infty$.

Each function $f_n(x)$ is continuous, hence integrable. The integral of $f_n(x)$ is just the area of the nth shaded triangle for every n; specifically

$$\int_0^1 f_n(x)dx = \int_0^{1/n} f_n(x)dx + \int_{1/n}^1 f_n(x)dx = \frac{1}{2}(2n)\left(\frac{1}{n}\right) + 0 = 1$$

Since for every n the integral has a fixed value of 1, the sequence of integrals is a constant sequence of real numbers that obviously converges to 1:

$$\lim_{n\to\infty} \int_0^1 f_n(x)dx = 1 \tag{8.14}$$

On the other hand, the pointwise limit of the sequence f_n is the zero function $f(x) = 0$. To see why, consider any fixed value $x = c$ in the interval $[0,1]$. Now, either $c = 0$ or $c > 0$. If $c = 0$, then $f_n(0) = 0$ by the definition of the sequence (see Figure 8.6). If $c > 0$, then for all indices n large enough that $1/n < c$ (or $n > 1/c$), c falls in the interval $[1/n, 1]$ where $f_n(c) = 0 = f(c)$. This implies that

$\lim_{n\to\infty} f_n(c) = 0$; and since c was an arbitrary value of x, we have actually proved that $\lim_{n\to\infty} f_n(x) = 0$ for every fixed value of x in $[0,1]$. In other words, $f_n \xrightarrow{p} 0$ on $[0,1]$.

But then the zero limit function $f(x) = 0$ is integrable as a constant function with integral

$$\int_0^1 f(x)dx = 0 \tag{8.15}$$

We see that (8.13) is not true for this sequence of integrable functions since by (8.14) and (8.15)

$$\lim_{n\to\infty} \int_0^1 f_n(x)dx = 1 \neq 0 = \int_0^1 \lim_{n\to\infty} f_n(x)dx$$

Note that the ε-index of the sequence $f_n(x)$ above is $N_\varepsilon = \lceil 1/c \rceil$ because as we argued above, $f_n(c) = 0$ for every index n that exceeds $\lceil 1/c \rceil$; so in particular, $f_n(c) < \varepsilon$ for every positive value of ε (notice that N_ε in this case is independent of ε). It follows that N_ε diverges to ∞ as $c \to 0^+$, consistent with the pattern we discovered earlier for the geometric power series and the inverse tangent sequence.

8.2.2 Uniform convergence

Is it possible to prevent the divergence of N_ε to ∞ through some modification so that nice things can happen, like the limit function be continuous or the equality in (8.13) hold?

The answer is yes, and the modification is often surprisingly easy.

Let's go back to the sequence of triangle-shaped functions and truncate the interval $[0,1]$ to $[a, 1]$ where a is fixed but may be as close to 0 as we like. Note that if $c \in [a, 1]$, then $c \geq a$; therefore, c cannot be chosen arbitrarily close to 0. Now, we examine the consequences of this modification.

As $n \to \infty$, the functions $f_n(x)$ approach the zero function pointwise on $[a, 1]$, but now the ε-index doesn't go to infinity; since $c \geq a$ for every fixed value c of x in $[a, 1]$, it follows that $1/c \leq 1/a$ and thus, $N \geq 1/a$ implies $N \geq 1/c$. We conclude that $N_\varepsilon = \lceil 1/a \rceil$, which is a constant for the interval $[a, 1]$.

Therefore, $f_n(x) = 0$ for every $n \geq N_\varepsilon$ and for all values of x in $[a, 1]$ because the nonzero triangular part of $f_n(x)$ has moved to the left of a. But since N_ε is a constant, it can't diverge to infinity when the value of x changes.

So now that N_ε doesn't diverge to ∞, what happens to the integrals? We again have $\int_a^1 f(x)dx = 0$ since the limit function $f(x)$ is the zero function on $[a, 1]$. Further, we just saw that $f_n(x) = 0$ for all x in $[a, 1]$ if $n \geq N_\varepsilon = \lceil 1/a \rceil$, so

$\int_a^1 f_n(x)dx = 0$ for all such large enough n. This means that the sequence of integrals is a sequence of constants that is eventually 0 from the index N_ε on:

$$\int_a^1 f_1(x)dx, ..., \int_a^1 f_{N_\varepsilon-1}(x)dx, \int_a^1 f_{N_\varepsilon}(x)dx, ... = I_1, ..., I_{N_\varepsilon-1}, 0, 0, ...$$

where $I_1, ..., I_{N_\varepsilon-1}$ are numbers less than 1 because this time the integrals do not represent the areas of *entire* shaded triangles (the tip of each triangle to the left of a is cut off). The above sequence ends in 0, so it converges to 0, which is the value of $\int_a^1 f(x)dx$. Therefore, (8.13) is true on $[a, 1]$ for every a, no matter how close to 0; mission accomplished!

What made the convergence in the preceding example work was the fact that on the truncated interval $[a, 1]$, the ε-index N_ε is independent of the value of the variable x, thus making it possible to keep the terms f_n of the sequence to be ε-close to the limit function f as $n \to \infty$. Whenever this is possible, the limit f of the sequence of functions retains many of the properties of the terms f_n of the sequence.

The next definition simply formalizes this property.

Uniform convergence and limit

A sequence of functions $f_n(x)$ *converges uniformly on a set S* of real numbers to a function $f(x)$, called the *uniform limit*, if (i) for every $\varepsilon > 0$, the ε-index N_ε does not depend on the value of x (it has the same value uniformly over all of S), and (ii) for all $n \geq N_\varepsilon$

$$|f_n(x) - f(x)| < \varepsilon \quad \text{for all } x \text{ in } S \tag{8.16}$$

We often use the notation: $f_n \xrightarrow{u} f$ on S.

A simple example of a uniformly convergent sequence is that in (8.9), which converges uniformly to $1/x$ on $(0, \infty)$ since N_ε didn't depend on x in this case.

We see in the definition of uniform convergence above the similarity to uniform continuity in Chapter 6; also see Theorem 310 below.

To help visualize the content of the definition of uniform convergence, we write the absolute value inequality in (8.16) in the equivalent form as two inequalities:

$$-\varepsilon < f_n(x) - f(x) < \varepsilon$$

or

$$f(x) - \varepsilon < f_n(x) < f(x) + \varepsilon$$

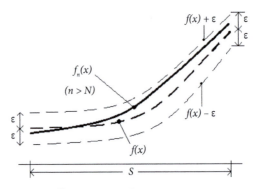

Fig. 8.7 Illustrating uniform convergence

The last pair of inequalities say that for every $n \geq N_\varepsilon$, the graph of $f_n(x)$ over the entire set S is contained within a strip of width 2ε that is centered at the limit curve $f(x)$. Figure 8.7 illustrates this situation.

Let's call the strip of width 2ε *the ε-strip centered at $f(x)$*. Looking back at Figures 8.5 and 8.6, it is easy to see that if we draw ε-strips centered at the limit curves in each figure, then the functions $f_n(x)$ can never be contained in the ε-strips *over the entire set S* no matter how large the index n is. It follows that convergence in those examples is not uniform.

Notice that every uniformly convergent sequence is automatically pointwise convergent, which we may remember as

$$f_n \xrightarrow{u} f \text{ on } S \text{ implies that } f_n \xrightarrow{p} f \text{ on } S.$$

The contrapositive of the above statement is also worth remembering:

If a sequence of functions f_n does not converge pointwise, then it does not converge uniformly.

However, in the several examples that we discussed above, we found that the *converse* of the above statement is false. In the case of triangle-shaped functions above, we can now say that that sequence converges to the zero function uniformly on $[a, 1]$ (no matter how close a is to 0) but only pointwise on $(0,1]$, since N_ε is unbounded. For the same reason, the tangent inverse sequence converges uniformly to the constant (and continuous) function $f(x) = \pi/2$ on $[a, \infty)$ for any choice of $a > 0$ but only pointwise on $(0, \infty)$.

Now, consider again the partial sums of the geometric power series and suppose that we truncate the interval $(-1, 1)$ to $[-a, a]$ where $a < 1$ is fixed and can be as close to 1 as we like. Then the value of r in (8.5) cannot approach 1 since $0 < r \leq$

$a < 1$, and $1 - a \leq 1 - r$. Remembering that $\ln x$ is an increasing function, taking logarithms does not affect inequality directions, so we have

$$\ln r \leq \ln a < 0 \quad \text{and} \quad \ln((1 - a)\varepsilon) \leq \ln((1 - r))\varepsilon < 0$$

Flipping the first pair now reverses the inequality:

$$\frac{1}{\ln r} \geq \frac{1}{\ln a}$$

But multiplying by a negative quantity also reverses the inequality, so

$$\left(\frac{1}{\ln r}\right) \ln((1 - a)\varepsilon) \leq \left(\frac{1}{\ln a}\right) \ln((1 - a)\varepsilon) \quad \text{and:}$$

$$\left(\frac{1}{\ln r}\right) \ln((1 - r)\varepsilon) \leq \left(\frac{1}{\ln r}\right) \ln((1 - a)\varepsilon) \quad \text{since } \frac{1}{\ln r} < 0$$

It follows that

$$\frac{\ln((1 - r)\varepsilon)}{\ln r} = \left(\frac{1}{\ln r}\right) \ln((1 - r)\varepsilon)$$

$$\leq \left(\frac{1}{\ln r}\right) \ln((1 - a)\varepsilon)$$

$$\leq \left(\frac{1}{\ln a}\right) \ln((1 - a)\varepsilon)$$

$$= \frac{\ln((1 - a)\varepsilon)}{\ln a}$$

So the constant ε-index

$$N_\varepsilon = \left\lceil \frac{\ln((1 - a)\varepsilon)}{\ln a} \right\rceil$$

is independent of any specific value of x and in particular, will not diverge to infinity. A similar calculation blocks the growth of N_ε in (8.7). We conclude that

The sequence $s_n(x)$ of partial sums of the geometric power series converges uniformly to the function $1/(1 - x)$ on the interval $[-a, a]$ if $0 < a < 1$ no matter how close a is to 1.

We show below that the above statement is what makes good things possible for the geometric power series.

Calculating the ε-index directly is often tedious, as the case of geometric power series above shows. It may be useful to have alternative ways of proving uniform convergence, like the next result.

Theorem 310 *Assume that* $f_n \xrightarrow{p} f$ *on a nonempty set S and define the quantity*[5]

$$\|f_n - f\| = \sup\{|f_n(x) - f(x)| : x \in S\}$$

Then $f_n \xrightarrow{u} f$ *on S if and only if*

$$\lim_{n \to \infty} \|f_n - f\| = 0 \tag{8.17}$$

Proof Note that for each n the quantity $\|f_n - f\|$ is a real number, so (8.17) gives the limit of a sequence of constants. If it holds, then there is a positive integer N_ε such that for each fixed $x = r \in S$

$$|f_n(r) - f(r)| \le \|f_n - f\| = |\|f_n - f\| - 0| < \varepsilon$$

for all $n \ge N_\varepsilon$. Since this is true regardless of the value of x, it follows that N_ε is the ε-index and independent of $x \in S$; therefore, $f_n \xrightarrow{u} f$.

Conversely, if $f_n \xrightarrow{u} f$, then for each $\varepsilon > 0$ the ε-index N_ε is independent of x; and if $n \ge N_\varepsilon$, then $|f_n(x) - f(x)| < \varepsilon$ for every $x \in S$. Therefore, $\|f_n - f\| = \sup\{|f_n(x) - f(x)| : x \in S\} \le \varepsilon$ for all $n \ge N_\varepsilon$; and since ε is arbitrary, (8.17) follows. ∎

To illustrate, we use Theorem 310 to prove that the sequence of partial sums of the geometric power series converges uniformly to $1/(1-x)$ on $[-a, a]$ if $0 < a < 1$. Note that

$$|s_n(x) - f(x)| = \left|1 + x + x^2 + \cdots + x^n - \frac{1}{1-x}\right|$$

$$= \frac{|(1 + x + x^2 + \cdots + x^n)(1-x) - 1|}{|1-x|}$$

$$= \frac{|x|^{n+1}}{|1-x|}$$

---

[5] The quantity $\|f\| = \sup\{|f(x)| : x \in S\}$ has the technical name *the sup-norm* or *the uniform norm*. The difference $\|f - g\|$ gives a measure of distance between two (bounded) functions f and g, and it often serves as a standard *metric* on spaces of bounded functions. We discuss another norm in the appendices.

Since $x \in [-a, a]$, it follows that $|x| \le a < 1$, and $|1 - x| \ge 1 - a$, so

$$|s_n(x) - f(x)| \le \frac{a^{n+1}}{1 - a}$$

The right-hand side of the above inequality doesn't depend on x, so

$$\left\| s_n(x) - \frac{1}{1 - x} \right\| = \sup \left\{ \left| s_n(x) - \frac{1}{1 - x} \right| : x \in [-a, a] \right\} \le \frac{a^{n+1}}{1 - a}$$

Finally, since $a < 1$, we have

$$\lim_{n \to \infty} \left\| s_n(x) - \frac{1}{1 - x} \right\| \le \frac{1}{1 - a} \lim_{n \to \infty} a^{n+1} = 0$$

Now Theorem 310 implies that s_n converges uniformly. This argument may not be much shorter than the earlier one using an ε-index calculation, but it has the advantage of working with the functions themselves rather than their indices, and in particular, not requiring the logarithm.

It is worth mentioning that as $a \to 1^-$ the fraction $a^{n+1}/(1 - a)$ diverges to ∞ for each fixed n, and the above argument breaks down. This is consistent with our earlier discussion involving the ε-index.

8.2.3 On preserving limits, integrals, and derivatives

We have seen that if a sequence of continuous functions $f_n(x)$ converges pointwise to a function $f(x)$, then $f(x)$ is not necessarily continuous. However, things look much better if the convergence is uniform.

Theorem 311 *If f_n is a sequence of functions that are continuous at a number $a \in S$, and $f_n \xrightarrow{u} f$ on S, then the uniform limit f is continuous at a.*

Proof For every $\varepsilon > 0$, we need to show that there is a $\delta > 0$ such that $|f(x) - f(a)| < \varepsilon$ if $x \in S$ and $|x - a| < \delta$. Note that, using the triangle inequality

$$|f(x) - f(a)| = |f(x) - f_n(x) + f_n(x) - f_n(a) + f_n(a) - f(a)|$$
$$\le |f(x) - f_n(x)| + |f_n(x) - f_n(a)| + |f_n(a) - f(a)|$$

The uniform convergence hypothesis implies that there is N (the ε-index) such that for all $n \ge N$

$$|f(x) - f_n(x)| \le \|f_n - f\| < \frac{\varepsilon}{3},$$
$$|f_n(a) - f(a)| \le \|f_n - f\| < \frac{\varepsilon}{3}$$

Further, since f_n is continuous for every n, it is continuous for $n = N$; so there is a $\delta > 0$ such that if $|x - a| < \delta$, then

$$|f_N(x) - f_N(a)| < \frac{\varepsilon}{3}$$

For this particular δ, we conclude that

$$|f(x) - f(a)| < \frac{\varepsilon}{3} + \frac{\varepsilon}{3} + \frac{\varepsilon}{3} = \varepsilon$$

Since ε was chosen arbitrarily, it follows that f is continuous at $x = a$.   ∎

The next corollary follows immediately since continuity on a set means continuity at every point of it.

Corollary 312 *If f_n is a sequence of continuous functions on a set S, and $f_n \overset{u}{\longrightarrow} f$ on S, then the uniform limit f is continuous on S.*

The contrapositive of Theorem 311 may be used to quickly tell if a converging sequence of functions does so uniformly. For instance, in the inverse tangent case above, we see that the convergence on any set of real numbers that contains 0 cannot be uniform since the limit function is not continuous at $x = 0$.

Using the definition of continuity, we can restate Theorem 311 in different ways. For example, by f_n or f being continuous at $x = a$, we mean

$$\lim_{x \to a} f_n(x) = f_n(a), \qquad \lim_{x \to a} f(x) = f(a)$$

But since $f_n \overset{u}{\longrightarrow} f$ as $n \to \infty$, it is true in particular that

$$\lim_{n \to \infty} f_n(a) = f(a)$$

Now combining things and using Theorem 311, we get

$$\lim_{n \to \infty} \left[\lim_{x \to a} f_n(x) \right] = \lim_{n \to \infty} f_n(a) = f(a) \tag{8.18}$$

Likewise, since for every $x \in S$

$$\lim_{n \to \infty} f_n(x) = f(x)$$

taking limit as $x \to a$ gives

$$\lim_{x \to a} \left[\lim_{n \to \infty} f_n(x) \right] = \lim_{x \to a} f(x) = f(a) \tag{8.19}$$

Comparing (8.18) and (8.19) gives the following result on interchange of limits, which is equivalent to Theorem 311.

Theorem 313 (*Interchanging limits*) *If f_n is a sequence of functions that are continuous at a number $a \in S$, and $f_n \xrightarrow{u} f$ on S, then*

$$\lim_{n\to\infty}\left[\lim_{x\to a} f_n(x)\right] = \lim_{x\to a}\left[\lim_{n\to\infty} f_n(x)\right] \tag{8.20}$$

More symmetrically (and generally to include left and right limits), if x_k is any sequence in S that converges to a as $k \to \infty$, then

$$\lim_{n\to\infty}\left[\lim_{k\to\infty} f_n(x_k)\right] = \lim_{k\to\infty}\left[\lim_{n\to\infty} f_n(x_k)\right] \tag{8.21}$$

If the convergence of $f_n(x)$ to $f(x)$ is only pointwise but *not uniform*, then Theorem 313 generally fails.

Consider once again the sequence of inverse tangent functions $f_n(x) = \tan^{-1}(nx)$. If x_k is any sequence that converges to 0 as $k \to \infty$, then $nx_k \to 0$ for each fixed n; so we have

$$\lim_{n\to\infty}\lim_{k\to\infty} f_n(x_k) = \lim_{n\to\infty}\lim_{k\to\infty}\tan^{-1}(nx_k) = \lim_{n\to\infty}\tan^{-1}(0) = \lim_{n\to\infty} 0 = 0$$

On the other hand, if $x_k \neq 0$ for all k, then $nx_k \to \infty$ for each fixed k as $n \to \infty$; and since $\tan^{-1} x \to \pi/2$ as $x \to \infty$, we have

$$\lim_{k\to\infty}\lim_{n\to\infty} f_n(x_k) = \lim_{k\to\infty}\lim_{n\to\infty}\tan^{-1}(nx_k) = \lim_{k\to\infty}\frac{\pi}{2} = \frac{\pi}{2}$$

So (8.21) is false for *every* sequence x_k of real numbers that converges to 0 and $x_k \neq 0$ for all k. For instance, if $x_k = 1/k \to 0$ as $k \to \infty$, then

$$\lim_{n\to\infty}\lim_{k\to\infty}\tan^{-1}\left(\frac{n}{k}\right) \neq \lim_{k\to\infty}\lim_{n\to\infty}\tan^{-1}\left(\frac{n}{k}\right)$$

We note in passing that Theorem 313 can be further generalized by replacing continuity with the existence of limits.

Next, let's consider the preservation of integrals. We take S to be an interval because we defined the (Riemann) integral on intervals.

Theorem 314 *If f_n is a sequence of integrable functions on an interval $[a, b]$, and $f_n \xrightarrow{u} f$, then the uniform limit f is also integrable on $[a, b]$ with*

$$\lim_{n\to\infty}\int_a^b f_n(x)dx = \int_a^b f(x)dx = \int_a^b \lim_{n\to\infty} f_n(x)dx \tag{8.22}$$

Note that the limit inside the last integral denotes a uniform limit not a point-wise one; also note that the limit outside the first integral is the limit of a sequence of numbers, say, $I_n = \int_a^b f_n(x)dx$.

Proof Let $\sigma_n = \|f_n - f\|$; and since by hypothesis $f_n \xrightarrow{u} f$, Theorem 310 implies that $\lim_{n \to \infty} \sigma_n = 0$. Further, for every $x \in [a, b]$

$$|f_n(x) - f(x)| \le \|f_n - f\| = \sigma_n$$

so for each fixed n

$$f_n(x) - \sigma_n \le f(x) \le f_n(x) + \sigma_n \tag{8.23}$$

If $P = \{x_0, x_1, \dots, x_N\}$ is an arbitrary partition of $[a, b]$, then for every x in each subinterval $[x_{j-1}, x_j]$

$$f(x) \ge \inf\{f_n(t) - \sigma_n : x_{j-1} \le t \le x_j\} = m_{j,n}$$

Therefore, for every $j = 1, 2, \dots, N$

$$m_j = \inf\{f(t) : x_{j-1} \le t \le x_j\} \ge m_{j,n}$$

It follows that

$$L(P, f) = \sum_{j=1}^{N} m_j \Delta x_j \ge \sum_{j=1}^{N} m_{j,n} \Delta x_j = L(P, f_n - \sigma_n)$$

Since this inequality is valid for every partition of $[a, b]$, it follows that

$$I_L(f) \ge I_L(f_n - \sigma_n) = \int_a^b [f_n(x) - \sigma_n]dx$$

where the last inequality holds because $f_n - \sigma_n$ is integrable on $[a, b]$. A similar argument shows that

$$I_U(f) \le \int_a^b [f_n(x) + \sigma_n]dx$$

so that

$$0 \le I_U(f) - I_L(f) \le \int_a^b 2\sigma_n dx = 2\sigma_n(b - a)$$

These inequalities hold for all n and $\sigma_n \to 0$, so we may conclude that

$$I_L(f) = I_U(f)$$

which means that f is integrable. Further, by the properties of integral and (8.23)

$$\left| \int_a^b f(x)\,dx - \int_a^b f_n(x)\,dx \right| \le \int_a^b |f(x) - f_n(x)|\,dx \le \sigma_n(b-a)$$

which is again true for all n. Letting $n \to \infty$ and recalling that $\sigma_n \to 0$, we obtain (8.22). ∎

Recalling that continuous functions are integrable and using Corollary 312 and Theorem 314, we obtain the following result.

Corollary 315 *If f_n is a sequence of continuous functions on an interval $[a, b]$, and $f_n \overset{u}{\longrightarrow} f$ on $[a, b]$, then (8.22) holds.*

The preceding results on preservation of continuity and integrability and the interchange of limits or of limits and integrals are often applied fruitfully in the context of infinite series that we discuss in the next section. But first, we must also discuss the interchanging limits with derivatives.

The natural question at this point is whether uniform convergence preserves differentiation too, and if it does, whether we can interchange the (uniform) limit and the derivative.

Specifically, if $f_n \overset{u}{\longrightarrow} f$ on a set S, and each f_n is differentiable at every point of S, can we say that f is differentiable too? Is it possible to interchange the limit and the derivative? The answer to the first of these questions is no, though we can do better for the second one.

An extreme example of an infinite sequence of differentiable functions that converges uniformly to a non-differentiable function is an infinite series defined by Weierstrass that we discuss below. A function $W(x)$ is defined as the sum of an infinite series of functions, i.e., the uniform limit of a sequence of partial sum functions $S_n(x)$ that are all differentiable on $(-\infty, \infty)$. But $W(x)$ itself is not differentiable at *any* number in $(-\infty, \infty)$.

For now, consider the sequence $f_n(x) = (1/n)\sin(nx)$ of differentiable functions on the set of all real numbers $(-\infty, \infty)$. For each fixed value of $x = c$, we have

$$\lim_{n\to\infty} f_n(c) = \lim_{n\to\infty} \frac{\sin(nc)}{n} = 0$$

because $-1 \le \sin(nc) \le 1$ for all n, so this sequence converges pointwise to the zero function $f(x) = 0$ on $(-\infty, \infty)$. It is easy to see that the convergence is actually uniform:

$$\|f_n - f\| = \sup\left\{\left|\frac{\sin(nx)}{n} - 0\right| : x \in (-\infty, \infty)\right\} \leq \frac{1}{n}$$

Hence, $\lim_{n\to\infty} \|f_n - f\| = 0$, and it follows that $f_n \xrightarrow{u} 0$ on $(-\infty, \infty)$. Note that the zero function is differentiable too.

On the other hand, looking at the derivatives

$$f'_n(x) = \frac{1}{n}\cos(nx)\frac{d}{dx}(nx) = \cos(nx)$$

we see that for a fixed value $x = c$, the sequence of derivatives $f_n'(c) = \cos(nc)$ does not converge for all choices of c. For instance, if $x = \pi$, then

$$f'_n(\pi) = \cos(n\pi) = \begin{cases} -1 & \text{if} \quad n \text{ is odd} \\ 1 & \text{if} \quad n \text{ is even} \end{cases}$$

So the sequence $f'_n(\pi)$ does not converge. We conclude that the sequence of derivatives $f'_n(x)$ doesn't converge pointwise on $(-\infty, \infty)$; and therefore, it can't converge uniformly.

But all is not lost; with proper conditions identified, it is still possible to interchange limits and derivatives, which is of crucial importance in the study of infinite series. To motivate the result on derivatives, let's go back to our series of functions above, namely, $f_n(x) = (1/n)\sin(nx)$, and instead consider the antiderivatives:

$$F_n(x) = -\frac{\cos(nx)}{n^2}$$

Now, because

$$\|F_n - F\| = \sup\left\{\left|-\frac{\cos(nx)}{n^2} - 0\right| : x \in (-\infty, \infty)\right\} \leq \frac{1}{n^2}$$

we may conclude that the sequence of antiderivatives converges uniformly to the zero function $F(x) = 0$. Further, since f_n converges uniformly to the zero function (which is differentiable), we can write $\lim_{n\to\infty} F'_n(x) = F'(x)$. It turns out that this is the most that can be said for derivatives. The hypotheses of the next theorem state the precise conditions for the interchange of limit with the derivative.

Theorem 316 *Let f_n be a sequence of differentiable functions on an interval $[a, b]$, and assume that their derivatives f'_n are continuous on $[a, b]$. If*

(i) *the sequence of derivatives f'_n converges uniformly on $[a, b]$, and*
(ii) *the sequence of constants $f_n(c)$ converges for some number $c \in [a, b]$,*

then the sequence f_n (antiderivatives of f'_n) converges uniformly on $[a, b]$ to a differentiable function f and $\lim_{n \to \infty} f'_n(x) = f'(x)$, i.e.,

$$\lim_{n \to \infty} \frac{df_n}{dx} = \frac{df}{dx} = \frac{d}{dx}\left(\lim_{n \to \infty} f_n\right)$$

Proof By hypothesis, f'_n is continuous for every n; so according to the fundamental theorem of calculus, with c as in hypothesis (ii)

$$f_n(x) = f_n(c) + \int_c^x f'_n(t)dt \tag{8.24}$$

Let g be the uniform limit of f'_n. Then by Corollary 315, the integral in (8.24) converges to the integral of g. Thus, for each $x \in [a, b]$

$$\lim_{n \to \infty} f_n(x) = \lim_{n \to \infty} f_n(c) + \lim_{n \to \infty} \int_c^x f'_n(t)dt$$

$$= \lim_{n \to \infty} f_n(c) + \int_c^x g(t)dt$$

This shows that the pointwise limit of f_n exists on $[a, b]$; let f be this pointwise limit, i.e.,

$$f(x) = \lim_{n \to \infty} f_n(x) = f(c) + \int_c^x g(t)dt \tag{8.25}$$

We now show that $f_n \xrightarrow{u} f$ on $[a, b]$.
Note that

$$|f_n(x) - f(x)| = \left| f_n(c) + \int_c^x f_n'(t)dt - f(c) - \int_c^x g(t)dt \right|$$

$$\le |f_n(c) - f(c)| + \int_c^x |f'_n(t) - g(t)|\, dt$$

By definition,

$$|f'_n(t) - g(t)| \le \|f'_n - g\|$$

so

$$\int_c^x |f'_n(t) - g(t)|\, dt \le \|f'_n - g\| \, |x - c| \le \|f'_n - g\| \, (b - a)$$

It follows that

$$|f_n(x) - f(x)| \leq \|f'_n - g\| (b - a)$$

and this inequality being true for all $x \in [a, b]$, we conclude that

$$\|f_n - f\| \leq \|f'_n - g\| (b - a)$$

Taking the limit and recalling that $f'_n \overset{u}{\longrightarrow} g$ on $[a, b]$, we obtain

$$0 \leq \lim_{n \to \infty} \|f_n - f\| \leq \lim_{n \to \infty} \|f'_n - g\| (b - a) = 0$$

This establishes that $f_n \overset{u}{\longrightarrow} f$ on $[a, b]$. Finally, (8.25) and the fundamental theorem of calculus imply that f is differentiable and

$$f'(x) = g(x) = \lim_{n \to \infty} f'_n(x) \qquad \blacksquare$$

The assumption that f'_n be continuous made it possible to use the fundamental theorem of calculus and antiderivatives, but this continuity of derivatives is not essential and can be dropped (the proof then becomes a little more technical). We don't discuss that generalization since it is not required for the rest of our work in this chapter.

Hypothesis (ii) in Theorem 316 seems strange, even though it was used in the proof. To see that the theorem may be false without it, consider $f_n(x) = \ln(nx)$ on $[1,2]$, and note that the sequence of derivatives

$$f'_n(x) = \frac{1}{nx} n = \frac{1}{x}$$

which as a constant sequence, converges (to $1/x$) as $n \to \infty$; and this convergence is (trivially) uniform. However, $\ln(nx)$ doesn't converge (uniformly or pointwise) on $[1,2]$ since for every fixed $c \in [1, 2]$, we have

$$\lim_{n \to \infty} \ln(nc) = \lim_{n \to \infty} \ln n + \ln c = \infty$$

8.2.4 Uniform convergence does not preserve lengths

The preceding discussion shows that uniform convergence preserves various important properties of functions that pointwise convergence fails to preserve. However, not every important property is preserved by uniform convergence; we

REAL ANALYSIS AND INFINITY   425

found that differentiability is not directly preserved, even though it was possible to find conditions that ensure differentiability indirectly to be able to interchange the limit with the derivative.

Another property that is not preserved is the length of a curve, and graphs of functions are typically curves.[6]

In this section, we study some sequences of functions that converge uniformly but whose lengths do not converge to the length of the limit function. The latter is an example of a function with *unbounded variation*, a topic that is studied in more advanced texts in real analysis. We study a specific sequence of functions that we can partially visualize using graphs. We study a similar function due to Weierstrass a little later.

In the introductory chapter, we discussed the "staircase curves" and found a discrepancy with the lengths of these curves.

The function corresponding to the nth staircase path is a step function that we may denote $f_n(x)$, and the diagonal is $f(x) = x$, all defined on the interval $[0,1]$. Then $f_n \xrightarrow{u} f$ on $[0, 1]$ because the difference between any point (x, y) on a staircase path and the point (x, x) on the diagonal is at most the length $1/n$ of each riser, which means $\|f_n - f\| = 1/n$. This clearly converges to 0 as $n \to \infty$.

But if l_n is the total length of f_n, then $l_n = 2$ for all n, and this constant sequence clearly doesn't converge to the length of the diagonal, $\sqrt{2}$.

In order to understand what is going on here, we recognize that there are many different types of paths that connect the points $(0,0)$ and $(1,1)$. The diagonal is one such path, and the zigzag staircase paths are among others. The staircase paths were constructed in such a way that in each step n, their length remained the same even as the formula or expression for the step function f_n was altered by adding more bumps.

We may connect the two points $(0,0)$ and $(1,1)$ using many other different types of paths of widely different lengths. Many of these paths are graphs of functions $f_n(x)$ that converge uniformly to the diagonal, or to any other fixed path that is expressed as a function $f(x)$. While these paths converge uniformly, their lengths form a sequence of real numbers l_n that may behave in an entirely unrelated fashion; l_n may converge to 2, or to $\sqrt{2}$, or any other number, or even not converge at all.

To illustrate this phenomenon,[7] we rotate the square and re-scale the diagonal's length to 1 rather than $\sqrt{2}$ to simplify the numbers (equivalent to considering a square with sides of length $1/\sqrt{2}$ instead of 1). The staircase paths now appear as continuous curves with sharp corners as shown in Figure 8.8 where four stages

---

[6] This material is not needed in the rest of the book, so you can omit this subsection and proceed to the next section without loss of continuity, so to speak.

[7] This failure of uniform convergence to preserve length shows up as a discontinuity in a function space; see Appendix 9.2.

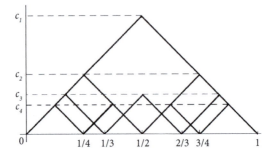

Fig. 8.8 Staircase curves: a different view

of the process, i.e., staircase functions having one, two, three, and four steps, are plotted together.

Note that the nth stage may still be represented as a function $f_n(x)$ on $[0,1]$, although the formula for f_n is now different than the original step function form.

The largest triangle in Figure 8.8 is just half of the original square, so its height is c_1. In the figure $c_1 = 1/2$, but we do not assign a specific length to c_1 in this discussion other than it be positive. Next, we split the interval $[0,1]$ in half and draw the two isosceles triangles shown: the one on the left has the interval $[0, 1/2]$ as its bottom side, and the one on the right has $[1/2, 1]$ as its bottom side. Each of these triangles has a height c_2; in general, we only assume that $0 < c_2 < c_1$; but in Figure 8.8, $c_2 = 1/4$, so the pairs of equal sides fall on the equal sides of the large triangle, consistent with the original staircase construction in Chapter 1. If $c_2 \neq 1/4$, then we have a zigzag path (no longer staircase shaped) whose sides are equal but not coincident with the sides of the larger triangle.

The third stage splits the interval $[0,1]$ into three equal pieces, and we get three triangles, each of height c_3; the fourth stage creates four triangles, each of height c_4, and so on. The general assumptions concerning the heights c_1, c_2, etc., are as follows:

$$c_1 > c_2 > c_3 > \cdots > 0 \quad \text{and} \quad c_n \to 0 \text{ as } n \to \infty \qquad (8.26)$$

Figure 8.8 illustrates the special case $c_n = 1/(2n)$. Let's consider the generic stage n. As we see in Figure 8.9, each of the n triangles has a side of length $1/n$ on the x-axis.

The other two sides are equal; and by the Pythagorean theorem, each of them has length

$$\sqrt{c_n^2 + \left(\frac{1}{2n}\right)^2} = \frac{1}{2n}\sqrt{(2nc_n)^2 + 1}$$

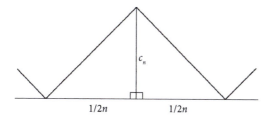

Fig. 8.9 A generic bump or triangle

There are $2n$ of these sides corresponding to the n triangles, so the total length of all of these sides (the length of the entire zigzag path) is

$$L_n = 2n \left[\frac{1}{2n} \sqrt{(2nc_n)^2 + 1} \right] = \sqrt{(2nc_n)^2 + 1} \qquad (8.27)$$

We are now in a position to take the limit as $n \to \infty$. We distinguish three possible cases:

Case 1: *If there is a number $c > 0$ such that $c_n = c/(2n)$ for every $n = 1, 2, 3, \ldots$* then

$$\lim_{n \to \infty} L_n = \sqrt{c^2 + 1} \qquad (8.28)$$

In particular, in the case shown in Figure 8.8 where $c = 1$ (originally, $c = \sqrt{2}$), we see that $\lim_{n \to \infty} L_n = \sqrt{2}$ (originally, 2).

But in fact, we can make the (fixed) length of the zigzag paths equal to any real number greater than 1 by adjusting the value of c, even as the paths converge uniformly to the interval $[0,1]$; for instance, to get paths of length $5/4 = 1.25$ at each stage, we set

$$\sqrt{c^2 + 1} = \frac{5}{4}$$

Solving this equation gives the value of c that gives the desired length:

$$c^2 + 1 = \frac{25}{16}$$

$$c = \sqrt{\frac{9}{16}} = \frac{3}{4} = 0.75$$

Figure 8.10 shows a generic triangular piece or bump of the zigzag path at stage n for different values of c.

This reduction in height reduces the lengths of the two equal sides enough for each triangle to reduce the total length from $\sqrt{2}$ to 1.25. If we choose $c > 1$, then

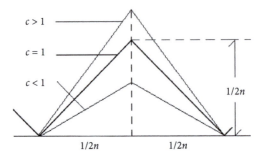

Fig. 8.10 Adjusting heights using the parameter c_n

with each partition we pull up, or extend the height of each triangle, so the total length gets larger too.

But with $c > 0$, the length $\sqrt{c^2 + 1}$ of every zigzag path is always greater than 1 no matter how small the value of c. It follows that the lengths of zigzag paths can never converge to the length of the base interval (originally, the diagonal) in this case.

Case 2: *If the sequence c_n converges to 0 fast enough that $nc_n \to 0$ as $n \to \infty$ (e.g., $c_n = 1/n^2$), then*

$$\lim_{n \to \infty} L_n = 1$$

This is clear from (8.27). In this case, as the zigzag paths flatten down to the interval $[0,1]$, and their lengths approach 1, the length of the interval $[0,1]$. There is no discrepancy between the lengths of the paths and the length of the limit curve in this case.

Case 3: *If the sequence c_n converges to 0 so slowly that $nc_n \to \infty$ as $n \to \infty$ (e.g., $c_n = 1/\sqrt{n}$), then*

$$\lim_{n \to \infty} L_n = \infty$$

Again, this is clear from (8.27). This case is especially puzzling because it implies that a (roughly) triangular region with a finite area[8] may have an arbitrarily long perimeter. This is reminiscent of a fractal like "Koch's snowflake" that has a finite area but an infinite perimeter; however, the limiting curve in this case isn't a fractal but a straight line segment, namely, the unit interval on the x-axis. So the limiting perimeter, being the perimeter of a triangle, is also finite, unlike a fractal. We can

---

[8] The area of the largest triangle in Figure 8.8 is 1/4, or half the area of the square of side $1/\sqrt{2}$.

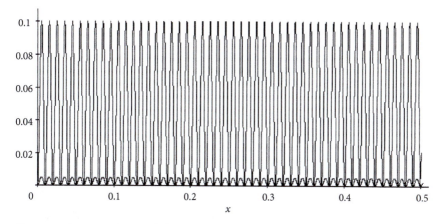

Fig. 8.11 Curves with very different variation

liken the curve to a fractal for large n because of the many bumps that it develops; but since this comparison works only for finite values of n, we don't have a true fractal for comparison. All bumps flatten out into the unit interval.

Figure 8.11 highlights the difference between Cases 1 and 3 above.

The tall curve, resembling a comb, shows the *left half* of the curve in Case 3 (from $x = 0$ to $x = 0.5$)[9] for $n = 100$ and $c_n = 1\sqrt{n}$. The little bumps at the bottom correspond to $c_n = 1/(2n)$. We see that the tall, comb-shaped curve in Figure 8.11 has a much longer total length than the much smaller curve at the bottom. The difference between these lengths grows even larger as n increases; but in the end, both processes lead to the interval $[0,1]$ on the x-axis, which has length 1.

A point worth mentioning before we wrap up this discussion is that the *non-smooth, jagged shape* of the zigzag paths is not responsible for the length discrepancy. We obtain the same results with very smooth (infinitely differentiable) curves too, a fact that we come across again in our discussion of the Weierstrass function below (defined as an infinite series). As an example, consider the sequence of smooth functions

$$f_n(x) = c_n \sin^2(n\pi x), \quad 0 \le x \le 1$$

where c_n is a decreasing sequence of positive numbers as in (8.26). Figure 8.12 illustrates four of these functions with $c_n = 1/(2n)$.

There is a standard formula in calculus for the lengths of these curves:

$$L_n = \int_0^1 \sqrt{1 + [f'_n(x)]^2}\,dx$$

[9] The right halves of both curves (from $x = 0.5$ to $x = 1$) are identical to what is shown in Figure 8.11.

The derivative $f'_n(x)$ is found using the chain rule:

$$f'_n(x) = 2n\pi c_n \sin(n\pi x) \cos(n\pi x)$$

$$= n\pi c_n \sin(2n\pi x)$$

Inserting this into the formula gives

$$L_n = \int_0^1 \sqrt{1 + \pi^2 n^2 c_n^2 \sin^2(2n\pi x)} \, dx, \quad n = 1, 2, 3, \ldots$$

This formula shows that if $nc_n \to 0$ as $n \to \infty$, then using the theory that we have already discussed in this section, we obtain

$$\lim_{n\to\infty} L_n = \int_0^1 \sqrt{1 + \pi^2 \lim_{n\to\infty} n^2 c_n^2 \sin^2(2n\pi x)} \, dx = \int_0^1 \sqrt{1 + 0} \, dx = 1$$

This result is similar to what we got for Case 2 above.

If nc_n does not converge to 0, then it is more difficult to deal with the integral directly due to a variety of technical issues. But we can talk about the other cases using a graphical argument and what have already learned about the staircase curves. If we superimposes the zigzag path in Figure 8.8 on the curves in Figure 8.12, we get what is shown in Figure 8.13.

Each straight line segment in Figure 8.13 is shorter than the piece of a smooth sine curve that joins its two ends. Therefore, the total length of the sine curve at each stage is larger than the total length the corresponding staircase curve. It follows that the lengths of the sine curves do not approach 1 as $n \to \infty$ if nc_n does

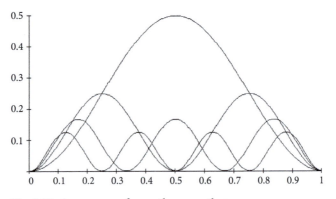

Fig. 8.12 A sequence of smooth curves that converges uniformly

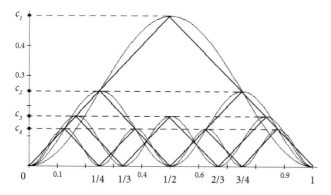

Fig. 8.13 Superimposing staircase curves on smooth ones

not converge to 0; further, if $nc_n \to \infty$, then the lengths of the sine curves actually diverge to infinity, nowhere near the length of the curve that they converge (uniformly) to, namely, the segment from 0 to 1 on the x-axis.

8.3 Infinite series of functions

The functions $f_n(x)$ in a sequence of functions can be added to generate an infinite series of functions similarly to the way we defined an infinite series of numbers earlier by adding a given sequence of constants a_n. In fact, since constants are actually constant functions $f_n(x) = a_n$ (independent of x), we may think of our earlier study as a very special case of what we discuss in this chapter.

Given an infinite sequence of functions over a given nonempty set S

$$f_1(x), f_2(x), \dots, f_n(x), \dots \qquad x \in S \qquad (8.29)$$

we define the corresponding sequence of partial sums:

$$s_1(x) = f_1(x)$$
$$s_2(x) = f_1(x) + f_2(x)$$
$$\vdots$$
$$s_n(x) = f_1(x) + f_2(x) + \cdots + f_n(x)$$
$$\vdots$$

It worth noticing here that if $f_n(x)$ is continuous, integrable, or differentiable for every n, then $s_n(x)$ also has the same properties because of the various rules that we previously discussed.

Infinite series of functions

Suppose that the sequence $s_n(x)$ of partial sums of the sequence in (8.29) converges pointwise (or uniformly) on S to a function $f(x)$. Then we write

$$\sum_{n=1}^{\infty} f_n(x) = \lim_{n\to\infty} s_n(x) = f(x) \qquad x \in S \qquad (8.30)$$

We say that *the infinite series on the left-hand side converges pointwise to $f(x)$ on S if $s_n \xrightarrow{p} f$, and likewise, the infinite series converges uniformly to $f(x)$ on S if $s_n \xrightarrow{u} f$.*

In our earlier discussion of the geometric power series, we proved the following:

$$\sum_{n=1}^{\infty} x^{n-1} = 1 + x + x^2 + \cdots = \frac{1}{1-x} \qquad \text{pointwise on } (-1, 1)$$

$$\sum_{n=1}^{\infty} x^{n-1} = 1 + x + x^2 + \cdots = \frac{1}{1-x} \qquad \text{uniformly on } [-r, r], \ 0 < r < 1$$

For the geometric power series, the functions being added are the power functions $f_n(x) = x^{n-1}$, with the limit function (or sum) $f(x) = 1/(1-x)$. We saw how the truncation of the interval $(-1, 1)$ to $[-r, r]$ led to uniform convergence by blocking the ε-index from diverging to infinity.

8.3.1 The Cauchy criterion and the Weierstrass test

For most infinite series of functions, a simple formula for $f(x)$ simply does not exist; the geometric power series is a rare exception to this rule. Convergence tests exist based on the *Cauchy convergence criterion*, which, as in the case of sequences and series of numbers, does not require knowing the limit.

Theorem 317 *(Cauchy criterion for uniform convergence) A sequence F_n of functions defined on a nonempty set S converges uniformly on S if and only if for every $\varepsilon > 0$, the ε-index N_ε is independent of the points in S and*

$$|F_n(x) - F_m(x)| < \varepsilon \qquad \text{for all } m, n \geq N_\varepsilon \text{ and for all } x \in S \qquad (8.31)$$

Equivalently, the sequence F_n converges uniformly on S if and only if

$$\|F_n - F_m\| \leq \varepsilon \qquad \text{for all } m, n \geq N_\varepsilon \qquad (8.32)$$

where as before, $\|F\| = \sup\{|F(x)| : x \in S\}$.

Proof First, assume that $F_n \xrightarrow{u} F$ on S where F is a given limit function. Then for every $\varepsilon > 0$, there is a positive integer N such that for all $n \geq N$

$$\|F_n - F\| < \frac{\varepsilon}{2}$$

Now if $m, n \geq N$, then for each $x \in S$

$$|F_n(x) - F_m(x)| \leq |F_n(x) - F(x)| + |F(x) - F_m(x)|$$
$$\leq \|F_n - F\| + \|F - F_m\|$$
$$< \varepsilon$$

Taking the supremum now gives

$$\|F_n - F_m\| \leq \varepsilon$$

Conversely, assume that the inequality (8.31), or equivalently, (8.32), holds. Then for each fixed $x \in S$

$$|F_n(x) - F_m(x)| \leq \|F_n - F_m\| \leq \varepsilon$$

so that the sequence of real numbers $F_n(x)$ is a Cauchy sequence; and as such, it converges to a real number[10] that we denote by $F(x)$. Since this is true for every $x \in S$, the set of numbers $F(x)$ defines a function F on S. By its construction, we have $F_n \xrightarrow{p} F$ on S, and it remains to show that the convergence is actually uniform.

Let $\varepsilon > 0$. Then by (8.32) for $m, n \geq N_\varepsilon$

$$\|F_n - F_m\| \leq \varepsilon$$

Then for all $x \in S$ if $n \geq N_\varepsilon$, then using the continuity of the absolute value function we obtain

$$|F_n(x) - F(x)| = \lim_{m \to \infty} |F_n(x) - F_m(x)| \leq \lim_{m \to \infty} \|F_n - F_m\| \leq \varepsilon$$

Now, taking the supremum gives

$$\|F_n - F\| \leq \varepsilon$$

which proves the uniform convergence.   ∎

---

[10] This is true by the Cauchy convergence criterion for sequences of numbers, which holds since $\mathbb{R}$ is complete.

An immediate corollary of the above theorem is obtained for infinite series where F_n is the sequence of partial sums of f_n. Specifically, suppose that $n > m$:

$$s_n(x) - s_m(x) = f_{m+1}(x) + \cdots + f_n(x)$$

so the Cauchy condition (8.32) can be written as

$$\|f_{m+1} + \cdots + f_n\| \leq \varepsilon \quad \text{for all } n > m \geq N_\varepsilon \tag{8.33}$$

Corollary 318 *(Cauchy criterion for uniform convergence of series) The infinite series $\sum_{k=1}^{\infty} f_k$ of functions f_k on a nonempty set S converges uniformly on S if and only if (8.33) holds for each $\varepsilon > 0$.*

The fact that in (8.33) we don't need to specify a limit function $f(x)$ makes the Cauchy criterion very useful. Also significant is the fact that Corollary 318 is an *if and only if*, or a *necessary and sufficient* condition, because we use *both* directions to prove the following useful convergence test[11] for infinite series of functions.

Theorem 319 *(Weierstrass test, or M test, for uniform convergence) Let f_n be a sequence of bounded functions on a nonempty set S, i.e.,*

$$\|f_k\| \leq M_k \quad \text{for } k = 1, 2, 3, \ldots \tag{8.34}$$

If the infinite series $\sum_{k=1}^{\infty} M_k$ of (non-negative) constants converges (i.e., $\sum_{k=1}^{\infty} M_k < \infty$), then the series of functions $\sum_{k=1}^{\infty} f_k$ converges uniformly on S.

Proof Let $\varepsilon > 0$. Thinking of M_k as constant functions, the series $\sum_{k=1}^{\infty} M_k$ converges uniformly; so by Corollary 318

$$M_{m+1} + \cdots + M_n \leq \varepsilon \quad \text{for all } n > m \geq N_\varepsilon$$

Further, for every $x \in S$

$$|f_{m+1}(x) + \cdots + f_n(x)| \leq |f_{m+1}(x)| + \cdots + |f_n(x)|$$
$$\leq \|f_{m+1}\| + \cdots + \|f_n\|$$
$$\leq M_{m+1} + \cdots + M_n$$
$$\leq \varepsilon$$

---

[11] Also known as the *Weierstrass M test*.

Taking the supremum gives

$$\|f_{m+1} + \cdots + f_n\| \leq \epsilon$$

for $n > m \geq N_\epsilon$. Now another application of Corollary 318 proves that $\sum_{k=1}^{\infty} f_k$ converges uniformly on S. ∎

To illustrate the utility of Theorem 319 (and indirectly, also of Cauchy's criterion), consider the following infinite series of functions:

$$\sum_{n=1}^{\infty} \frac{x^3 + n}{x + n^3} \qquad x \in [0, c] \qquad (8.35)$$

where c is a fixed, positive, real number. To see if this series converges uniformly on the interval $[0, c]$, we check each term of the sum:

$$f_n(x) = \frac{x^3 + n}{x + n^3}$$

to see if it is bounded by some number M_n. Notice that

$$x^3 + n \leq c^3 + n \qquad \text{since } x \leq c \qquad (8.36)$$

$$x + n^3 \geq 0 + n^3 = n^3 \qquad \text{since } x \geq 0$$

Therefore,

$$f_n(x) \leq \frac{n + c^3}{n^3} = \frac{1}{n^2} + \frac{c^3}{n^3}$$

The last expression on the right-hand side is a good candidate for our M_n (it depends only on n). In fact, if we define it to be M_n for each n, then

$$\sum_{n=1}^{\infty} M_n = \sum_{n=1}^{\infty} \left(\frac{1}{n^2} + \frac{c^3}{n^3} \right) = \sum_{n=1}^{\infty} \frac{1}{n^2} + \sum_{n=1}^{\infty} \frac{c^3}{n^3}$$

The sum $\sum_{n=1}^{\infty} 1/n^2$ converges as a p-series with power $p = 2 > 1$, and the sum $\sum_{n=1}^{\infty} 1/n^3$ is likewise a convergent p-series. Since $\sum_{n=1}^{\infty} 1/n^3 < \infty$, we conclude that $c^3 \sum_{n=1}^{\infty} 1/n^3 < \infty$ too; so as a whole, $\sum_{n=1}^{\infty} M_n < \infty$. Now, by Theorem 319, the series in (8.35) converges uniformly on $[0, c]$.

8.3.2 Integration and differentiation term by term of function series

Looking back at the definition of infinite series of functions, we see that the idea of taking limits, derivatives, or integrals is a familiar one: we use the results that we already found for sequences of functions and simply apply them to the sequences of partial sums of a series. We begin with continuity.

Corollary 320 *Suppose that the series of functions $\sum_{k=1}^{\infty} f_k$ converges uniformly to a function f on a nonempty set S. If f_k is continuous at a point $a \in S$ for every k, then f is continuous at a. If all f_k are continuous on S, then f is continuous on S.*

Proof Note that the partial sums $s_n = \sum_{k=1}^{n} f_k$ are continuous at a. Since by assumption $s_n \xrightarrow{u} f$, Theorem 311 implies that f is continuous at a. The last statement is obvious now. ∎

This is useful information that is not obvious; for example, we don't have formulas for functions that the series in (8.35) converge to, but we know that the limit function must be continuous because every term of the series is a continuous function, and the series converges uniformly.

We actually get a little bit more mileage out of the continuity theorem above because $f(x)$ is continuous on S if and only if it is continuous at every x in S. To illustrate, consider the fact that the geometric power series converges only pointwise on the interval $(-1, 1)$, but its limit $1/(1 - x)$ is continuous on all of this interval. We recall that the series converges uniformly on intervals of type $[-r, r]$ if $0 < r < 1$. Now, pick any value of x in $(-1, 1)$, say, $x = c$. We can then choose r so that $-r < c < r$ and still have $0 < r < 1$. Since this is possible for every real number x in $(-1, 1)$, it follows that the limit is continuous on all of $(-1, 1)$. This observation is not limited to the geometric power series; it extends to other series and their uniform limits since it depends only incidentally on the geometric power series or its limit.

Next, we consider the integral of a series of functions.

Corollary 321 *(Term by term integration) Assume that the series of functions $\sum_{k=1}^{\infty} f_k$ converges uniformly to a function f on an interval $[a, b]$. If f_k is integrable on $[a, b]$, then f is integrable on $[a, b]$ and*

$$\sum_{k=1}^{\infty} \int_a^b f_k(x)\,dx = \int_a^b f(x)\,dx = \int_a^b \sum_{k=1}^{\infty} f_k(x)\,dx \qquad (8.37)$$

Proof The sum of a (finite) number of integrable functions is integrable by the linear property of the Riemann integral, so the partial sum $s_n(x)$ is integrable on $[a, b]$ for every n and

$$\int_a^b s_n(x)dx = \int_a^b f_1(x)dx + \cdots + \int_a^b f_n(x)dx = \sum_{k=1}^n \int_a^b f_k(x)dx \qquad (8.38)$$

Next, since the sequence $s_n(x)$ converges uniformly by hypothesis, $\lim_{n\to\infty} s_n(x) = f(x)$ is integrable by Theorem 314, and we can move the limit outside the integral to get

$$\int_a^b f(x)dx = \int_a^b \lim_{n\to\infty} s_n(x)dx = \lim_{n\to\infty} \int_a^b s_n(x)dx \qquad (8.39)$$

Now the last sum on the right in (8.38) is the partial sum of a sequence of constants; therefore, taking the limit as $n \to \infty$ gives an infinite series (of numbers):

$$\lim_{n\to\infty} \int_a^b s_n(x)dx = \lim_{n\to\infty} \left(\sum_{k=1}^n \int_a^b f_k(x)dx \right)$$

$$= \sum_{n=1}^\infty \int_a^b f_n(x)dx$$

that converges to $\int_a^b f(x)dx$ by (8.39) and proves the validity of (8.37).   ∎

To illustrate the above theorem, consider the geometric power series with alternating signs. This converges uniformly by the Weierstrass test (Theorem 319) on any interval $[-r, r]$ if $0 < r < 1$:

$$1 - x + x^2 - x^3 + \cdots = \frac{1}{1+x}$$

Let u be any number[12] in the interval $(-1, 1)$, and integrate the above series term by term from 0 to u to get

$$\int_0^u 1dx - \int_0^u xdx + \int_0^u x^2dx - \int_0^u x^3dx + \cdots = \int_0^u \frac{1}{1+x}dx$$

After carrying out the integration, we obtain the interesting result:

$$u - \frac{u^2}{2} + \frac{u^3}{3} - \frac{u^4}{4} + \cdots = \ln(1+u), \qquad u \in (-1, 1)$$

---

[12] As long as $u \neq \pm 1$, we can find r between 0 and 1 such that the interval $[0, u]$ (or $[u, 0]$ if u is negative) is contained in $[-r, r]$.

We show later that this is indeed the Taylor series for $\ln(1 + u)$ centered at the origin.

Finally, there is also a corollary corresponding to Theorem 316 that lays out the conditions under which we can take the derivative of a series term by term. The proof is left as an exercise.

Corollary 322 *Let f_k be a sequence of differentiable functions on an interval $[a, b]$, and assume that their derivatives f'_k are continuous on $[a, b]$. If*

(i) the derivative series $\sum_{k=1}^{\infty} f'_k$ converges uniformly on $[a, b]$,

(ii) the series of constants $\sum_{k=1}^{\infty} f_k(c)$ converges for some number $c \in [a, b]$,

then the sequence $\sum_{k=1}^{\infty} f_k$ converges uniformly on $[a, b]$ to a differentiable function f and

$$\frac{d}{dx} \sum_{k=1}^{\infty} f_k(x) = f'(x) = \sum_{k=1}^{\infty} f'_k(x)$$

To illustrate the above corollary, consider the series

$$\sum_{k=1}^{\infty} \frac{\sin(kx)}{2^k} \tag{8.40}$$

where $x \in (-\infty, \infty)$. The nth partial sum of this series is

$$s_n(x) = \sum_{k=1}^{n} \frac{\sin(kx)}{2^k}$$

and taking the derivative of this finite sum gives

$$s_n'(x) = \sum_{k=1}^{n} \frac{k \cos(kx)}{2^k}$$

To use Corollary 322, we show that this sequence converges uniformly on $(-\infty, \infty)$, and we do this using the Weierstrass test (Theorem 319). Note that for every $k \geq 1$ and all $x \in (-\infty, \infty)$

$$\left| \frac{k \cos(kx)}{2^k} \right| = \frac{k |\cos(kx)|}{2^k} \leq \frac{k}{2^k}$$

Let $M_k = k/2^k$ and note that

$$\sum_{k=1}^{\infty} M_k = \sum_{k=1}^{\infty} \frac{k}{2^k} < \infty$$

The convergence may be proved quickly via either the ratio test or the root test (you may find the exact value in Exercise 348). Theorem 319 now implies that the series

$$\sum_{k=1}^{\infty} \frac{k\cos(kx)}{2^k}$$

converges uniformly on $(-\infty, \infty)$. This shows that Condition (i) in Corollary 322 is satisfied on any interval $[a, b]$ of real numbers. Further, (ii) is satisfied trivially at $x = 0$ since $\sin(kx) = 0$ if $x = 0$. Now Corollary 322 implies that series in (8.40) converges uniformly on every interval I that contains 0, and for all $x \in I$

$$\frac{d}{dx}\sum_{k=1}^{\infty} \frac{\sin(kx)}{2^k} = \sum_{k=1}^{\infty} \frac{k\cos(kx)}{2^k} \tag{8.41}$$

Note that (8.41) is actually true for all real numbers x since every x is in some interval that contains 0.

Worth mentioning also is that although we do not know of any closed form formulas for the limits or sums of the series on either side of (8.41), if we define $f(x)$ to be the series in (8.40), then $f'(x)$ is given by (8.41). For certain values of x we can obtain exact numerical values, while generally we can approximate these values to a desired level of accuracy by adding enough terms of the series. For instance, if $x = \pi$, then

$$f(\pi) = \sum_{k=1}^{\infty} \frac{1}{2^k} \sin(k\pi) = 0$$

while

$$f'(\pi) = \sum_{k=1}^{\infty} \frac{k}{2^k} \cos(k\pi) = \sum_{k=1}^{\infty} \frac{(-1)^k k}{2^k} = -\frac{2}{9}$$

The last number is calculated in Exercise 348. So even though we don't have a formula for $f(x)$, we can tell that $f(\pi) = 0$, and f is decreasing at $x = \pi$. You can similarly calculate the exact values of $f(0)$ and $f'(0)$ and tell if f is increasing or decreasing at $x = 0$ (note also that $f(x + 2\pi) = f(x)$ for all values of x, so f is a periodic function with period 2π).

8.3.3 Weierstrass's continuous yet nowhere differentiable function

It is not hard to imagine a continuous curve or function that fails to have a derivative (or be non-differentiable) at some points; the absolute value function $|x|$ is an example. By joining several such v-shaped curves, we can create a sawtooth shaped curve that has a large number of points where a derivative does not exist; there may be even an infinite number of points, say, at every integer. But it is hard to imagine a continuous curve that doesn't have a derivative at *any* point.

In fact, is impossible to *visualize* such a curve because the only known examples are obtained as limits of infinite sequences of simpler curves so as to have infinitely many sharp corners in every bounded interval. We discussed one case involving the sequence of functions $c_n \sin^2(n\pi x)$ above when showing that uniform convergence doesn't preserve lengths of curves.

On the other hand, when we go back to our definition of continuity and derivatives in Chapter 6, a moment's reflection shows that there is no reason why a continuous curve has to have derivatives at *any point* of its domain. The limitation of our ability to visualize such continuous curves is not a valid reason for ruling out their existence. However, historically this was not as clear a point as it may seem nowadays.

When analysis was not yet mature and understanding calculus relied on visual intuition, it was hard to accept the existence of what are now called *nowhere differentiable continuous functions*. For instance, in an 1839 calculus text by J. L. Raabe, it was taken for granted that every continuous function can have at most a *finite* number of points where it is *not* differentiable in any bounded interval.

It is actually not hard to come up with a continuous function that has no derivatives at an infinite number of points in some bounded interval, no matter how tiny. For example, take the interval $[0,1]$ and the sequence of numbers $1, 1/2, 1/3, \ldots$ For each positive integer n, set up an isosceles triangle having the segment between $1/n$ and $1/(n+1)$ on the x-axis as its base and a height of 1 that is reached at the midpoint of the interval $[1/(n+1), 1/n]$. Now define a function $f(x)$ so that its value at any x is the vertical distance from the x-axis to the side of the triangle that sits above x (to ensure continuity, we define $f(0) = 0$). The graph of $f(x)$ is just the top sides of all the triangles; notice that this graph is a continuous (polygonal) curve that has an infinite number of sharp corners between 0 and 1. However, this curve also contains an infinite number of line segments, and each of these segments is smooth, so our $f(x)$ is differentiable at most points of $[0,1]$.

It was not until 1872 that Weierstrass constructed a function that was continuous on all of $(-\infty, \infty)$ but didn't have a derivative at *any point*. This function was defined as the limit of a trigonometric series similar to that in (8.40). We now discuss Weierstrass's function.

Consider the following trigonometric series:

$$\sum_{k=0}^{\infty} c^k \cos(m^k \pi x) \quad \text{where } 0 < c < 1, \text{ and } m \text{ is a positive integer} \qquad (8.42)$$

Because

$$\left| c^k \cos(m^k \pi x) \right| = c^k \left| \cos(m^k \pi x) \right| \leq c^k$$

and $\sum_{k=0}^{\infty} c^k = 1/(1 - c)$ is a convergent geometric series, the Weierstrass test for uniform convergence implies that the series in (8.42) converges uniformly on all of $(-\infty, \infty)$ to some function, say, $W(x)$, which is continuous on $(-\infty, \infty)$ since each term $c^k \cos(m^k \pi x)$ of the series is continuous on $(-\infty, \infty)$. But Weierstrass showed that if we choose m to be an *odd integer* large enough that

$$m > \frac{3\pi + 2}{2c} \tag{8.43}$$

then $W(x)$ fails to have a derivative at any real number x. Notice that since $c < 1$, we have

$$\frac{3\pi + 2}{2c} > \frac{3\pi + 2}{2} \simeq 5.71$$

so the *least* odd integer for which Weierstrass's inequality in (8.43) is valid is $m = 7$.

To simplify the notation in what follows, we set $c = 1/2$ and prove the following result (some graphical illustrations follow the proof).

Theorem 323 *(Weierstrass) Define the function*

$$W(x) = \sum_{k=0}^{\infty} \frac{\cos(m^k \pi x)}{2^k} \tag{8.44}$$

where m is an odd integer such that

$$m > 3\pi + 2 \quad (m \geq 13) \tag{8.45}$$

Then $W(x)$ is continuous on $(-\infty, \infty)$, but it is not differentiable at any point of $(-\infty, \infty)$.

Proof The proof is technical but not difficult. Since we have already seen that $W(x)$ is continuous, we now proceed to show that its difference ratio does not have a limit at any real number. We do this by finding a sequence h_n of (positive) real numbers that converges to 0 such that

$$\lim_{n \to \infty} \left| \frac{W(x + h_n) - W(x)}{h_n} \right| = \infty \quad \text{for all } x \in (-\infty, \infty) \tag{8.46}$$

The nth partial sum of the series in (8.44) is

$$S_n(x) = \sum_{k=0}^{n-1} \frac{\cos(m^k \pi x)}{2^k}$$

Let $R_n(x) = W(x) - S_n(x)$ be the remainder of the series and notice that for all positive integers n

$$\frac{W(x + h_n) - W(x)}{h_n} = \frac{R_n(x + h_n) - R_n(x)}{h_n} + \frac{S_n(x + h_n) - S_n(x)}{h_n}$$

Taking the absolute value and using the alternate version of the triangle inequality, we obtain

$$\left| \frac{W(x + h_n) - W(x)}{h_n} \right| \geq \left| \frac{R_n(x + h_n) - R_n(x)}{h_n} \right| - \left| \frac{S_n(x + h_n) - S_n(x)}{h_n} \right| \qquad (8.47)$$

To complete the proof of the theorem, we show that the difference on the right-hand side above goes to infinity for a properly chosen sequence h_n of positive real numbers that converges to 0. To this end, we find an upper bound for the difference quotient of S_n and a lower bound for that of R_n.

For S_n, the mean value theorem gives a number t between 0 and h_n such that

$$\frac{\cos(m^k \pi(x + h_n)) - \cos(m^k \pi x)}{h_n} = -m^k \pi \sin(m^k \pi(x + t))$$

so

$$\left| \frac{S_n(x + h_n) - S_n(x)}{h_n} \right| \leq \sum_{k=0}^{n-1} \left| \frac{\cos(m^k \pi(x + h_n)) - \cos(m^k \pi x)}{2^k h_n} \right|$$

$$\leq \sum_{k=0}^{n-1} \frac{m^k \pi}{2^k}$$

The last sum is a geometric progression with value

$$\sum_{k=0}^{n-1} \frac{m^k \pi}{2^k} = \pi \sum_{k=0}^{n-1} \left(\frac{m}{2} \right)^k = \frac{\pi[(m/2)^n - 1]}{(m/2) - 1} < \frac{2\pi}{m - 2} \left(\frac{m}{2} \right)^n$$

so we have the upper bound below:

$$\left| \frac{S_n(x + h_n) - S_n(x)}{h_n} \right| \leq \frac{2\pi}{m - 2} \left(\frac{m}{2} \right)^n \qquad (8.48)$$

Next, we obtain a lower bound for the difference ratio of R_n and in the process also determine the proper sequence h_n. A key observation is to write the real number $m^n x$ as

$$m^n x = j_n + \delta_n$$

where j_n is a positive integer, and $\delta_n \in [-1/2, 1/2)$. Now define

$$h_n = \frac{1 - \delta_n}{m^n}$$

Note that $1/2 < 1 - \delta_n \leq 3/2$, so

$$\frac{2}{3} m^n \leq \frac{1}{h_n} < 2m^n \tag{8.49}$$

Now, for $k \geq n$

$$m^k \pi(x + h_n) = m^{k-n} m^n \pi(x + h_n)$$
$$= m^{k-n} \pi(m^n x + (1 - \delta_n))$$
$$= m^{k-n} \pi(j_n + 1)$$

Since m^{k-n} is odd, and k_n is an integer, it follows that

$$\cos(m^k \pi(x + h_n)) = \cos(m^{k-n} \pi(j_n + 1)) = (-1)^{j_n+1} \tag{8.50}$$

Further,

$$\cos(m^k \pi x) = \cos(m^{k-n} m^n \pi x)$$
$$= \cos(m^{k-n} \pi(j_n + \delta_n))$$
$$= \cos(m^{k-n} \pi j_n + m^{k-n} \pi \delta_n)$$

Using the trigonometric sum identity $\cos(u + v) = \cos u \cos v - \sin u \sin v$, and noting that $\sin(m^{k-n} \pi j_n) = 0$, we obtain

$$\cos(m^k \pi x) = \cos(m^{k-n} \pi j_n) \cos(m^{k-n} \pi \delta_n) = (-1)^{j_n} \cos(m^{k-n} \pi \delta_n) \tag{8.51}$$

From (8.50) and (8.51) we have

$$\cos(m^k \pi(x + h_n)) - \cos(m^k \pi x) = (-1)^{j_n+1}[1 + \cos(m^{k-n} \pi \delta_n)]$$

It follows that

$$\left|\frac{R_n(x+h_n)-R_n(x)}{h_n}\right| = \left|\sum_{k=n}^{\infty}\frac{\cos(m^k\pi(x+h_n))-\cos(m^k\pi x)}{2^k h_n}\right|$$

$$= \left|\frac{(-1)^{j_n+1}}{h_n}\sum_{k=n}^{\infty}\frac{[1+\cos(m^{k-n}\pi\delta_n)]}{2^k}\right|$$

$$= \frac{1}{h_n}\sum_{k=n}^{\infty}\frac{1+\cos(m^{k-n}\pi\delta_n)}{2^k}$$

$$\geq \frac{1}{h_n}\frac{1+\cos(\pi\delta_n)}{2^n}$$

Note that $\cos(\pi\delta_n) \geq 0$ since $\delta_n \in [-1/2, 1/2)$, so by (8.49)

$$\left|\frac{R_n(x+h_n)-R_n(x)}{h_n}\right| \geq \frac{1}{h_n}\frac{1}{2^n} \geq \frac{2}{3}\left(\frac{m}{2}\right)^n$$

Now we go back to (8.47) and use this inequality and (8.48) to get

$$\left|\frac{W(x+h_n)-W(x)}{h_n}\right| \geq \frac{2}{3}\left(\frac{m}{2}\right)^n - \frac{2\pi}{m-2}\left(\frac{m}{2}\right)^n = \left(\frac{2}{3}-\frac{2\pi}{m-2}\right)\left(\frac{m}{2}\right)^n$$

By (8.45), $m > 2$ and

$$\frac{2}{3} - \frac{2\pi}{m-2} > 0$$

so (8.46) follows, and the proof is complete.   ∎

Weierstrass's inequality, which can also be written as

$$mc > 1 + \frac{3\pi}{2}$$

is a sufficient condition for the non-existence of derivatives, but it is not necessary. Looking at the series

$$\sum_{n=0}^{\infty} c^n \cos(a^n\pi x)$$

the English mathematician G. H. Hardy (1877–1947) showed that it is nowhere differentiable as long as

$$ac \geq 1 \tag{8.52}$$

where a is any (positive) *real* number, and $0 < c < 1$. Aside from being more general and looking more natural than 8.43, Hardy's inequality 8.52 makes it much easier to conduct a graphical exploration.

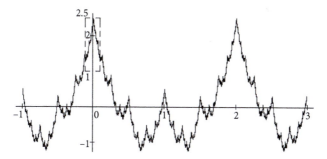

Fig. 8.14 A partial sum of the Hardy–Weierstrass series

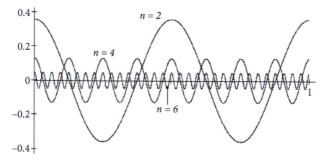

Fig. 8.15 Three terms of the Hardy–Weierstrass series

Let's explore the Hardy–Weierstrass series above graphically for specific values of a and c. If $a = 2$ and $c = 0.6$, Hardy's inequality (8.52) is satisfied and we obtain

$$H(x) = \sum_{n=0}^{\infty} 0.6^n \cos(2^n \pi x) \qquad (8.53)$$

Graphing any stretch of $H(x)$ properly is impossible, so we take a look at some of the partial sums of the series.

Figure 8.14 shows the partial sum $S_{12}(x)$ over the interval $[-1, 3]$. Even with these few terms of the infinite series in (8.53), we see a spiky, non-smooth graph. In Figure 8.15, we see three of the 13 individual terms of this partial sum.

To get a better sense of what is going on in Figure 8.15, we zoom in on a portion of it near $x = 0$; the result is shown in Figure 8.16, which represents a 20-fold magnification (plus a 10-fold increase in resolution to improve the image).

We still see a spiky curve at this magnification; if this was indeed the graph of $H(x)$ (the entire infinite series), then further magnifications would continue to show a spiky curve, essentially resembling the portion that we magnified. This self-similarity would be preserved with further zooming in on the graph of $H(x)$. Figure 8.17 shows a 10-fold magnification of the portion near the tip in Figure 8.16.

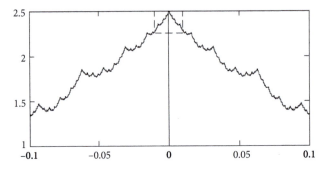

Fig. 8.16 Zooming on the selected portion in Figure 8.14

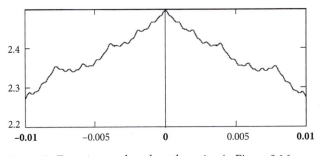

Fig. 8.17 Zooming on the selected portion in Figure 8.16

The partial sums of $H(x)$ are similar to the limit of the sequence of sinusoidal smooth versions of the zigzag paths that we studied above in the sense that both limits in both cases are curves whose lengths are not limits of the lengths of the curves in the sequence. The underlying concept of variation is studied rigorously in more advanced texts on real analysis.

8.4 Power series expansions of functions

One of the most powerful concepts in calculus is being able to express functions as infinite series. Given a function $f(x)$, we may express it as a power series consisting of the power functions x, x^2, etc., if it is differentiable infinitely often; otherwise, we may express $f(x)$ as a different series of functions, such as a Fourier series consisting of the trigonometric functions $\sin kx$ and $\cos kx$. We discuss these latter series after studying the power series.

8.4.1 Power series and their convergence

We start with a formal definition of power series.

A **power series** is an infinite series of (translated or shifted) power functions:

$$\sum_{n=0}^{\infty} c_n(x - c)^n = c_0 + c_1(x - c) + c_2(x - c)^2 + \cdots \tag{8.54}$$

where c and c_n are real numbers, and x is a variable in a set S of real numbers. The number c is the *center* of the power series, and the numbers c_n are its *coefficients*.

Notice that the standard notation for power series starts it at the *zeroth term* $a_0(x - c)^0$; defining $(x - c)^0 = 1$ we see that the zeroth term of every power series is always a number c_0, i.e., a constant. The *nth term* is the power function $f_n(x) = c_n(x - c)^n$ so the theory that we developed for series of functions applies to this type of series too.

If $c = 0$, then the power series in (8.54) takes a simpler-looking form:

$$\sum_{n=0}^{\infty} c_n x^n = c_0 + c_1 x + c_2 x^2 + \cdots \tag{8.55}$$

which is a power series centered at zero. Given the simpler form of the power series (8.55), why bother with any other center? The answer is, for practical reasons; we need the greater flexibility for the power series expansions of functions like the natural logarithm $\ln x$ that are not defined at $x = 0$.

The geometric power series is the simplest example of a power series centered at zero because all of its coefficients are equal: $c_n = a$ for all $n \geq 0$ (we set $a = 1$ earlier to simplify the notation without missing anything of significance). The most general version of the geometric power series is

$$\sum_{n=0}^{\infty} a(x - c)^n = a + a(x - c) + a(x - c)^2 + \cdots$$

This has a nonzero center c and (common) coefficient $c_n = a$. Recall from our earlier discussion that this series converges if $-1 < x - c < 1$, i.e.,

$$c - 1 < x < c + 1 \qquad \text{or} \qquad x \in (c - 1, c + 1)$$

For this range of values of x, the following formula is valid:

$$a + a(x - c) + a(x - c)^2 + \cdots = \frac{a}{1 - (x - c)} = \frac{a}{c + 1 - x}$$

Notice that for $x \in (c - 1, c + 1)$, the geometric series converges *absolutely*, i.e.,

$$\sum_{n=0}^{\infty} |a| |x - c|^n < \infty \quad \text{for } x \in (c - 1, c + 1)$$

so we may call $(c - 1, c + 1)$ an interval of absolute convergence. This concept extends to all power series.

Finding the range of values of x for which the general power series in (8.54) converges requires a different, more general method. Of course, *every power series converges at its center $x = c$* trivially because all terms of the series (except maybe the first) drop out.

Beyond the center, we may use the root test or the ratio test. Recall that these tests were extensions of the concept of common ratio to non-constant ratios. We can use these tests with slight modification to allow for the fact that the variable x has a range of values that needs to be specified.

Theorem 324 *(Convergence of power series)*

   (a) The power series $\sum_{n=0}^{\infty} c_n (x - c)^n$ converges absolutely if

$$|x - c| < R \quad \text{or } x \in (c - R, c + R)$$

   where

$$R = \frac{1}{\limsup_{n \to \infty} \sqrt[n]{|c_n|}} \tag{8.56}$$

   if the denominator is nonzero. If $\limsup_{n \to \infty} \sqrt[n]{c_n} = 0$, then we set $R = \infty$; also if $\limsup_{n \to \infty} \sqrt[n]{c_n} = \infty$, then we take $R = 0$.
   (b) The power series diverges if $|x - c| > R$.
   (c) The power series converges uniformly if $|x - c| \le r$, i.e., if $x \in [c - r, c + r]$ for every $r \in (0, R)$.

Proof (a) According to Theorem 150 in Chapter 5, the power series converges (absolutely) if

$$\limsup_{n \to \infty} \sqrt[n]{|c_n| |x - c|^n} < 1 \tag{8.57}$$

and it diverges if the reverse inequality holds. Simplifying the inequality in (8.57), we obtain

$$\limsup_{n \to \infty} \sqrt[n]{|c_n|} |x - c| < 1 \quad \text{or} \quad |x - c| < \frac{1}{\limsup_{n \to \infty} \sqrt[n]{|c_n|}}$$

If $\limsup_{n\to\infty} \sqrt[n]{|c_n|} = 0$, then the first inequality above reduces to $0 < 1$, which is always true for all $x \in (-\infty, \infty)$. This is equivalent to setting $R = \infty$ in (8.56).

(b) Theorem 150 in Chapter 5 implies that if $|x - c| > R$ in (8.56), then the power series diverges.

(c) To prove the uniform convergence, we use Weierstrass's test (Theorem 319). If $0 < r < R$, then

$$\limsup_{n\to\infty} \sqrt[n]{|c_n|} = \frac{1}{R^k} < \frac{1}{r^k}$$

This implies that there is a positive integer N such that $\sqrt[n]{|c_n|} \le 1/b$ for $n \ge N$ where b is a number such that $1/R < 1/b < 1/r$. Therefore, if $|x - c| \le r$, then for $n \ge N$

$$|c_n(x - c)^n| \le \frac{1}{b^n} r^n = \left(\frac{r}{b}\right)^n$$

Note that $\sum_{n=1}^{\infty} (r/b)^n < \infty$ since $r < b$, so by Weierstrass's M test, the power series converges uniformly on the interval $[c - r, c + r]$.   ∎

Theorem 324 suggests the following definitions.

Radius of convergence The real number R is the *radius of convergence* of the power series. If $R = 0$, then the power series converges only at its center $x = c$.

Interval of absolute convergence. The open interval $(c - R, c + R)$, where the power series converges absolutely (pointwise), is the *interval of absolute convergence*. If $R = \infty$, then the power series converges for all real numbers.

To illustrate the above concepts, consider the power series

$$\sum_{n=0}^{\infty} [2 + (-1)^n]^n (x - 1)^n = 1 + (x - 1) + 3^2(x - 1)^2 + \cdots \tag{8.58}$$

In this series, the coefficients are

$$c_n = [2 + (-1)^n]^n = \begin{cases} 1 & \text{if } n \text{ is odd} \\ 3^n & \text{if } n \text{ is even} \end{cases}$$

It follows that

$$\sqrt[n]{|c_n|} = \begin{cases} 1 & \text{if } n \text{ is odd} \\ 3 & \text{if } n \text{ is even} \end{cases}$$

and therefore,

$$\limsup_{n\to\infty} \sqrt[n]{|c_n|} = 3$$

Therefore, $R = 1/3$, and the interval of convergence of the power series is $(2/3, 4/3)$. Further, the series converges uniformly in closed subintervals of $(2/3, 4/3)$, such as the interval $[0.67, 1.33]$.

For another example, consider the power series

$$\sum_{n=0}^{\infty} \frac{x^{3n}}{2^n} = 1 + \frac{x^3}{2} + \frac{x^6}{2^2} + \cdots \qquad (8.59)$$

Here the power $3n$ seems at odds with the general power series notation, but we may think of the above series as a standard power series in which $c_k = 0$ if $k \neq 3n$. The zero terms don't contribute to the sum of the series, but *they can affect its radius of convergence.*

To see how, we go back to the inequality in (8.57); for this series, it takes the form

$$\limsup_{n\to\infty} \sqrt[n]{|c_n| |x - c|^{3n}} < 1$$

or

$$|x - c|^3 \limsup_{n\to\infty} \sqrt[n]{|c_n|} < 1$$

From this we obtain

$$|x - c|^3 < \frac{1}{\limsup_{n\to\infty} \sqrt[n]{|c_n|}}$$

or

$$|x - c| < \frac{1}{\sqrt[3]{\limsup_{n\to\infty} \sqrt[n]{|c_n|}}}$$

Therefore, the radius of convergence in this problem is given by the formula

$$R = \frac{1}{\sqrt[3]{\limsup_{n\to\infty} \sqrt[n]{1/2^n}}} = \frac{1}{\sqrt[3]{1/2}} = \sqrt[3]{2}$$

and this gives the interval of absolute convergence as $(-\sqrt[3]{2}, \sqrt[3]{2})$.

We may independently check this conclusion using the geometric power series formula, since the series in (8.59) can be written as

$$\sum_{n=0}^{\infty}\left(\frac{x^3}{2}\right)^n = 1 + \left(\frac{x^3}{2}\right)^1 + \left(\frac{x^3}{2}\right)^2 + \cdots$$

For each fixed x, this is a geometric series with common ratio $x^3/2$, so it converges to the function

$$\frac{1}{1 - x^3/2} = \frac{2}{2 - x^3}$$

provided that $|x^3/2| < 1$, which holds for $x \in (-\sqrt[3]{2}, \sqrt[3]{2})$, the same as what we obtained using Theorem 324.

The ratio test method

We saw in Chapter 5 that the ratio test works efficiently when the coefficients involve factorials. The same is true for power series.

To use the ratio test, we assume that $c_n \neq 0$ for all n. In this case, for each fixed value of x, we must have

$$\lim_{n\to\infty}\left|\frac{c_{n+1}(x - c)^{n+1}}{c_n(x - c)^n}\right| = \lim_{n\to\infty}\left|\frac{c_{n+1}}{c_n}\right||x - c| < 1 \qquad (8.60)$$

We assume for simplicity that the limit above exists and define

$$\rho = \lim_{n\to\infty}\left|\frac{c_{n+1}}{c_n}\right|$$

Then either $\rho = 0$, in which case (8.60) is true for all real values of x, or $\rho \neq 0$, in which case the inequality

$$|x - c| < \frac{1}{\rho} \qquad (8.61)$$

guarantees that the power series converges. The above inequality may be written as

$$c - \frac{1}{\rho} < x < c + \frac{1}{\rho}$$

The number $1/\rho$ defines the range of values of x before and after c for which the power series converges. If we define $R = 1/\rho$, then we see that the power series converges if

$$c - R < x < c + R$$

Also notice that if $x < c - R$ or $x > c + R$, then the ratio test implies that the series diverges.

In our future calculations, we are interested in finding R rather than $\rho = 1/R$. So we modify the calculation to yield the value of R directly by defining

$$R = \lim_{n \to \infty} \left| \frac{c_n}{c_{n+1}} \right| \tag{8.62}$$

Consider the following power series as an illustration:

$$\sum_{n=0}^{\infty} n(x-1)^n = (x-1) + 2(x-1)^2 + 3(x-1)^3 + \cdots$$

Here we have $c_n = n$ so that

$$R = \lim_{n \to \infty} \left| \frac{c_n}{c_{n+1}} \right| = \lim_{n \to \infty} \frac{n}{n+1} = 1$$

Therefore, the radius of convergence is $R = 1$, and the interval of absolute convergence is $(0, 2)$.

For another illustration, consider the series

$$\sum_{n=0}^{\infty} \frac{(x-c)^n}{n!} = 1 + (x-c) + \frac{(x-c)^2}{2} + \frac{(x-c)^3}{6} + \cdots \tag{8.63}$$

where the center c is any fixed real number. Here $c_n = 1/n!$ so that

$$R = \lim_{n \to \infty} \left| \frac{c_n}{c_{n+1}} \right| = \lim_{n \to \infty} \frac{1/n!}{1/(n+1)!} = \lim_{n \to \infty} \frac{(n+1)n!}{n!} = \lim_{n \to \infty} (n+1) = \infty$$

Since $R = \infty$, the interval of (absolute) convergence of (8.63) is $(-\infty, \infty)$, regardless of the value of c. We find below that with $c = 0$, (8.63) gives the power series expansion of the function e^x.

The series in (8.63) is a "nice" extreme case where convergence is global. At the other extreme, consider

$$\sum_{n=0}^{\infty} n!(x-c)^n = 1 + (x-c) + 2(x-c)^2 + 6(x-c)^3 + \cdots$$

In this case, $c_n = n!$, and we find that

$$R = \lim_{n \to \infty} \left| \frac{c_n}{c_{n+1}} \right| = \lim_{n \to \infty} \frac{n!}{(n+1)!} = \lim_{n \to \infty} \frac{1}{n+1} = 0$$

This means that the series converges only when $x = c$.

Convergence or divergence at the endpoints

The interval of convergence of the power series in (8.54) is $(c - R, c + R)$. If R is a positive real number, then we showed that the series converges absolutely over this interval and diverges outside it. But *what happens at the endpoints $c \pm R$?*

At the endpoints, both the root test and the ratio test fail, so the series may converge or diverge; if it converges, then the series may converge conditionally or absolutely.

To illustrate, consider the series

$$\sum_{n=0}^{\infty} \frac{x^n}{(n+1)^2} = \frac{1}{1^2} + \frac{x}{2^2} + \frac{x^2}{3^2} + \frac{x^3}{4^2} + \cdots \tag{8.64}$$

Here we have $c_n = 1/(n+1)^2$ and $c_{n+1} = 1/(n+2)^2$ so

$$R = \lim_{n \to \infty} \left| \frac{c_n}{c_{n+1}} \right| = \lim_{n \to \infty} \frac{(n+2)^2}{(n+1)^2} = \lim_{n \to \infty} \left(\frac{n+2}{n+1} \right)^2 = \left(\lim_{n \to \infty} \frac{n+2}{n+1} \right)^2 = 1^2 = 1$$

So the interval of absolute convergence is $(-1, 1)$. The endpoints in this case are ± 1, and we check each separately:

$$x = 1: \qquad \sum_{n=0}^{\infty} \frac{1^n}{(n+1)^2} = \frac{1}{1^2} + \frac{1}{2^2} + \frac{1}{3^2} + \frac{1}{4^2} + \cdots$$

This series is a p-series with $p = 2$, which converges. Therefore, the power series (8.64) converges (absolutely) when $x = 1$. Next, consider $x = -1$:

$$x = -1: \qquad \sum_{n=0}^{\infty} \frac{(-1)^n}{(n+1)^2} = \frac{1}{1^2} - \frac{1}{2^2} + \frac{1}{3^2} - \frac{1}{4^2} + \cdots$$

This is an alternating series that converges absolutely, as we saw with $x = 1$. So the power series (8.64) converges when $x = -1$ too, and the interval of convergence is $[-1, 1]$, with convergence being absolute at all points, including both of the endpoints.

For comparison, consider the similar power series

$$\sum_{n=0}^{\infty} \frac{x^n}{n+1} = 1 + \frac{x}{2} + \frac{x^2}{3} + \frac{x^3}{4} + \cdots$$

Here we have $c_n = 1/(n+1)$ and $c_{n+1} = 1/(n+2)$, so

$$R = \lim_{n \to \infty} \left| \frac{c_n}{c_{n+1}} \right| = \lim_{n \to \infty} \frac{(n+2)}{(n+1)} = 1$$

So the interval of absolute convergence is $(-1, 1)$. The endpoints in this case are again ±1, and we check each separately:

$$x = 1: \qquad \sum_{n=0}^{\infty} \frac{1^n}{n+1} = \frac{1}{1} + \frac{1}{2} + \frac{1}{3} + \frac{1}{4} + \cdots$$

This series is the harmonic series that diverges. We conclude that the power series diverges when $x = 1$. Next, consider $x = -1$:

$$x = -1: \qquad \sum_{n=0}^{\infty} \frac{(-1)^n}{n+1} = \frac{1}{1} - \frac{1}{2} + \frac{1}{3} - \frac{1}{4} + \cdots$$

This is an alternating series that converges conditionally. So the power series converges when $x = -1$. We conclude that the interval of convergence is $[-1, 1)$, with convergence being conditional at $x = -1$ and absolute on $(-1, 1)$.

8.4.2 Continuity and integrability

We have seen how a power series converges pointwise (absolutely or conditionally) in its interval of convergence and possibly at the endpoints of that interval. Further, it converges uniformly on closed subintervals of the open interval $(c - R, c + R)$ of absolute convergence. This latter fact and the theory discussed earlier imply the following two important results. I leave their formal proofs as an exercise.

Corollary 325 *(Continuity) Assume that a power series $\sum_{n=0}^{\infty} c_n(x - c)^n$ has radius of convergence $R > 0$. Then it converges uniformly to a function $f(x)$ on each closed interval $[c - r, c + r]$ for every real number $r \in [0, R)$. Further, $f(x)$ is continuous on the entire interval $(c - R, c + R)$.*

For example, we showed earlier that the series

$$\sum_{n=0}^{\infty} \frac{x^n}{n!} \tag{8.65}$$

which is the series in (8.63) with $c = 0$, converges absolutely on $(-\infty, \infty)$; so by Corollary 325, its limit or sum (namely, e^x as we show below) is continuous on $(-\infty, \infty)$.

Corollary 326 *(Term by term integration of power series) Assume that a power series $\sum_{n=0}^{\infty} c_n(x - c)^n$ has radius of convergence $R > 0$ and converges absolutely to* *a function $f(x)$ on $(c - R, c + R)$. If $a, b \in (c - R, c + R)$, and $a < b$, then $f(x)$ is integrable on $[a, b]$ and*

$$\int_a^b \left(\sum_{n=0}^{\infty} c_n(x-c)^n \right) dx = \int_a^b f(x)dx = \sum_{n=0}^{\infty} \int_a^b c_n(x-c)^n dx$$

The power series expansion of the natural logarithm

The geometric power series $\sum_{n=0}^{\infty}(x-c)^n$ has an interval of convergence $(c-1, c+1)$ that contains the center c. Now, if u is any number in the interval of convergence, then by Corollary 326

$$\int_c^u \frac{1}{1-(x-c)}dx = \sum_{n=0}^{\infty} \int_c^u (x-c)^n dx = \sum_{n=0}^{\infty} \frac{(u-c)^{n+1}}{n+1}$$

$$\ln 1 - \ln(c+1-u) = \sum_{n=0}^{\infty} \frac{1}{n+1}[(u-c)^{n+1} - 0^{n+1}]$$

$$\ln(c+1-u) = -\sum_{n=0}^{\infty} \frac{(u-c)^{n+1}}{n+1} \qquad c-1 < u < c+1$$

In particular, if $c = 0$, then

$$\ln(1-u) = -\sum_{n=0}^{\infty} \frac{u^{n+1}}{n+1} = -u - \frac{u^2}{2} - \frac{u^3}{3} - \cdots \qquad -1 < u < 1 \qquad (8.66)$$

To obtain a power series for $\ln x$ itself, set $x = 1 - u$ in (8.66) to get

$$\ln x = \sum_{n=0}^{\infty} \frac{-(1-x)^{n+1}}{n+1} = \sum_{n=0}^{\infty} \frac{(-1)^n(x-1)^{n+1}}{n+1} \qquad 0 < x < 2$$

We now check the endpoints 0 and 2. If $x = 0$, then the series

$$\sum_{n=0}^{\infty} \frac{-1}{n+1} = -1 - \frac{1}{2} - \frac{1}{3} - \cdots$$

is just the negative of the harmonic series; and therefore, the power series diverges at $x = 0$. At $x = 2$, we have

$$\sum_{n=0}^{\infty} \frac{-(-1)^{n+1}}{n+1} = \sum_{n=0}^{\infty} \frac{(-1)^n}{n+1} = 1 - \frac{1}{2} + \frac{1}{3} - \frac{1}{4} + \cdots$$

which converges (conditionally) by the alternating series test from Chapter 5.

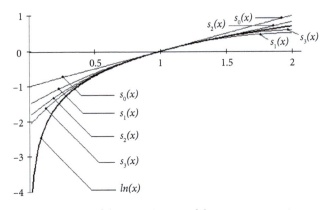

Fig. 8.18 Some of the partial sums of the series converging to *lnx*

Putting this information together, we obtain the power series expansion for the natural logarithm as

$$\ln x = \sum_{n=0}^{\infty} \frac{(-1)^n (x-1)^{n+1}}{n+1} = \frac{x-1}{1} - \frac{(x-1)^2}{2} + \frac{(x-1)^3}{3} - \cdots \qquad 0 < x \le 2$$

$$(8.67)$$

This series is the *standard expansion of* $\ln x$ *as a power series centered at* $c = 1$, the midpoint of the interval of convergence $(0, 2]$.

Figure 8.18 shows the graph of $\ln x$ (thick curve) and its approximation using the first four partial sums $s_0(x)$, $s_1(x)$, $s_2(x)$, $s_3(x)$ of the series in (8.67).

Power series like (8.66) that are centered at zero are sometimes used for translated or shifted versions of $\ln x$ when it simplifies calculations.

It is important to remember that *the power series in (8.67) does not converge to anything outside the interval* $(0, 2]$ *and therefore, it is not equal to* $\ln x$ *even though* $\ln x$ *is well-defined for* $x > 2$. However, using the established properties of logarithms in Chapter 7, we can use the series in (8.67) to obtain infinite series for $\ln x$ for $x > 2$, and thus to approximate the values of logarithm throughout its domain using infinite series; see the exercises.

Inverse tangent power series and the Madhava–Gregory–Leibniz series for π

We have met the inverse tangent function $\tan^{-1} x$ before in different contexts. Now we discuss its power series and the historical significance that is attached to the infinite series that is obtained when $x = 1$. In the process, we discover another interesting application of term by term integration of power series.

Let's start with the series

$$1 - u^2 + u^4 - u^6 + \cdots = \frac{1}{1 + u^2}$$

that we obtained earlier with the help of the geometric power series. This power series converges in the interval $(-1, 1)$. If x is any number in this interval, then integrating the above series gives us

$$\int_0^x (1 - u^2 + u^4 - u^6 + \cdots)du = \int_0^x \frac{1}{1 + u^2} du$$

$$\left[u - \frac{u^3}{3} + \frac{u^5}{5} - \frac{u^7}{7} + \cdots \right]_{u=0}^{u=x} = \int_0^x \frac{1}{1 + u^2} du$$

$$x - \frac{x^3}{3} + \frac{x^5}{5} - \frac{x^7}{7} + \cdots = \int_0^x \frac{1}{1 + u^2} du \qquad (8.68)$$

But the integral on the right-hand side is $\tan^{-1} x$, so we have the expansion

$$x - \frac{x^3}{3} + \frac{x^5}{5} - \frac{x^7}{7} + \cdots = \tan^{-1} x \qquad -1 \le x \le 1 \qquad (8.69)$$

We now check the endpoints $x = \pm 1$. At $x = 1$, the left-hand side of (8.69) gives

$$1 - \frac{1}{3} + \frac{1}{5} - \frac{1}{7} + \cdots$$

which converges by the alternating series test.[13] On the other side of (8.69), $\tan^{-1} 1 = \pi/4$ since $\tan(\pi/4) = 1$; therefore,

$$1 - \frac{1}{3} + \frac{1}{5} - \frac{1}{7} + \cdots = \frac{\pi}{4} \qquad (8.70)$$

The above discussion is similar to the case of $\ln 2$ earlier in more ways than one, including the fact that they are really simple alternating series that generally converge to irrational numbers.

The series in (8.70) attracted special attention over the course of history as a simple way of representing π; it appeared in one of Leibniz's publications around 1676 and became known as "Leibniz's series." But it had already appeared in a 1668 publication of the Scottish mathematician James Gregory (1638–1675), so the name Gregory–Leibniz series has also been used.

---

[13] The same is true for $x = -1$ where the series converges to $-\pi/4$.

Later, the series was discovered in works of the old Kerala school in India, founded by Madhava of Sangamagrama (1340–1425). It seems then that the series in (8.70) was known over 200 years before Gregory's publication, so it may be called the *Madhava–Gregory–Leibniz series*.

The series in (8.70) converges too slowly to be of practical use for calculating π. We can certainly improve on the convergence rate by using another angle; for instance, if we use $\theta = \pi/6$, then $\tan\theta = 1/\sqrt{3}$ is in the interior of the interval of convergence rather than at the end, and the series that we obtain is

$$\frac{1}{\sqrt{3}} - \frac{(1/\sqrt{3})^3}{3} + \frac{(1/\sqrt{3})^5}{5} - \frac{(1/\sqrt{3})^7}{7} + \cdots = \tan^{-1}\frac{1}{\sqrt{3}}$$

$$\frac{1}{\sqrt{3}} - \frac{1}{\sqrt{3}}\frac{1}{3(3)} + \frac{1}{\sqrt{3}}\frac{1}{5(3)^2} - \frac{1}{\sqrt{3}}\frac{1}{7(3)^3} + \cdots = \frac{\pi}{6}$$

If we pull out $1/\sqrt{3}$ and multiply by 6, we get the following series for π:

$$\pi = \frac{6}{\sqrt{3}}\left(1 - \frac{1}{3(3)} + \frac{1}{5(3)^2} - \frac{1}{7(3)^3} + \cdots\right)$$

Madhava actually used 21 terms of the above series to obtain a value of π that was accurate to 11 decimal places:

$$\pi = 3.1415926535922\ldots$$

Other series that involve π are also known that not only converge fast, but they *do not involve irrational numbers* like $\sqrt{3}$; we discuss one such series later in this chapter that involves $\sin^{-1} x$.

8.4.3 Taking the derivative of a power series

We have seen at times that certain power series have known functions associated with them, like $1/(1-x)$, $\tan^{-1} x$, $\ln x$, or e^x. However, we have yet to show that the power series in (8.65) in fact converges to e^x. The general method for doing this that applies to all other known, infinitely differentiable functions like $\sin x$ and $\cos x$ is the method of Taylor series expansion that we discuss below.

The key idea behind Taylor expansion is being able to take the derivative of a power series term by term. If we take the derivative of a power series term by term, then we obtain

$$\frac{d}{dx}\sum_{n=0}^{\infty} c_n(x-c)^n = \frac{d}{dx}\left(c_0 + c_1(x-1)^1 + c_2(x-1)^2 + c_3(x-1)^3 + \cdots\right)$$

$$= 0 + c_1 + 2c_2(x-1)^1 + 3c_3(x-1)^2 + 4c_4(x-1)^3 + \cdots$$

$$= \sum_{n=1}^{\infty} nc_n(x-c)^{n-1}$$

It remains to prove that if $f(x) = \sum_{n=0}^{\infty} c_n(x-c)^n$ then the last series above indeed equals the derivative $f'(x)$ and also determines the range of values of x for which this happens.

Theorem 327 (*Term by term differentiation*) *Assume that the power series* $\sum_{n=0}^{\infty} c_n(x-c)^n$ *has radius of convergence R and that it converges to a function* $f(x)$, *i.e.,*

$$f(x) = \sum_{n=0}^{\infty} c_n(x-c)^n, \quad x \in (c-R, c+R)$$

Then the power series $\sum_{n=1}^{\infty} nc_n(x-c)^{n-1}$ *also has radius of convergence R and equals* $f'(x)$, *i.e.,*

$$f'(x) = \sum_{n=1}^{\infty} nc_n(x-c)^{n-1}, \quad x \in (c-R, c+R) \tag{8.71}$$

Proof First, we write the series in (8.71) in the following equivalent form that starts it from $n = 0$:

$$\sum_{n=0}^{\infty} (n+1)c_{n+1}(x-c)^n \tag{8.72}$$

Now we use the coefficients $c'_n = (n+1)c_n$ in Theorem 324 and obtain

$$\limsup_{n\to\infty} \sqrt[n]{|c'_n|} = \lim_{n\to\infty} \sqrt[n]{n+1}\limsup_{n\to\infty} \sqrt[n]{|c_n|}$$

We find the limit on the right-hand side using the l'Hospital rule; note that

$$\lim_{n\to\infty} \ln\sqrt[n]{n+1} = \lim_{n\to\infty} \frac{\ln(n+1)}{n} = \lim_{n\to\infty} \frac{1}{n+1} = 0$$

which yields

$$\lim_{n\to\infty} \sqrt[n]{n+1} = e^0 = 1$$

Therefore,

$$\limsup_{n\to\infty} \sqrt[n]{|c'_n|} = \limsup_{n\to\infty} \sqrt[n]{|c_n|} = \frac{1}{R}$$

which implies that the series converges absolutely on $(c - R, c + R)$ and uniformly on $[c - r, c + r]$ for all $r \in (0, R)$. Now, since the series in (8.72), and thus, the one in (8.71), converges at $x = c$, we conclude that by Corollary 322, $f(x)$ is differentiable on $[c - r, c + r]$; and its derivative is given as in (8.71). Finally, $f'(x)$ is defined at all $x \in (c - R, c + R)$ since for all such x, there is $r \in (0, R)$ such that $r \geq |x|$, i.e., $x \in [c - r, c + r]$.   ∎

For illustration, consider the following function defined by the infinite series in (8.65):

$$E(x) = \sum_{n=0}^{\infty} \frac{x^n}{n!} = 1 + \frac{x}{1!} + \frac{x^2}{2!} + \frac{x^3}{3!} + \cdots$$

We saw earlier that this formula defines a continuous $E(x)$ on $(-\infty, \infty)$ and by

$$E'(x) = \sum_{n=1}^{\infty} (n)\frac{x^{n-1}}{n!} = \sum_{n=1}^{\infty} \frac{x^{n-1}}{(n-1)!} = \sum_{n=0}^{\infty} \frac{x^n}{n!} = E(x)$$

which is what we might expect if $E(x) = e^x$.

A useful feature of the above theorem is that it can be applied repeatedly to obtain power series for derivatives of a higher order.

By repeatedly applying the above theorem and taking derivatives, we see that if $f(x) = \sum_{n=0}^{\infty} c_n(x - c)^n$ where the power series has a positive radius of convergence R, then all higher-order derivatives of $f(x)$ exist on the same interval of convergence $(c - R, c + R)$, and they are found by taking the derivative of the original power series term by term. For the second derivative

$$f''(x) = \frac{d}{dx} \sum_{n=1}^{\infty} nc_n(x - c)^{n-1}$$

$$= \frac{d}{dx} \left[c_1 + 2c_2(x - 1)^1 + 3c_3(x - 1)^2 + 4c_4(x - 1)^3 + \cdots \right]$$

$$= 0 + (1)(2)c_2 + (2)(3)(x - 1)^1 + (3)(4)(x - 1)^2 + \cdots$$

$$= \sum_{n=2}^{\infty} n(n - 1)c_n(x - c)^{n-2}$$

Continuing this procedure k times leads to the following general formula:

$$f^{(k)}(x) = \sum_{n=k}^{\infty} n(n-1)(n-2)\cdots(n-k+1)c_n(x-c)^{n-k} \qquad (8.73)$$

As we saw in the proof of Theorem 327, it is possible to write the derivative formulas in the standard power series notation, which starts from $n = 0$. Since the summation index is a dummy variable (it is only a placeholder), we can use any letter; so using n again to keep formulas looking uniform, we get for the second derivative

$$f''(x) = \sum_{n=0}^{\infty} (n+2)(n+1)c_{n+2}(x-c)^n$$

This process when continued k times gives

$$f^{(k)}(x) = \sum_{n=0}^{\infty} c_n'(x-c)^n \quad \text{with } c_n' = (n+k)(n+k-1)\cdots(n+1)c_{n+k} \qquad (8.74)$$

In the form (8.74), we see more easily that all higher derivatives are also ordinary power series.

8.5 The Taylor series expansion of functions

In the discussion of power series above we noted that in some cases, these series converged on some open interval to a known function, like $1/(1-x)$, $\ln x$, or $\tan^{-1} x$. In this section, we turn to the issue of starting from a given function $f(x)$ and finding the power series that may converge to it on some open interval.

8.5.1 Infinitely differentiable functions and power series

Being able to take the derivative of a power series repeatedly and indefinitely makes the function that is represented by a power series have a special property.

Infinitely differentiable functions If a power series converges to a function $f(x)$, then $f(x)$ has derivatives of all orders defined on its interval of convergence. We say that such a function is *infinitely differentiable*. The set of all functions that are infinitely differentiable on an open interval (a, b) is denoted $C^{\infty}(a, b)$. A function that has a power series with radius of convergence $R > 0$ is thus an element of $C^{\infty}(c - R, c + R)$.

For example, the function $1/(1 - x)$ is infinitely differentiable on the interval $(-1, 1)$ where the geometric power series converges to it. Therefore, $1/(1 - x) \in C^\infty(-1, 1)$. Similarly, $\ln x$ is infinitely differentiable on $(0, 2)$, so $\ln x \in C^\infty(0, 2)$.

On the other hand, $f(x) = x^{4/3}$ is defined on $(-\infty, \infty)$ and has a derivative $f'(x) = (4/3)x^{1/3}$ that is also defined on $(-\infty, \infty)$; however, $f''(x) = (4/9)x^{-2/3}$ is not defined on $(-\infty, \infty)$ because $f'(0)$ cannot be defined; so $x^{4/3}$ is differentiable on $(-\infty, \infty)$, but only once. We may say that $x^{4/3} \in C^1(-\infty, \infty)$, but $x^{4/3} \notin C^2(-\infty, \infty)$.

More generally, the power function

$$f(x) = x^{n+1/3}$$

where n is a positive integer is differentiable n times on $(-\infty, \infty)$ but no more at $x = 0$; each time that we take the derivative, we reduce n by one until after n times, only the power $1/3$ is left. Taking the derivative again leads to a negative power $-2/3$ and division by 0 at $x = 0$. It follows that this type of function does not expand in the form of a power series. On the other hand, it is not hard to show that $x^{n+1/3} \in C^\infty(0, \infty)$.

We mention for reference that the notation C^n typically refers to n times *continuously* differentiable functions, i.e., functions f where $f^{(n)}$ is continuous. Here the derivative of $x^{4/3}$ is $(4/3)x^{1/3}$, which is continuous; and its second derivative $(4/9)x^{-2/3}$ does not exist at 0, and thus is not continuous there. There are functions whose derivatives exist but are not continuous. For instance,

$$f(x) = x^2 \sin \frac{1}{x} \quad \text{if } x \neq 0, \quad f(0) = 0 \tag{8.75}$$

has a derivative given by the formula

$$f'(x) = 2x \sin \frac{1}{x} - \cos \frac{1}{x}$$

if $x \neq 0$ and $f'(0) = 0$ (see Exercise 354). Since $f'(x)$ above has no limit at $x = 0$, it follows that f' is not continuous at 0. The function in (8.75) is therefore differentiable on $(-\infty, \infty)$, but $f \notin C^1(-\infty, \infty)$.

A basic but important type of function that is infinitely differentiable is a polynomial function that we met earlier when discussing algebraic numbers. By taking the derivatives repeatedly, we see that a polynomial of degree n such as

$$P(x) = x^n + a_1 x^{n-1} + a_2 x^{n-2} + \cdots + a_{n-1}x + a_n$$

is infinitely differentiable on $(-\infty, \infty)$, i.e., $P \in C^\infty(-\infty, \infty)$ with $P^{(k)}(x) = 0$ for all $k > n$.

Infinitely differentiable functions may seem (and are) very special; yet, remark-ably, we will soon discover that most of the familiar functions including e^x, $\sin x$, $\cos x$, etc., are all of this type.

A simple but important consequence of the formula (8.74) is the following observation: if we set $x = c$ in that formula, then we obtain

$$f^{(k)}(c) = (0 + k)(0 + k - 1) \cdots (0 + 1)c_{0+k} = k!c_k$$

because all higher-degree terms contain the difference $x - c$, and they drop out. Solving for c_k gives the formula in the following result.

Corollary 328 (*Coefficients formula*) *Suppose that $f(x)$ has a power series of radius $R > 0$, so that*

$$f(x) = \sum_{n=0}^{\infty} c_n(x - c)^n, \quad x \in (c - R, c + R)$$

Then the coefficients c_n are given by the formula

$$c_n = \frac{f^{(n)}(c)}{n!}, \quad n = 1, 2, 3, \ldots \tag{8.76}$$

Also note that $c_0 = f(c)$.

The significance of the formula in (8.76) is that using it we can go backward, *starting from a function and building its power series by calculating the coefficients.* One more piece of information ensures that the power series that we get using this process is *the* power series for the function.

Corollary 329 (*Uniqueness of the power series for a C^∞ function*) *Two power series, $\sum_{n=0}^{\infty} c_n(x - c)^n$ and $\sum_{n=0}^{\infty} d_n(x - c)^n$ are both power series for the same function $f(x)$ on an interval $(c - R, c + R)$ if and only if $c_n = d_n$ for all $n \geq 0$.*

Proof First, if $c_n = d_n$, then the two power series are equal and thus represent the same function. Conversely, if the two series are both power series for $f(x)$, then by (8.76)

$$c_n = \frac{f^{(n)}(c)}{n!} \quad \text{and} \quad d_n = \frac{f^{(n)}(c)}{n!}$$

so $c_n = d_n$ for $n \geq 1$. Further, $c_0 = f(c) = d_0$ also, and the proof is complete. ∎

To illustrate how this new information is used, we find the power series (cen-tered at $c = 0$) for the natural exponential function $f(x) = e^x$.

We know that

$$f'(x) = e^x, \quad f''(x) = e^x, \quad f'''(x) = e^x, \quad \text{and so on}$$

In short, $f^{(n)}(x) = e^x$. Therefore, $f^{(n)}(0) = e^0 = 1$ for all $n \geq 0$, and (8.76) gives $c_n = 1/n!$ These are the coefficients of the power series:

$$e^x = \sum_{n=0}^{\infty} \frac{x^n}{n!} = 1 + \frac{x}{1!} + \frac{x^2}{2!} + \frac{x^3}{3!} + \cdots \tag{8.77}$$

whose interval of convergence we earlier found to be $(-\infty, \infty)$.

Since by Corollary 329, the power series for e^x is unique on $(-\infty, \infty)$, the formula in (8.77) gives the power series centered at 0 for e^x.

As a by-product, setting $x = 1$ in (8.77) gives the following series for the irrational number e:

$$e = 1 + \frac{1}{1!} + \frac{1}{2!} + \frac{1}{3!} + \cdots \tag{8.78}$$

This series converges to the irrational value of e rather quickly; the first 7 terms give $e \simeq 2.71806$, which is correct to three decimal places.

Cauchy's function and the existence of power series

Let f be a C^∞ function f and consider the problem of finding an infinite series that converges to it in some open interval. We did this successfully for the exponential function e^x above. The crucial aspect of e^x in that discussion is that the higher derivatives of e^x are easy to find, and as a result, we could prove that the corresponding series converged.

If we want to generalize the above process for e^x to an arbitrary C^∞ function f, then we need to find formulas or expressions for the higher derivatives $f^{(n)}$ at the point c. This is difficult for all but the simplest functions; and if we cannot do it generally, then we need to answer an even more basic question:

How do we even know that there is a power series expansion for f that converges to it on some open interval (not just at the center c)?

This is a subtle point: the knowledge that a function f has higher derivatives of all possible orders at some point $x = c$ (i.e., $f^{(n)}(c)$ is defined for all n) is not sufficient to guarantee the existence of a power series expansion. Without a formula for $f^{(n)}(c)$, we cannot tell whether a series of the type in Corollary (328) converges to f on some open interval.

This issue concerned Cauchy too, and he managed to find an example of a C^∞ function f whose power series doesn't converge to f on any open interval. This meant that the answer to the reverse question that we asked above requires some additional theory that guarantees the existence of a power series expansion on an open interval.

Before discussing that theory, we take a look at Cauchy's function. Consider

$$f(x) = e^{-1/x^2} \text{ if } x \neq 0, \quad \text{and } f(0) = 0$$

First we note that if x_n is a sequence that converges to 0, and $x_n \neq 0$, then $1/x_n^2$ diverges to ∞, so

$$\lim_{n \to \infty} e^{-1/x_n^2} = \lim_{n \to \infty} \frac{1}{e^{1/x_n^2}} = 0$$

It follows that f is continuous at $x = 0$. Further, if $x \neq 0$, then using derivative rules we find that

$$f'(x) = \frac{2}{x^3} e^{-1/x^2}$$

and if $x = 0$, then

$$f'(0) = \lim_{x \to 0} \frac{f(x) - f(0)}{x - 0} = \lim_{x \to 0} \frac{e^{-1/x^2}}{x}$$

We show that the last limit from either right or left converges to 0 using l'Hospital's rule and the substitution $r = 1/x$ as follows:

$$\lim_{x \to 0^+} \frac{e^{-1/x^2}}{x} = \lim_{r \to \infty} \frac{r}{e^{r^2}} = \lim_{r \to \infty} \frac{1}{2re^{r^2}} = 0$$

A similar calculation leads to a limit of 0 if $x \to 0^-$. Hence the limit exists, and it is 0, i.e., $f'(0) = 0$. Using this type of straightforward (though somewhat tedious) calculation gives

$$f''(x) = \left(-\frac{6}{x^4} + \frac{4}{x^6} \right) e^{-1/x^2} \text{ if } x \neq 0, \quad \text{and } f''(0) = 0$$

and so on. Using mathematical induction, we can show that (left as an exercise)

$$f^{(n)}(x) = P\left(\frac{1}{x}\right) e^{-1/x^2} \text{ if } x \neq 0, \quad \text{and } f^{(n)}(0) = 0$$

where P is a polynomial of degree $3n$.

In particular, Cauchy's function is infinitely differentiable on $(-\infty, \infty)$, but $f^{(n)}(0) = 0$ for all n. It follows that the standard power series for f converges to the zero function on $(-\infty, \infty)$; however, it converges to f only at $x = 0$ and not any open interval that contains 0.

8.5.2 Taylor polynomials and approximation

We now return to the question of whether a given function $f(x)$ can be expanded in the form of a power series that converges to it on some open interval. To determine the proper conditions for the existence of such a power series, we begin with the partial sums.

For series of the type discussed in Corollary (328), the partial sums are given a name.

Taylor polynomials and the Taylor series. Assume that a function $f(x)$ is n-times differentiable on an open interval (a, b), and let $c \in (a, b)$. The polynomial

$$T_n(x) = \sum_{k=0}^{n} \frac{f^{(k)}(c)}{k!}(x - c)^k$$

is the *Taylor polynomial of order n* centered at c.

If $f \in C^{\infty}(a, b)$, and the power series

$$\sum_{k=0}^{\infty} \frac{f^{(k)}(c)}{k!}(x - c)^k$$

has radius of convergence $R > 0$, then it is the *Taylor series* of f centered at c.

The polynomials and the power series above are named after the English mathematician Brook Taylor (1685–1731). In the special case $c = 0$, the series is sometimes called *Maclaurin series*.[14]

With regard to the definitions of Taylor polynomials and series, we must remember two points: first, a Taylor polynomial of order n need not have *degree n; the degree is actually less than or equal to the order*, and the equality holds if $f^{(n)}(c) \neq 0$.

Second, the Taylor series of $f(x)$ as defined above need not converge to $f(x)$ on (a, b). We give the appropriate conditions that ensure the important convergence property in what follows.

Since partial sums of a series approximate the value of a (convergent) series, it is clear that Taylor polynomials can serve in the same capacity. But we may use Taylor polynomials for approximating a function $f(x)$ only if we settle the

---

[14] After the Scottish mathematician Colin Maclaurin (1698–1746).

issue of whether a Taylor series converges to the function f on an open interval. Without this assurance, we have no reason to think that Taylor polynomials actually approximate $f(x)$ at any point (except trivially at the center).

As we might guess, the Maclaurin series of a polynomial $P(x)$ is $P(x)$ itself; you can prove this fact in Exercise 357. So we may think of a polynomial as its own Taylor series about 0. If the center is not 0, then the Taylor series of a polynomial is a different polynomial (but still a finite sum).

We earlier found the Maclaurin series for e^x by explicitly calculating all the higher derivatives of e^x. We now do a similar calculation for the Maclaurin series for $f(x) = \sin x$. We find the function value and its higher derivatives at 0:

$$f(0) = \sin 0, \quad f'(x) = \cos x, \text{ so } f'(0) = \cos 0 = 1$$
$$f''(x) = -\sin x, \text{ so } f''(0) = -\sin 0 = 0$$
$$f'''(x) = -\cos x, \text{ so } f'''(0) = -\cos 0 = -1$$
$$f^{(4)}(x) = \sin x, \text{ so } f^{(4)}(0) = \sin 0 = 0$$

Since $f^{(4)}(x) = f(x)$, the set of values: $0, 1, 0, -1$ repeats every 4 steps. So all of the coefficients of the power series can be calculated:

$$c_0 = \frac{f^{(0)}(0)}{0!} = \frac{f(0)}{1} = 0, \qquad c_1 = \frac{f'(0)}{1!} = \frac{1}{1} = 1$$
$$c_2 = \frac{f''(0)}{2!} = \frac{0}{2} = 0, \qquad c_3 = \frac{f'''(0)}{3!} = -\frac{1}{3!}$$

The next four coefficients cycle the same four numerators, but the denominators simply contain larger factorials:

$$c_4 = \frac{f^{(0)}(0)}{4!} = \frac{f(0)}{4!} = 0, \qquad c_5 = \frac{f'(0)}{5!} = \frac{1}{5!}$$
$$c_6 = \frac{f''(0)}{6!} = \frac{0}{6!} = 0, \qquad c_7 = \frac{f'''(0)}{7!} = -\frac{1}{7!}$$

The pattern is evident: all even-indexed coefficients are 0, and they do not appear in the power series, while the odd coefficients alternate in sign. Therefore, the power series is[15]

$$\sin x = x - \frac{x^3}{3!} + \frac{x^5}{5!} - \frac{x^7}{7!} + \cdots = \sum_{n=0}^{\infty} \frac{(-1)^n x^{2n+1}}{(2n+1)!} \tag{8.79}$$

With the complete series at hand, the radius of convergence may be found using the ratio test. To simplify calculations, we factor out x and substitute $u = x^2$ to get

---

[15] The variable x here and elsewhere for the trigonometric functions is measured in radians.

$$x \sum_{n=0}^{\infty} \frac{(-1)^n x^{2n}}{(2n+1)!} = x \sum_{n=0}^{\infty} \frac{(-1)^n u^n}{(2n+1)!}$$

The coefficients of the last series are $c_n = (-1)^n/(2n+1)!$ so

$$R = \lim_{n\to\infty} \left| \frac{c_n}{c_{n+1}} \right| = \lim_{n\to\infty} \left| \frac{(-1)^n (2(n+1)+1)!}{(-1)^{n+1}(2n+1)!} \right| = \lim_{n\to\infty} \frac{(2n+3)!}{(2n+1)!} = \infty$$

To see how the limit diverges to infinity, notice that $(2n+3)! = (2n+3)(2n+2)(2n+1)!$ so that

$$\frac{(2n+3)!}{(2n+1)!} = \frac{(2n+3)(2n+2)(2n+1)!}{(2n+1)!} = (2n+3)(2n+2)$$

Therefore, the radius of convergence of the power series with u is $R = \infty$. Of course, $u = x^2$ is never negative, but it can be infinitely large, so x can be any real number, that is, with values ranging in $(-\infty, \infty)$; conveniently, this also happens to be the domain of $\sin x$, so we conclude that the Taylor series of $\sin x$ centered at 0 converges to $\sin x$ on its domain $(-\infty, \infty)$.

Figure 8.19 illustrates the graph of $\sin x$ (the thick curve) and the first four (distinct) Taylor polynomials (the indicated partial sums).

We clearly see in Figure 8.19 that the approximations improve over larger stretches of the x-axis with the increasing degree of the polynomial. The partial sum $s_9(x)$, namely, the Taylor polynomial of order 9, is visually indistinguishable in Figure 8.19 from $\sin x$ on the interval $[-\pi, \pi]$ a full period of $\sin x$.

With a power series at hand, using the theory of power series that we discussed earlier, we can use the series to estimate integrals.

To illustrate, consider $f(x) = \sin(x^2)$. There is no known antiderivative for this function, but we may easily obtain the Taylor series for it by simply replacing x by x^2 in (8.79):

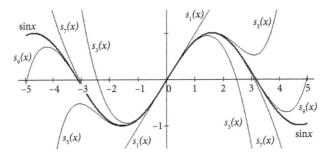

Fig. 8.19 Several Taylor polynomials for $\sin x$

$$\sin(x^2) = x^2 - \frac{(x^2)^3}{3!} + \frac{(x^2)^5}{5!} - \frac{(x^2)^7}{7!} + \cdots$$

$$= x^2 - \frac{x^6}{3!} + \frac{x^{10}}{5!} - \frac{x^{14}}{7!} + \cdots \qquad (8.80)$$

Suppose we retain the first three nonzero terms to get the Taylor polynomial (of degree 10) as an approximation for $\sin(x^2)$:

$$\sin(x^2) \simeq x^2 - \frac{x^6}{3!} + \frac{x^{10}}{5!}$$

Since this is the Taylor polynomial about 0, we get better estimates for integrals taken over intervals that surround 0. For instance, integrating from 0 to 1 (remember, in radians!), we have

$$\int_0^1 \sin(x^2)dx \simeq \int_0^1 \left(x^2 - \frac{x^6}{3!} + \frac{x^{10}}{5!} \right) dx$$

$$= \left[\frac{x^3}{3} - \frac{x^7}{3!(7)} + \frac{x^{11}}{5!(11)} \right]_{x=0}^{x=1}$$

$$= \frac{1}{3} - \frac{1}{6}\left(\frac{1}{7}\right) + \frac{1}{120}\left(\frac{1}{11}\right)$$

$$\simeq 0.310281$$

where the last number was rounded to six decimal places. But how good an approximation is this number?

The series (8.80) is an alternating series for which we found an error bound earlier in Chapter 5. For the particular problem here, the approximation error in using the first three terms of the series is less than the magnitude of the fourth term, that is,

$$\left| -\int_0^1 \frac{x^{14}}{7!}dx \right| = \left| \frac{x^{15}}{7!(15)} \right|_{x=0}^{x=1} = \left| \frac{1}{7!(15)} \right| < 0.000093$$

So although we don't know the exact value of the integral $\int_0^1 \sin(x^2)dx$, we can confidently declare that our answer 0.310281 is off from the exact value by less than 0.000093. If this is not small enough for some purpose, then we simply add more terms of the series until the error is within acceptable bounds.

Now that we know the series in (8.79) converges to the function $\sin x$ for every real number x, we can use its partial sums, namely, the Taylor polynomials, to approximate the value of $\sin x$. For example, to get an approximation of $\sin 1$ to a certain level of accuracy, we may use as many terms as necessary to accomplish the task. For instance, we can be sure that the Taylor polynomial of order 7 with $x = 1$ (radian)

$$\sin 1 \approx 1 - \frac{1}{3!} + \frac{1}{5!} - \frac{1}{7!}$$

is a valid approximation of the number $\sin 1$.

As with the integral problem above, we can tell how many terms are necessary to *make certain* that we get an approximation that is accurate to four decimal places using the properties of alternating series. But this approach doesn't work for non-alternating series like the one in (8.77) for e^x; for instance, how many terms of that series do we need to approximate $e = e^1$ to 16 decimal places?

8.5.3 The remainder, approximation error, and convergence of Taylor polynomials

The answer to the question about the accuracy of approximation is related to the issue of the convergence of the Taylor series for a given function $f(x)$. If the series does converge to the function $f(x)$, then the Taylor polynomial gives a valid approximation to the function $f(x)$. The *error* in such an approximation is called the *remainder*, defined simply as

$$R_n(x) = f(x) - T_n(x) \tag{8.81}$$

Notice that

$$\lim_{n \to \infty} T_n(x) = f(x) \quad \text{if and only if} \quad \lim_{n \to \infty} R_n(x) = 0$$

Therefore, the Taylor series at a point x converges (as the limit of its partial sums) to $f(x)$ if and only if $R_n(x)$ converges to 0. We formalize this observation as follows.

Lemma 330 Let $f \in C^\infty(a, b)$, and $c \in (a, b)$. Then for every $x \in (a, b)$

$$f(x) = \sum_{n=0}^{\infty} \frac{f^{(n)}(c)}{n!}(x - c)^n$$

if and only if $\lim_{n \to \infty} R_n(x) = 0$.

The formula

$$f(x) = f(c) + \frac{f'(c)}{1!}(x - c) + \frac{f''(c)}{2!}(x - c)^2 + \cdots$$

$$+ \frac{f^{(n)}(c)}{n!}(x - c)^n + R_n(x)$$

is known as *Taylor's formula with remainder*.

Lagrange's form of the remainder

The above result is not very useful if we don't know much about the remainder $R_n(x)$. There are two common formulas for the remainder that we now discuss. The first is attributed to Joseph Lagrange (1736–1813) We prove it using Cauchy's generalized mean value theorem from Chapter 6.

Theorem 331 *(Taylor remainder, Lagrange's differential form) Let f be differentiable $n+1$ times on an open interval (a, b), and let $c \in (a, b)$. Then for every $x \in (a, b)$, there is a number r between x and c such that*

$$R_n(x) = \frac{f^{(n+1)}(r)}{(n + 1)!}(x - c)^{n+1} \tag{8.82}$$

Proof   First, observe from the definition of remainder in (8.81) that

$$R_n(c) = f(c) - T_n(c) = 0$$

since $T_n(c) = f(c)$ for all n. Further, $R'_n(x) = f'(x) - T'_n(x)$ with

$$T'_n(x) = \frac{f'(c)}{1!} + \frac{2f(c)}{2!}(x - c) + \cdots + \frac{nf^{(n)}(c)}{n!}(x - c)^{n-1}$$

so that $T'_n(c) = f'(c)$ and therefore, $R'_n(c) = f'(c) - T'_n(c) = 0$. Taking additional derivatives we obtain

$$R_n^{(k)}(c) = f^{(k)}(c) - T_n^{(k)}(c) = 0 \tag{8.83}$$

for $k = 1, 2, \ldots, n$. Next, define

$$g(x) = (x - c)^{n+1} \tag{8.84}$$

Then $g(x)$ is differentiable for all $x \in (a, b)$, and $g'(x) = (n + 1)(x - c)^n$ so $g'(c) = 0$. Similarly, $g^{(k)}(c) = 0$ for $k = 0, 1, \ldots, n$, but

$$g^{(n+1)}(c) = (n + 1)! \tag{8.85}$$

Let $x \in (a, b)$ be fixed, $x \neq c$. By the generalized mean value theorem (Theorem 208), there is a number r_1 between x and c such that

$$\frac{R_n(x)}{g(x)} = \frac{R_n(x) - R_n(c)}{g(x) - g(c)} = \frac{R'_n(r_1)}{g'(r_1)} \tag{8.86}$$

Note that the interval between r_1 and c is contained in the interval between x and c. Because of (8.83), we may apply Theorem 208 again to find a number r_2 between r_1 and c such that

$$\frac{R'_n(r_1)}{g'(r_1)} = \frac{R'_n(x) - R'_n(c)}{g'(x) - g'(c)} = \frac{R_n''(r_2)}{g''(r_2)} \tag{8.87}$$

Now (8.86) and (8.87) imply that for some r_2 between x and c

$$\frac{R_n(x)}{g(x)} = \frac{R_n''(r_2)}{g''(r_2)}$$

Continuing in this way using Theorem 208, we eventually obtain the equality

$$\frac{R_n(x)}{g(x)} = \frac{R_n^{(n+1)}(r_{n+1})}{g^{(n+1)}(r_{n+1})} \tag{8.88}$$

for a number r_{n+1} between x and c. Now using (8.81) and (8.85) in (8.88) and setting $r = r_{n+1}$, it follows that

$$\frac{R_n(x)}{g(x)} = \frac{f^{(n+1)}(r)}{(n+1)!}$$

and given (8.84), this is equivalent to (8.82).   ∎

Using the above theorem we can get an estimate for the approximation error of e^x by its Taylor polynomial of order n. In this case,

$$|R_n(x)| = \frac{e^r |x|^n}{(n+1)!}$$

For $x = 1$ and center $c = 0$, we have $0 < r < 1$, so $e^r < e^1 < 3$ and therefore,

$$|R_n(1)| \leq \frac{3}{(n+1)!}$$

So to get an approximation of $e = e^1$ that is correct to 16 decimal places, we choose the order n large enough that

$$\frac{3}{(n+1)!} < 0.5 \times 10^{-17}$$

i.e.,

$$(n + 1)! > 6 \times 10^{17}$$

The least n that satisfies the above inequality is $n = 19$, so the Taylor polynomial of order 19, or the first 19 terms of series for e^x, give us an estimate

$$e \approx 1 + \frac{1}{1!} + \frac{1}{2!} + \cdots + \frac{1}{19!}$$

that is accurate to 16 (or more) decimal places.

Lemma 330 and Theorem 331 together give the following result that tells us under what conditions we can expect the Taylor polynomials $T_n(x)$ to converge to a given function $f(x)$ on some open interval.

Theorem 332 (*Convergence of Taylor polynomials*) *Let $f \in C^\infty(a, b)$ and $c \in (a, b)$. If r is such that $[c - r, c + r] \subset (a, b)$, and there is $M > 0$ such that for all positive integers n*

$$|f^{(n)}(x)| \le M^n \quad \text{for all } x \in [c - r, c + r] \tag{8.89}$$

then the Taylor polynomials $T_n(x)$ converge to $f(x)$ for every x in $[c - r, c + r]$, i.e.,

$$f(x) = \sum_{n=0}^{\infty} \frac{f^{(n)}(c)}{n!}(x - c)^n \quad \text{for all } x \in [c - r, c + r]$$

Proof If (8.89) holds, then by (8.82)

$$|R_n(x)| \le \frac{M^{n+1}}{(n + 1)!}|x - c|^{n+1} \le \frac{(Mr)^{n+1}}{(n + 1)!}$$

Now by (5.40) in Theorem 151 (Chapter 5), we have

$$0 \le \lim_{n \to \infty} |R_n(x)| \le \lim_{n \to \infty} \frac{(Mr)^{n+1}}{(n + 1)!} = 0$$

for all $x \in [c - r, c + r]$. Now the conclusion follows from Lemma 330. ∎

As a quick application of Theorem 332, we prove the convergence of Taylor polynomials to each of the functions e^x and $\sin x$. Earlier we used the theorem on term by term differentiation of power series to derive the Taylor series for these functions.

If $f(x) = e^x$, then $f^{(n)}(x) = e^x$ for all n; so, given that e^x is an increasing function, (8.89) holds on an interval of type $[-r, r]$ if $M = e^r$. Then $e^r > 1$ since $r > 0$; so $M > 1$ and

$$|f^{(n)}(x)| \leq e^r = M \leq M^n$$

for all n as is required.

For $\sin x$, all derivatives are among the following functions: $\pm \cos x, \pm \sin x$. All of these have absolute value of at most 1, so choosing $M = 1$ satisfies (8.89).

Infinity hidden in a single argument

Let's take a break from formal discussion to point out the remarkable bundling of infinity within the Lagrange's form of the remainder. Lemma 330 and Theorem 331 show that every C^∞ function f can be written as

$$f(x) = \sum_{k=0}^{n} \frac{f^{(k)}(c)}{k!}(x - c)^k + \frac{f^{(n+1)}(r)}{(n + 1)!}(x - c)^{n+1} \tag{8.90}$$

for all x in an open interval as long as the remainder converges to 0 as $n \to \infty$. The remainder term in (8.90) is then equal to the infinite series

$$\sum_{k=n+1}^{\infty} \frac{f^{(k)}(c)}{k!}(x - c)^k \tag{8.91}$$

This series contains the "soul of $f(x)$" in a manner of speaking, since the Taylor polynomial is only an approximation of f that matches the value of $f(x)$ exactly only at $x = c$ (unless f is a polynomial, of course). Nevertheless, this essential aspect of f is contained in the single remainder term, which looks a lot like the very first term of the series in (8.91). What then is special about this particular term?

Let's examine the difference between the remainder term in (8.90) and the first term in (8.91):

$$\frac{(x - c)^{n+1}}{(n + 1)!}[f^{(n+1)}(r(x)) - f^{(n+1)}(c)], \quad r(x) \text{ is between } x \text{ and } c$$

This quantity is 0 (there is no difference) if $x = c$, which implies that $r(x) = c$ as well. But if $x \neq c$, then the above difference is literally the difference between f and its nth Taylor polynomial T_n.

The difference $f^{(n+1)}(r(x)) - f^{(n+1)}(c)$ accounts for the essence of $f(x)$ as a (transcendental) function somehow just by the way the quantity $r(x)$ varies as a function of x.

Note also that different values of the center c or of the order n result in a different remainder, and thus in a different r. For a given function f, the quantity r depends not only on x but on n and c as well: $r = r(x, c, n)$. By fixing c and n, we can focus on the variation of r as a function of x.

In this sense, $r(x, c, n)$ bundles and conceals the infinity that lies deep within and characterizes a transcendental function. It's role is loosely reminiscent of the ε-index $N_\varepsilon(x)$ of a sequence of functions that highlighted the difference between pointwise convergence and uniform convergence. For a sequence x_n, the index is its argument by definition, which could just as well be written $x_{N_\varepsilon} = x(N_\varepsilon)$, and the quantity $r(x, c, n)$ also appears as an argument (in $f^{(n+1)}$).

For a sequence of functions that converge pointwise, $N_\varepsilon(x)$ may become infinitely large depending on the value of x for a fixed ε, thus clearly pointing to the infinity that is hidden in pointwise converging sequences that doesn't exist for the uniformly converging one. In the case of Taylor series, the variation of $r(x, c, n)$ within $f^{(n+1)}$ encodes the way the Taylor polynomials converge to f.

Under conditions of Theorem 332, the magnitude of $f^{(n+1)}$ is suppressed by the fraction $(x - c)^{n+1}/(n + 1)!$ and therefore, the variation of $r(x, c, n)$ has no major effects on the outcome, namely, the convergence of Taylor polynomials to f. But for a function like Cauchy's e^{-1/x^2} the variation of $r(x, c, n)$ is noticeable, as it causes spikes in the value of $f^{(n+1)}$ that are not suppressed enough and occur close to the origin for large n. Figure 8.20 shows a few of the higher derivatives of e^{-1/x^2} to illustrate these high variations in the derivative values near the origin.

The left panel of Figure 8.20 shows e^{-1/x^2} (thick curve) and its higher derivatives $f^{(1)}, f^{(2)}, f^{(3)}$ in a neighborhood of 0. Notice that the third derivative cannot be fit completely in the frame here. The right panel illustrates portions of $f^{(3)}$ and $f^{(4)}$; on this scale, the third derivative is the squat thick curve, while the tall curve is the fourth derivative.

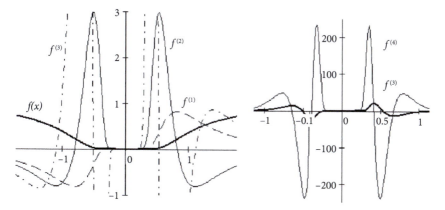

Fig. 8.20 Variations in higher derivatives of Cauchy function near 0

For all C^∞ functions, the variation of $r(x, c, n)$ seems to indicate the nature of a transcendental function at a deep level; but we notice its effects only when the Taylor series fails to converge to f.

Cauchy's integral form of the remainder

Another useful form of the remainder is due to Cauchy, which we discuss next. This form is slightly more restricted than Lagrange's form in that the derivative $f^{(n+1)}$ must be assumed integrable. This restriction is rarely a practical impediment.

Theorem 333 *(Taylor remainder, Cauchy's integral form) Let f be differentiable $n+1$ times on an open interval (a, b), and let $c \in (a, b)$. If $f^{(n+1)}$ is Riemann integrable on every closed and bounded subinterval of (a, b), then*

$$R_n(x) = \frac{1}{n!} \int_c^x f^{(n+1)}(t)(x - t)^n dt, \quad x \in (a, b) \tag{8.92}$$

Proof The proof is based on mathematical induction from Chapter 2 and integration by parts (Corollary 258, Chapter 7). For $n = 1$, we have by definition

$$R_1(x) = f(x) - f(c) - f'(c)(x - c)$$

Using the fundamental theorem of calculus we get

$$R_1(x) = \int_c^x f'(t)dt - \int_c^x f'(c)dt$$

$$= \int_c^x [f'(t) - f'(c)]dt$$

Next, using integration by parts with $u = f'(t) - f'(c)$ and $dv = dt$ we obtain

$$du = f''(t)dt, \quad v = t - x$$

where we included a "constant of integration" x that we need in the next step. Now,

$$R_1(x) = [f'(t) - f'(c)](t - x)\big|_c^x - \int_c^x f''(t)(t - x)dt$$

$$= \int_c^x f''(t)(x - t)dt$$

This proves (8.92) for $n = 1$. Next, we assume that (8.92) holds for $n = k$ and prove it for $n = k + 1$.

$$R_{k+1}(x) = f(x) - T_{k+1}(x)$$

$$= f(x) - T_k(x) - \frac{f^{(k+1)}(c)}{(k+1)!}(x-c)^{k+1}$$

$$= R_k(x) - \frac{f^{(k+1)}(c)}{(k+1)!}(x-c)^{k+1}$$

By the induction hypothesis, we may express $R_k(x)$ using (8.92), and as in the case $n = 1$, write

$$R_{k+1}(x) = \frac{1}{k!}\int_c^x f^{(k+1)}(t)(x-t)^k dt - \frac{f^{(k+1)}(c)}{k!}\int_c^x (x-t)^k dt$$

$$= \frac{1}{k!}\int_c^x [f^{(k+1)}(t) - f^{(k+1)}(c)](x-t)^k dt$$

Next, we use integration by parts with

$$u = f^{(k+1)}(t) - f^{(k+1)}(c), \quad dv = (x-t)^k dt$$

$$du = f^{(k+2)}(t)dt, \quad v = -\frac{1}{k+1}(x-t)^{k+1}$$

to obtain

$$R_{k+1}(x) = \frac{1}{k!}\left[-\frac{f^{(k+1)}(t) - f^{(k+1)}(c)}{k+1}(x-t)^{k+1}\Big|_c^x + \int_c^x \frac{f^{(k+2)}(t)}{k+1}(x-t)^{k+1}dt\right]$$

$$= \frac{1}{(k+1)!}\int_c^x f^{(k+2)}(t)(x-t)^{k+1}dt$$

and complete the proof by induction. ∎

We illustrate an application of the integral form of the remainder in our discussion of the binomial series below.

The irrationality of the number e

Let's now apply our knowledge of the remainder term to prove that the base e of the natural exponential is irrational.

We start with the observation that there is an integer N such that $e < N$; of course, we can pick $N = 3$, the least possible integer; but the value of N is not important, and we can simply assume its existence as a consequence of the Archimedean property of real numbers.

From our discussion above, for every real number x and each n there is $r(x, n)$ between 0 and x such that

$$e^x = 1 + x + \frac{x^2}{2!} + \cdots + \frac{x^n}{n!} + \frac{e^r x^{n+1}}{(n+1)!}$$

For $0 < r < x \le 1$, we have $e^r < e < N$, and it follows that

$$0 < e - \left(2 + \frac{1}{2!} + \cdots + \frac{1}{n!}\right) < \frac{N}{(n+1)!}$$

Note that the above inequality is true for all positive integers n.

Now, suppose that e is rational, so that $e = k/m$ in reduced form, i.e., the integers k and m are relatively prime and share no common factors. Then

$$0 < \frac{k}{m} - \left(2 + \frac{1}{2!} + \cdots + \frac{1}{n!}\right) < \frac{N}{(n+1)!}$$

Next, find an integer n that has m as a factor and is larger than N, say, $n = mN$. Multiplying the above inequalities by the factorial of this number yields

$$0 < \frac{kn!}{m} - n!\left(2 + \frac{1}{2!} + \cdots + \frac{1}{n!}\right) < \frac{n!N}{(n+1)!}$$

$$0 < \frac{kn!}{m} - 2n! - \frac{n!}{2} - \cdots - n - 1 < \frac{N}{n+1}$$

Notice that there is something wrong with the above inequalities; on the right-hand side, the fraction is less than 1, but what is sandwiched between this fraction and 0 is another integer. But this is impossible, as there are no integers between 0 and 1. The only way to avoid this contradiction is if e is not rational, i.e., if e is irrational!

8.5.4 The binomial series and its applications

We proved the following equality in Chapter 5 (in Theorem 151) using mathematical induction:

$$(a + b)^n = a^n + \binom{n}{1}a^{n-1}b + \binom{n}{2}a^{n-2}b^2 + \cdots + \binom{n}{n}b^n$$

$$= \sum_{k=0}^{n} \frac{n(n-1)(n-2)\cdots(n-k+1)}{k!}a^{n-k}b^k$$

This equality known as *the binomial theorem* is valid for all positive integers n.

If n is not a positive integer, then the above equality is not valid; but it turns out that for all real values of p, the binomial expression $(a + b)^p$ may be expressed as a power series. We now discuss this series and some of its applications.

Deriving the binomial series as a power series

We begin with the function

$$f(x) = (1 + x)^p$$

where the constant p is a real number, and calculate the derivatives:

$$f'(x) = p(1 + x)^{p-1},$$
$$f''(x) = p(p - 1)(1 + x)^{p-2},$$
$$f'''(x) = p(p - 1)(p - 2)(1 + x)^{p-3},$$

$$\vdots$$

$$f^{(n)}(x) = p(p - 1)(p - 2) \cdots (p - n + 1)(1 + x)^{p-n}$$

$$\vdots$$

Notice that if p is not a positive integer, then the above list of derivatives is infinite. At $x = 0$,

$$f^{(n)}(0) = p(p - 1)(p - 2) \cdots (p - n + 1)$$

These and $f(0) = 1$ give the sequence of Taylor polynomials for $(1 + x)^p$

$$T_n(x) = 1 + \sum_{k=1}^{n} \frac{p(p - 1) \cdots (p - k + 1)}{k!} x^k$$

$$= 1 + \frac{p}{1!}x + \frac{p(p - 1)}{2!}x^2 + \cdots + \frac{p(p - 1) \cdots (p - n + 1)}{n!}x^n$$

The obvious question now is if these polynomials converge to $(1+x)^p$ over some open interval. There are two ways of answering this question; one is by showing that the remainder $R_n(x)$ converges to 0, and the other is by using the power series methods that we discussed earlier. We do it *both* ways for the sake of comparison and illustration.

First, let us consider the limit $\lim_{n \to \infty} T_n(x)$, which is the power series

$$1 + \sum_{n=1}^{\infty} \frac{p(p-1) \cdots (p-n+1)}{n!} x^n = 1 + \frac{p}{1!} x + \frac{p(p-1)}{2!} x^2 + \cdots \qquad (8.93)$$

To find the radius of convergence R for this power series with coefficients $c_n = p(p-1) \cdots (p-n+1)/n!$ we use the ratio test:

$$R = \lim_{n \to \infty} \left| \frac{c_n}{c_{n+1}} \right|$$

$$= \lim_{n \to \infty} \left| \frac{p(p-1) \cdots (p-n+1)(n+1)!}{p(p-1) \cdots (p-n+1)(p-n)n!} \right|$$

$$= \lim_{n \to \infty} \left| \frac{n+1}{p-n} \right|$$

Finding the last limit is straightforward:

$$R = \left| \lim_{n \to \infty} \frac{n+1}{p-n} \right| = \left| \lim_{n \to \infty} \frac{1 + 1/n}{-1 + p/n} \right| = 1$$

Therefore, the interval of convergence of the series in (8.93) is $(-1, 1)$. Convergence at the end-points depends on the value of p, as we see below. In particular, if $p < 0$, then $x = -1$ is not even in the domain of the function $(1 + x)^p$.

Note that the series in (8.93) is a power series, and further, it is the limit of the partial sums $T_n(x)$ of $(1 + x)^p$ for every $x \in (-1, 1)$. So by the uniqueness of power series (Corollary 329), it is the Taylor series for $(1 + x)^p$:

$$(1 + x)^p = 1 + \sum_{n=1}^{\infty} \frac{p(p-1) \cdots (p-n+1)}{n!} x^n, \qquad x \in (-1, 1) \qquad (8.94)$$

We refer to the above series as the *binomial series*.

In particular, if $p = n$, a positive integer, then

$$(1 + x)^n = 1 + \frac{n}{1!} x + \frac{n(n-1)}{2!} x^2 + \cdots + \frac{n(n-1) \cdots (1)}{n!} x^n \qquad (8.95)$$

where all terms after the last one listed above drop out since the coefficients of all the remaining terms of the series contain a factor $(n - n)$ and are therefore zeros. The coefficients are none other than the binomial coefficients, so (8.95) may be written as

$$(1 + x)^n = 1 + \binom{n}{1}x + \binom{n}{2}x^2 + \cdots + \binom{n}{n}x^n \qquad (8.96)$$

In this form, we see that if we set $x = b/a$, or a/b if $b > a$ so that x is in the interval of convergence $(-1, 1)$, and then multiply the sum by a^n (or b^n as the case may be), we obtain the binomial theorem for integer powers mentioned above.

Noteworthy is the fact that if $x = \pm 1$ ($b = \pm a$), then the equality in (8.96) remains valid, so (8.96) holds for $x \in [-1, 1]$.

Also noteworthy is the observation that we have here a proof of the binomial theorem (in Theorem 151) without using induction!

Next, set $p = -1$ in (8.94) to get

$$(1 + x)^{-1} = 1 + \frac{(-1)}{1!}x + \frac{(-1)(-2)}{2!}x^2 + \frac{(-1)(-2)(-3)}{3!}x^3 + \cdots$$

$$\frac{1}{1 + x} = 1 - x + x^2 - x^3 + \cdots$$

This is the geometric series; in fact, by changing x to $-x$ above, we obtain the original geometric power series as a special case of the binomial series (8.93). This also means that *all other series* that we obtained from the geometric power series, including those for $\ln x$ and $\tan^{-1} x$, can be considered direct consequences of the binomial series! Only now we have a series with a much wider range of applicability.

Binomial series as a Taylor series with remainder

Before exploring the binomial series further, let's prove that it is the Taylor series for $(1 + x)^p$ by showing that the remainder $R_n(x)$ converges to zero as $n \to \infty$. For convenient reference, we state this as a theorem.

Theorem 334 *(Convergence of the binomial series) For every real number p, and $x \in (-1, 1)$*

$$\lim_{n \to \infty} R_n(x) = \lim_{n \to \infty} \left[(1 + x)^p - 1 + \sum_{k=1}^{n} \frac{p(p - 1) \cdots (p - k + 1)}{k!} x^k \right] = 0 \qquad (8.97)$$

That is, the series in (8.94) converges to $(1 + x)^p$ on $(-1, 1)$.

Proof Note that if $p = 0$, we get a constant function $(1 + x)^p = 1$, which leads to $T_n(x) = 1$ for all n and x; so $R_n(x) = 0$ for $n > 1$, and we have a trivial case (in fact, if p is any non-negative integer, then $R_n(x)$ is the zero function for $n > p$).

We assume that $p \neq 0$. To reduce notational clutter we define

$$c(p, n) = p(p - 1) \cdots (p - n), \quad n = 0, 1, 2, 3, \ldots$$

The integral form of the remainder in (8.92) with $f(x) = (1 + x)^p$ and (8.94) gives

$$R_n(x) = \frac{1}{n!} \int_c^x f^{(n+1)}(t)(x - t)^n dt$$

$$= \frac{1}{n!} \int_0^x c(p, n)(1 + t)^{p-n-1}(x - t)^n dt$$

$$= \frac{c(p, n)}{n!} \int_0^x (1 + t)^{p-n-1}(x - t)^n dt \qquad (8.98)$$

Next, since $R_n(0) = 0$, we consider the remaining cases as follows:
Case 1: $-1 < x < 0$. In this case, we rewrite (8.98) as

$$R_n(x) = (-1)^{n+1} \frac{c(p, n)}{n!} \int_x^0 \left(\frac{t - x}{1 + t}\right)^n (1 + t)^{p-1} dt$$

and note that for $-1 < x \leq t \leq 0$

$$0 \leq \frac{t - x}{1 + t} \leq -x = |x|$$

Therefore, if $-1 < x \leq t \leq 0$, then $0 < 1 + t \leq 1$ and for all n,

$$0 \leq \left(\frac{t - x}{1 + t}\right)^n \leq |x|^n$$

With this in hand, we go back to (8.98) to get

$$|R_n(x)| \leq \frac{c(p, n)}{n!} \left| \int_x^0 |x|^n (1 + t)^{p-1} dt \right|$$

$$= \frac{c(p, n)}{n!} |x|^n \left| \frac{(1 + t)^p}{p} \right|_x^0$$

$$= \left| \frac{1 - (1 + x)^p}{p} \right| \frac{c(p, n)}{n!} |x|^n$$

The first factor does not change as $n \to \infty$, so we prove that the limit of the remaining terms goes to 0. To that end, note that if

$$a_n = \frac{c(p, n)}{n!} |x|^n$$

then

$$\begin{aligned}
\frac{a_{n+1}}{a_n} &= \frac{c(p, n+1)|x|^{n+1}}{(n+1)!} \frac{n!}{c(p, n)|x|^n} \\
&= \frac{p(p-1)\cdots(p-n)(p-n-1)|x|}{(n+1)p(p-1)\cdots(p-n)} \\
&= \frac{(p-n-1)|x|}{n+1}
\end{aligned}$$

Thus,

$$\left| \frac{a_{n+1}}{a_n} \right| = |x| \left| 1 - \frac{p}{n+1} \right| \leq |x| \left(1 + \frac{|p|}{n+1} \right)$$

If we choose N large enough that (since $|x| < 1$)

$$|x| \left(1 + \frac{|p|}{N+1} \right) \leq |x|[1 + (1 - |x|)] = |x|^2$$

then to the decreasing nature of $|p|/(n+1)$,

$$\left| \frac{a_{n+1}}{a_n} \right| \leq |x|^2 < 1 \quad \text{for all } n \geq N$$

Now $n = N + k$ with $k = n - N$ so that

$$\left| \frac{a_n}{a_N} \right| = \left| \frac{a_{N+1}}{a_N} \right| \left| \frac{a_{N+2}}{a_{N+1}} \right| \cdots \left| \frac{a_{N+k}}{a_{N+k-1}} \right| \leq |x|^{2k}$$

so that

$$|a_n| \leq |x|^{2(n-N)} |a_N| \quad \text{for all } n \geq N$$

It now follows that

$$\lim_{n \to \infty} \frac{c(p, n)}{n!} |x|^n = \lim_{n \to \infty} |a_n| = 0$$

and since

$$0 \leq |R_n(x)| = \left| \frac{1 - (1+x)^p}{p} \right| |a_n|$$

the squeeze theorem implies that $\lim_{n \to \infty} |R_n(x)| = 0$.

Case 2: $0 < x < 1$. The argument for this case similar to that for Case 1 and therefore, left as an exercise. ∎

Binomial series: Illustration and application

An obvious application of the binomial series is to estimate roots of rational numbers; a few centuries ago when there were no computers or calculators, being able to estimate roots quickly was a major issue. Newton was familiar with the binomial series before it became well known. Speedy publication or wide communication of ideas and results was not possible those days, so he managed to impress many smart people of his time by doing seemingly impossible calculations.

To illustrate this application of the binomial series, let's use it to estimate $\sqrt{2.5}$. The "obvious" substitution $x = 1.5$ in $(1 + x)^p$ with $p = 1/2$ doesn't work because 1.5 is not in the interval of convergence $(-1, 1)$. But

$$\sqrt{2.5} = \sqrt{\frac{5}{2}} = \left(\frac{5}{2}\right)^{1/2} = \left(\frac{2}{5}\right)^{-1/2} = \left(1 - \frac{3}{5}\right)^{-1/2}$$

and $3/5 = 0.6$ is well within the interval of convergence of the binomial series for $(1 + x)^{-1/2}$.

Consider the case $p = -1/2$ in (8.94):

$$(1 + x)^{-1/2} = 1 + \frac{-1/2}{1!}x + \frac{(-1/2)(-3/2)}{2!}x^2 + \frac{(-1/2)(-3/2)(-5/2)}{3!}x^3 + \cdots$$

$$\frac{1}{\sqrt{1 + x}} = 1 - \frac{1}{1!2}x + \frac{(1)(3)}{2!2^2}x^2 - \frac{(1)(3)(5)}{3!2^3}x^3 + \cdots \qquad (8.99)$$

or more compactly,

$$\frac{1}{\sqrt{1 + x}} = \sum_{n=0}^{\infty} \frac{(-1)^n(3)\cdots(2n - 1)}{n!2^n}x^n$$

We use the series in (8.99) with $x = -0.6$. If we use just the first four terms, then we obtain the approximation

$$\sqrt{2.5} \simeq 1 - \frac{1}{1!2}(-0.6) + \frac{(1)(3)}{2!2^2}(-0.6)^2 - \frac{(1)(3)(5)}{3!2^3}(-0.6)^3$$

$$= 1 + 0.3 + \frac{3(0.36)}{8} + \frac{1.5(0.36)}{8} = 1.5025$$

which is not a bad estimate of $\sqrt{2.5}$ whose value rounded to four decimal places is 1.5811. Of course, to improve the accuracy of approximation, we need only use a Taylor polynomial of higher order.

The inverse-sine series and an efficient way to approximate π

The binomial series is one of the most versatile power series known. We use it now to derive the Taylor series for the inverse-sine function $\sin^{-1} x$.

Substituting $-u^2$ for x in (8.99) doesn't change the interval of convergence, since $-u^2$ is in the interval $(-1, 1)$ if and only if x is. We obtain the following equality:

$$\frac{1}{\sqrt{1 - u^2}} = 1 - \frac{1}{1!2}(-u^2) + \frac{(1)(3)}{2!2^2}(-u^2)^2 - \frac{(1)(3)(5)}{3!2^3}(-u^2)^3 + \cdots$$

$$\frac{1}{\sqrt{1 - u^2}} = 1 + \frac{1}{1!2}u^2 + \frac{(1)(3)}{2!2^2}u^4 + \frac{(1)(3)(5)}{3!2^3}u^6 + \cdots \qquad (8.100)$$

The left-hand side of (8.100) happens to be the derivative of $\sin^{-1} u$

$$\frac{d}{du}\sin^{-1} u = \frac{1}{\sqrt{1 - u^2}}$$

as claimed. Now, if x is in $(-1, 1)$, then we may integrate (8.100) term by term to obtain

$$\int_0^x \frac{1}{\sqrt{1 - u^2}}du = \left[u + \frac{u^3}{1!2(3)} + \frac{(1)(3)u^5}{2!2^2(5)} + \frac{(1)(3)(5)u^7}{3!2^3(7)} + \cdots \right]_{u=0}^{u=x}$$

$$\sin^{-1} x = x + \frac{x^3}{1!2(3)} + \frac{(1)(3)x^5}{2!2^2(5)} + \frac{(1)(3)(5)x^7}{3!2^3(7)} + \cdots \qquad (8.101)$$

This is the standard, inverse-sine, Maclaurin series that we can write more compactly as

$$\sin^{-1} x = \sum_{n=0}^{\infty} \frac{(1)(3)\cdots(2n - 1)x^{2n+1}}{n!2^n(2n + 1)}, \qquad -1 < x < 1$$

I note in passing that this series shows that $\sin^{-1} x$ is an odd function. Let's use (8.101) to approximate π one more time!

Note that $\sin(\pi/6) = 1/2$ so that $\sin^{-1}(1/2) = \pi/6$. Now we insert $x = 1/2$ in (8.101) and use four of its terms to obtain

$$\frac{\pi}{6} \simeq \frac{1}{2} + \frac{(1/2)^3}{1!2(3)} + \frac{(1)(3)(1/2)^5}{2!2^2(5)} + \frac{(1)(3)(5)(1/2)^7}{3!2^3(7)}$$

$$= \frac{1}{2} + \frac{1}{2^4(3)} + \frac{3}{2^8(5)} + \frac{5}{2^{11}(7)}$$

Multiplying by 6 and doing the arithmetic gives the following approximation of π rounded to four decimal places:

$$\pi \simeq 3.1412$$

The difference between this number and the actual value of π is less than 0.0004. Using just 18 terms of (8.101) with $x = 1/2$ gives the estimate

$$\pi \simeq 3.14159265358959$$

which is already better than the one using 21 terms of the inverse tangent series earlier. Further, this new approximation is *rational*, i.e., it does not include irrational factors like $\sqrt{3}$.

It is worth remembering that *both the inverse sine and the inverse tangent series are integrals of power series that we can easily obtain from the binomial series.*

8.6 Trigonometric series and Fourier expansions: A glance

In this section, we take a brief look at trigonometric series and their well-known special cases, the Fourier series. The theory of Fourier series and integrals represents a vast and powerful machinery that requires its own book. We do not enter it in a substantial way in this section; our purpose is limited and twofold: first, to discuss briefly an alternative to power series to which the general theory of the earlier sections applies; second, use the results to find, among other things, the exact numerical value of the p-series

$$\sum_{n=1}^{\infty} \frac{1}{n^2}$$

which happens to be also the value of the Riemann ζ function at 2, i.e., $\zeta(2)$.

We have already discussed one prominent trigonometric series, namely, the one that converges uniformly to Weierstrass's function $W(x)$ that we proved to be continuous yet nowhere differentiable. That series dramatically highlights all of the differences between power series and trigonometric series and shows just how much more interesting material beyond power series infinite series of functions can produce.

In our discussion of power series, we found that they may converge only to functions that are infinitely differentiable. If $f(x) = \sum_{n=0}^{\infty} c_n(x - c)^n$ with radius of convergence $R > 0$, then $f(x)$ is basically the limit of a sequence of *polynomials* $T_n(x) = \sum_{k=0}^{n} c_k(x - c)^k$ on the interval $(c - R, c + R)$. Every $T_n(x)$ and the limit $f(x)$ are C^{∞} functions.

Functions having a Taylor series are even more special: if $f(x)$ has a Taylor series, then its analytic nature throughout the interval of convergence of the Taylor series is fully determined by the list of numbers $f(c), \dot{f}(c), f''(c), \ldots, f^{(n)}(c), \ldots$ calculated at a *single point* $x = c$.

Weierstrass's function is defined by an infinite trigonometric series of type

$$\sum_{n=0}^{\infty} a^n \cos(m^n \pi x) \tag{8.102}$$

which is also the limit of a sequence of C^∞ functions, namely, the partial sums

$$s_n(x) = \sum_{k=0}^{n} a^k \cos(m^k \pi x), \quad n = 1, 2, 3, \ldots$$

on $(-\infty, \infty)$. But as we saw earlier, the limit $W(x)$ is not even once differentiable at *any* point.

Trigonometric series are not generally as extreme as Weierstrass's, and they don't need to be; the functions that we encounter in practice are generally piecewise smooth, i.e., they have continuous derivatives except at a finite number of points. A simple example is $|x|$, which is differentiable on the set $(-\infty, \infty)$ of real numbers except at $x = 0$. We show later in this section that it can be expanded on a closed bounded interval as a trigonometric series. For these functions, which are not infinitely differentiable, trigonometric series succeed where power series fail.

8.6.1 Trigonometric series

Consider a function such as that shown in Figure 8.21.

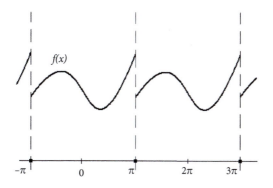

$f(x)$

$-\pi \qquad 0 \qquad \pi \qquad 2\pi \qquad 3\pi$

Fig. 8.21 A discontinuous periodic function

Such a function is not even continuous, but because it is periodic, we may represent it using infinite series of (also periodic) trigonometric functions. Trigonometric series that are most commonly used in practice to represent functions are the following:

$$\sum_{n=0}^{\infty} [a_n \cos(nx) + b_n \sin(nx)] = a_0 + a_1 \cos(x) + b_1 \sin(x) + \cdots \qquad (8.103)$$

whose coefficients a_n and b_n are real numbers. Recall that $\cos 0 = 1$ and $\sin 0 = 0$, so when $n = 0$, we always get the constant a_0 for the first term. For this reason, (8.103) is commonly written as

$$a_0 + \sum_{n=1}^{\infty} [a_n \cos(nx) + b_n \sin(nx)] \qquad (8.104)$$

Note that the Weierstrass series in (8.102) is a trigonometric series that is *not of the above type*.

The Weierstrass test for uniform convergence (Theorem 319) is used to determine if in particular a trigonometric series such as (8.104) converges uniformly.

Notice that

$$|a_n \cos(nx) + b_n \sin(nx)| \leq |a_n \cos(nx)| + |b_n \sin(nx)|$$

by the triangle inequality; so because $|\cos(nx)| \leq 1$, and $|\sin(nx)| \leq 1$, we may conclude that

$$|a_n \cos(nx) + b_n \sin(nx)| \leq |a_n| + |b_n|$$

So applying the Weierstrass test with $M_n = |a_n| + |b_n|$ yields the following result.

Theorem 335 *Assume that a_n and b_n are sequences of real numbers such that $\sum_{n=1}^{\infty} (|a_n| + |b_n|) < \infty$. Then the trigonometric series in (8.104) converges uniformly on $(-\infty, \infty)$ to a continuous function $f(x)$ of period 2π.*

To illustrate, consider the series

$$\sum_{n=0}^{\infty} \left[\frac{1}{2^n} \cos(nx) + \frac{(-1)^n}{3^n} \sin(nx) \right] = 1 + \frac{1}{2} \cos x - \frac{1}{3} \sin x + \cdots \qquad (8.105)$$

The series of coefficients converges absolutely since

$$\sum_{n=0}^{\infty} \frac{1}{2^n} = 2, \qquad \sum_{n=0}^{\infty} \frac{1}{3^n} = \frac{3}{2}$$

Therefore, the series in (8.105) converges to a continuous function $f(x)$ of period 2π on $(-\infty, \infty)$. We can derive a number of facts about the function $f(x)$ using the series and the theory that we discussed earlier.

For example, we can find certain values of $f(x)$ such as the following:

$$f(0) = \sum_{n=0}^{\infty}\left[\frac{1}{2^n}\cos 0 + \frac{(-1)^n}{3^n}\sin 0\right] = \sum_{n=0}^{\infty}\frac{1}{2^n} = 2 = f(2\pi k), \ k \in \mathbb{Z}$$

or

$$f(\pi) = \sum_{n=0}^{\infty}\left[\frac{1}{2^n}\cos(n\pi) + \frac{(-1)^n}{3^n}\sin(n\pi)\right] = \sum_{n=0}^{\infty}\frac{(-1)^n}{2^n} = \frac{2}{3}$$

Also, since continuous functions are Riemann integrable, and the series converges uniformly, we may also calculate the integral of f by integrating the series in (8.105) term by term:

$$\int_0^{\pi} f(x)dx = \sum_{n=0}^{\infty}\left[\frac{1}{2^n}\int_0^{\pi}\cos(nx)dx + \frac{(-1)^n}{3^n}\int_0^{\pi}\sin(nx)dx\right]$$

$$= \pi + \sum_{n=1}^{\infty}\left[\frac{1}{2^n n}[\sin(n\pi) - \sin 0] - \frac{(-1)^n}{3^n n}[\cos(n\pi) - \cos 0]\right]$$

$$= \pi + \sum_{n=1}^{\infty}\frac{(-1)^n}{3^n n}[1 - (-1)^n] = \pi + \sum_{n=1}^{\infty}\frac{(-1)^n}{3^n n} - \sum_{n=1}^{\infty}\frac{1}{3^n n} \qquad (8.106)$$

The integral is thus given by series of constants. In general, this the most that we can expect; and to obtain a numerical answer, we may approximate the series of constants. But in the case of (8.106), both series can be calculated using the integral of the geometric power series to obtain exact values; for the last series consider

$$\sum_{n=0}^{\infty}\frac{x^n}{3^n} = \sum_{n=0}^{\infty}\left(\frac{x}{3}\right)^n = \frac{3}{3 - x}$$

Integrating this series gives

$$\int_0^1 \sum_{n=0}^{\infty}\frac{x^n}{3^n}dx = 3\int_0^1\frac{1}{3 - x}dx$$

$$\sum_{n=0}^{\infty}\int_0^1\frac{x^n}{3^n}dx = -3\int_3^2\frac{1}{u}du$$

$$\sum_{n=0}^{\infty}\frac{1}{3^n(n + 1)} = 3(\ln 3 - \ln 2)$$

Now notice that

$$\sum_{n=1}^{\infty}\frac{1}{3^n n}=\sum_{n=0}^{\infty}\frac{1}{3^{n+1}(n+1)}=\frac{1}{3}\sum_{n=0}^{\infty}\frac{1}{3^n(n+1)}=\ln 3-\ln 2$$

In Exercise 366, you can likewise calculate the other series in (8.106) using a similar geometric power series to obtain the exact value of the integral:

$$\int_0^{\pi}f(x)dx = \pi+2\ln 3-3\ln 2$$

As the above example shows, we typically do not have a formula for the limit function of a trigonometric series and have to depend on convergence results that we proved earlier in order to obtain analytical information about the limit functions, or to meaningfully use partial sums to approximate the values of the limit function.

8.6.2 Fourier series and their coefficients

We discovered earlier that it is possible to expand a given function as a Taylor series about a given center. In this section, we consider expanding functions in the form of trigonometric series.

Recall that a function $f(x)$ is periodic of period p if $f(x + p) = f(x)$ for all x. The functions $\cos(nx)$ and $\sin(nx)$ in (8.104) have period 2π for every value of the positive integer n, but we can generalize (8.104) to a series with an arbitrary period using a standard method. A slight modification of (8.104) gives:

$$a_0 + \sum_{n=1}^{\infty}\left(a_n\cos\frac{n\pi x}{L}+b_n\sin\frac{n\pi x}{L}\right) \tag{8.107}$$

where L is a fixed positive real number. For each fixed value of n notice that

$$\cos\frac{n\pi(x+2L)}{L}=\cos\left(\frac{n\pi x}{L}+2n\pi\right)=\cos\frac{n\pi x}{L}$$

so $\cos(n\pi x/L)$ has period $2L$. Similarly, $\sin(n\pi x/L)$ has period $2L$, so the series in (8.107) has period $p = 2L$. In particular, if we set $L = \pi$, then this series reduces to (8.104).

Now suppose that a function $f(x)$ with period $p = 2L$ is given, and we want to find the numbers a_n, b_n for which we can say that

$$f(x) = a_0 + \sum_{n=1}^{\infty} \left(a_n \cos \frac{n\pi x}{L} + b_n \sin \frac{n\pi x}{L} \right), \quad -L \le x \le L \qquad (8.108)$$

Notice that we need only consider the equality in (8.108) for x in the interval $[-L, L]$ since all other values of x are taken care of by periodicity. For instance, in Figure 8.21, we see that $f(2\pi) = f(0)$ where 0 is in the interval $[-L, L]$ with $L = \pi$ even though 2π is not in $[-L, L]$.

Recall that in the case of Taylor series, we calculated the coefficients c_n using the higher derivatives of $f(x)$. To calculate the numbers a_n and b_n in (8.108), we use a property of the sine and cosine functions known as "orthogonality."[16] If we multiply both sides of (8.108) by $\cos(m\pi x/L)$ where m is any unspecified positive integer then we get

$$f(x) \cos \frac{m\pi x}{L} = a_0 \cos \frac{m\pi x}{L} + \sum_{n=1}^{\infty} \left(a_n \cos \frac{m\pi x}{L} \cos \frac{n\pi x}{L} + b_n \cos \frac{m\pi x}{L} \sin \frac{n\pi x}{L} \right)$$

Next, we integrate the above from $-L$ to L; the term by term integration of the series on the right-hand side is possible if, for instance, the series in (8.108) converges uniformly to $f(x)$ on $[-L, L]$ (see Theorem 335):

$$\int_{-L}^{L} f(x) \cos \frac{m\pi x}{L} dx = a_0 \int_{-L}^{L} \cos \frac{m\pi x}{L} dx+ \qquad (8.109)$$
$$+ \sum_{n=1}^{\infty} \left(a_n \int_{-L}^{L} \cos \frac{m\pi x}{L} \cos \frac{n\pi x}{L} dx \right.$$
$$\left. + b_n \int_{-L}^{L} \cos \frac{m\pi x}{L} \sin \frac{n\pi x}{L} dx \right)$$

The first integral on the right-hand side is easy to evaluate by the substitution rule:

$$\int_{-L}^{L} \cos \frac{m\pi x}{L} dx = \frac{L}{m\pi} (\sin(m\pi) + \sin(m\pi)) = 0$$

---

[16] Orthogonal is synonymous with perpendicular. Recall from calculus that two vectors are perpendicular if their "dot product" is zero. In vector spaces of functions, two functions are orthogonal if their "inner product" is zero. Integrals are used to define a type of inner product that is especially important in physics. The concept of inner product is discussed in most standard analysis textbooks that discuss metric spaces.

The integrals inside the summation symbol can be evaluated using the *orthogonality formulas* for trigonometric functions.

$$\int_{-L}^{L} \cos \frac{m\pi x}{L} \cos \frac{n\pi x}{L} dx \int_{-L}^{L} \sin \frac{m\pi x}{L} \sin \frac{n\pi x}{L} dx = \begin{cases} L, & \text{if } m = n \\ 0, & \text{if } m \neq n \end{cases} \qquad (8.110)$$

and

$$\int_{-L}^{L} \sin \frac{m\pi x}{L} \cos \frac{n\pi x}{L} dx = 0 \quad \text{for all } m, n \qquad (8.111)$$

The derivation of the orthogonality relations is straightforward. For example, to derive (8.111), we use the trigonometric identity

$$2 \sin u \cos v = \sin(u + v) + \sin(u - v)$$

to obtain

$$\int_{-L}^{L} \sin \frac{m\pi x}{L} \cos \frac{n\pi x}{L} dx = \frac{1}{2} \int_{-L}^{L} \sin \frac{(m + n)\pi x}{L} dx + \frac{1}{2} \int_{-L}^{L} \sin \frac{(m - n)\pi x}{L} dx$$

Next, we may calculate each of the integrals using the substitution method, or use the fact that $\sin(ax)$ is an odd function for every nonzero real number a, so integrating it over a symmetric interval $[-L, L]$ gives a net value of zero (you may check this result by actually calculating each integral). This completes the derivation of (8.111).

The derivation of (8.110) is similar but has one extra case. First, if $m \neq n$, then using the trigonometric identity

$$2 \cos u \cos v = \cos(u + v) + \cos(u - v)$$

we obtain

$$\int_{-L}^{L} \cos \frac{m\pi x}{L} \cos \frac{n\pi x}{L} dx = \frac{1}{2} \int_{-L}^{L} \cos \frac{(m + n)\pi x}{L} dx + \frac{1}{2} \int_{-L}^{L} \cos \frac{(m - n)\pi x}{L} dx$$

If k is any integer nonzero, then

$$\int_{-L}^{L} \cos \frac{k\pi x}{L} dx = \frac{L}{k\pi} \int_{-k\pi}^{k\pi} \cos u \, du = 0$$

since $\sin(k\pi) = 0$ for every integer k. Finally, if $m = n \geq 1$, then

$$\int_{-L}^{L} \cos\frac{m\pi x}{L} \cos\frac{n\pi x}{L} dx = \int_{-L}^{L} \cos^2\frac{n\pi x}{L} dx$$

In this case, since

$$\cos^2\frac{n\pi x}{L} = \frac{1}{2}\left(1 + \cos\frac{2n\pi x}{L}\right)$$

we obtain

$$\int_{-L}^{L} \cos^2\frac{n\pi x}{L} dx = \frac{1}{2}\int_{-L}^{L}\left(1 + \cos\frac{2n\pi x}{L}\right) du$$

Next, we substitute $u = 2n\pi x/L$ to get

$$\int_{-L}^{L} \cos^2\frac{n\pi x}{L} dx = \frac{L}{4n\pi}\int_{-2n\pi}^{2n\pi} [1 + \cos(u)] \, du = L$$

This completes the proof of the first equality in (8.110); the second is proved in a similar way and is left as an exercise.

Now, using the orthogonality equalities in (8.109) gives

$$\int_{-L}^{L} f(x) \cos\frac{m\pi x}{L} dx = a_m L$$

Dividing by L and recalling that m was an arbitrary positive integer, this results in the following formula for a_n:

$$a_n = \frac{1}{L}\int_{-L}^{L} f(x) \cos\frac{n\pi x}{L} dx, \quad n \neq 0 \qquad (8.112)$$

If $n = 0$, then $\cos(n\pi x/L) = 1$ and (8.109) with $m = 0$ we get

$$\int_{-L}^{L} f(x)dx = a_0 \int_{-L}^{L} 1 dx = a_0(2L)$$

$$a_0 = \frac{1}{2L}\int_{-L}^{L} f(x)dx \qquad (8.113)$$

To find the formula for b_n, we multiply by $\sin(m\pi x/L)$ and integrate; similar calculations to the above for the case $n \neq 0$ ultimately give the formula

$$b_n = \frac{1}{L}\int_{-L}^{L} f(x) \sin\frac{n\pi x}{L} dx \qquad (8.114)$$

Formulas (8.112)–(8.114) were derived by Fourier (as well as by Euler earlier).

Fourier series expansion and coefficients

The numbers a_n, b_n are determined by calculating the integrals in (8.112)–(8.114), and they are called the *Fourier coefficients*. With these numbers, the trigonometric series in (8.108) is called the *Fourier series expansion* of $f(x)$.

Convergence issues, namely, if the Fourier series converges in some sense to the function $f(x)$, are important questions in the theory of Fourier series. We discuss them briefly later. For now, we find the Fourier expansion for the function $f(x) = |x|$ on the interval $[-1, 1]$, which with $L = 1$ takes the form

$$|x| = a_0 + \sum_{n=1}^{\infty} [a_n \cos(n\pi x) + b_n \sin(n\pi x)], \quad -1 \leq x \leq 1$$

We calculate the coefficients a_n, b_n. First, using (8.113), we get the value of a_0:

$$a_0 = \frac{1}{2} \int_{-1}^{1} |x| dx = \frac{1}{2} \left[\int_{-1}^{0} (-x)dx + \int_{0}^{1} x\, dx \right] = 1$$

Next, to calculate a_n where $n \neq 0$, we use (8.112):

$$a_n = \frac{1}{1} \int_{-1}^{1} |x| \cos(n\pi x)dx = \int_{-1}^{0} (-x) \cos(n\pi x)dx + \int_{0}^{1} x \cos(n\pi x)dx$$

Each of the integrals on the right-hand side is found using integration by parts:

$$\int_{0}^{1} x \cos(n\pi x)dx = \frac{x}{n\pi} \sin(n\pi x)\Big|_{0}^{1} - \frac{1}{n\pi} \int_{0}^{1} \sin(n\pi x)dx$$

$$= 0 - \frac{1}{n\pi} \left[-\frac{1}{n\pi} \cos(n\pi x) \right]_{x=0}^{x=1}$$

$$= \left(\frac{1}{n\pi} \right)^2 [\cos(n\pi) - \cos 0]$$

Since $\cos(n\pi) = (-1)^n$ we obtain

$$\int_{0}^{1} x \cos(n\pi x)dx = \frac{(-1)^n - 1}{n^2 \pi^2}$$

Note that $\int_{-1}^{0}(-x)\cos(n\pi x)dx = \int_{0}^{1} u\cos(n\pi u)du$ by a change of variable $u = -x$, so

$$a_n = \frac{2[(-1)^n - 1]}{n^2\pi^2}$$

Next, using integration by parts again and proceeding as in the previous case, we find that

$$b_n = \frac{1}{1}\int_{-1}^{1}|x|\sin(n\pi x)dx = \int_{-1}^{0}(-x)\sin(n\pi x)dx + \int_{0}^{1} x\sin(n\pi x)dx = 0$$

Therefore, we have the Fourier series expansion for $-1 \le x \le 1$:

$$|x| = \frac{1}{2} + \sum_{n=1}^{\infty}\frac{2[(-1)^n - 1]}{n^2\pi^2}\cos(n\pi x) = \frac{1}{2} - \frac{4}{\pi^2}\cos(\pi x) - \frac{4}{9\pi^2}\cos(3\pi x) - \cdots$$

$$(8.115)$$

Figure 8.22 illustrates the first two partial sums of the above series, $s_1(x)$ in the left-hand side panel, and $s_3(x)$ in the other panel; we see that $s_3(x)$ looks more like $|x|$ than $s_1(x)$.

Figure 8.23 shows the partial sum $s_{99}(x)$, which is visually indistinguishable from $|x|$ unless we zoom on some part of it, like near the origin as shown in the right-hand side panel.

A note about periodicity

Notice that the series in (8.115) for $|x|$ is periodic with period 2, although $|x|$ clearly is not periodic. Therefore, the series converges to $|x|$ only on the interval $[-1, 1]$.

If we create a periodic function out of $|x|$ by placing over each adjacent interval of length 2 the copy that is defined over $[-1, 1]$, then the resulting sawtooth

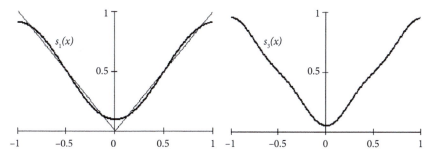

Fig. 8.22 The first two partial sums of the Fourier series for $|x|$

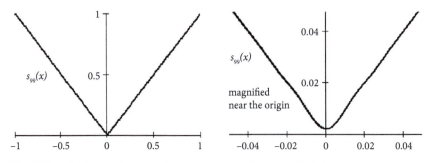

Fig. 8.23 A higher-order partial sum of the Fourier series for $|x|$

function is indeed periodic, and the series in (8.115) converges to it uniformly on $(-\infty, \infty)$.

A solution of the Basel problem and the value of $\zeta(2)$

Earlier, using the integral test, we found that the following series converges:

$$\sum_{n=1}^{\infty} \frac{1}{n^2} = \frac{1}{1^2} + \frac{1}{2^2} + \frac{1}{3^2} + \cdots \tag{8.116}$$

But the integral test does not give the value of the number, namely, $\zeta(2)$, to which the series in (8.116) converges. We can use the Fourier series theory to find out the exact value of $\zeta(2)$.

Finding the sum of the series in (8.116) was first posed as a problem by the Italian mathematician Pietro Mengoli in 1650. It was solved by Euler in 1734, who proved that the sum was $\pi^2/6$. The proof was ingenious, although the rigorous proof of Euler's arguments was not completed until 100 years later when Weierstrass provided the last missing piece, namely, that Euler's representation of $\sin x$ as an infinite product was, in fact, valid. Mengoli's problem has since come to be known as the *Basel problem*, named after Euler's home town.

To find the value of the sum of the series in (8.116), we begin by setting $x = 1$ in (8.115) to get

$$|1| = \frac{1}{2} - \frac{4}{\pi^2}\cos\pi - \frac{4}{9\pi^2}\cos(3\pi) - \frac{4}{25\pi^2}\cos(5\pi) - \cdots$$

Since cosine of an odd multiple of π is -1, we end up with

$$1 = \frac{1}{2} + \frac{4}{1^2\pi^2} + \frac{4}{3^2\pi^2} + \frac{4}{5^2\pi^2} + \cdots$$

Subtracting $1/2$ and multiplying by $\pi^2/4$ gives the equality

$$\frac{\pi^2}{8} = \frac{1}{1^2} + \frac{1}{3^2} + \frac{1}{5^2} + \cdots \qquad (8.117)$$

This shows that the sum of the reciprocals of the squares of all *odd* natural numbers is $\pi^2/8$. This result is rather striking, no less so than the solution of the Basel problem itself.

To find the value of $\zeta(2)$, we use (8.117) and a little algebraic manipulation: we split the (absolutely) convergent series in (8.116) into two as follows:

$$\zeta(2) = \left(1 + \frac{1}{3^2} + \frac{1}{5^2} + \cdots\right) + \left(\frac{1}{2^2} + \frac{1}{4^2} + \frac{1}{6^2} + \cdots\right) \qquad (8.118)$$

We found the value of the first series in (8.117); the second series can be written as

$$\frac{1}{2^2} + \frac{1}{4^2} + \frac{1}{6^2} + \cdots = \frac{1}{4}\left(1 + \frac{1}{2^2} + \frac{1}{3^2} + \cdots\right)$$

It follows that the value of the second series in (8.118) is $(1/4)\zeta(2)$. Inserting what we have calculated so far in (8.118) gives

$$\zeta(2) = \frac{\pi^2}{8} + \frac{1}{4}\zeta(2)$$

We solve for $\zeta(2)$ to obtain

$$\zeta(2) = \frac{\pi^2}{6}$$

which completes our task!

Even and odd functions

Notice that the trigonometric series for $|x|$ involves only the cosine functions, with all the sine terms dropping out. The reason for this is simply that $f(x) = |x|$ is an even function: $f(-x) = f(x)$. The cosine function terms are all even too since for all n

$$\cos(n\pi(-x)) = \cos(n\pi x)$$

However, sine functions are odd, i.e., $f(-x) = -f(x)$ because

$$\sin(n\pi(-x)) = -\sin(n\pi x)$$

So if there is a nonzero sine term ($b_n \neq 0$ for some $n \geq 1$), then the function $f(x)$ that is represented by the series in (8.108) cannot be even. A similar observation verifies that an odd function cannot have a cosine term ($a_n \neq 0$ for some $n \geq 0$). These observations justify the following statement.

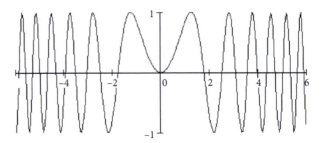

Fig. 8.24 The non-periodic, even function $\sin(x^2)$

Fourier series for odd and even functions

If $f(x)$ is an *even* function, then its Fourier series (and all of its partial sums) contain only the cosine terms; such a series where $b_n = 0$ for all $n \geq 1$ is called a *Fourier cosine series*. If $f(x)$ is an *odd* function, then its Fourier series (and all of its partial sums) contain only the sine terms; such a series where $a_n = 0$ for all $n \geq 0$ is called a *Fourier sine series*.

For example, we can readily verify that the function $f(x) = \sin(x^2)$ is even; in spite of the sine, f is neither odd nor periodic; see Figure 8.24.

This function is continuous on $(-\infty, \infty)$, and its Fourier *cosine* series converges uniformly on every interval of length 2π, like $[-\pi, \pi]$.

On the other hand, $\sin(x^3)$ is odd, and it has a Fourier sine series. Calculating the coefficients of the Fourier series in both cases is difficult due to the integration requirement.

Discontinuous functions

We have seen that *finding the Fourier coefficients requires integration rather than differentiation.* Therefore, functions that can be expanded into a Fourier series need not even be continuous.

To illustrate, consider a simple discontinuous function like the following step function on the interval $[-1, 1]$:

$$f(x) = \begin{cases} 2, & \text{if } x \geq 0 \\ 0, & \text{if } x < 0 \end{cases} \qquad (8.119)$$

The Fourier coefficients in this case are easy to calculate. From (8.112)–(8.114) with $L = 1$ and noting that $f(x) = 0$ when $x < 0$, we obtain

$$a_0 = \frac{1}{2} \int_{-1}^{1} f(x)dx = \frac{1}{2} \int_{0}^{1} 2dx = 1$$

$$a_n = \frac{1}{1} \int_{-1}^{1} f(x) \cos(n\pi x)dx = \int_{0}^{1} 2 \cos(n\pi x)dx = 0$$

$$b_n = \frac{1}{1} \int_{-1}^{1} f(x) \sin(n\pi x)dx = \int_{0}^{1} 2 \sin(n\pi x)dx = \frac{2[1 - (-1)^n]}{n\pi}$$

So we have the Fourier series expansion[17]

$$f(x) = 1 + \sum_{n=1}^{\infty} \frac{2[1 - (-1)^n]}{n\pi} \sin(n\pi x) = 1 + \frac{4}{\pi} \sin(\pi x) + \frac{4}{3\pi} \sin(3\pi x) + \cdots$$

for a function that is clearly discontinuous at $x = 0$. Figure 8.25 shows the partial sums $s_1(x)$ (dashed) and $s_{11}(x)$ (solid).

In Figure 8.25, it is evident that the trigonometric polynomial $s_{11}(x)$ is a better estimate of $f(x)$ (displayed as thick lines). But even that is not a good estimate; we can improve upon our estimate by using larger partial sums. Figure 8.26 shows the graph of $s_{50}(x)$, which is really the same as $s_{49}(x)$; we see that the approximation has improved substantially.

But we also notice in Figure 8.26 that as we approach the point of discontinuity at $x = 0$ from either side, the graph of $s_{50}(x)$ deviates noticeably from the graph of $f(x)$ in the form of larger and more rapid oscillations. The same discrepancies exist at the two endpoints. Using even larger partial sums does improve the approximation

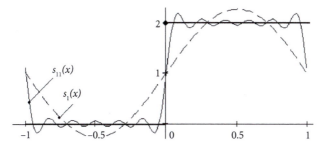

Fig. 8.25 Two partial sums of the Fourier series around a jump

---

[17] Notice that this is *not* a sine series despite the absence of explicit cosine terms because the constant term 1 is a cosine term corresponding to $n = 0$. Also observe that $f(x)$ here is neither an odd function nor an even one.

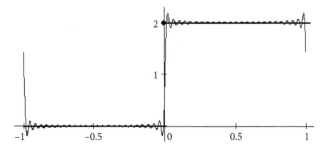

Fig. 8.26 A higher-order partial sum of the Fourier series near a jump

in the middle section, but it does not remove the discrepancies near $x = 0, \pm 1$, which persist no matter how large a partial sum we use.

> **The Gibbs phenomenon.** *The persistent discrepancy is not limited to this particular function $f(x)$ and appears in the series for other discontinuous functions. It even has a name: the "Gibbs phenomenon," named after the American scientist Josiah Willard Gibbs who discovered it in 1899. The issue had been noted and explained in 1848 by the English mathematician Henry Wilbraham, but Gibbs seems to have been unaware of Wilbraham's work.*

Evidently, something important is going on that shows up as discrepancies that we observe in Figure 8.26. If we attribute the discrepancy at $x = 0$ to the discontinuity somehow, then why does the same thing seem to be happening at the endpoints $x = \pm 1$ where there are no discontinuities in $f(x)$?

A partial understanding comes from the nature of uniform convergence. If a trigonometric series converges uniformly, then we know from earlier discussion that its limit function must be continuous. It follows that *if a trigonometric series converges to a (periodic) function that is not continuous, then the convergence is not uniform*. This non-uniform convergence shows up as the Gibbs phenomenon, like the discrepancies that we see in Figure 8.26.

8.7 The Weierstrass approximation theorem

In this concluding section of the chapter, we prove another remarkable result by Weierstrass. To see why this result is remarkable, we go back to the Taylor series where we showed that a C^∞ function $f(x)$ for x in a closed and bounded interval $[a, b]$ may be expressed as a uniform limit of the sequence of Taylor polynomials $T_n(x)$ centered about some point c in $[a, b]$.

If $f(x)$ is continuous but not a C^∞ function, then the Taylor series cannot be defined for it. We might in this case conclude that such a function cannot be the uniform limit of a sequence of polynomials, and we need to expand $f(x)$ as a Fourier series using trigonometric sums rather than polynomials. This approach is a viable alternative, but the conclusion that polynomials don't work is false.

As long as we stay with a closed and bounded interval $[a, b]$, it is indeed possible to express any function that is continuous on $[a, b]$ as a uniform limit of a sequence of polynomials. This is basically what Weierstrass proved.

We may compare this fact with the *density* of rational numbers among the real numbers, which meant essentially that every real number can be approximated by a sequence of rational numbers.

If we replace "rational number" with "polynomial" and "real number" with "continuous function," then the approximation theorem implies that *polynomials are dense in the space of continuous functions on* $[a, b]$ with respect to the uniform (or sup-norm) metric (an extension of the concept of absolute value to spaces of functions). Further pursuit of this idea can be found in a more advanced analysis book.

We present a proof of Weierstrass's theorem given by Serguei N. Bernstein (1880–1968) because Bernstein's proof actually supplies the approximating polynomial sequence.

Given a continuous function $f(x)$, the corresponding *Bernstein polynomial* of degree n is defined as

$$P_n(x) = \sum_{k=0}^{n} f\left(\frac{k}{n}\right) \binom{n}{k} x^k (1-x)^{n-k}, \qquad x \in [0, 1] \qquad (8.120)$$

The above definition is based on the interval $[0,1]$, but the generalization to arbitrary intervals is straightforward, and we explain how later.

The dependence of P_n on f is through the coefficients $f(k/n)$, which are obtained by partitioning the interval $[0,1]$ into k subintervals of equal length with endpoints at

$$0, \frac{1}{n}, \frac{2}{n}, \ldots, \frac{n-1}{n}, 1 \qquad (8.121)$$

and then calculating the values of $f(x)$ at these $n+1$ partition points. The resulting numbers give the coefficients $f(k/n)$. The remaining factors are independent of f.

For example, if $f(x) = \cos(3x)$, then the Bernstein polynomial of degree 2 is defined by partitioning the interval $[0,1]$ into two subintervals with endpoints 0, 1/2, 1; then calculating $\cos 0$, $\cos(3/2)$, and $\cos 3$; and finally filling in the rest of the expression in (8.120):

$$P_2(x) = \sum_{k=0}^{2} \cos\left(\frac{3k}{2}\right)\binom{2}{k}x^k(1-x)^{2-k}$$

$$= (1-x)^2 + \cos\left(\frac{3}{2}\right)\binom{2}{1}x(1-x) + \cos\left(3\right)x^2$$

$$= \left(1 + \cos 3 - 2\cos\frac{3}{2}\right)x^2 - 2\left(1 - \cos\frac{3}{2}\right)x + 1$$

It is useful to compare the polynomials in (8.120) with the Taylor polynomials $T_n(x)$ to see in what ways they are different. For $\cos(3x)$ the Taylor polynomial of degree 2 (order 2 or 3) is

$$T_2(x) = 1 - \frac{(3x)^2}{2!} = 1 - \frac{9}{2}x^2$$

Clearly we have two very different types of polynomials. In general, the essential differences between the Taylor polynomials T_n and the polynomials P_n are as follows:

(i) To define T_n, we need to know only one value of $f(x)$ at the center $x = c$; but to define P_n, we need to know the values of $f(x)$ for all $n + 1$ points that are listed in (8.121).

(ii) To define T_n, we need to also know the values of all the higher derivatives $f^{(k)}(c)$ for $k = 1, 2, ..., n$, but P_n requires no derivative values of f.

From these observations, we gather that when we use Bernstein's polynomials, the $n+1$ bits of information that are needed are the function values at $n+1$ points; whereas a Taylor polynomial of order n requires the value of the function and its first n higher-order derivatives, also $n + 1$ bits of information.

It follows that for a continuous function $f(x)$ that is not C^∞ we cannot construct a convergent Taylor series, but a sequence of Bernstein polynomials $P_1, P_2, P_3, ...$ is well-defined.

The question is whether this sequence converges uniformly to f on a closed and bounded interval. If so, then P_n can be used to estimate the values of $f(x)$ for x in-between the points in (8.121) to an arbitrary degree of accuracy. The approximation theorem below proves the uniform convergence, so we indeed have a uniform way of estimating the value of $f(x)$ over a closed and bounded interval.

Before proving the approximation theorem, it is convenient to first collect a few preliminary facts in the form of a lemma.

Lemma 336 For every number $x \in [0,1]$ and each positive integer n, the following are true:

(a)
$$\sum_{k=0}^{n} \binom{n}{k} x^k (1-x)^{n-k} = 1 \tag{8.122}$$

(b)
$$\sum_{k=0}^{n} \frac{k}{n} \binom{n}{k} x^k (1-x)^{n-k} = x \tag{8.123}$$

(c) If $n \geq 2$, then

$$\sum_{k=0}^{n} \frac{k(k-1)}{n(n-1)} \binom{n}{k} x^k (1-x)^{n-k} = x^2 \tag{8.124}$$

(d) If $n \geq 2$, then

$$\sum_{k=0}^{n} \left(x - \frac{k}{n} \right)^2 \binom{n}{k} x^k (1-x)^{n-k} = \frac{x(1-x)}{n} \tag{8.125}$$

Proof (a) The identity in (8.122) is quickly obtained by setting $a = x$ and $b = 1-x$ in Theorem 151 (binomial theorem, Chapter 5).

(b) and (c) may be derived from (a); see Exercise 373.

(d) We simplify the left-hand side of (8.125):

$$\sum_{k=0}^{n} \left(x - \frac{k}{n} \right)^2 \binom{n}{k} x^k (1-x)^{n-k} = \sum_{k=0}^{n} \left(x^2 - \frac{2k}{n} x + \frac{k^2}{n^2} \right) \binom{n}{k} x^k (1-x)^{n-k}$$

Now using (8.122) and (8.123), we can simplify the right-hand side of the above equality to obtain

$$\sum_{k=0}^{n} \left(x - \frac{k}{n} \right)^2 \binom{n}{k} x^k (1-x)^{n-k} = x^2 - 2x^2 + \sum_{k=0}^{n} \frac{k^2}{n^2} \binom{n}{k} x^k (1-x)^{n-k} \tag{8.126}$$

Next, pulling out an n from the last term on the right-hand side above, multiplying by $n-1$, and using (8.123) and (8.124), we obtain

$$\sum_{k=0}^{n} \frac{k^2}{n^2} \binom{n}{k} x^k (1-x)^{n-k} = \frac{n-1}{n} \sum_{k=0}^{n} \frac{k^2}{n(n-1)} \binom{n}{k} x^k (1-x)^{n-k}$$

$$= \frac{n-1}{n} \sum_{k=0}^{n} \frac{k(k-1)+k}{n(n-1)} \binom{n}{k} x^k (1-x)^{n-k}$$

$$= \frac{n-1}{n} \left(x^2 + \frac{x}{n-1} \right)$$

Inserting the last quantity on the right-hand side above into the right-hand side of (8.126) and simplifying the resulting expression proves (8.125). ∎

Theorem 337 *(The Weierstrass approximation theorem) Let $f(x)$ be a continuous function for all x in a closed and bounded interval $[a, b]$. For every $\varepsilon > 0$, there is a polynomial $P(x)$ such that*

$$|f(x) - P(x)| < \varepsilon \quad \text{for all } x \in [a, b] \tag{8.127}$$

In the sup-norm notation,

$$\|f - P\| < \varepsilon$$

Proof   First we prove the theorem for the interval $[0,1]$ and then extend the result to the given interval $[a, b]$.

Recall from the uniform continuity theorem that f is uniformly continuous on $[0,1]$; so for a given $\varepsilon > 0$, there is a $\delta > 0$ such that for all $u, v \in [0, 1]$

$$|f(u) - f(v)| < \frac{\varepsilon}{2} \quad \text{if } |u - v| < \delta \tag{8.128}$$

Further, the boundedness theorem, or the extreme value theorem, imply that f is bounded by some constant $M > 0$, i.e.,

$$|f(x)| \leq M \quad \text{for all } x \in [0, 1] \tag{8.129}$$

Now we show that $P(x) = P_n(x)$ for n that is sufficiently large, then (8.127) holds.

For each $x \in [0, 1]$ and k such that $1 \leq k \leq n$ if $|x - k/n| < \delta$, then by (8.128), $|f(x) - f(k/n)| < \varepsilon/2$. Otherwise, if $|x - k/n| \geq \delta$, then

$$\left(x - \frac{k}{n} \right)^2 \geq \delta^2 \tag{8.130}$$

by (8.129) and the triangle inequality

$$\left| f(x) - f\left(\frac{k}{n} \right) \right| \leq |f(x)| + \left| f\left(\frac{k}{n} \right) \right| \leq 2M$$

so using (8.130)

$$\left| f(x) - f\left(\frac{k}{n} \right) \right| \leq \frac{2M}{\delta^2} \delta^2 \leq \frac{2M}{\delta^2} \left(x - \frac{k}{n} \right)^2$$

Therefore, for $0 \leq k \leq n$

$$\left| f(x) - f\left(\frac{k}{n} \right) \right| < \frac{\varepsilon}{2} + \frac{2M}{\delta^2} \left(x - \frac{k}{n} \right)^2 \tag{8.131}$$

Now by (8.122)

$$f(x) = \sum_{k=0}^{n} f(x) \binom{n}{k} x^k (1 - x)^{n-k}$$

so

$$f(x) - P_n(x) = \sum_{k=0}^{n} \left[f(x) - f\left(\frac{k}{n}\right) \right] \binom{n}{k} x^k (1 - x)^{n-k}$$

Using the triangle inequality, (8.131), (8.122), and (8.125), in that order

$$|f(x) - P_n(x)| \le \sum_{k=0}^{n} \left| f(x) - f\left(\frac{k}{n}\right) \right| \binom{n}{k} x^k (1 - x)^{n-k}$$

$$\le \sum_{k=0}^{n} \left| \frac{\varepsilon}{2} + \frac{2M}{\delta^2} \left(x - \frac{k}{n} \right)^2 \right| \binom{n}{k} x^k (1 - x)^{n-k}$$

$$= \frac{\varepsilon}{2} + \frac{2M}{\delta^2} \sum_{k=0}^{n} \left(x - \frac{k}{n} \right)^2 \binom{n}{k} x^k (1 - x)^{n-k}$$

$$= \frac{\varepsilon}{2} + \frac{2M}{\delta^2 n} x(1 - x)$$

Therefore, if $0 \le x \le 1$, then

$$|f(x) - P_n(x)| \le \frac{\varepsilon}{2} + \frac{2M}{\delta^2 n}$$

Now we must choose n large enough that the second fraction on the right-hand side above is less than $\varepsilon/2$. This is possible (by the Archimedean property of $\mathbb{R}$); in fact, if we choose

$$n > \frac{2M}{\delta^2 \varepsilon}$$

then n is large enough that (8.127) holds, as we had claimed, and the proof for the interval $[0,1]$ is complete.

To complete the proof, consider the original interval $[a, b]$ and the function

$$g(t) = a + (b - a)t, \quad t \in [0, 1]$$

Then g is continuous bijection of $[0,1]$ onto $[a, b]$. Thus, the composition $f \circ g$ is continuous on $[0, 1]$; and by the above argument, there is a Bernstein polynomial P_n that uniformly approximates $f \circ g$; i.e., for each $\varepsilon > 0$

$$|f(g(t)) - P_n(t)| < \varepsilon, \quad t \in [0, 1] \tag{8.132}$$

Now since g is onto $[a, b]$, for each $x \in [a, b]$ there is $t \in [0, 1]$ such that $a + (b - a)t = x$ or

$$t = \frac{x - a}{b - a}$$

Consider the polynomial

$$P(x) = P_n\left(\frac{x - a}{b - a}\right), \quad x \in [a, b]$$

Then (8.132) shows that (8.127) holds with this $P(x)$, and the proof is complete. ∎

An interesting corollary of Theorem 337 is the following:

Corollary 338 *The continuous, nowhere differentiable function $W(x)$ can be uniformly approximated by polynomials on every closed and bounded interval, e.g., $[-\pi, \pi]$, which contains one period of $W(x)$.*

Whereas previously $W(x)$ could be the uniform limit of trigonometric series only, now we can think of it also as a uniform limit of a sequence of polynomials. Both the trigonometric partial sums and the polynomials are sequences of C^∞ functions that converge uniformly to the nowhere differentiable function $W(x)$ on closed and bounded intervals.

Figure 8.27 illustrates approximations of $\cos(3x)$ over the interval $[0,1]$ by the polynomial $P_6(x)$ in the left-hand panel and the Taylor polynomial $T_6(x)$ in the right-hand panel. In both panels, $\cos(3x)$ is the dashed curve.

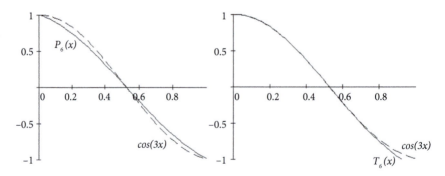

Fig. 8.27 Approximations by Bernstein polynomial vs. Taylor polynomial

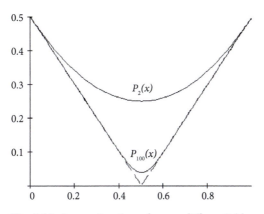

Fig. 8.28 Approximation of a non-differentiable function by Bernstein polynomials

We now calculate a Bernstein polynomial for $f(x) = |x - 1/2|$, which is continuous on $[0,1]$ but not differentiable at $x = 1/2$.

$$P_2(x) = \sum_{k=0}^{2} \left| \frac{k}{2} - \frac{1}{2} \right| \binom{2}{k} x^k (1-x)^{2-k}$$

$$= \frac{1}{2}(1)(1-x)^2 + 0 + \frac{1}{2}(1)x^2$$

$$= x^2 - x + \frac{1}{2}$$

Figure 8.28 shows the graph of $P_2(x)$ as well the graph of $f(x)$ (dashed). For comparison, we also see a computer-generated graph of $P_{100}(x)$ that more closely approximates $f(x)$.

We close by pointing out that the Weierstrass approximation theorem is not true if the interval is not closed or not bounded. We can trace the restriction on the interval to the uniform continuity theorem in the proof above. Thus, for example, we cannot apply Theorem 337 to the function $1/x$ on the interval $(0, 1]$ or the function e^x on $[0, \infty)$.

8.8 Exercises

Exercise 339 *To obtain a more extreme example, modify the sequence of f_n in (8.14) and Figure 8.6 so that the height of triangle n is n^2 instead of $2n$. Show that the sequence of integrals changes to*

$$\int_0^1 f_1(x)dx, \quad \int_0^1 f_2(x)dx, \ldots, \quad \int_0^1 f_n(x)dx, \ldots \;=\; \frac{1}{2}, 1, \frac{3}{2}, 2, \ldots, \frac{n}{2}, \ldots$$

which diverges to infinity. Nevertheless, show that the pointwise limit is still the zero function, so its integral is also still 0.

Exercise 340 *Consider the sequence of power functions $f_n(x) = x^n$. Note that f_n is differentiable, hence continuous on $(-\infty, \infty)$ for every n.*

(a) *Prove that f_n converges pointwise to the zero function $f(x) = 0$ for $x \in [0, 1)$.*

(b) *Calculate the ε-index N_ε for the interval $[0, 1)$ and show that it diverges to infinity as $x \to 1^-$.*

(c) *Prove that on the interval $[0, 1]$ the sequence f_n converges pointwise to the discontinuous function*

$$f(x) = \begin{cases} 0 & \text{if } 0 \le x < 1 \\ 1 & \text{if } x = 1 \end{cases}$$

but f_n does not converge pointwise on the interval $[-1, 1]$.

(d) *If a is a fixed real number in $(0, 1)$, then no matter how close a is to 1, the sequence f_n converges uniformly on $[-a, a]$ to the zero function. Prove that the ε-index N_ε is a constant on such an interval.*

Exercise 341

(a) *Prove that the inverse tangent sequence $f_n(x) = \tan^{-1}(nx)$ converges uniformly to the constant function $f(x) = \pi/2$ on the interval $[a, \infty)$ if $a > 0$ no matter how close a may be to 0. Decide whether you would use Theorem 310 in this case or use a formula for the ε-index following the discussion in the text.*

(b) *Prove that*

$$\lim_{n\to\infty} \lim_{k\to\infty} \tan^{-1}(nx_k) = \lim_{k\to\infty} \lim_{n\to\infty} \tan^{-1}(nx_k) = \frac{\pi}{2}$$

for every convergent sequence x_k in $[a, \infty)$. You may start by assuming that $x_k \to b$ for some unspecified number $b \ge a$.

(c) *Let $f_n(x) = \tan^{-1}(n \sin x)$. Explain why f_n is differentiable (hence continuous) on the interval $[0, \infty)$ for every n, and find the pointwise limit $f(x)$.*

Are there values of a such that this sequence converges uniformly to f on $[a, \infty)$?

Exercise 342

(a) Use Theorem 310 to prove that the sequence $f_n(x) = x/n$ converges uniformly to the constant function $f(x) = 0$ on the interval $[-b, b]$ for every positive real number b, no matter how large.

(b) Show that this sequence converges pointwise but not uniformly on $(-\infty, \infty)$. Calculate the formula for N_ε to show that it depends on x on $(-\infty, \infty)$ but not on $[-b, b]$.

Exercise 343

(a) Show that the sequence $f_n(x) = nx/(n + x)$ converges pointwise but not uniformly to the constant function $f(x) = x$ on $[0, \infty)$.

(b) Use Theorem 310 to prove that the convergence is uniform on the interval $[0, b]$ for every positive real number b, no matter how large.

(c) Find the pointwise limit g of the sequence of functions $g_n(x) = nx/(1 + nx)$ on $[0, \infty)$. Does g_n converge to g uniformly on $[0, \infty)$? If not on $[0, \infty)$, then how about $[0, b]$ for some $b > 0$? What about $(0, \infty)$? Explain.

Exercise 344 Prove that each of the following series converges uniformly on the indicated set:

$$(a) \quad \sum_{n=1}^{\infty} \frac{2x + 1}{x + n^2}, \quad x \in [0, b], \text{ for any } b > 0$$

$$(b) \quad \sum_{n=1}^{\infty} \frac{\sin(nx)}{n\sqrt{n}}, \quad x \in (-\infty, \infty)$$

Exercise 345 Prove that the infinite series in Exercise 344(a) converges pointwise on $[0, \infty)$, and let $f(x)$ be the sum of the series. Prove that $f(x)$ is continuous on $[0, \infty)$ using the fact that the number b in Exercise 344(a) can be as large as we like.

Exercise 346

(a) Prove that the geometric power series

$$1 + x^2 + x^4 + x^6 + \cdots = \frac{1}{1 + x^2}$$

converges uniformly on the interval $[-r, r]$ for every $r \in (0, 1)$.

(b) Use the series in (a) and term by term integration to obtain a power series for the inverse tangent function $\tan^{-1} x$ for $x \in (-1, 1)$.

Exercise 347 Prove Corollary 322.

Exercise 348 Consider the geometric power series $1 + x + \cdots = 1/(1-x)$ for $x \in (-1, 1)$.

(a) Prove that this series is integrable and differentiable term by term on the interval $[-r, r]$ for all $r \in (0, 1)$, and establish the formulas

$$\sum_{k=1}^{\infty} \frac{(-1)^{k+1} r^k}{k} = r - \frac{r^2}{2} + \frac{r^3}{3} - \frac{r^4}{4} + \cdots = \ln(r+1)$$

$$\sum_{k=1}^{\infty} k r^{k-1} = 1 + 2r + 3r^2 + 4r^3 + \cdots = \frac{1}{(1-r)^2}$$

(b) Use the formulas in (a) to prove the following equalities:

$$\sum_{k=1}^{\infty} \frac{(-1)^{k+1}}{k 2^k} = \frac{1}{2} - \frac{1}{2(2^2)} + \frac{1}{3(2^3)} - \frac{1}{4(2^4)} + \cdots = \ln 3 - \ln 2$$

$$\sum_{k=1}^{\infty} \frac{k}{2^k} = \frac{1}{2} + \frac{2}{2^2} + \frac{3}{2^3} + \frac{4}{2^4} + \cdots = 2$$

$$\sum_{k=1}^{\infty} \frac{(-1)^k k}{2^k} = -\frac{1}{2} + \frac{2}{2^2} - \frac{3}{2^3} + \frac{4}{2^4} + \cdots = -\frac{2}{9}$$

Exercise 349 Find the radius and the interval of convergence for each of the following power series:

(a) $\displaystyle\sum_{n=0}^{\infty} \frac{n}{2^n} (x-1)^n$ (b) $\displaystyle\sum_{n=0}^{\infty} 2^n (x-1)^n$ (c) $\displaystyle\sum_{n=0}^{\infty} 2^{n^2} (x-1)^n$

Exercise 350 By re-casting the series $\sum_{n=0}^{\infty} 2^n (x-1)^n$ as a geometric series, find a formula for its sum. Also find a formula for $\sum_{n=1}^{\infty} n 2^n (x-1)^{n-1}$ by writing it in terms of the derivative of a geometric series. Use similar ideas to find a formula for each of the following series, and specify the interval of convergence:

(a) $\displaystyle\sum_{n=1}^{\infty} \frac{n}{2^n} (x-1)^{n-1}$,   (b) $\displaystyle\sum_{n=1}^{\infty} \frac{2^n}{n} (x-1)^n$

Exercise 351 *When paired with the properties of the logarithm, the series (8.67) or its equivalent centered at 0 can be used to estimate the numerical values of the logarithms of (positive) real numbers. If $x > 1$, the property $\ln(1/x) = -\ln x$ extends the usefulness of (8.67) to all positive real numbers. For instance,*

$$\ln 2 = -\ln \frac{1}{2}$$

and the series for $\ln(1/2)$ in (8.67) converges faster than the alternating harmonic series above for $\ln 2$ because $1/2$ is in the interval of absolute convergence, and also it is closer to the center than 2 is.

(a) *Estimate the following logarithms using the first four terms of the series in (8.67):*

$$\ln 3 \qquad \ln 4.5$$

(b) *If you were to use a few terms of the series in (8.67) to estimate $\ln 15$, which of the following methods do you think gives better values, using the same number of terms of the series in each case?*

$$\ln 15 = -\ln \frac{1}{15}$$

$$\ln 15 = \ln 3 + \ln 5 = -\ln \frac{1}{3} - \ln \frac{1}{5}$$

You can check by adding the first four terms of (8.67) and compare the answers to what a calculator gives for $\ln 15$. Can you explain why the second way gives consistently better results no matter how many terms of the series you use? Figure 8.18 offers some insights!

Exercise 352 *Prove Corollaries 325 and 326.*

Exercise 353 *Find the radius and interval of convergence of each of the following power series:*

(a) $\displaystyle\sum_{n=0}^{\infty} \frac{x^{2n}}{3^n}$ 　　(b) $\displaystyle\sum_{n=0}^{\infty} \frac{x^{3n}}{n!}$ 　　(c) $\displaystyle\sum_{n=0}^{\infty} \frac{x^{n^2}}{2^n}$

Note that the formula in Theorem 324 is not applicable in the stated form; see the derivation for the series (8.59) in the text. Consider using the ratio test with factorial coefficients.

Exercise 354 *Use the definition of derivative to prove that $f'(0) = 0$ for the function*

$$f(x) = x^2 \sin \frac{1}{x} \quad \text{if } x \neq 0, \quad f(0) = 0$$

Exercise 355 *Let x_n and y_n be sequences of non-negative real numbers with $x = \lim \sup_{n \to \infty} x_n \leq \infty$, and $y = \lim \sup_{n \to \infty} y_n \leq \infty$.*

(a) $\lim \sup_{n \to \infty}(x_n y_n) \leq xy$ *assuming that xy is not a $0 \cdot \infty$ form.
Note that the case where $x = \infty$ or $y = \infty$ is the easy case; also note that
the inequality is generally strict; consider $x_n = 2 + (-1)^n$ and $y_n = x_{n+1}$.*
(b) *If x_n has a limit, and $x = \lim_{n \to \infty} x_n > 0$, then $\lim \sup_{n \to \infty}(x_n y_n) = xy$.*

Exercise 356

(a) *Complete the calculations for the derivatives of Cauchy's function; to derive
the zero value of the derivative at $x = 0$, you need to prove that for all n*

$$\lim_{x \to 0} \frac{1}{x^n} e^{-1/x^2} = 0$$

(b) *Relevant to the Taylor series, prove that there is no single number $M > 0$
such that for all n*

$$f^{(n)}(x) \leq M^n \quad \text{for all } x$$

*This can be deduced directly from a general formula for the nth deriva-
tive, which can then be used to calculate the local maxima and minima of
$f^{(n-1)}(x)$.*

Exercise 357

(a) *Prove that the Maclaurin series of the polynomial $P(x) = a_0 + a_1 x + a_2 x^2 + \cdots + a_n x^n$ is $P(x)$ itself.*
(b) *Calculate the Taylor polynomial of order 4 for the polynomial $x^2 - 3x + 2$
centered at $c = -1$.*
(c) *Prove that the Taylor series of the polynomial in (a) about any center c is
a polynomial of degree n (if $a_n \neq 0$).*

Exercise 358

(a) *Find the Taylor polynomial of order 4 centered at $c = \pi/2$ for $\sin x$; repeat
for $c = \pi/4$.*
(b) *Find the Taylor polynomial of order 4 centered at $c = \ln 2$ for e^x; repeat for
$c = -\ln 2$.*

Exercise 359

(a) *By finding the coefficients as we did in the text for* $\sin x$, *show that the Maclaurin series for* $\cos x$ *is the following, with interval of convergence* $(-\infty, \infty)$:

$$\cos x = 1 - \frac{x^2}{2!} + \frac{x^4}{4!} - \frac{x^6}{6!} + \cdots = \sum_{n=0}^{\infty} \frac{(-1)^n x^{2n}}{(2n)!} \qquad (8.133)$$

(b) *Alternatively, derive the above series expansion for* $\cos x$ *the quick way by taking the derivative of the series in (8.79) in the text term by term.*

Exercise 360 *If we substitute* $-x^2$ *for* u *in the Taylor series for* e^u *centered at 0, then we obtain the power series*

$$1 - \frac{x^2}{1!} + \frac{x^4}{2!} - \frac{x^6}{3!} + \cdots = \sum_{n=0}^{\infty} \frac{(-1)^n x^{2n}}{n!} \qquad (8.134)$$

(a) *Prove that the above series in fact converges to* e^{-x^2} *on* $(-\infty, \infty)$ *and is therefore its Taylor series. Note that this can be done either using Theorem 331 or by proving that the series in (8.134) converges on* $(-\infty, \infty)$ *using the ratio test, and its partial sums are the Taylor polynomials of* e^{-x^2} *centered at 0.*

(b) *Explain why the series in (8.134) can be integrated term by term to give a power series for the integral*

$$\int_0^a e^{-x^2} dx$$

Then use the first four terms of your power series to give an estimate of the integral for $a = 1$. *What is the maximum error of this approximation? Note that this can be found using either the remainder term or the alternating series properties.*

Exercise 361 *Use (8.82) to determine what order Taylor polynomial for* e^x *ensures an estimate of* $\sqrt{e} = e^{1/2}$ *that is accurate to 16 or more decimal places.*

Exercise 362 *Prove that* e^2 *is irrational using an argument similar to the one we used in the text to prove that* e *is irrational. Explain why the irrationality of* e *and* e^2 *imply that* $\sqrt{e}$ *and* $e^{2/3}$ *must also be irrational.*

Exercise 363

(a) Find the Taylor polynomials $T_n(x)$ for $n = 1, 2, 3, 4$ for $f(x) = e^{\sin x}$ centered at 0.

(b) Use $T_4(1/2)$ to obtain an estimate of $e^{\sin(1/2)}$. Also find an upper bound for the error of this approximation using the remainder.

(b) Prove that the Taylor polynomials converge to a power series for $e^{\sin x}$ on $(-\infty, \infty)$. It is not necessary to obtain the complete Taylor series.

Exercise 364 Complete the proof of Theorem 334 by proving Case 2.

Exercise 365

(a) Find the binomial series for $\sqrt{1 + x}$.

*(b) Use the first four terms of your series in (a) to estimate $\sqrt{1.1}$.

(c) Determine how accurate your estimate in (a) is by calculating an error bound using a method that we have discussed earlier.

Exercise 366 Calculate the other series of constants in (8.106).

Exercise 367 Let the Weierstrass function $W(x)$ be defined as the uniform limit of the trigonometric series

$$W(x) = \sum_{k=0}^{\infty} c^k \cos(m^k \pi x)$$

where $0 < c < 1$, and m is an odd positive integer that satisfies the inequality in (8.43).

(a) Show that $W(x)$ is periodic with period 2 and calculate the values of $W(0)$, $W(1/2)$, and $W(1)$.

(b) Recall that $W'(x)$ is not defined for any value of x. But explain why $W(x)$ is integrable, and show that

$$\int_{-1}^{1} W(x)dx = \int_{0}^{1} W(x)dx = 0$$

Also prove that for all $r \in [-1, 1]$

$$\left| \int_{0}^{r} W(x)dx \right| \leq \frac{1}{1 - c}$$

Exercise 368 (*a*) *Prove that the following trigonometric series converges uniformly on* $(-\infty, \infty)$ *to a continuous function with period* 2π:

$$\sum_{n=0}^{\infty} \frac{\sin^n(nx)}{2^n}$$

(*b*) *Let* $f(x)$ *be the function that the series in* (*a*) *converges to. Calculate the exact value of* $f(x)$ *for* $x = 0, \pi/2$.

Exercise 369 (*a*) *Prove the second orthogonality identity in* (8.110).
(*b*) *Derive the formula for* b_n *in* (8.114).

Exercise 370 (*a*) *Determine the Fourier series for* $f(x) = |x|$ *on the interval* $[-2, 2]$ ($L = 2$).
(*b*) *Determine the Fourier series for* $f(x) = x$ *on the interval* $[-1, 1]$ ($L = 1$).

Exercise 371 (*a*) *Determine the Fourier series for* $f(x) = x^2/2$ *on the interval* $[-\pi, \pi]$.
(*b*) *Use your series to prove that*

$$\sum_{n=1}^{\infty} \frac{(-1)^n}{n^2} = \frac{\pi^2}{12}$$

Exercise 372 *Find the Fourier expansion of the discontinuous function* $f(x)$ *in* (8.119) *over the interval* $[-2, 2]$.

Exercise 373 *Prove identities* (8.123) *and* (8.124) *in Lemma 336.*
Note: First show that

$$\binom{n-1}{k-1} = \frac{k}{n}\binom{n}{k} \qquad \text{if } 1 \le k \le n$$

$$\binom{n-2}{k-2} = \frac{k(k-1)}{n(n-1)}\binom{n}{k} \qquad \text{if } 2 \le k \le n$$

Next, replace n *by* $n - 1$ *in* (8.122) *and multiply by* x *to obtain* (8.123). *Use a similar procedure to prove* (8.124).

Exercise 374 *For each of the functions below, determine the explicit formulas for the Bernstein polynomial $P_2(x)$ with $x \in [0, 1]$:*

$$(a) \quad \sqrt{x} \qquad\qquad (b) \quad \ln(x + 1)$$

Exercise 375 *Consider the function $f(x) = |x - 1/2|$ that is mentioned in the text. Use the proof of the Weierstrass approximation theorem to determine P_n explicitly for some n large enough that $|f(x) - P_n| < 1/4$.*

9
Appendices

9.1 Appendix: Cantor's construction, additional detail

The material in this section supplements the construction of real numbers that we discussed in Chapter 4.

Theorem 376 *If a sequence q_n of rational numbers converges to a rational number q, then q_n is a Cauchy sequence.*

Proof First using the triangle inequality (Lemma 54 in Chapter 3), we have

$$|q_n - q_m| = |q_n - q + q - q_m| \leq |q_n - q| + |q - q_m|$$

Since $q_n \to q$, we can say that for every (rational) $\varepsilon > 0$, there is an integer index N large enough that both $|q_n - q| < \varepsilon/2$ and $|q - q_m| < \varepsilon/2$ if $m, n > N$. For such large index values then it is true that

$$|q_n - q_m| < \frac{\varepsilon}{2} + \frac{\varepsilon}{2} = \varepsilon$$

As ε can be arbitrarily small, we conclude that $|q_n - q_m| \to 0$, and q_n is Cauchy. ∎

Theorem 1 can be stated equivalently (in contrapositive) as

If a sequence is not Cauchy, then it does not converge.

This shows why we need to consider *Cauchy* sequences in building the real numbers. The next result tells us what happens if a Cauchy sequence of rational numbers does *not* converge to 0.

Theorem 377 *Let q_n be a sequence of rational numbers that does not converge to 0. Then the sequence is either eventually positive or eventually negative.*

Proof Because q_n does not converge to zero, there is $\varepsilon_0 > 0$ such that for every positive integer K, there is $n \geq K$ such that $|q_n| \geq \varepsilon_0$. But q_n is also Cauchy,

Real Analysis and Infinity. Hassan Sedaghat, Oxford University Press.
© H. Sedaghat (2022). DOI: 10.1093/oso/9780192895622.003.0009

so for each $\varepsilon > 0$, there is an index N such that $|q_n - q_m| < \varepsilon$ if $m, n > N$. Now suppose that q_n and q_m have opposite signs (they are on opposite sides of 0) for a pair of indices $m, n > N$. Then without loss of generality, assume that $q_m < 0 \leq q_n$. Then

$$|q_n - q_m| = q_n - q_m \geq q_n$$

By taking $K = N$, we may also assume that $q_n = |q_n| \geq \varepsilon_0$. Now if we choose the arbitrary ε to be less than ε_0, say, $\varepsilon = \varepsilon_0/2$, then we arrive at

$$\varepsilon_0 \leq q_n \leq |q_n - q_m| < \varepsilon = \frac{\varepsilon_0}{2}$$

which is impossible. Thus q_n and q_m must be on the same side of zero eventually, as claimed.   ∎

The next result states another property of all Cauchy sequences. The proof assumes that the sequence is rational, but this assumption has no bearing on the proof.

Theorem 378 *Every Cauchy sequence is bounded.*

Proof We start the proof by recalling that a sequence q_n is bounded if there is a rational number $M > 0$ such that $|q_n| \leq M$ for every index n. Now suppose that q_n is a (rational) Cauchy sequence. Then there is an index N large enough that $|q_n - q_m| < 1$ for all $m, n > N$ (pick $\varepsilon = 1$). In particular, for $m = N + 1$, we get

$$|q_n - q_{N+1}| < 1 \qquad \text{if } n > N$$

Since we can write, using the triangle inequality,

$$|q_n| = |q_n - q_{N+1} + q_{N+1}| \leq |q_n - q_{N+1}| + |q_{N+1}| < 1 + |q_{N+1}|$$

and this is true for every $n > N$, it follows that for all $n \geq 1$

$$|q_n| \leq \max\{|q_1|, |q_2|, ..., |q_N|, 1 + |q_{N+1}|\}$$

To complete the proof, we need only define M to be the quantity to the right side of the above inequality.   ∎

Both of the preceding theorems and their proofs are valid for all real numbers; but since the real numbers are not yet defined, we assume that all numbers are rational.

$[\mathcal{C}]$ is an algebraic field.

Consider the set $[\mathcal{C}]$ of all equivalence classes of rational Cauchy sequences as described in the text and define two operations: for each pair of equivalence classes $[q_n]$ and $[q'_n]$

$$[q_n] + [q'_n] = [q_n + q'_n] \qquad \text{(addition)}$$

$$[q_n][q'_n] = [q_n q'_n] \qquad \text{(multiplication)}$$

On the right-hand sides of the above equalities, we have the equivalence classes of the sum or product of two rational Cauchy sequences. Specifically, given any pair of sequences $q_n = q_1, q_2, q_3, \ldots$ and $q'_n = q'_1, q'_2, q'_3, \ldots$ in $[\mathcal{C}]$, we have

$$q_n + q'_n = (q_1, q_2, q_3, \ldots) + (q'_1, q'_2, q'_3, \ldots) = q_1 + q'_1, q_2 + q'_2, q_3 + q'_3, \ldots$$

$$q_n q'_n = (q_1, q_2, q_3, \ldots)(q'_1, q'_2, q'_3, \ldots) = q_1 q'_1, q_2 q'_2, q_3 q'_3, \ldots$$

Since we are dealing with equivalence classes rather than individual sequences, it is necessary to ensure that these operations are well-defined functions in the sense that *the result of addition or multiplication does not depend on which sequences in the equivalence classes we choose to add or multiply.*

Theorem 379 *If p_n and p'_n are Cauchy sequences of rationals such that $[q_n] = [p_n]$ and $[q'_n] = [p'_n]$, then $[q_n + q'_n] = [p_n + p'_n]$ and $[q_n q'_n] = [p_n p'_n]$. Therefore, addition and multiplication in $[C]$ are well-defined.*

Proof Note that $[q_n] = [p_n]$ implies $|q_n - p_n| \to 0$ as $n \to \infty$; similarly, from $[q'_n] = [p'_n]$, it follows that $|q'_n - p'_n| \to 0$ as $n \to \infty$. Therefore, for every (rational) $\varepsilon > 0$, there is a large enough index N such that $|q_n - p_n| < \varepsilon/2$, and $|q'_n - p'_n| < \varepsilon/2$.

$$|(q_n + q'_n) - (p_n + p'_n)| = |(q_n - p_n) + (q'_n - p'_n)| \le |q_n - p_n| + |q'_n - p'_n| < \frac{\varepsilon}{2} + \frac{\varepsilon}{2} = \varepsilon$$

where the first equality is implied by the triangle inequality. Since ε can be arbitrarily small, we have shown that

$$|(q_n + q'_n) - (p_n + p'_n)| \to 0 \quad \text{as } n \to \infty$$

This means that the sequences $q_n + q'_n$ and $p_n + p'_n$ are equivalent, which is saying the same thing as $[q_n + q'_n] = [p_n + p'_n]$. For products, a little more work is needed; we start by writing

$$|q_n q'_n - p_n p'_n| = |q_n q'_n - p_n q'_n + p_n q'_n - p_n p'_n|$$

$$= |(q_n - p_n)q'_n + p_n(q'_n - p'_n)|$$

$$\le |(q_n - p_n)q'_n| + |p_n(q'_n - p'_n)| \qquad \text{(triangle inequality)}$$

$$= |q_n - p_n||q'_n| + |p_n||q'_n - p'_n|$$

Both of the quantities $|q'_n|$ and $|p_n|$ are bounded by Theorem 3, so there is (rational) $M > 0$ such that $|q'_n| \le M$, and $|p_n| \le M$. Further, we know that $|q_n - p_n| \to 0$, and $|q'_n - p'_n| \to 0$; so there is an index N large enough that for $n > N$

$$|q_n - p_n| < \frac{\varepsilon}{2M}, \qquad |q'_n - p'_n| < \frac{\varepsilon}{2M}$$

Therefore, for all $n > N$

$$|q_n q'_n - p_n p'_n| \le \frac{\varepsilon}{2M}M + \frac{\varepsilon}{2M}M = \varepsilon$$

which, because ε can be arbitrarily small, is saying the same thing as $|q_n q'_n - p_n p'_n| \to 0$ as $n \to \infty$. Therefore, $[q_n q'_n] = [p_n p'_n]$. ∎

Okay, now we have two well-defined operations of addition and multiplication in $[\mathcal{C}]$. The next thing to do is verify that these are field operations; this means that it is necessary to verify that the nine field properties hold. This is a task that is more tedious than difficult.

The commutative property: Since addition and multiplication are commutative in the field $\mathbb{Q}$, we have

$$[q_n] + [q'_n] = [q_n + q'_n] = [q'_n + q_n] = [q'_n] + [q_n]$$
$$[q_n][q'_n] = [q_n q'_n] = [q'_n q_n] = [q'_n][q_n]$$

Hence, addition and multiplication are also commutative in $[\mathcal{C}]$.

The associative property: Since addition and multiplication are associative in the field $\mathbb{Q}$, we have

$$[q_n] + ([q'_n] + [q'_n]) = [q_n] + [q'_n + q'_n]$$
$$= [q_n + (q'_n + q'_n)]$$
$$= [(q_n + q'_n) + q'_n]$$
$$= ([q_n] + [q'_n]) + [q'_n]$$

And similarly,

$$[q_n]([q'_n][q'_n]) = [q_n][q'_n q'_n]$$
$$= [q_n(q'_n q'_n)]$$
$$= [(q_n q'_n)q'_n]$$
$$= ([q_n][q'_n])[q'_n]$$

Hence, addition and multiplication are also associative in $[\mathcal{C}]$.

The distributive property: Since addition and multiplication have the distributive property in the field $\mathbb{Q}$, we have

$$[q_n]([q'_n] + [q''_n]) = [q_n][q'_n + q''_n]$$
$$= [q_n(q'_n + q''_n)]$$
$$= [q_nq'_n + q_nq''_n]$$
$$= [q_nq'_n] + [q_nq''_n]$$
$$= [q_n][q'_n] + [q_n][q''_n]$$

Hence, addition and multiplication also have the distributive property in $[\mathcal{C}]$.

The additive identity or zero element: Consider $[0]$, namely, the equivalence class of the constant sequence $0, 0, 0, \dots$ Since for every $[q_n]$ in $[\mathcal{C}]$,

$$[q_n] + [0] = [q_n + 0] = [q_n]$$

it follows that $[0]$ is the zero element of $[\mathcal{C}]$. Note also that for every $[q_n]$

$$[q_n][0] = [q_n 0] = [0]$$

The multiplicative identity: Consider $[1]$, namely, the equivalence class of the constant sequence $1, 1, 1, \dots$ Since for every $[q_n]$ in $[\mathcal{C}]$,

$$[q_n][1] = [q_n 1] = [q_n]$$

it follows that $[1]$ is the multiplicative identity in $[\mathcal{C}]$.

The additive inverses, or negatives: If to each sequence $q_1, q_2, q_3, \dots$ in a class $[q_n]$, we add the sequence of negatives $-q_1, -q_2, -q_3, \dots$ we obtain the zero sequence $0, 0, 0, \dots$ Further, the sequence of negatives is again Cauchy since for all indices m, n

$$|(-q_m) - (-q_n)| = |-q_m + q_n| = |q_m - q_n|$$

So we simply define the additive inverse (the negative) of the class $[q_n]$ as

$$-[q_n] = [-q_n]$$

to obtain the desired result: $[q_n] + (-[q_n]) = [q_n + (-q_n)] = [0]$.

The multiplicative inverses, or reciprocals: If we multiply a sequence of *nonzero* rational numbers $q_1, q_2, q_3, \dots$ by the sequence of reciprocals of its terms $1/q_1, 1/q_2, 1/q_3, \dots$ we obtain the unit sequence $1, 1, 1, \dots$ So it makes sense to define $[q_n]^{-1} = [1/q_n]$. But in every *class* $[q_n]$, there are sequences with lots of zero terms. In fact, if we take any sequence that is not equivalent to zero (say, 2, 2, 2, ...) and add any number of zeros to the beginning of it, then both the original sequence and the one starting with zeros say, (0, 0, 0, 0, 2, 2, 2, ...) are in the same equivalence class ([2]) because their tail ends certainly approach each other (actually meet). But of course, the rational number 0 has no reciprocal, so we need to resolve this issue of sequences containing zeros. It turns out that as long as $[q_n] \neq [0]$, every sequence in $[q_n]$ has at most a *finite* number of zeros, which makes the issue easy to resolve.

First, observe that if a Cauchy sequence $q_1, q_2, q_3, \ldots$ of rationals is in the class $[0]$, i.e., it is equivalent to the constant sequence $0, 0, 0, \ldots$ then q_n approaches 0. This means that for every (rational) $\varepsilon > 0$, there is an index N such that $|q_n| = |q_n - 0| < \varepsilon$ for $n > N$. From this we infer that if $q_1, q_2, q_3, \ldots$ is *not* equivalent to $0, 0, \ldots$ then there is a (rational) $\varepsilon_0 > 0$ such that $|q_n| = |q_n - 0| \geq \varepsilon_0$ infinitely often (i.e., for every positive integer N there is an index $n_0 \geq N$ such that $|q_{n_0}| \geq \varepsilon_0$). Since we are assuming that q_n is Cauchy, there is an index N such that $|q_m - q_n| < \varepsilon_0/2$ for $m, n > N$. Now, with $m = n_0$ and using the triangle inequality in the form (3.14), it follows that for all $n > N$

$$|q_n| \geq |q_{n_0}| - |q_{n_0} - q_n| > \varepsilon_0 - \frac{\varepsilon_0}{2} = \frac{\varepsilon_0}{2}$$

In summary

If the sequence $q_1, q_2, q_3, \ldots$ is not in the class $[0]$, then there is a rational number $\varepsilon_0 > 0$ and an index N such that $|q_n| \geq \varepsilon_0$ for all $n > N$. In particular, $q_n \neq 0$ for $n > N$.

Now if $[q_n] \neq [0]$, then the above statement applies to *every* sequence in $[q_n]$, including the sequence $\tilde{q}_n$ such that

$$\tilde{q}_n = \begin{cases} q_n & \text{if } n > N \\ 1 & \text{if } n \leq N \end{cases}$$

Note that $[\tilde{q}_n] = [q_n]$, and further, *none* of the terms of $\tilde{q}_n$ are zeros. So we can define the multiplicative inverse of $[q_n]$ as (you may verify in Exercise 380 that the reciprocal sequence is Cauchy)

$$[q_n]^{-1} = [\tilde{q}_n]^{-1} = \left[\frac{1}{\tilde{q}_n}\right]$$

The above definition does not apply to one equivalence class in $[\mathcal{C}]$, namely, $[0]$. Although $[0]$ does contain sequences with no zero terms, like $1, 1/2, 1/3, \ldots$ it also contains sequences that are zeros infinitely often, like $1, 0, 1/2, 0, 1/3, 0, \ldots$ This means that

$$[0] = [1, 0, 1/2, 0, 1/3, 0, \ldots] = [1, 1/2, 1/3, \ldots]$$

But $1, 0, 1/2, 0, 1/3, 0, \ldots$ does not have a reciprocal sequence, while $1, 1/2,$ $1/3, \ldots$ does, so it is not possible to define $[0]^{-1}$ unambiguously. It is also worth noticing that the reciprocal sequence of $1, 1/2, 1/3, \ldots$ namely, $1, 2, 3, \ldots$ is *not* Cauchy and thus not a member of any equivalence class in $[\mathcal{C}]$ to begin with!

The preceding discussion verifies that $[\mathcal{C}]$ is indeed an algebraic field.

Exercise 380 *Suppose that* $q_1, q_2, q_3, \ldots$ *is Cauchy and not in* $[0]$, *i.e., not equivalent to* 0. *Assuming that* $q_n \neq 0$ *for all* n, *show that the reciprocal sequence* $1/q_1, 1/q_2, 1/q_3, \ldots$ *is Cauchy.*

$[\mathcal{C}]$ is a totally ordered field.

Recall that the set $\mathbb{Q}$ of rational numbers is totally ordered. If p and q are rational numbers, then saying that $p < q$ is equivalent to saying that $q - p > 0$. This is the way we introduce the ordering of equivalence classes in $[\mathcal{C}]$ since we know that a difference of equivalence classes $[q_n] - [p_n]$ is just the equivalence class $[q_n - p_n]$.

With this in mind, we start with what it means to say something is "positive" in $[\mathcal{C}]$.

The equivalence class $[q_n]$ of a rational Cauchy sequence $q_1, q_2, q_3, \ldots$ is *positive* if there is (rational) $\varepsilon > 0$ and an index N such that $q_n \geq \varepsilon$ for all $n > N$.

I had to use a positive lower bound ε here to ensure that q_n does not converge to 0. For example, $1/n > 0$ for all n, but $q_n = 1/n$ is equivalent to the zero sequence. So its equivalence class is $[0]$, but it would be counterproductive to declare $[0]$ positive! Also starting from an index N is necessary, since as I mentioned above there are infinitely many sequences with a finite number of zeros in each member of $[q_n] \neq [0]$.

To turn the above definition into an ordering $\prec$ in $[\mathcal{C}]$, we start by defining

$[0] \prec [q_n]$ if $[q_n]$ is positive.

Now if $q'_1, q'_2, q'_3, \ldots$ is *any* other sequence in $[q_n]$, then $[q'_n] = [q_n]$; so for the definition to make sense, there must be a (rational) $\varepsilon' > 0$ and an index N' such that $q'_n \geq \varepsilon'$ for $n > N'$. Since $q'_1, q'_2, q'_3, \ldots$ is equivalent to $q_1, q_2, q_3, \ldots$ there is an index N_0 such that $|q'_n - q_n| < \varepsilon/2$ for all $n > N_0$. This means that $-\varepsilon/2 < q'_n - q_n < \varepsilon/2$ for $n > N_0$; and since $q_n \geq \varepsilon$ for $n > N$, if I pick N' to be the larger of N_0 and N; then for all $n > N'$ it is the case that

$$q'_n = q'_n - q_n + q_n > -\frac{\varepsilon}{2} + \varepsilon = \frac{\varepsilon}{2}$$

If I pick $\acute{\varepsilon} = \varepsilon/2$, then I have shown that $q'_n > \acute{\varepsilon}$ for $n > N'$, which was necessary to make $\prec$ well-defined. Now let us extend the relation $\prec$ to all pairs in $[\mathcal{C}]$ as follows:

$$[p_n] \prec [q_n] \text{ if } [0] \prec [q_n] - [p_n], \text{ (i.e., } [q_n - p_n] \text{ is positive).}$$

To prove that $\prec$ makes $[\mathcal{C}]$ a totally ordered field, we need to establish the following: for all classes $[p_n]$ and $[q_n]$ in $[\mathcal{C}]$

1. If $[p_n] \prec [q_n]$, then $[p_n] + [r_n] \prec [q_n] + [r_n]$ for every class $[r_n]$ in $[\mathcal{C}]$.
2. If $[p_n] \prec [q_n]$, then $[p_n][r_n] \prec [q_n][r_n]$ for every class $[r_n]$ in $[\mathcal{C}]$ such that $[0] \prec [r_n]$.
3. If $[p_n] \neq [q_n]$, then either $[p_n] \prec [q_n]$ or $[q_n] \prec [p_n]$ (not both).
4. The relation $\prec$ is transitive.

To prove Item 1 above, it is necessary to show that

$$[0] \prec ([q_n] + [r_n]) - ([p_n] + [r_n])$$

Using the field properties of $[\mathcal{C}]$, the right-hand side works out to $[q_n] - [p_n] = [p_n - q_n]$. This is assumed to be positive in Item 1, so we are done.

To prove Item 2, we must show that

$$[0] \prec [q_n][r_n] - [p_n][r_n]$$

The field properties and the definitions of addition and multiplication reduce the problem to proving

$$0 \prec ([q_n] - [p_n])[r_n] = [q_n - p_n][r_n] = [(q_n - p_n)r_n]$$

So it remains to show that there is an ε_0 and an index N such that $(q_n - p_n)r_n \geq \varepsilon_0$ for all $n \geq N$. By the positivity assumptions, there are $\varepsilon_1, \varepsilon_2 > 0$ and indices N_1, N_2 such that $q_n - p_n \geq \varepsilon_1$ for $n \geq N_1$ and $r_n \geq \varepsilon_2$ for $n \geq N_2$. So now simply choose $\varepsilon_0 = \varepsilon_1 \varepsilon_2$ and N to be the greater of N_1 and N_2 to conclude that $[(q_n - p_n)r_n]$ is positive and finish the proof.

Next, I prove Item 3. Assume that $[p_n] \neq [q_n]$. This means that $q_n - p_n$ is a rational Cauchy sequence that does not converge to 0; so by Theorem 2, $q_n - p_n$ is eventually positive or eventually negative and stays some positive distance ε_0 above or below zero. If eventually positive, then $[q_n - p_n]$ is positive and thus $[p_n] \prec [q_n]$. If eventually negative, then then $[p_n - q_n]$ is positive and thus $[q_n] \prec [p_n]$.

Finally, I show that $\prec$ is transitive, Item 4. Suppose that there are classes $[p_n], [q_n],$ and $[r_n]$ in $[\mathcal{C}]$ such that $[p_n] \prec [q_n]$ and $[q_n] \prec [r_n]$. Then

$$[0] \prec [q_n] - [p_n] \quad \text{and} \quad [0] \prec [r_n] - [q_n]$$

Using the field properties of $[\mathcal{C}]$ and Item 1 above, we have

$$[0] = [0] + [0] \prec [r_n] - [q_n] + [q_n] - [p_n] = [r_n] - [p_n]$$

It follows that $[p_n] \prec [r_n]$, and transitive property is proved.

The above discussion proves that $[\mathcal{C}]$ is a totally ordered field under the relation $\prec$ and thus also under its reflexive extension $\preceq$.

Rational numbers are dense in the set of real numbers.

We have shown so far that the set $[\mathcal{C}]$ of equivalence classes of rational Cauchy sequences is a totally ordered field that contains a copy of the set $\mathbb{Q}$ of all rational numbers in the sense that each rational number q is uniquely identified with the equivalence class $[q]$ of the constant sequence $q, q, q, \ldots$ Verifying this uniqueness is a good exercise; the preceding sections contain enough details to draw on here.

Exercise 381 *Show that if p, q are distinct rational numbers, then $[p] \neq [q]$; in fact, show that the inequality $p < q$ in $\mathbb{Q}$ is equivalent to the inequality $[p] \prec [q]$ in $[\mathcal{C}]$.*

This copy of $\mathbb{Q}$ turns out to be a very important subset in calculus and beyond. To explore this further, it is helpful to blur the precise nature of elements in $[\mathcal{C}]$; now that we have some structure for $[\mathcal{C}]$, we can use a simpler (even if more vague) notation for more effective progress moving forward. We start by formally defining the set of real numbers:

The Real Numbers: Each equivalence class of rational Cauchy sequences in $[\mathcal{C}]$ is a *real number*. We use the common notation $\mathbb{R}$ instead of $[\mathcal{C}]$ for the *set of all real numbers*.

We have thus shown that $\mathbb{R}$ is a totally ordered field. We now identify the rational numbers with the equivalence classes of constant rational sequences so as to simply say that $\mathbb{Q}$ is a subset of $\mathbb{R}$.

The concepts of positive and negative are easy to define. A real number $r = [q_n]$ is *negative* if $[q_n] \prec [0]$ and *positive* if $[0] \prec [q_n]$. Identifying $[0]$ with the rational number 0 and the relations $\prec$ and $\preceq$ with the usual orderings $<$ and $\leq$ of the rationals, we simply write $r < 0$ and $r > 0$ to indicate whether the real number r is negative or positive.

The extension of the concept of absolute value to $\mathbb{R}$ is also straightforward:

$$|r| = \begin{cases} r & \text{if } r \geq 0 \\ -r & \text{if } r < 0 \end{cases}$$

From this definition, you can see that $|r| \geq 0$ (i.e., $[0] \leq |[q_n]|$) no matter what real number r we choose. It is well worth a mention that the triangle inequality readily extends to the real numbers now, since the proof of that theorem simply considered the signs of rational numbers and not their nature as rationals. Now we are ready to state an important result about the real numbers.

Theorem 382 *(density of rational numbers). For every pair of real numbers x and y such that $x < y$, there is a rational number q such that $x < q < y$.*

Proof Suppose that $x = [q_n]$ and $y = [r_n]$ where q_n and r_n are Cauchy sequences of rationals. We are given that $[q_n] \prec [r_n]$ or equivalently, $[r_n - q_n] \succ [0]$. Therefore, there is a (rational) $\varepsilon_0 > 0$ and index N such that $r_n - q_n \geq \varepsilon_0$ for all $n \geq N$. On the other hand, q_n is a Cauchy sequence, so there is an index N_0 such that $|q_n - q_m| < \varepsilon_0$ for all $m, n > N_0$. If I define K to be the larger of N and $N_0 + 1$ and the rational number $q = q_K$, then

$$q_n < q + \varepsilon_0 \leq r_n \quad \text{for all } n \geq K$$

In particular, $q_n < q < r_n$ for all $n \geq K$, which translates to $[q_n] < [q] < [r_n]$, or equivalently, $x < q < y$. ∎

This theorem shows that every real number can be approximated by rational numbers, and so it should come as no surprise; we in fact defined a real number essentially as something that is approximated by a sequence of rational numbers. The important difference is that now we can actually associate a sequence with a real number rather than the other way around. For instance, suppose that x is a real number and we want to find a sequence of rationals that approximates x. We begin with some rational number q_1 and then find a rational q_2 that is between q_1 and x. Clearly, q_2 is closer to x than q_1 was, in the sense that

$$|q_2 - x| < |q_1 - x|$$

We may go on to a rational q_3 that is between q_2 and x and so on, creating a sequence $q_1, q_2, q_3, \ldots$ in this way that converges to x. Such a sequence is Cauchy by Theorem 1 (this theorem, like all of the Theorems 1–3, readily extends to real numbers) so $x = [q_n]$.

The set of real numbers is complete.

We now come to the most important property of real numbers, and the reason for their invention: the set $\mathbb{R}$ of real numbers is complete, or put differently, $\mathbb{R}$ has no gaps or holes. First, it is convenient to have the extensions of Theorems 1–3 listed for easy reference. The proofs are essentially the same as before; I present the proof of one theorem in order to set the stage for the completeness theorem below. As in the previous section, we take advantage of all our work above and use the common notation for rational numbers instead of the more accurate but cumbersome equivalence class notation.

Theorem 383 *If a sequence x_n of real numbers converges to a real number x, then x_n is a Cauchy sequence.*

This theorem can be stated equivalently (in contrapositive form) as the following: *If a sequence is not Cauchy, then it does not converge.*

Proof It is given that for every (real, or rational) $\varepsilon > 0$ there is an index N such that if $n > N$, then $|x_n - x| < \varepsilon/2$. So for $m > N$

$$|x_n - x_m| = |x_n - x + x - x_m| \leq |x_n - x| + |x - x_m| < \frac{\varepsilon}{2} + \frac{\varepsilon}{2} = \varepsilon$$

It follows that x_n is a Cauchy sequence.   ∎

Theorem 384 *Let x_n be a sequence of real numbers that does not converge to 0. Then the sequence is either eventually positive or eventually negative.*

Theorem 385 *Every Cauchy sequence of real numbers is bounded.*

It is worth mentioning the technical point that in proving statements about the real numbers, the numbers ε or ε_0 in all proofs can be chosen real or be kept rational since by Theorem 5, if ε is real, then a rational ε' can be chosen that is arbitrarily close to it, and either smaller or larger, as needed.

A very useful result about sequences is the following.

Theorem 386 *Every nondecreasing sequence of real numbers that is bounded from above is Cauchy.*

Proof Suppose that x_n is a nondecreasing sequence in $\mathbb{R}$ that is bounded from above, i.e., there is a fixed real number M such that

$$x_n \leq M \quad \text{for every } n = 1, 2, 3, \ldots \tag{9.1}$$

We show that if x_n is *not* Cauchy, then we get a contradiction. That will be grounds for declaring the sequence Cauchy. If x_n is not Cauchy, then there must be some positive ε_0 and an increasing sequence of indices n_k

$$n_1 < n_2 < n_3 < \cdots$$

such that

$$x_{n_{j+1}} - x_{n_j} \geq \varepsilon_0 \quad \text{for every } j = 1, 2, 3, \ldots \tag{9.2}$$

The absolute value was not required above since x_n is nondecreasing so that

$$x_{n_j} \leq x_{n_j+1} \leq x_{n_j+2} \leq \cdots \leq x_{n_{j+1}}$$

Now using (9.2) I get

$$(x_{n_{j+1}} - x_{n_j}) + (x_{n_j} - x_{n_{j-1}}) + \cdots + (x_{n_2} - x_{n_1}) \geq \varepsilon_0 + \varepsilon_0 + \cdots + \varepsilon_0$$

$$x_{n_{j+1}} - x_{n_1} \geq j\varepsilon_0$$

$$x_{n_{j+1}} \geq x_{n_1} + j\varepsilon_0$$

There are infinitely many j, so $j\varepsilon_0$ gets arbitrarily large as j does (this is a consequence of the Archimedean property that is discussed in the text). Therefore, there is J large enough that $J\varepsilon_0 > M - x_{n_1}$. This means that

$$x_{n_{J+1}} \geq x_{n_1} + J\varepsilon_0 > M$$

This contradicts the hypothesis (9.1); to avoid this, x_n must be Cauchy. ∎

The next result is the *converse* of Theorem 6. Up to now, when dealing with Cauchy sequences of rational numbers, we could only assert that they accumulate around a certain value. If that value was not rational, then we had reached a gap or hole in the set of rationals. The next theorem fills such gaps and says more than just Cauchy sequences of rational numbers converge to real numbers.

Theorem 387 *(completeness) Every Cauchy sequence of real numbers converges to a real number.*

Proof Suppose that r_n is a Cauchy sequence of *real* numbers, i.e., $r_n \in \mathbb{R}$ for every index n. First we dispense with a trivial case: if there is an index N such that $r_n = r$ for all $n \geq N$ (i.e., if r_n is eventually constant), then clearly r_n converges to r and we are done.

So let us assume that r_n is not eventually constant. Then for each index n, there is another $n' > n$ such that $r_{n'} \neq r_n$. We may assume that n' is the smallest or least such index. Now by the density of rationals (Theorem 5), there is a rational number q_n between r_n and $r_{n'}$; in other words, q_n is closer to each of r_n and $r_{n'}$ than the two real numbers are to each other. In particular,

$$|r_n - q_n| < |r_n - r_{n'}|$$

Although I have blurred the notation in favor of better readability, it is worth remembering that each term of r_n is an equivalence class $[p_k]_n$ of rational Cauchy sequences; and likewise, each term of the rational sequence q_n is the equivalence class of a constant sequence; i.e., q_1 is $[q_1, q_1, q_1, \ldots]$, and so on.

So now we know that there is a sequence (at least one) of *rational* numbers $q_1, q_2, q_3, \ldots$ whose terms are sandwiched in between the terms of the original sequence r_n. Let us verify that all such rational sequences are Cauchy; let $\varepsilon > 0$ and note that by the triangle inequality

$$\begin{aligned} |q_n - q_m| &= |q_n - r_n + r_n - r_m + r_m - q_m| \\ &\leq |q_n - r_n| + |r_n - r_m| + |r_m - q_m| \\ &< |r_{n'} - r_n| + |r_n - r_m| + |r_m - r_{m'}| \end{aligned}$$

Since r_n is Cauchy by assumption, there is an index N such that if $m, n > N$, then $|r_n - r_m| < \varepsilon/3$. This inequality is valid for $|r_{n'} - r_n|$ and $|r_m - r_{m'}|$ too because $m' > m$ and $n' > n$ by the construction above. Therefore, if $m, n > N$, then

$$|q_n - q_m| < \frac{\varepsilon}{3} + \frac{\varepsilon}{3} + \frac{\varepsilon}{3} = \varepsilon$$

and it follows that q_n is Cauchy. Therefore, its equivalence class $[q_n]$ is a real number; let us call it r. It remains to show that r_n converges to r. With ε and N defined above, we can state (with $m = N + 1$) that $|r_n - r_{N+1}| < \varepsilon/3$, and $|q_{N+1} - r_{N+1}| < \varepsilon/3$. Further, since r is the equivalence class of q_n by taking N large enough, we can ensure that $|q_{N+1} - r| < \varepsilon/3$ also. Therefore, using the triangle inequality again,

$$\begin{aligned} |r_n - r| &= |r_n - r_{N+1} + r_{N+1} - q_{N+1} + q_{N+1} - r| \\ &\leq |r_n - r_{N+1}| + |r_{N+1} - q_{N+1}| + |q_{N+1} - r| \\ &< \frac{\varepsilon}{3} + \frac{\varepsilon}{3} + \frac{\varepsilon}{3} = \varepsilon \end{aligned}$$

This verifies that r_n converges to r and concludes this proof.   ∎

9.2 Appendix: Discontinuity in a space of functions

In this appendix, we explain the sudden changes in lengths of the zigzag functions in Chapters 1 and 8 by looking at the lengths of curves as functions defined on a space of curves.

Consider the length of a curve as a function L that maps a function $f(x)$ to a number, namely, the length of its graph. In other words, *the domain of L is a set of* *ordinary functions* not a set of points on the x-axis or in a space with finitely many dimensions. For example, if $f(x) = \sqrt{1 - x^2}$ whose graph is the upper half-circle of radius 1 for $-1 \leq x \leq 1$, then $L(f) = \pi$ is just half the circumference of the full circle of radius 1. Or if $f(x) = 2x$ whose graph is a straight line, then the length of the stretch of this line from $x = 0$ to $x = 1$ is easily found using the Pythagorean theorem as $L(f) = \sqrt{5}$.

To understand the nature of the discontinuity in lengths and relate it to the usual definition of continuity in Chapter 6, we need to understand spaces of functions. These are spaces in which individual points are real-valued functions, i.e., subsets of the infinite product of the set of real numbers with itself. These sets are technically complicated and unsuitable for any substantial discussion in this book. However, I give a brief, simplified discussion of one such space here to explain the discontinuity that we observe in the lengths of zigzag paths in terms of the usual notions of continuity and limits.

Consider the set of all continuous functions (in the sense of Chapter 6) on the interval $[0,1]$, and let's denote this set of functions by the symbol $C[0, 1]$. The zigzag paths and the smooth sine functions that we encountered in Chapter 6 are all continuous functions on $[0,1]$ and therefore, points or elements in $C[0, 1]$. Since every function in $C[0, 1]$ is also an ordinary real-valued function on $[0,1]$, we see that each function in $C[0, 1]$ is an element of the infinite product $\mathbb{R}^{[0,1]}$.[1] This is analogous to identifying every vector in the usual three-dimensional space $\mathbb{R}^3$ as a point of that space. The main difference is that the space of functions $C[0, 1]$ has infinite dimension.

Norm.

Just as vectors have lengths (their magnitudes), so can the functions in $C[0, 1]$. But unlike $\mathbb{R}^3$, many different length concepts are possible for $C[0, 1]$. Each turns out to be useful for a different purpose. The one that we discuss here is one of the most important in analysis because it is associated with uniform convergence.

Given any function f in $C[0, 1]$ and any number x_0 in $[0,1]$, the absolute value $|f(x_0)|$ gives us the magnitude of $f(x_0)$ for the specific number x_0. This magnitude is different for different numbers x_0 in $[0,1]$. So let's consider $|f(x)|$ for *all* numbers x in $[0,1]$; the largest, or maximum, of these numbers is a finite number that we denote by $\|f\|$. In symbols

$$\|f\| = \max\{|f(x)| : x \in [0, 1]\}$$

We call $\|f\|$ the *max norm*[2] of f, which represents the size or magnitude of f relative to this norm. It so happens that $\|f\|$ has the same basic properties as the

---

[1] $C[0, 1]$ turns out to be a very small subset of $\mathbb{R}^{[0,1]}$, which has cardinal number $2^{2^{\aleph_0}}$ compared to the cardinal number of $C[0, 1]$, which is a "mere" $2^{\aleph_0}$.

[2] The standard term is the *sup norm*, where as usual "sup" is short for *supremum*, or the least upper bound. Recall that this term is synonymous with maximum when the largest value is attained at a specific number.

vector magnitude; in particular, it satisfies the important triangle inequality: for every pair of functions f, g in $C[0, 1]$

$$\|f + g\| \leq \|f\| + \|g\|$$

Recall that because of continuity, the extreme value theorem implies that there is at least one number x^* in $[0,1]$ such that $\|f\| = |f(x^*)|$.

Calculating the norm of a function in $C[0, 1]$ is straightforward. For example, $\|x^2\|$ is the maximum value of x^2 over the interval $[0,1]$. This maximum occurs at $x = 1$ and $|1^2| = 1$, so $\|x^2\| = 1$. Can you figure out $\|2x\|$ the same way?

More generally, we simply find the absolute value and if necessary, use calculus to find the norm. For example, the maximum value of $\sin 2x$ is 1, which occurs at $x = \pi/4 \simeq 0.785$ where the derivative $2\cos 2x$ is 0. Therefore, $\|\sin 2x\| = 1$. Can you figure out $\|\sin x\|$? Keep in mind that $0 \leq x \leq 1$.

Distance and the uniform metric.

Just as for vectors $\vec{v}$ and $\vec{u}$ in $\mathbb{R}^3$, we define the distance between $\vec{v}$ and $\vec{u}$ as the magnitude of the difference $\vec{v} - \vec{u}$; we define *the distance between two functions f, g* in $C[0, 1]$ to be *the norm* of their difference, that is,

$$\text{distance between } f \text{ and } g \text{ is } \|f - g\|$$

As we saw above, this distance can often be calculated using standard methods of calculus. For example, the distance between $f(x) = x$ and $g(x) = x^3$ is

$$\|x - x^3\| = \max\{|x - x^3| : x \in [0, 1]\}$$

Figure 9.1 shows that the difference $x - x^3$ is non-negative for all x in $[0,1]$; if we define $h(x) = x - x^3$, then the maximum value of $h(x)$ occurs where $h'(x) = 0$, that is, for the value of x that satisfies

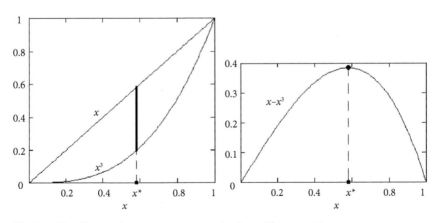

Fig. 9.1 The distance between two curves in the uniform metric

$$h'(x) = 1 - 3x^2 = 0$$

Solving the last equation gives a solution $x^* = 1/\sqrt{3} \simeq 0.58$ in the interval $[0,1]$; inserting this number in $x - x^3$ gives the distance between the two curves:

$$\left\| x - x^3 \right\| = h\left(x^*\right) = \frac{1}{\sqrt{3}} - \left(\frac{1}{\sqrt{3}}\right)^3 = \frac{2}{3\sqrt{3}} \simeq 0.38$$

The set $C[0,1]$ of continuous functions endowed with the distance function above is an example of a *metric space*. "Metric" is synonymous with distance. The max norm distance here is also called the *uniform metric* for the following reason. Recall that a sequence of functions f_n in $C[0,1]$ converges uniformly to a function f in $C[0,1]$ (or $f_n \overset{u}{\longrightarrow} f$) if for every $\varepsilon > 0$, there is a positive integer N such that $|f_n(x) - f(x)| < \varepsilon$ for all $n > N$ *regardless of the choice of x in $[0,1]$*. It follows that

$$\|f_n - f\| = \max\{|f_n(x) - f(x)| : x \in [0,1]\} < \varepsilon \qquad \text{if } n > N \qquad (9.3)$$

Therefore, we can conveniently define uniform convergence in terms of the uniform metric as

$$f_n \overset{u}{\longrightarrow} f \quad \text{if} \quad \|f_n - f\| \to 0 \quad \text{as } n \to \infty \qquad (9.4)$$

In words, *the sequence of functions f_n converges to f uniformly if the distance between f_n and f as points in $C[0,1]$ goes to zero.*

Notice that $\|f_n - f\|$ is just a sequence of real numbers, which is easier to deal with than a sequence of functions.

Functionals.

Before we can discuss continuity of functions on a metric space like $C[0,1]$, we need to define what we mean by functions on such a space. To keep technical details at a minimum, I consider only functions from $C[0,1]$ into the real numbers $\mathbb{R}$. These types of functions are called *functionals* because their range is the set of real numbers.[3] The norm or length of a function is an example of a functional because for every f in $C[0,1]$, its norm $\|f\|$ is a real number.

We also worked with another important type of functional earlier, namely, *the integral $\int_a^b f(x)dx$*. We define the integral functional $I : C[0,1] \to \mathbb{R}$ as

$$I(f) = \int_0^1 f(x)dx$$

---

[3] The range of a functional does not have to be $\mathbb{R}$; the set of complex numbers is also a useful range for various purposes.

We may readily calculate the value of I for a variety of functions using the Fundamental Theorem of Calculus. For example,

$$I(x^n) = \int_0^1 x^n dx = \frac{x^{n+1}}{n+1}\Big|_{x=0}^{x=1} = \frac{1}{n+1}$$

for all positive integer values of n. Here the functional I assigns to each function x^n the real number $1/(n+1)$; in particular, $I(x) = 1/2$, $I(x^2) = 1/3$, and so on. Similarly,

$$I(e^x) = \int_0^1 e^x dx = e^1 - e^0 = e - 1$$

A simple interpretation of the integral functional is to think of it as giving the *average value* of a function. For example, $I(x^2) = 1/3$ is the average value of x^2 on the interval $[0,1]$, and $I(e^x) = e - 1$ is the average value of e^x over $[0,1]$.

The integral belongs to an important class of functionals known as *linear functionals*. A functional $F : C[0, 1] \to \mathbb{R}$ is linear if for every pair of functions f, g in $C[0, 1]$ and every pair of real numbers a, b it is true that

$$F(af + bg) = aF(f) + bF(g) \qquad (9.5)$$

The integral I is a linear functional (recall the properties of integral that we discussed in Chapter 7), but the norm is not linear. To see why not, let $F(f) = \|f\|$ so that $F(x - x^3) = \|x - x^3\|$. We showed above that $F(x - x^3) = 2/3\sqrt{3}$. On the other hand,

$$F(x) - F(x^3) = \|x\| - \|x^3\| = 1 - 1 = 0 \neq \frac{2}{3\sqrt{3}}$$

which shows that (9.5) is not satisfied (with $a = 1$, $b = -1$, $f(x) = x$ and $g(x) = x^3$).

Continuous functionals.

A functional $F : C[0, 1] \to \mathbb{R}$ is *continuous at a point f* in $C[0, 1]$ relative to the max norm, if for every sequence f_n in $C[0, 1]$ that converges uniformly to f we have $\lim_{n \to \infty} F(f_n) = F(f)$. Notice that this limit is just the limit of a sequence of real numbers. If F is continuous at every point f in $C[0, 1]$, then F is a *continuous functional*.

The integral I is a continuous functional because for every continuous function f and every sequence of continuous functions f_n that converges uniformly to f the theory of Chapter 8 gives

$$\lim_{n \to \infty} I(f_n) = \int_0^1 \lim_{n \to \infty} f_n(x) dx = \int_0^1 f(x) dx = I(f)$$

With regard to the first equality, recall that in the case of uniform convergence, we can move the limit in or out of the integral.

The norm itself is another continuous functional; this is quickly proved using the triangle inequality. We are more interested here in the *length functional*, which we consider next.

Lengths of curves are discontinuous functionals.

The graph of each continuous function on the interval [0,1] is a continuous curve. For each such curve *f* we define $L(f)$ to be the length of *f*. To ensure that L is a functional, it is necessary that the graph of every function *f* have a finite, well-defined length that is a real number. For $C[0, 1]$ this is technically complicated to deal with, so we limit our attention to a subset of $C[0, 1]$ for which it is easy to see that L is well-defined.

A function in $C[0, 1]$ is *smooth* if it is *continuously differentiable* at every number in [0,1]; so *f* in $C[0, 1]$ is smooth if its derivative $f'(x)$ exists *and* is continuous at every number *x* in [0,1]. In Chapter 8, we used the notation $C^1[0, 1]$ for the set of all smooth functions in $C[0, 1]$.

The set $C^1[0, 1]$ is a proper subset of $C[0, 1]$ because there are functions that are continuous but not smooth; a simple example is the absolute value function $|x|$, whose derivate at $x = 0$ cannot be defined continuously. An extreme example of a continuous function that is *not smooth at even a single point* is Weierstrass's function that we discussed in Chapter 8.

On the other hand, $C^1[0, 1]$ contains most functions that we find in calculus: all polynomials, as well as the trigonometric functions $\sin x$ and $\cos x$, and the exponential and logarithmic functions are all smooth on their domains.

The set $C^1[0, 1]$ endowed with the uniform metric is a metric space. It is called a *subspace* of $C[0, 1]$ because the uniform metric on $C^1[0, 1]$ is inherited from the larger space $C[0, 1]$.

In calculus, it is shown that if *f* is smooth, then its length can be defined using an integral; specifically, if *f* is in $C^1[0, 1]$, then

$$L(f) = \int_0^1 \sqrt{1 + [f'(x)]^2}\, dx$$

If $f'(x)$ is continuous, then so is the square root expression inside the integral. It follows that the integral is well-defined. This formula shows that L is a well-defined functional on $C^1[0, 1]$.

But is L continuous?

Our discussion of the zigzag paths shows that L is not continuous relative to the uniform metric. To prove this, the functions

$$f_n(x) = \frac{\sin^2(n\pi x)}{2n}$$

are smooth functions on the interval [0,1] with continuous derivatives $(\pi/2)\sin(2n\pi x)$. Therefore, f_n is a point in $C^1[0, 1]$ for every *n*. Using (9.3) or (9.4), we see that the sequence f_n converges uniformly to the zero function $f(x) = 0$ on $[0,1]$ because the maximum value of $\sin^2(n\pi x)$ is simply 1 so that

$$\|f_n - 0\| = \left\|\frac{\sin^2(n\pi x)}{2n}\right\| = \frac{1}{2n} \to 0 \quad \text{as } n \to \infty$$

so $f_n \xrightarrow{u} 0$. The length of the zero function on the interval $[0,1]$ is

$$L(0) = \int_0^1 \sqrt{1 + 0}\,dx = 1$$

However, as we showed in our discussion of zigzag path in Chapter 8 (see the discussion surrounding Figure 8.13 in Chapter 8), for every positive integer n

$$L(f_n) = \int_0^1 \sqrt{1 + \frac{\pi^2}{4}\sin^2(2n\pi x)}\,dx \geq \int_0^1 \sqrt{2}\,dx > 1$$

It follows that the length functional L is not continuous on the space $C^1[0,1]$ relative to the uniform metric. Therefore, *the drop in the lengths of zigzag paths in the limit that we observed earlier may be attributed to the discontinuity of the length functional!*

This is analogous to the jump discontinuity for ordinary functions that we discussed in Chapter 6, but the domain is now a metric space of functions; so it is not possible to visualize the discontinuity using an ordinary graph.

Before closing, it is worth mentioning that other norms can be defined on $C^1[0,1]$, and some of those may actually be sensitive to variations in function values. One such norm is

$$\|f\|_1 = \max\{|f(x)| : x \in [0, 1]\} + \max\{|f'(x)| : x \in [0, 1]\}$$

This norm defines a metric or distance on $C^1[0, 1]$ that takes into account the derivative, or the variation in the function as well as its raw values. It is therefore a refinement of the uniform norm. Specifically, if f and g are any two functions in $C^1[0, 1]$, then the distance between them is

$$\|f - g\|_1$$

Since derivatives are slopes of curves, the greater the oscillations, the larger the distance in the above norm is. Figures 8.7 and 9.2 illustrate this point.

We see that the difference in values between $f_n(x)$ and $f(x)$ is smaller than ε throughout; however, the derivatives $f'_n(x)$ and $f'(x)$ can be very different from each other, especially at the points where $f_n(x)$ and $f(x)$ intersect since at these

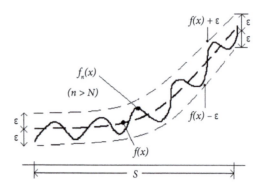

Fig. 9.2 Curves that are uniformly close in value but not in variation

points, the slopes of these curves may be different from each other. Near an intersection point x, the difference

$$|f'(x) - g'(x)|$$

is much larger than ε, and this leads to a large value of $\|f - g\|_1$ overall.

Let's take a look at the norms of the earlier sine curves (with arbitrary c_n). Notice that

$$\|f_n\|_1 = \max\left\{\left|c_n \sin^2(n\pi x)\right| + \left|n\pi c_n \sin(2n\pi x)\right| : x \in [0,1]\right\} \geq n\pi c_n$$

since the maximum value of the sine function $\sin(2n\pi x)$ is 1. Therefore, the sequence f_n does not converge to the zero function (or any other function) in the new norm. In fact, we see that the new norm can be infinitely large if nc_n goes to infinity. The anomaly of zigzag paths does not occur if we measure the distance between each path and the diagonal of the square using the new, refined norm!

9.3 Numbered Elements in every chapter

Chapter 1:

Exercises 1–3

Chapter 2:

Theorems 4–12
Lemma 13
Corollary 14
Theorem 15

Corollaries 16, 17
Exercises 18-53

Chapter 3:

Lemma 54
Theorem 55
Corollary 56
Theorem 57–61
Lemma 62, 63
Theorem 64
Corollary 65
Theorem 66–68
Corollary 69
Exercises 70–97

Chapter 4:

Theorem 98
Corollary 99
Theorem 100–104
Corollary 105
Theorem 106
Corollary 107
Theorem 108–109
Corollary 110
Theorem 111–112
Corollary 113
Theorem 114–115
Corollary 116
Exercises 117–138

Chapter 5:

Theorem 139
Lemma 140
Theorem 141–142
Corollary 143–144
Lemma 145
Theorem 146
Corollary 147
Theorem 148
Corollary 149
Theorem 150–154

Chapter 6:

Chapter 7:

References and Further Reading

David Bressoud, *A Radical Approach to Real Analysis*, Mathematical Association of America, Washington DC, 1994

T. G. Faticoni, *The Mathematics of Infinity: A Guide to Great Ideas*, Wiley, New York, 2006

Patrick M. Fitzpatrick, *Advanced Calculus*, 2nd ed., Thomson, Brooks-Cole, Belmont, CA, 2006

George Gamov, *1, 2, 3 ... Infinity: Facts and Speculations of Science*, Viking-Compass, New York, 1962

Luke Heaton, *A Brief History of Mathematical Thought*, Robinson-Little, Brown, London, 2015

J. M. Henle and E. M. Kleinberg, *Infinitesimal Calculus*, MIT Press, Cambridge, 1979

David Hilbert and S. Cohn-Vossen, *Geometry and the Imagination*, 2nd ed., Chelsea Publishing Company, New York, 1952

Morris Klein, *Calculus: An Intuitive and Physical Approach*, 2nd ed., Dover Publications, New York, 1998

Erwin Kreyszig, *Introductory Functional Analysis with Applications*, Wiley, New York, 1978

Serge Lang, *Analysis I*, Addison-Wesley, Reading, MA, 1968

Abraham Robinson, *Non-standard Analysis*, Princeton University Press, Princeton, 1996

Bertrand Russell, *Introduction to Mathematical Philosophy*, Martino Fine Books, Eastford, CT, 2017 (originally published in 1919, George Allen and Unwin, LTD, London)

Hassan Sedaghat, *Achieving Infinite Resolution: A Gentle Look at the Role of Infinity in Calculus*, Amazon, 2020

Manfred Stoll, *Introduction to Real Analysis*, Addison-Wesley, Reading, MA, 1997

Patrick Suppes, *Axiomatic Set Theory*, Dover Publications, New York, 1972

Terence Tao, *Analysis I*, 3rd ed., Hindustan Book Agency, New Delhi, 2014

Terence Tao, *Analysis II*, 3rd ed., Hindustan Book Agency, New Delhi, 2014

William F. Trench, *Introduction to Real Analysis*, Pearson Education, New York, 2003

Index